Writing Embodiment in Victorian Microscopy

Writing Embodiment in Victorian Microscopy

Beautiful Mechanism

MEEGAN KENNEDY

OXFORD
UNIVERSITY PRESS

OXFORD
UNIVERSITY PRESS

Great Clarendon Street, Oxford, OX2 6DP,
United Kingdom

Oxford University Press is a department of the University of Oxford. It furthers the University's objective of excellence in research, scholarship, and education by publishing worldwide. Oxford is a registered trade mark of Oxford University Press in the UK and in certain other countries

Published in the United States of America by Oxford University Press
198 Madison Avenue, New York, NY 10016, United States of America

British Library Cataloguing in Publication Data
Data available

Library of Congress Control Number: 2024943328

ISBN 9780198939009

DOI: 10.1093/9780198940623.001.0001

Printed and bound in the UK by
Clays Ltd, Elcograf S.p.A.

Acknowledgements

It is perhaps not surprising that my fascination with "vastness" and "difficulty" has itself expanded into a multi-year project. The book has been long in the making, and I owe much to the generous colleagues and friends who've accompanied me along that winding path.

This project grows out of many years of happy burrowing around in libraries, archives, and museums. I started thinking about these ideas while browsing in the stacks at Brown University's Rockefeller Library when I was working on a different project, but over the course of my research I've benefitted from research access to a number of institutions and the guidance of their staff. I am especially grateful to the U.S. National Library of Medicine and the Huntington Library for their assistance in securing images republished here. I enjoyed a fruitful summer trawling the collections at Oxford University, especially the Museum and Library of the History of Science (with special thanks to Lucy Blaxland and Jennifer Dumbelton), and a memorable spring at the Huntington Library in San Marino, California (with thanks to Daniel Lewis). I'm also indebted to the generous assistance of staff at the Florida State University Libraries, the John Hay Library at Brown University, the British Library, the Countway Library at Harvard University, the Royal College of Surgeons of England, the Quekett Microscopical Club, the London School of Hygiene and Tropical Medicine, the Bill Douglass Collection at Exeter University, and the Whipple Museum at Cambridge University. The Wellcome Library provided valuable access to periodical resources as well as print books. I'm grateful for the assistance of Trenton Streck-Havill at the National Museum of Health and Medicine in Washington, D.C. and Bernie Alie at York County (Maine) Community College—who went above and beyond to help me access particular sources during the Covid pandemic. And I appreciate the work of staff at the Merrill Library in Yarmouth, Maine and the Bowdoin College Library, for their help in accessing needed books and providing a welcoming place to work.

Research for this book was carried out with support from a number of funders, for which I am most grateful. The most consequential work on the book occurred during a National Endowment for the Humanities summer stipend and yearlong fellowship, and during a Huntington-Linacre Exchange Fellowship in Oxford and a Dibner Fellowship in the History of Science and Technology at the Huntington Library in California. Florida State University has also supported my work with sabbaticals, summer support, and funding for the work of research, travel, and publication. This has been a three-chair book; my thanks to Eric

Walker, Gary Taylor, and Andrew Epstein, and to Peggy Wright-Cleveland, Aline Kalbian, and Sam Huckaba, for their encouragement and assistance in securing internal and external funding for this project.

I very much appreciate the guidance and support of Eleanor Collins, Jack Mcnichol, and Imogene Haslam at Oxford University Press, as well as Jothi Alysia Stephenson, Chris Bessant, Helen Taylor, and everyone who helped in the publication process.

The book is better for the many conversations I've enjoyed with smart and generous people. My thanks to the many friends and interlocutors at conferences like NAVSA, MLA, and HSS whose questions helped me develop my argument. I also appreciate having had the opportunity to discuss my ideas with audiences at Bowdoin College, the Birkbeck Centre for Nineteenth-Century Studies at the University of London, the Georgia Colloquium for Eighteenth- and Nineteenth-Century Studies at the University of Georgia, and the Werkmeister Literature and Evolution Workshop at FSU; my thanks to all who helped to facilitate these conversations. Members of the Boston Victorianist Writing Group offered valuable commentary on the project during my NEH year. Others who have helped make this book better include Jay Clayton, Maria Frawley, and Peter Melville Logan, Amongst FSU colleagues and students, I owe special thanks to Candace Ward, Leigh Edwards, Fritz Davis, Anne Coldiron, Michael Ruse, Barry Faulk, Judith Pascoe, Lindsey Eckert, Molly Marotta, Holly Horner, Carley Robertson, and Rachel Duke, for many fruitful conversations. Outside FSU, Martha Bayless, Cathy Komisaruk, Sophie Weeks, Julie Couch, Nina Markov, and Piers Hale helped me keep moving ahead on the project. I'm also indebted to Cindy Holmes for her steady good sense and support. These friends and colleagues remind me why I entered this profession, and why I stay in it.

The book has benefitted from the practical assistance of several family members. Bay read early chapters; Caty and Jeff consulted on Excel and statistics; and Scott helped me ship Victorian microscopy books home from England; I so appreciate these efforts in support of my work. I am also grateful for all those who have lived with, and cheered on, this book for so long. My mother and father showed me as a child what it looks like to be curious, wonder-filled, and painstaking, and they found Victorian microscopes, slides, and books for me to work with; Gail and Dick offered quiet work spaces or food and companionship, as needed; Austin brought good humor and conversation; and Skipper, Dory, and James took me on many helpful "thinking walks." Most of all, thank you to Scott, Caty, and Bay, for your unwavering patience and support. The best example of "counting the uncountable" might be in trying to enumerate the myriad instances of your good cheer and encouragement.

Contents

II. BEAUTY AND WONDER

III. DIFFICULTY AND ERROR

List of Figures

PART I

THEORY, HISTORY, AND CULTURES OF VICTORIAN MICROSCOPY

1
Embodiment in Microscopical Print and Practice

Microscope, difficulties of, 184; charms of, 197.
Index entry in Philip Henry Gosse, *A Naturalist's Rambles on the Devonshire Coast*, 1853

Introduction

Victorian readers of Philip Henry Gosse's 1853 guidebook *A Naturalist's Rambles on the Devonshire Coast* might have noticed mixed messages about the microscope: is it a complicated tool or a delightful toy? The book certainly suggests the latter; it helped spark the mid-century fashion of beachcombing, as middle-class Victorians filled their leisure hours with the study of mollusks, algae, and other seaside creatures.[1] But in citing both the "difficulties" and "charms" of microscopy, Gosse acknowledges the various and unsettled meanings of this technology at mid-century, when the curious might ramble the hills with a pocket lens, dissect flowers under a simple (single-lens) microscope, entertain friends with a projection microscope, or scrutinize unicellular organisms through one of the newly powerful compound (multi-lens) microscopes.[2]

In fact, the conjunction of wonder and skepticism seen in Gosse's index shapes British microscopy throughout the century and across a broad range of texts, whether written for hospital, laboratory, schoolroom, parish hall, or parlor. The phycologist Margaret (Mrs. Alfred) Gatty, for example, in her popular 1872 guide to *British Sea-weeds*, warns that the "delicacy of eye and judgment" in viewing a microscopical object is "not so easily taught or acquired," even as she praises the microscope as "one of those desirable possessions" needed to experience fully the "delight" of studying seaweeds.[3] Her references to desire and delight, delicacy and difficulty imply different approaches to microscopical vision, but as the examples above show, they often appear in the same text, sometimes on the same page. These expressions of wonder and skepticism reproduce persistent Victorian ideas about microscopical practice as a process that both aroused affective energies and engaged a necessary discipline of eye, hand, and body. This book focuses on such tropes—particular literary metaphors, narratives, and concepts, adapted from old forms or newly developed—to study how writers sought to establish microscopy as a difficult but valuable contribution to Victorian society. Whether depicting microscopy as difficult or delightful, these literary forms describe an embodied practice evoked through the interrelation of observer, microscope, and object. I will show how Victorian texts work with such tropes not just to promote microscopical observation but also to regulate the entanglement of those very different bodies. These tropes shape every level of microscopical interest and proficiency, and they illuminate wider patterns of Victorian embodiment, affect, and scientific practice.

A suggestive early episode in this story begins in January 1851, when John Quekett stood in the Histological Theatre of the Royal College of Surgeons (RCS). With posted sketches of microscopic structure behind him and a gleaming Ross compound microscope before him, he began a course of introductory lectures on "normal and pathological microscopical anatomy" for members of the RCS—experienced medical men.[4] Quekett exemplified the rising science of microscopy. He was at age thirty-five serving as the Assistant Conservator of the RCS Museum under Richard Owen. Colleagues brought him specimens to examine, seeking his opinion as a skilled microscopist.[5] His RCS lectures would be reprinted in the *Medical Times* and would attract glowing reviews when republished in book form—and he was presenting them at a pivotal moment in Victorian microscopy.

After the development in the 1830s of compound instruments that could reduce optical errors, the microscope offered new potential for scientific investigators. Through the 1840s, some medical men in Britain began to encourage students and colleagues to use it in medicine as well. While some investigators were eager to explore the powers of this technology, many medical practitioners were slow to adopt it in practice. In 1841, the Scottish physician and physiologist John Hughes Bennett, a booster of the new technology, complained in a published lecture that even "men who keep pace with the advance of science" were likely to learn about microscopic structures only from "descriptions and drawings," and that "by far the greater number of practitioners and anatomists in Britain, and the whole mass of students, have no idea of them at all."[6] By 1848, in his *Practical Treatise on the Microscope*, Quekett more optimistically argued that "the Microscope has now become indispensable as an educational instrument" for "the medical profession," and that "perhaps, by far the most numerous class of microscopic observers" were medical men.[7]

But while the microscope may have been energizing physiological research, especially on the Continent, British doctors were slow to adopt it in their diagnosis and practice. In contrast to Germany, where medical microscopy was integrated into the curriculum of every medical school by mid-century, fewer than half of London medical schools featured a course in microscopy by the 1850s.[8] So to open his first lecture of the 1851 series, even though he was speaking to an audience already experienced in the medical arts, Quekett needed to make the practical argument that knowledge of histology, enabled by "that powerful auxiliary the microscope," would bring "many advantages" to medical practitioners and anchor "the successful practice of the healing art."[9] And while scientific and medical uses of microscopy did increase in the 1850s, naysayers persisted. An 1851 *Quarterly Review* article, for example, grumbled that "[t]he microscope is a good and useful servant, but it has often been abused," resulting in the "too confident assertion of results." The article swept Quekett himself into the fray, calling out his consulting work for its role in some "hasty blunders." Although the article was unsigned, internal evidence pointed to its author: Quekett's colleague (and superior) Owen.[10]

Owen had built his researches into comparative anatomy in part on microscopical findings; indeed, he had served as the first President of the Microscopical Society of London. Addressing the fledgling Society in 1841, he asserted that "the microscope is…essential to the full and true interpretation" of fossilized plants and animals.[11] In other settings, however, he was more critical, alleging that advocates for microscopy (like Quekett) exaggerated and abused its powers. In the Preface to his massive *Odontography* (1840), Owen briefly acknowledges the "value" of microscopical analysis of fossilized teeth but writes much more on his concerns about the technology: "the difficulty of testing" or replicating observations, the misguided analysis of isolated fragments, and the "abuse" (a repeated term) due to ambitious observers' "ignorance of the mode and limits of its application."[12] While Owen's work with the instrument helped raise its status at this critical moment, statements like these helped perpetuate skepticism about microscopy.

Significantly, when Quekett revised his first lecture for publication in 1852, he rewrote his introduction to directly address the uncertain status of microscopy at the time. He concedes that "the microscope, except as a mere toy, fell into disuse during the greater part of the eighteenth century." (This was not quite true, but it was a widespread perception in Victorian Britain.)[13] In contrast, he explains, the microscope in 1852 boasts "such a magnifying power" and "such clearness of definition" that it could ground the new field of histology, a "most interesting and important field of research," currently "being gradually evolved by the labours of microscopic physiologists." Indeed, he proclaims, microscopy could yield access to previously unimagined truths, for "in organized beings nature works out her most secret processes by structures far too minute for observation, unless with the microscope."[14] Quekett's printed lectures counter familiar dismissals of the microscope (an instrument *in decline*, misused as a *mere toy*) by adopting other familiar tropes of wonder (*the natural philosopher probing Nature's hidden secrets* and *seeing the invisible*) and of skepticism (the need for *patient labor*) in order to legitimate the practice of microscopical science. Quekett's language here implicitly acknowledges the difficulties he was facing. Nineteenth-century promoters of the microscope in Britain had to overcome longstanding perceptions that it was a useless gadget, a toy for dilettantes, or a degrading tool for examining vermin and other filth. But these critiques of the microscope receded as writers drew upon a range of literary tropes to circulate alternative metaphors and narratives about modern microscopy as a challenging but rewarding pursuit. These tropes powered the remarkable story of the nineteenth-century microscope, which by the end of the century had become a widely enjoyed amusement, a respected scientific instrument, and a useful tool in agriculture, industry, and medicine.

This book examines how Victorian microscopists like Quekett, Gosse, and Gatty used various tropes about microscopy to manage the uptake and use of the instrument. Their expressions of wonder and skepticism helped promote microscopy as an embodied practice entangling the observer, the apparatus (the microscope and its associated tools), and the object of study in a network of optical, mechanical, cognitive, and affective interchanges. These tropes of embodiment shape microscopical observation through every level of interest and proficiency: working microscopists (skilled and committed observers working at the microscope for research or amusement), hobbyists with more sporadic and limited engagement with the instrument, even passive viewers of public microscope shows.

Microscopists were not unique in this regard; other nineteenth-century investigators also used "literary technology" to promote and direct developing sciences. For example, as Simon Schaffer shows, William Herschel coined new astronomical terms that strategically "required appeal to networks of familiar concepts, and extended that network in one direction rather than another."[15] However, such efforts were particularly needful in mid-century microscopy, due to the difficulty of use and relative newness of the achromatic compound technology, and a lingering suspicion of microscopists' claimed ability to reveal marvels within ordinary things.

This book joins a lively and longstanding scholarly conversation on the relations of nineteenth-century science and literature—in particular, on the forms of scientific writing and how those forms shaped scientific practice. I build here on a rich tradition of scholarship examining specific literary patterns as tools for scientific work, such as (for example) Ralph O'Connor's and Adelene Buckland's work on narrative in geology. As Buckland explains, "geologists elaborated new *literary* forms with which they would explain, interpret, order, describe, argue about, and bring into existence" Victorian geological science; she argues that "geology was *written* into existence...as much as it was found, discovered, collected, mapped, or modeled."[16] One could say the same about Victorian microscopy, although the tropes I study often center on the observer's visual and physical access to objects of study, which was (as I will show) particularly fraught, both precarious and overdetermined.

This book thus examines how Victorians used competing stories about wonder and skepticism to defend and refine microscopical practice—how these stories prescribe particular embodied relations between observer, apparatus, and object, and discourage others. They refute unhelpful narratives—for example, replacing the old saw *seeing is believing* with inspirational stories about *revelations* or skeptical ones about the *trained eye*. They push back against longstanding dismissals of microscopical work as trivial, unreliable, or even disgusting by reframing it as a moral practice of self-discipline and wonder. These tropes convey complex ideas concisely, often using colorful or memorable phrases, to form a more persuasive and transferrable knowledge. They implicitly establish a normative practice by endorsing some interchanges (optical, mechanical, cognitive, affective) over others. They model certain forms of work and preferred responses to the image for readers to learn and embody in turn. Overall, the tropes I examine here nudge microscopists into preferred forms of embodiment, both skeptical and wonder-filled. They argue that microscopical vision demands certain forms of patient, humble labor that, properly executed, yield a deeply rewarding engagement with the world, even as this work forces us to confront the unreliability of our senses and cognition, the limits on our skill and knowledge, and the unsettling depths of minute organic life around and within us. They leverage a productive tension, formed in the convergence of affective flows and physical practices, to direct the expansion of Victorian microscopy. (A list of the tropes discussed in this book is provided in Appendix III.)

Some of these tropes were borrowed from other settings, such as the *narrative of natural history* or the astronomical metaphor of a *personal equation* correcting for individual bias. Others revised familiar literary or philosophical conventions—such as the concepts of *beauty* and *the sublime*—into versions specific to microscopy. And still others developed out of the unique promises and challenges of microscopy, such as the image of a *diatom as a Gothic rose window* or a *drop of water as a teeming world*, the notion of *the vastness of the very small*, the fantasy of *microscopical fairylands*, or the promise of *the repulsive revealed as wonderful*. Humorous tropes such as *diatomaniacs* mockingly weighed in on Victorian disputes over the proper work of microscopy. The notion of a *trained eye* or *educated hand* met the challenges of microscopical work by encouraging novices to train in a disciplined practice. Although these tropes did not always achieve their intended effect, many circulated widely between practitioners and non-practitioners and became conventional frames for Victorian microscopical experience.

I analyze how these tropes enlist observers and audiences in what feminist theorist Karen Barad calls "material-discursive" practices, where cultural elements (like the language and image patterns that make up microscopical discourse) and material ones (like the bodies of observers, instruments, and objects) help configure one another.[17] Over decades of use, these widely circulated tropes helped build an influential material-discursive network: they emerged out of microscopists' accumulated experience even as they helped produce and regularize Victorian microscopical culture. The practice of observing minute things offers a valuable window into nineteenth-century debates over what we know and how we know it; how to research, teach, and learn; how to talk about what you know and don't know; how to form improving habits; how to test the quality of what we make, eat, and use; where illness comes from; what it means to be modern, or a man of science, or British; what happened eons ago, or yesterday; what is life itself, and faith; and how we fit within a complex, curious, and sometimes threatening natural world. This book pursues a broad-ranging and detailed analysis of Victorians' attempts to use microscopy and its metaphors to organize human experience—to observe and make sense of their bodies and their surroundings, and to form communities engaged in this purpose. Microscopy offered Victorians a model for how to study the world around them, and how to feel about it all.

Most of the scholarship on nineteenth-century microscopy has focused on its relation to debates about visuality—not surprisingly, since the instrument overtly relies on, extends, and challenges human vision.[18] Vision and perspective have long symbolized cultural mastery. Victorians valued the eye for its role in gathering and authenticating knowledge, but they also worried about its limitations and errors. Investigators increasingly scrutinized vision and the microscope as images became ever more important to, and contested in, nineteenth-century scientific practice. Microscopists developed a variety of strategies to safeguard reliable observations even as their faith in the instrument and in human perception was challenged by the work of Hermann von Helmholtz, Ernst Abbe, and others. The instrument presents views disjunct from one another and so distant from the macroscopic world that microscopic vision symbolizes and also requires imagination, simply to comprehend what is seen. In texts from this period, the microscope mimics, aids, extends, supplements, and replaces human vision but can also damage eye and image. The microscope thus offers Victorians a technological and worldly creation that reveals and celebrates—but also revises, supplants, and possibly injures—the natural and spiritual one.

This book contributes to the "visual turn" in scholarship in that it examines this Victorian fascination with optical instruments and the production of images; indeed, the microscope might seem an ideal site from which to examine the supposed hegemony of vision during this period. But scholars like Jonathan Crary have shown how nineteenth-century theories of vision were grounded in the body and brain. Although nineteenth-century texts do sometimes describe microscopy in terms suggesting a simple dualism of mind and body, seer and seen, they often articulate a more complex view informed by contemporary work in physiology and psychology. Chapter 5 demonstrates how the ideas underlying Victorian microscopical discourse track with the work of a group of "physiological psychologists," who discuss the human senses and cognition as embodied, unstable phenomena in dynamic interrelation with the material and social environment. Overall, I show that Victorians understood vision and cognition—especially microscopical vision and related feats of intellection—to be deeply embodied and contingent on an array of environmental conditions including unstable physiological and affective states. My field of interest thus encompasses not just the observing body but the whole system of relationships between observer, apparatus, object, text, and audience, which produces microscopical vision as a (sometimes vexed) collaboration between bodies. This book thus endorses the wider field of analysis demonstrated by scholars like the historian of science Tiffany Watt Smith, whose work on the "flinch" shows that the "emotional and embodied dimensions of experimental practice" are often neglected by historians but were openly discussed by Victorian investigators.[19] Similarly, the literary critic Cannon Schmitt has urged colleagues to develop "a way to take seriously the affective, visceral, animal, and familial modalities of knowing that inform or underpin the production of human knowledge about the natural world."[20] This book works within that radically expanded field of study.

The embodiment thesis here builds on a longstanding historical tradition examining the details of scientific practice. Here that tradition shapes my discussion of how Victorian microscopists were asked to attend to diverse forms of embodiment in practical terms, although these terms encompassed not only physical but also affective and cognitive actions. Observers were instructed in the careful placement of head, torso, arms, hands, and feet and the coordinated movement of their eyes and fingers in relation to the apparatus and object; and they were advised on the feelings that could or should be aroused by microscopical work. I use the term "entangled" to describe this joint role of human and non-human bodies in creating microscopical experience and knowledge. As the archaeologist Ian Hodder explains,

> Entanglement as a term aims to allow a materialism but embedded within the social, the historical, the contingent.... The interlacings of the web are composed of bundles of material, ecological, economic, social, ideational, emotional and cognitive processes. Human movements to and away from things are made possible by, but are also bounded or channeled by, these bundles of material and immaterial interactions and dependences.[21]

Victorians portrayed microscopy as a practice requiring human operators to work within a configuration that enacts and constrains their observations. While not discounting the effects of human actions, they describe agency as dispersed across an interdependent structure of humans and non-humans. Discussions of microscopical practice repeatedly place the human observer and the microscopical object as elements of the observational apparatus, both optically and physically, even as they describe the physical and apparently psychological foibles of individual observers, microscopes, lenses, and objects. This book examines how Victorian microscopists constructed communities of practice by attempting to manage observers' embodied participation in this system and to direct the flows circulating through it.

Victorians' discussions of this integrative model present a suggestive resonance with scholarly writing on actor-network theory, postphenomenology, agential realism, disability studies, affect theory, and other contemporary methodologies that argue for embodiment and distributed agency in human–technology interrelations.[22] These approaches have shaped my own work with this material, especially in helping me recognize entanglement in Victorian ideas about the interrelation of multiple actors in any microscopical observation. Benjamin Morgan similarly draws on affect theory and other approaches to argue that, for Victorians, "aesthetic experience is a forceful, bodily encounter of a person and a pattern," where the observer's agency is understood "as a distributed property of a network of humans and nonhumans rather than as the secure possession of a self-governing subject." Morgan identifies a trend in Victorian aesthetics, an "exteriorization of mind, consciousness, and the self into networks of matter, sensations, and objects," influenced by the general cultural familiarity with recent work in physiological psychology. This "outward turn" characterizes not just the writing about art that Morgan studies but the microscopical work I examine here as well.[23]

However, Victorian writing on microscopy demonstrates important differences from these contemporary theorists. Most important, the nineteenth-century writers examined here assume the independent pre-existence of observer, apparatus, and object even as they frame these as elements linked or even constituted in a system of related flows. They often embrace the positivist goal of "mastery," whether of objects or bodies of knowledge, and the scientific ideal of mechanical objectivity, which, as Lorraine Daston and Peter Galison explain, called for the conscious self-denial, even "self-annihilation," of the observer's subjectivity in hopes of automatizing the act of observation.[24] Victorian microscopists often frame these goals as unattainable. The aspirational ideal of mechanical objectivity was both necessary and impossible to fully achieve in microscopy precisely because the instrument presents in heightened form these issues of entangled agency and embodiment in human–technology interrelations. While Victorian microscopists still consider agency and identity to be aspects of pre-existing independent subjects, they see agency as playing out within an embodied system, and the observer as part of the apparatus. They often describe agency as dispersed across and constituted through this system and discuss how agency and identity are directed and shaped by the entangled practice of observer, apparatus, and microscopical object. They debate whether the figure of the *microscopist* is formed by work at the microscope, or play. They emphasize that an observer's identity as investigator or hobbyist develops through one's actions with the cooperative or recalcitrant apparatus and object; they consider how a particular microscope's identity (as elite instrument, student's instrument, or rubbish) develops and even changes over years of use. The agency and identity of observer, apparatus, and object also shift over the course of many observations; the work leaves material traces that form and reform all participants. The acquisition of *tact*—a sensitive manual dexterity in handling the microscopical apparatus—shows how bodies can be enhanced through careful practice. The oft-noted fatigue of eye, arm, and torso reminds us that human bodies are also worn down by the work, their performance degraded. Instruments, too, become worn or damaged with much use. The objects studied, of course, are compressed, sectioned, dissected, killed; Victorians noted how even basic techniques like staining could change structure.

Finally, for these writers, microscopical objects with defined attributes do exist in the world (the holy grail of *true structure*) but knowledge of those attributes can emerge only through an embodied system of observation, only unevenly, and never definitively. While the very term *true structure* suggests stasis, working microscopists consistently frame those coveted images in the context of development (evolution and life history), systems (environment), and practices (their own investigative work). Further, they demonstrate their awareness that—as suggested by the sciences of optics, physiology, and psychology as well as decades of microscopical reports—observation is often misleading, always embodied, always partial and flawed, and always contingent. Worse, conflicting observations can both be true. As I will show, some commentators insist that even clearly "wrong" observations are true in that they produce an image corresponding to the particular constraints of the system as configured. The instrument is resistant, often deceptive, and always right. The tropes I study here express and seek to direct this fraught embodiment of observer, apparatus, and object, even as they also celebrate the history of microscopical work and its frustrations, fears, and joys. Microscopical tropes endorse particular forms of embodiment by modeling and soliciting specific emotions and practices within this economy of actions. While these tropes all describe a visual event and invoke an embodied practice, they can be distinguished as primarily concerned with affective flows, or primarily concerned with optical, cognitive, and mechanical (physical) linkages. I differentiate these tropes under the general headings of wonder and skepticism. It is important to remember that, although I discriminate between them for the sake of argument, tropes of wonder and of skepticism were often used in the same texts to describe the same experiences. In fact, the overarching narrative endorsed by many Victorian microscopists was that skeptical practice is closely imbricated with wonder and delight: it is necessary for, inspired by, and yields those rewards.

Wonder and Skepticism

Victorian microscopists frame *wonder* as an expression of the observer's affective response to particular qualities of the microscopical image, especially beauty, delicacy, minuteness, mystery, color, or a sense of expansive landscape or infinite worlds. Tropes of wonder typically voice this affective response in exaggerated, romantic prose that solicits a similar response from the reader. Writers offer to enlist observers and audiences alike in shared feelings that are aroused and mediated by the microscopical image and by these tropes. The sympathy and identification with microscopical objects reported by viewers, especially at projection microscope shows, can also evoke and enhance a sense of wonder.[25]

In this book, I use the term *wonder* broadly in order to identify a group of affective states frequently invoked by Victorian microscopists.[26] Philip Fisher influentially defines wonder as "a sudden experience of an extraordinary object that produces delight," an experience that is "unreflective and immediate" and sparked by an unexpected view of "precious transience."[27] However, many related varieties of affective experience shade off from this; as Sophia Vasalou notes, "concepts like surprise, astonishment, amazement, wonder, and awe, belong to a family of terms that cannot always be sharply distinguished."[28] And as Lorraine Daston and Katharine Park remind us, even wonder itself, like other emotions, can exhibit through varieties of "felt experience," where "the felt substance of an emotion depends to a significant degree on the company it keeps."[29] In this book, I study how Victorians spoke and wrote about the kinds of microscopical wonder. A microscopical object can evoke, for example, admiration (wonder as marveling and awe, as in the trope of *revelations*), doubt and uncertainty (wonder as interest and curiosity, as in the *mystery* of microscopical *worlds*), utter incredulity (wonder as amazement, as in the *vastness of the very small*), or playful imaginativeness (wonder as fancy, as in the *Oriental tale*).[30] These tropes counter competing narratives of disgust or fear associated with microscopy. Other tropes of wonder, such as *invisible life surrounds us*, may be used when writers leverage disgust or fear to other ends, perhaps to show the need for microscopical investigation, or simply to attract the attention of readers or viewers. I work here with a "big tent" definition of wonder, one that can reach from the sublime to the whimsical, because it allows me to capture a fuller range of the affective experience that Victorian microscopists experienced in relation to their work—and hoped to evoke in their readers. This broad definition provides a frame large enough to see diverse forms of wonder in relation,

William Notcutt, a dissenting chemist, botanist, and popularizer of science, remarks that the microscope "brings before us marvels so strange, and apparently incredible, that all the inventions of romance and fiction are rendered trivial and tame" by contrast.[31] Despite these extravagant claims, many tropes of wonder describe serious stances that are not just congruent with but even supportive of philosophical aims. For example, microscopists express *admiration*, typically when viewing a beautifully structured natural object, but also when appreciating beautifully rendered scientific work. Admiration is the most common form of wonder found in research texts (reports, technical manuals) for investigators. Even the *Handbook for the Physiological Laboratory*, a notorious 1873 guidebook that roused the ire of anti-vivisectionists, incorporates occasional notes of admiration, referencing "a very beautiful network of pale protoplasm," "the most beautiful muscular nerve endings," "labyrinths of capillaries of surpassing beauty," and the like.[32] Observers also express *awe* in encountering the sublime, such as the apparently infinite recursion of minute life or the vast numbers of foraminifera making up geological formations; and *reverence* in considering the intricacy and vastness of the minute world, seemingly demonstrating the power and patience of a Creator God. Microscopists additionally convey wonder in lighter forms echoing literary romance. They express *astonishment*, as when projection microscopists and their audiences compare the scene within a drop of water to the phantasmagoria, and *whimsy*, as when Gosse plunges into a fanciful, fictional dreamscape or an Oriental tale or Arabella Buckley weaves descriptions of microscopical fairylands. Writers usually invoke tropes of wonder in describing the observer's responses, sparked by what they see under the lens; but they can also acknowledge the affective life of the object and even—as in the tropes *microscope as companion* or *microscope as pet*—the instrument as well.

Even within these traditions, the writer's credibility could be damaged by any excessive or naïve wonder, or wonder expressed in injudicious terms or in an inappropriate (textual or physical) space. For example, the Scottish physician Martin Barry claimed to have discovered a microscopical spiral form present in all organic tissue. In the early 1840s, Barry's speculations were critiqued for their grandiosity and excessive wonder; they were damned as "the monstrous products of a fertile imagination," displaying "far too much the stamp of a lively fancy."[33] Barry's exuberance marks him out here for censure.

At the same time, theologies of nature brought credibility and meaning to wonder in microscopical work; and most forms of wonder likely helped recruit new adherents with their promise of microscopical pleasures. Given the frequency of these tropes in microscopical writing, especially in popularizing books, articles, and performances, Victorian readers would have expected wonder to be a common, ordinary occurrence when using a microscope; the wonderful is an everyday event in this setting. Advocates for the microscope juxtapose such promised visions with a skeptic's call for *patient labor* or warning of *counterfeit revelations*.

Microscopical whimsy and wonder may inspire fantastical writing, but it is important to distinguish this type of playful imaginative act, which contemporaries often termed *fancy*, from Victorian naturalists' vision of the *scientific imagination*, which over the course of the century became recognizable as an intellectual act and framed as judicious speculation, not capricious fantasizing. The distinction is often framed in terms of control and restraint and became especially cogent at mid-century: the imagination was valuable in part because supervision and restraint prevented it from running into extremes. In *Mental Physiology*, Carpenter explains that "whilst the *Imagination* works on the sense of beauty, of grandeur, of sublimity, or, by way of contrast, on that of aversion, awe, or even terror,—Fancy plays with the grotesque, the whimsical, the ridiculous, in short, with the fanciful; its function being rather to amuse by its superficial 'caprices' and 'strange conceits,' than to appeal to the deeper nature."[34]

By mid-century speculation and hypothesis, if well grounded in empirical observation and applied judiciously, were touted as elements of scientific work. John Herschel's *Preliminary Discourses on the Study of Natural Philosophy* (1830) and William Whewell's *Philosophy of the Inductive Sciences* (1840) laid out guidelines for the use of hypotheses in empirical investigation. So by 1850, John Hughes Bennett argued against Barry's spiral formations using a very different tone: Bennett explained that "instead of deducing theories from well observed facts," Barry had presented "supposed facts...invented to support theories." The problem was not that Barry speculated, but that his speculations needed better grounding in sound observation.[35]

By the late 1860s, the physicist John Tyndall was advocating even more strongly for the scientific imagination. In his controversial 1874 Belfast Address to the British Association for the Advancement of Science (BAAS), Tyndall asserted the materialist basis of science and claimed the authority of scientific investigation over material domains—while he also famously rejects naïve empiricism in favor of a judicious speculation, especially when our physical tools falter. In fact, he posed imagination as a kind of physical instrument, specifically as yet another supplement to the microscope, saying that "it is plain that beyond the present outposts of microscopic enquiry lies an immense field for the exercise of the speculative power."[36] He considered that this power required careful management, of course; but, he argued, when "[b]ounded and conditioned by cooperant Reason, imagination becomes the mightiest instrument of the physical discoverer."[37] Tyndall here advocates for a new approach to scientific materialism, one shaped by the findings of physiological psychology, which can and must strategically leverage the instrumental power of an informed imagination.[38]

In his 1874 *Mental Physiology*, Carpenter similarly endorses the use of imaginative mechanisms in science. He says, "It is a very limited view of the Imagination to regard its operation as restricted to works of Art." Rather, he argues, "Its highest exercise is put forth in the discovery of great fundamental Truths, as well in Science as in Art."[39] In fact, Carpenter defends science by associating it with imagination. He hotly rebuts John Ruskin's attacks on science, saying,

> To represent Science and Poetry as antagonistic, and to maintain (as Mr. Ruskin continually does) that the Poet sees into Nature both *more deeply* and *more truly* than the Scientific inquirer, simply shows this able writer's ignorance of what science really is. The Truths which the man of Science and the Poet respectively discover, are of different orders, but cannot be antagonistic.[40]

Carpenter's heated words here contribute to the same cultural shift that is more famously registered in Tyndall's *Essays on the Use and Limit of the Imagination in Science* (1870) and 1874 BAAS address in Belfast.

Identifying wonder as a form of affective experience usefully allows us to see its relation to the primarily cognitive tradition of scientific imagination. The affective experience ("Wonderful!") sometimes but not always occurs together with the intellectual one ("I wonder..."). By recognizing Victorian wonder as an affective experience distinct from (but not necessarily conflicting with) the intellectual work of speculation and the scientific imagination, I can examine microscopists' interest in affective flows apart from contemporary debates over the value or risk of fancy and speculation. A wonder-filled experience may inspire fancies to explain it; conversely, imaginative work often rouses wonder in creator and audience. When Victorian microscopists warn against speculation, fancy, or unrestrained imagination, they do not necessarily critique wonder (admiration, awe, reverence, astonishment, or whimsy). And although the physiological psychologists argued that thinking and feeling are entwined, Victorian microscopists did not expect that the scientific imagination would always engage wonder.

While romantic tropes of wonder engage affective flows, skeptical tropes like *patient labor* and the *personal equation*—realist in their detached and doubtful tone—seek to manage primarily optical, mechanical, and cognitive relations between observer, apparatus, and object instead. Skeptical tropes usually adopt a moderating, deflationary, realist discourse, but they do not necessarily suppress the influence of the observer's mind and body as might be suggested by the theory of mechanical objectivity. Instead, microscopists use these tropes to direct action, to recruit the observer into particular practices. Such directives appear in optical diagrams, illustrations of apparatus, advice on manipulating the instrument, instructions on setting up a workspace, warnings about confirmation bias or the dangers of fatigue, and more. The skeptical voice responds to the challenges of embodied observation, where the eye, mind, and body of the observer are acknowledged as integral elements of the optical and physical system of the instrument; it acknowledges and seeks to curtail potential sources of error; it recognizes areas of uncertainty; it invokes method. Skepticism underwrites the protocols developed in the period to safeguard the work of empirical science, disciplining that work through a disinterested practice of precise description (with measurement, detail, and specialist language), record-keeping, and confirmation. Skeptical tropes model the observer's commitment to thorough, meticulous, and objective work—in short, the observer's rigor.

Today the figure of "the skeptic" hints of the carping killjoy, but Victorian microscopists portrayed skepticism as expressing an ethic of care. The physician Lionel Beale, for instance, repeatedly warns that careless, foolish observation and hasty speculation mislead the colleagues—current and future investigators—who may rely upon it. Such work thus also imperils the progress of microscopical science.[41] Skepticism and method, that is, seek to entrain observers in practices that support one's peers and the community of microscopists as a whole.

Wonder and skepticism describe different approaches to the microscopical world, but they are not incompatible with one another; indeed, they might both arise in response to uncertainty. In fact, the affective engagement of wonder can spark a skeptical investigation. A number of writers on wonder ask if, in pushing someone toward investigation, wonder risks extinguishing wonder. Fisher sees wonder as "the origin of intellectual life," placing it as the instigating factor of a narrative where the "pleasure [of wonder] induces, converts into, and then sustains investigation."[42] But although wonder thus spurs us to convert the exciting and unknown into the familiar and known, he argues that aspects of the "process of explanation" can actually reanimate wonder, by recharging "the same triggering mechanisms that brought on attention in the first place."[43] Indeed, skeptical microscopical investigation can often prompt new outbreaks of wonder, by providing better access to new views of the world. The opposite of wonder is not, then, objectivity or skepticism but indifference. Similarly, skepticism does not counter wonder but rather opposes hasty, careless, or partial work. It fosters a more disciplinary practice, asking microscopists to recognize how observation is shaped by optical, mechanical, and cognitive forces such as the limits of the instrument or the eye, their own bodily fatigue or clumsiness, and the human tendency to see what you hope or expect to see. Even working microscopists generally resisted any extreme or unrestrained skepticism that could endanger this empirical, inductive science. They embraced instead the many tools and strategies that could mitigate doubt enough to enable productive work. This book will show how Victorian observers used tropes of wonder and skepticism in this way to define and promote a burgeoning microscopical culture.

Our definitions of wonder and skepticism affect how we understand debates within microscopical communities. For example, Carpenter has been discussed as a foe of wonder because he warns microscopists to report only "what they actually *witness*," not "what they *believe* and *infer*."[44] But this advice, which draws on contemporary research showing the effects of human perception and cognition on empirical truth, reflects a suspicion not of wonder but of imagination.[45] Indeed, if we define wonder as affective response (not conflating it with imagination), it is readily visible in Carpenter's text. For instance, he titles his book with a nod to the microscope's "revelations" and frequently refers to the beauty of microscopical objects, encouraging microscopical observers to admire what they do see. He also refers to theologies of nature: in the same Introduction where he warns novices about undisciplined belief, he invokes the "operations of Creative Wisdom," which are "wonderfully displayed" in microscopical objects, and which "conspicuously solicit the attention of the Student of Nature by the beauty of their aspect or the elaborateness of their organization."[46] Overall, working with a definition of wonder as an affective engagement often expressed in romantic prose, I find not a strong division between writers advocating for or eschewing wonder, but rather a range of discursive spaces more or less welcoming to various forms of wonder and skepticism, and speakers more or less willing to engage in them. Broadening the critical gaze to comprehend both the wonder and the skepticism in Victorian microscopy can, I believe, provide a transformative perspective on the negotiations that helped form scientific roles and communities in the period.

Some literary forms, such as the *narrative investigation* that I examine later in this book, allow the microscopical writer to manipulate wonder and skepticism in productive tension. And some narratives, such as the microscopical fairyland, the fictional soirée, and articulated practice, create virtual storyspaces where the observer's body is represented interacting with the instrument or image, enlisting the reader in a compelling form of virtual witnessing. I hope also to surface the pleasure that animates both wonder and skepticism, whether it derives from enjoyment in studying beautiful or complex objects or satisfaction in a discipline meticulously practiced; for Victorian writers repeatedly describe how work with the microscopic world engenders curiosity, appreciation, and delight in the observer.

Two Cultures

Victorians used a stunning variety of optical instruments for both work and play. Some instruments developed in the course of optical experiments (such as the kaleidoscope or stereoscope) were widely adopted as toys; the microscope, too, enabled both serious study and sensational entertainment. Similarly, the camera could be used for scientific or artistic aims or as a diversion. Just as components of one visual tool (such as a lens or eyepiece) appeared in others as well, the landscape of Victorian visual instruments shows a significant overlap between the functions of research, instruction, and amusement. But, with some exceptions, contemporary scholarly fields often reinscribe a two-cultures view of Victorian microscopy. Scholarship on original investigators understandably emphasizes how expert practitioners develop a discipline of skeptical empirical method, building systems of technique and communities of practice to compensate for possible sources of error. Indeed, Lorraine Daston has noted how, in the history and philosophy of science generally, "the skeptical tinge of modern epistemology" results in scholarship "much more heavily weighted toward the wary avoidance of errors through experimental tests and evidentiary argument than toward the eager pursuit of the new."[47] At the same time, scholarship on "popularizers" or the performance of science often focuses on how wonder works in these texts.

There are good reasons for the prevalence of the two-cultures model; it reflects real differences in microscopical practice and a long tradition of plain speech in the scientific investigation of nature. After all, Thomas Sprat, in his 1667 *History of the Royal Society*, specifies that the natural philosopher "endeavours rather to know, than to admire."[48] The scholarly separation of wonder and skepticism also reflects the work of historians of science like Daston, Park, and Peter Galison, who have published influential histories of wonder in the early modern period and objectivity in the eighteenth and nineteenth centuries. In particular, Daston and Galison's analysis of mechanical objectivity emphasizes the nineteenth-century observer's attempt to exclude any subjective influence over observations. The ideal of mechanical objectivity, they say, "battles the general, all-too-human tendencies to aestheticize, anthropomorphize, judge, interpret, or in any other way 'tamper' with the givens of nature."[49]

The divergence of scholarship on wonder and skepticism also follows from Greg Myers' distinction between a *narrative of nature* (describing natural objects without mentioning the scientific observer) and a *narrative of science* (foregrounding a theoretical scientific argument).[50] In response, Barbara Gates and Ann Shteir propose two nineteenth-century precursors, both wonder-filled: the *narrative of natural theology* and the *narrative of natural history*. The tradition of natural theology encourages studying the natural world as evidence of design and a Creator (see Part II). The *narrative of natural theology* as defined by Gates and Shteir teaches natural theology by using dialogue within a domestic scene.[51] The *narrative of natural history* often uses fanciful prose to encourage a detailed study of nature.[52] Scholars do note the realist elements of these nineteenth-century genres, but they tend to emphasize their romantic approach. One exception derives from Amy King's work on a realist narrative related to the above that she calls *reverent natural history*, which uses "*detail*, *dilation*, and *minuteness*" to explore "the divine in the commonplace."[53] Victorian microscopical texts frequently reference all these nineteenth-century genres.

The two-cultures model has enabled influential and productive analyses, but the inadvertent result of the focus on local differences is to establish formal borders cutting across what was at the time a very fluid landscape and to suggest that microscopists had developed a more rigid generic code than actually existed. In fact, Victorian writers frequently tap multiple genres, and microscopical texts are no different.

This book bridges different strands of critical inquiry by studying how Victorians saw tropes of wonder and skepticism not as incommensurate responses to the natural world but as complementary approaches across a variety of settings, from the journal article to the instruction manual, the soirée to the solar microscope show. Many Victorian microscopists worked across multiple settings. Individual writers negotiated these approaches in different ways, using the strategies of separation, juxtaposition or counterbalancing, and combination to manage the productive tension of wonder and skepticism within any single text (see Part IV).

It is possible to identify generic trends as well. I read microscopical genres as community-building efforts, where writers with similar aims can declare themselves to the world, foster a sense of shared practice, and attract and enculturate new members in part by how they address wonder and skepticism in the print record. Writers used wonder and skepticism differently in different genres and for different audiences, to address the local demands of these various settings. Of course, most Victorian texts are not internally homogeneous but draw from multiple genres, even if one predominates; microscopists' use of wonder and skepticism can vary depending not only on a text's general aims and audience but also from moment to moment. By examining the circulation of these tropes, we can track the development of distinct communities of practice. For example, popularizations of microscopy such as lectures or articles for broad audiences do emphasize the marvelous, but they often authorize these marvels with elements of skeptical method such as quantification.

Unsurprisingly, articles in scientific journals rarely turn to sensationalism or whimsy. However, many do admire beauty, and they can do so fulsomely. Scientific texts also use romantic tropes such as *counting the uncountable* or *microscopical vistas* that help observers channel their own experiences of the visceral, affective force of the microscopic image, identify with the agency of the instrument and object, and evoke admiration, awe, reverence, astonishment, or whimsy from the reader. For example, as I will show in Part II, they may describe feeling a vertiginous plunge into the vastness of the very small, a pang of shock and sympathy upon seeing a water-flea devoured by a water-lion, or a thrilling sense that they could reach out and physically manipulate a particularly clear image. Wonder and skepticism thus work together to direct embodied observation even within the most elite ranks of microscopical culture. Finally, although tropes of wonder and skepticism helped microscopists identify and form distinct communities, they also helped construct a broadly shared understanding that microscopy had become a challenging yet rewarding practice.

My argument about microscopy builds on scholarship about the relation between Victorians and their machines. While scholars like Anson Rabinbach have largely focused on the industrial, steam-powered machines that exemplified the dynamic nature of nineteenth-century technological change, my interest is in a scientific tool, extant since around 1600, that moves very little and only under human power. Still, my reading of microscopy expands upon Tamara Ketabgian's claim for the affective power of "human-mechanical amalgams and assemblages." Ketabgian emphasizes the conflictual nature of the human–machine hybrid in industrial settings, innately strained by "prosthetic conflict, fusion, and control."[54] The microscope, a scientific and domestic instrument, does stage operators' ever-present struggles with the innate difficulty of vision and the recalcitrance of the instrument and object. However, the battle is often framed as a challenge of the self: one must train the eye, hand, and mind to meet these complications. Microscopists, like pianists in Ketabgian's reading, form "communities of feeling" (and, I would add, meaning) based on "sympathy and technique."[55] They meet the challenge and hope to reap a reward as much spiritual as scientific.

Metaphor and Microscopy

This book argues that Victorians used particular tropes to shape and direct their embodied experiences with the microscope and its objects. I use the broader definition of *trope* as figurative language or plot device generally, reserving *metaphor* to describe a kind of figurative language that associates two unlike objects (including simile). The tropes I examine take various forms, but many of them use metaphor. As George Lakoff and Mark Johnson have shown, our everyday life is deeply inflected by common metaphors that structure our thinking. They argue, "The most fundamental values in a culture will be coherent with" its metaphors and the concepts they represent—although "individuals, like groups, vary" in their responses.[56]

Scientific metaphors do more than express scientific ideas; they shape scientific thinking and help scientists advocate for particular scientific models of the world. There is a distinguished history of scientific metaphor; the seventeenth-century natural philosophers Robert Hooke and Robert Boyle both famously argued that, despite concerns about literary language, well-chosen metaphor acted as a kind of figurative microscope, dilating upon significant detail.[57] Boyle explains, "Proper Comparisons do the Imagination almost as much Service, as Microscopes do the Eye," because both enlarge the object of investigation and make them more "easy to be apprehended."[58] This formulation of the trope *metaphor as lens* has persisted over centuries, framing both metaphor and magnifier as valuable tools.

Anson Rabinbach and Laura Otis have focused books on particular scientific metaphors such as motors, networks, membranes, and borders, but my work here is closer to that of Andrew Reynolds (2018), who has surveyed the variety of metaphors used to describe a particular object (the cell) and inspire research on the topic. He examines the cell described as an empty chamber, building block, citizen individual within a polis, factory, or chemical laboratory. I am doing something similar with the microscope, which served not just as an object of study but also as a scientific tool, educational aid, and toy. Victorians circulated tropes about the microscope in order to claim specific cultural roles for it and shape microscopical practice. These tropes call readers into—interpolate them into—a network of objects, affects, and actions.

Reynolds defines metaphor as "the *conceptual organization* of a target domain" (in his case, the cell) that operates by adopting the structure of another domain (the polis, factory, etc.).[59] He emphasizes that metaphors do more than just aid in communication or teaching; they help researchers think through problems and build testable hypotheses. The best metaphors allow scientists to "make successful predictions and . . . discover objectively real similarities" between the domains being compared. In fact, like Boyle, he sees metaphor as an important scientific tool, arguing that "how to talk and to think about cells has always been just as important an issue as how to physically investigate them by means of material technologies and techniques."[60] Reynolds discriminates between *metaphor as lens* and *metaphor as tool*—although of course a lens is a kind of tool. Specifically, he posits metaphor as "the 'third lens' through which we see cells, the first two being the objective and ocular lenses of a compound microscope." He notes that "researchers often use multiple, even contrary, metaphors," a strategy like choosing "different magnifying lenses of a microscope" for different tasks.[61] Similarly, Pablo Maurette identifies a "haptic style" in prose, which "organizes space in idiosyncratic ways by means of rhetorical figures such as metaphor, simile, and metonymy, by reducing distance, by zooming into details and blurring contexts and backgrounds." Like Reynolds, Maurette suggests an analogy between figurative language and microscopical vision, with its interest in scale, detail, and texture.[62]

The trope of *metaphor as a microscope* rests upon the use of microscopes to enable careful visual and mental scrutiny. In fact, a complementary trope—*microscopy as scrutiny*—is often used to describe a sharp, skeptical inspection and analysis. This accounts for the use of "microscope" to describe texts of social or aesthetic critique, such as Catharine Quigley's poem *The Microscope; or Village Flies* (1819). Cornelius Tuthill says of his short-lived satirical periodical in 1820 that other names were considered—"*the Instructor, the Guardian*, or the *Connoisseur…the Observer, the Spectator* or *the Looker-on*," before *The Microscope* was chosen "on account of its enabling one to examine objects *nearer* than can be done…by the naked eye."[63] Algernon Charles Swinburne elaborates at length on the trope of *microscopy as scrutiny* in his screed *Under the Microscope* (1872), which responds to a vicious reviewer by figuring critics as vermin, himself as an entomologist, and his book as a great microscope putting these insects under the lens in an "examination of…[this] class of the infinitely little," the "varieties of the critical tribe."[64]

Microscopy can also figure as a very specific kind of scrutiny: reading. Emily Steele Elliott's children's book *Under the Microscope; or, "Thou shalt call me My Father"* (1861) uses the trope *microscopy as reading* to give a point to her moral tale. At a Sunday school outing for factory children, the clergyman Mr. Vaughan brings a microscope to the picnic scene, showing the eager children the glory of a "fly's wing," "spider's claw," and "beetle wing." Then he shows them what looks like a "speck of sand." Under a stronger lens, this object is discernible as a microphotograph of the Lord's Prayer. A "lense of yet higher power" is needed to enable them to make out the words with ease. A teacher explains the analogy to a doubting student: "we can't understand" the words of the prayer, "or read them right by ourselves," until God "opens our eyes by putting [the prayer] *under the microscope*." In this case, the trope *microscopy as reading* serves a religious goal: the teacher explains, "there is often need of a higher magnifying power," which she identifies as "*the love of Christ*." She concludes, "In care and gladness, joy and sorrow, we must just look *under the microscope*…that we may be enabled to read aright our Father's name."[65] Other titles use the term in the same way, such as William Sewall's book *The Microscope of the New Testament* (1878), where "microscope" refers to a close hermeneutical reading that dilates upon and magnifies (glorifies) the Word of God. The persistence of this cluster of tropes—*metaphor as lens, microscopy as scrutiny, microscopy as reading*—shows how the microscope was considered a valuable instrument of thought and interpretation as well as vision.

Reynolds examines how metaphors shape explanation and research; in this book, I pursue a similar interest in how metaphor sets up particular narratives and discourages others. To take an example discussed by Reynolds, the metaphor *cell as factory* enables narratives about the working of mechanism, where the metaphor *cell as empty chamber* does not. By enabling particular favored narratives, metaphors implicitly encourage the research programs that explore or extend these narratives over others that ignore or conflict with them—though sites of conflict may also spark research. The tropes I examine here encourage certain narratives about microscopy and discourage others, encourage particular forms of practice and discourage others.

Victorian writers respected this work of metaphor—its ability to yoke far-distant ideas, highlighting certain structures and asking readers to dilate upon those qualities. Like Carpenter, whose defense of scientific imagination was noted above, influential thinkers like Alexander Bain and George Henry Lewes advocated for the judicious use of scientific metaphor as well as scientific imagination and wonder. In *The Emotions and the Will* (1859), Bain, who was a linguist as well as mathematician and psychologist, applauds the use of scientific metaphor, attributing its power to its elements of wonder—its delightful appeal to feeling. He comments, "Those identifications of likeness in remote objects, and under deep disguises strike the mind with an effect of surprise, brilliancy, exhilaration, or charm." He points to "the feeling excited by a flash of identification between things never regarded as like before," and remarks, "we are arrested, startled, and excited into a pleasing wonderment." Indeed, "our attention is awakened and held fast upon the objects with a sort of temporary fascination." Bain claims that scientific metaphors function like imagination and plot-interest to help lighten the difficult task of understanding for students and investigators alike, catching our interest and holding it steady on the object for a deeper gaze. In fact, he argues, scientific metaphors provide valuable intellectual leverage through the pleasure they afford, by sparking this delightful sense of surprise and recognition. In the process of "scientific discovery," he explains,

> the sudden shock of wonder is accompanied with a marked degree of the pleasure of *rebound*, the lightening of an intellectual burden, or the solving of a difficulty that formerly weighed on the mind.... The labour of intellectual comprehension is reduced by every new discovery of likeness; and the first feeling of this gives a rush of delight, the delight we feel when we are relieved of some long-standing burden, or discharged from a laborious obligation.[66]

Bain explains these benefits over and over again, insisting that metaphorical "strokes of identity" offer uniquely powerful tools to clarify difficult subjects, as in example metaphors like lungs as bellows and infectious matter as yeast. He repeatedly refers to physical properties—weight and rebound, the spring of "surprise" and "intellectual relief." This lift-effect, he suggests, accounts for their power, as the "mind, labouring and struggling to understand what is difficult, is suddenly illuminated by the help of the well-chosen analogy, and enjoys the buoyant reaction." This pleasant surprise buoys the mind up into understanding with "the lightening sense of a deliverance from toil."[67] Like so many other elements in Bain's work, metaphor proves to be an embodied form of cognition.

For Bain, metaphor may be even necessary for comprehension. He warns that simply "having a thing before the eyes is not seeing, far less knowing." Students should rather seek "[t]he happy comparison,—by classification, analogy, or simile,—and the pointed contrast" in order to "vivify the mind" and recognize those structures that "formerly slept before the open eyes."[68] Here he pushes back against the convention that doing is better than reading, since (he believed) memory and imagination ignite the same train of nervous action that occurs with physical experience.

Carpenter also defines comparison as central to scientific work. In *Principles of Mental Physiology* (1874), he praises the "power of bringing together the *like* in the midst of the *unlike*." Like Bain, he points to the "peculiar feeling of *pleasure*" that often accompanies an apt metaphor. More cautious than Bain, though, Carpenter argues for a "judicious use of metaphor": it "adds force [and] ornamental variety to the diction" if appropriately used.[69]

In contrast, Lewes, who took up physiology only after years of experience as a literary writer and editor, was cautious about the overextension of scientific metaphors. In *The Physical Basis of Mind* (1877), he challenges the metaphor of *the body as a machine*, a trope that would gain particular traction in the later nineteenth century. He asserts that there is a "profound difference between ORGANISMS AND MACHINES," and that comparisons between "steam-engines and animal organisms" ignore "all that specially distinguishes vital activity." For instance, in an organism "all are parts of one sympathetic whole" where "[n]ot a nerve is stimulated, nor a muscle moved, but the entire organism is affected."[70] (For similar reasons he disapproved of the metaphors of the *nervous system as telegraph or railway*.)[71] But while Lewes refuses to identify humans as machines, he does not deny that we share significant features with them. His concern is essentially a linguistic one: the body–machine metaphor identifies partial congruence between organisms and machines, but it has been extended beyond its usefulness, obscuring important qualities of living form such as the integration of parts into "one sympathetic whole."

However, even as Lewes protests against the misuse of imagination, he uses metaphor extensively. He even turns to metaphor to criticize researchers' overreliance on imagination, which (he claims) accounts for the slow pace and poor quality of microscopical research on neuroanatomy, much of which was rapidly discarded as inaccurate. He complains that "the statements confidently repeated in text-books and monographs are very often for the most part only ingenious guesses, in which Observation is to Imagination what the bread was to the sack in Falstaff's tavern bill."[72] Lewes's frustration here is not with imaginative work generally, as the literary reference demonstrates, but with its misuse—with unwarranted speculation, taken as scientific fact. Indeed, he turns to a series of increasingly elaborated metaphors to make this point. First, he says, hypothesis "may be used as a finger-post, not as a foundation. It may suggest a direction in which truth may be sought; it cannot take the place of Observation. It may link together scattered facts, it must not take the place of a fact." He asserts that "[h]ypothesis is indispensable" to scientific work, just as "[w]e are glad of corks until we have learned to swim." However, he warns, an unverified hypothesis can lead us astray, like "a torchbearer sent out to look for a missing child in one direction, while the child was wandering in another."[73] To make his argument against unbridled speculation, Lewes leads the reader through a succession of increasingly fanciful metaphors: hypothesis as Falstaff's sack, a finger-post, swimming-corks, and a torchbearer seeking a lost child. In this passage, then, he uses one form of imaginative work to control another.

Lewes again uses a series of metaphors to describe the judicious use of imagination in this famous statement from *Foundation of a Creed* (1874). He says, "Science is fertile not because it is a tank but because it is a spring. The grandest discoveries...have revealed by the telescope of Imagination what the microscope of Observation could never have seen, although it may afterwards be employed to verify the vision."[74] His metaphors here may be memorable, but their message has sometimes been misunderstood. As with the passage examined just above, this is not an unrestricted endorsement of imagination. Critics often omit the last phrase, but Lewes emphasizes here the importance of (microscopical) verification as a check on speculation—even as he uses a striking metaphor about microscopy to make this point, one that he makes repeatedly. In *Physical Basis of Mind*, for example, he exhorts "medical and psychological students" to take up the microscope to counter the "positive statements, and delusive diagrams" of some psychologists; only microscopical work can "correct...precipitate conclusions" by following out "organic connections not previously suspected."[75] In these examples, the trope *microscope as scrutiny* brings its skeptical force to bear upon unwarranted speculation. Overall, Lewes charges investigators to respect metaphor and imagination as powerful scientific tools that—like a telescope or microscope—require care in use. In Chapter 8, I will discuss the trope *beautiful mechanism* as a reference to human and microscopical bodies, but it could just as easily describe metaphor itself: aesthetically pleasing but also providing a potent—sometimes, perhaps, too potent—cognitive leverage.

A Note on Terms

In this book I often speak of *microscopy* or a *microscopist*. These terms from the period may seem clunky or overly broad now, but they are actors' terms describing a particular historical moment. During the time period examined by this book, a wide-ranging field of endeavor grew up centered on the microscope and its diverse uses, and on the figure of the microscopist, whether that be a doctor at the bedside or a tourist at the seaside, who knew how to manage its complexities. By the end of the century, the microscope had gained stature as a scientific instrument, but popular microscopical clubs and societies were declining and the general identifier *microscopist* was losing ground to more specific roles such as *bacteriologist* and *petrologist*.

I use the contemporary terms *observer* when speaking primarily of the act of looking through the microscope, and *operator* when discussing the required fiddling with instrument, slide, and apparatus. A microscopist would ordinarily act as both operator and observer. In some cases, as in soirée microscopy or mass microscopy using projection instruments, an operator worked the microscope and *viewers* passively looked at the images.

I also frequently use a term, *microscopical*, which was much more common in the Victorian period than it is now. The Rev. J. B. Reade, a founding member of the Microscopical Society of London, humorously rebuffed the suggestion of "the Microscopic Society of London" as the new group's name, suggesting that they use the term "microscopical" instead to prevent "the possibility of ourselves being mistaken for microscopic objects."[76] The editors of the *Quarterly Journal of Microscopical Science*, introducing the journal in 1853, say:

> It is, perhaps, hardly necessary to apologise for the title of the Journal, as the term "Microscopical," however objectionable in its origin, has acquired a conventional meaning by its application to Societies having the cultivation of the use of the Microscope in view, and so fully expresses the objects of the Journal, that it immediately occurred as the best understood term to employ. It will undoubtedly be a Journal of Microscopy and Histology; but the first is a term but recently introduced into our language, and the last would give but a contracted view of the objects to which the Journal will be devoted.[77]

Microscopists were still considering this question in 1886.[78] Victorians did not consistently use the term, but it makes a useful distinction. In this book I preserve this distinction, using "microscopical" (despite its "objectionable" etymology from both Greek and Latin and its stuffy overtones) to denote "having to do with microscopy." I reserve "microscopic" for references to minute objects.

Victorians debated the appropriate classification of many of these objects, and they refer to them using a mixture of folk terms and scientific nomenclature. In their use of scientific nomenclature, they sometimes use italics and capitalization irregularly or use names that are no longer in use. I retain the terms used in the original passage, providing clarification where needed.

Chapter Review

The book is divided into four parts, each of which includes three to six chapters. Part I introduces important historical contexts for the book. It surveys microscopical roles from investigator to viewer, emphasizing how individuals might move between these roles and between different communities; and it previews some of the major sources for the book.

Part II focuses on beauty and wonder in Victorian microscopy, showing how Victorians employed romantic tropes of wonder in both scientific and popular microscopy. These tropes model and solicit particular affective flows across the material system, from observer to wriggling animalcule and back. While it is of course not possible to determine the actual emotions experienced by Victorian microscopists, their texts clearly privilege particular formulations of admiration, awe, reverence, astonishment, and whimsy. They help frame the emotional experiences of microscopy within established epistemological structures such as natural theology or the sublime; they harness the affective charge of conventional fantasy worlds as presented by the fairytale, Oriental tale, or phantasmagoria. These tropes present the felt experiences of microscopical observation in acceptable forms and seek to evoke similar experiences in readers. Writers thus authorize their own work and enlist readers in microscopical communities, a goal that they often explicitly stated. In an argument informed by the work of Victorian physiological psychologists and contemporary affect theorists—that is to say, an understanding of emotion as embodied, dynamic, relational, and modulated by culture—I show how in nineteenth-century microscopy these affective flows were mediated by both literary and technological structures.

Part III of this book demonstrates how skeptical tropes help working microscopists manage concerns about embodied observation and the linkage of observer, apparatus, and object. Where romantic tropes solicit and guide affective flows, these skeptical tropes promote particular optical, mechanical, and cognitive practices in microscopical work. I analyze these concerns in the context of nineteenth-century discourses around how perception, attention, and fatigue contribute to discovery, learning, and error. Although most of the scholarship on skepticism in microscopy focuses on scientific texts, I show how important it is in instructional writing and how popularizers could also use skeptical tropes performatively in recreative settings.

The fourth and final part of the book studies diverse approaches used by microscopical writers to channel the productive tensions that arise when wonder and skepticism appear in the same text. I show how writers separate, juxtapose, and combine these elements in a range of texts and examine virtual storyspaces in microscopical writing. Appendices at the end of this book provide a list of abbreviations, a timeline of developments in microscopical technology and culture, a list of microscopical tropes discussed in this book, and graphs showing trendlines in nineteenth-century microscopical publishing, as well as a glossary of common microscopical terms. Appendices available online offer an analysis of library catalogues showing the holdings of microscopical texts and a list of "wonders of the microscope" titles, available at meegankennedy.net.

The Promises and Ends of Microscopy

Microscopy has lost much of its vitality as a field category in the modern world. The discipline has been folded into the practice of science more generally; few twenty-first-century scientists would define themselves primarily as a *microscopist*. However, the term provides a valuable opportunity to study Victorian science and culture through one of the most iconic instruments of the period. Reading the nineteenth century through its use of the microscope productively disrupts our usual field categories, making visible forgotten patterns of Victorian work, knowledge, and expression.

The microscope achieved a remarkable status in nineteenth-century culture, rivaling the railway and telegraph as an emblem of modernity and enjoying an astonishing diversity of applications. It is "multistable" (in Don Ihde's term) in that it sustains different meanings in different contexts and uses. Buoyed by the technical improvements proposed by Joseph Jackson Lister and William Tulley in 1830, the compound microscope became one of the century's most iconic and versatile tools. It was an affordable, reliable instrument for students; a familiar diagnostic resource for practicing physicians; a necessary tool for research in botany, physiology, and other fields; a foundation of cell theory and bacteriology; and a symbol of scientific rigor. The simple microscope continued to provide important service for dissections, while specialty microscopes were developed for diverse professions. Microscopy provides a valuable window into important cross-disciplinary debates of the period on topics such as natural selection, sanitation, vitalism, and germ theory. It offers a unique vantage point on Victorian scientific culture, since it served at once as a specialized study and an element of almost every other science.

The microscope illuminated not just scientific subjects but also religious, popular, and domestic ones with compelling exhibits that large Victorian telescopes could not match. Versions of the instrument entertained general audiences in the educational play of bourgeois families, the earnest instruction of Mechanics'

Institutes, and the spectacular images projected in crowded lantern shows. Its images were a staple demonstration for texts and lectures in natural theology and natural history. It provided income for street exhibitors and a penny peep into nature for their customers. It was touted as adjudicator in businesses, kitchens, customs houses, and courts. The microscope could be a scientific tool, a fascinating hobby, an educational plaything, a mass entertainment. Although this book attempts to give a sense of the remarkable breadth and richness of Victorian microscopical practices, it does not claim to be a comprehensive survey but rather a map of the territory.

The microscope is not alone in its outsize effect on modern culture. For example, photography and cinema (two other lens-based technologies) also reshaped Victorians' experience of visuality and continue to structure our own understanding of the world. Indeed, as Isobel Armstrong points out, John Ruskin's critique of microscopy was part of a larger opposition to a scopic modernity aligned with mass visuality and the cultivation of spectacle.[79]

But microscopy offered a unique and influential version of modernity. Modernist artists and writers famously critiqued their Victorian forebears, but microscopic vision, with its inherently alienated, fragmented, ungrounded, defamiliarized, abstract, and private image, in many ways prefigures the modernist aesthetic. The microscope concentrates in one portable instrument both a sublime promise of extraordinary access to other living worlds and a forceful reminder of the irreducible individuality of human perception. The Aestheticist art critic Walter Pater likely draws on investigations executed with the help of the microscope—in particular, Lewes's and Spencer's writing on physiological research—when in 1873 he famously figures "our physical life" as in constant movement: "a perpetual motion of…the wasting and repairing of the lenses of the eye, the modification of the tissues of the brain by every ray of light and sound." And Pater echoes the concerns of the late-Victorian microscopist when he urges readers to a "sharp and eager observation" of "impressions unstable, flickering, inconsistent," through "momentary acts of sight and passion and thought"—or when he warns that "the whole scope of observation is dwarfed into the narrow chamber of the individual mind."[80] In short, microscopical skepticism, found across a wide range of Victorian texts, endorses the ironic, experimental, relativist, and demystifying elements that would become important features of high modernism. Microscopical images, founded on often fleeting glimpses, evoke (as Baudelaire says of modernity) "the ephemeral, the fugitive, the contingent."[81] Roberta McGrath argues that, with nineteenth-century microscopy, "the body was revealed as a puzzling mosaic of cells and strangely mutating structures. If the overall picture was lost, then it returned in a sequence of incomplete and splintered details."[82] This sense of lost organic unity describes not only the human body, born into processes of growth and decay, but also many aspects of the natural and built Victorian environment as seen under the lens.

Against this alienating vision of endlessly fragmenting detail, Victorian microscopists set the promise that insight and understanding, though perhaps not an ultimate knowledge, may be achieved through prolonged focus on a single image set out in a circle of light, with the steady labor required of original microscopical work. They assure readers that, by training the eye, the hand, and the judgment, the disciplined student may achieve expertise. They propose skeptical and wonder-filled tropes to direct this process, inviting observers' bodies and minds into particular interrelations with the microscope and its world. The modernists mocked their grandfathers' love of awe and spectacle, their earnest diligence; but Victorians' work with the microscope helped them navigate many of the challenges of the nineteenth century. By negotiating the visionary and cautionary language of microscopy, these writers strove to navigate many of the great scientific, religious, and ethical debates of the period, and to explore their own ambiguous position as humans in a rapidly changing world. After all, as Robert Browning puts it, "a man's reach should exceed his grasp": the microscope provides a site from which Victorians could try—and come to terms with—the limits of the observing subject. Microscopical practice, as laid out in the texts examined here, powerfully informs the observer of the deeply embodied and contingent nature of all work and the provisional status of all knowledge; and in this way it shapes a Victorian observer whom we can recognize as modern.

Writing Embodiment in Victorian Microscopy: Beautiful Mechanism. Meegan Kennedy, Oxford University Press.
 DOI: 10.1093/9780198940623.003.0001

2
A Brief History of Victorian Microscopy

> The history of the microscope, like that of nations and arts, has had its brilliant periods, in which it shone with uncommon splendour, and was cultivated with extraordinary ardour; and…intervals marked with no discovery…in which the science seemed to fade away, or at least to lie dormant.
>
> Jabez Hogg, *The Microscope*, 1854

> The last fifteen years have sufficed to elevate the compound microscope…[to be] the most important instrument ever bestowed by art upon the investigator of nature.
>
> Andrew Ross, "Microscope," 1839

Introduction

While *Writing Embodiment in Victorian Microscopy* is not a history, it takes historical change seriously as a necessary foundation on which to build an argument about microscopical culture.[1] I focus on the middle decades of the nineteenth century (1830–80), when a series of inflection points marked important shifts in British microscopy. Microscopy had featured in natural philosophy since the seventeenth century and had become a fashionable pursuit during the eighteenth century. Microscopical attractions attracted curious crowds in the **1830s**. For example, Philip Carpenter's *Microcosm* exhibition of lucernal and solar microscopes at 24 Regent Street ran from 1827 to 1833, and the Adelaide Gallery, featuring the brilliant and lively projections of the new oxyhydrogen microscope, powered by limelight, opened in 1832.

But these early instances of microscopy centered on the simple (single-lens) microscope, due to imaging problems in the compound instrument, which offered a wider field of vision and could be more powerful, due to its use of two widely separated lenses: the eyepiece screwed into the top of the body tube and the objective at the other end. However, the compound microscope suffered from spherical and chromatic aberrations causing deformed, blurry, and discolored images.[2] A well-corrected compound design had been achieved in eighteenth-century telescopes but was hampered in microscopes by the smaller size of their lenses.[3] Over the early decades of the century, many studied the problem of an achromatic objective for the compound microscope, including Louis Vincent Chevalier, Joseph von Fraunhofer, and Giovanni Battista Amici, as well as C. R. Goring, Andrew Pritchard, John Dollond, David Brewster, William Hyde Wollaston, John Herschel, and others. Amici experimented with a reflecting microscope design (using a mirror in place of a lens); Wollaston invented a doublet lens (two lenses glued together); Brewster and Pritchard suggested lenses made of diamond or sapphire. In 1830, Joseph Jackson Lister published his work with the optician William Tulley on a compound objective that was well corrected for both spherical and chromatic aberration.[4]

Most investigators used simple (single-lens) microscopes before Lister and Tulley developed the compound (multiple-lens) objectives that could provide sharp, well-corrected images.[5] However, many microscopists continued to use simple microscopes such as a dissecting microscope, or used an older compound microscope. Even as late as 1877, readers of the *English Mechanic* debated whether a serious worker needed a state-of-the-art compound microscope. This periodical addressed a broad readership of technology enthusiasts, including many of modest means, so this debate (which recurred in its pages) may say as much about the periodical as the instrument. However, this example also shows that, by this date, a range of competent instruments were available, not all of them wildly expensive; and it serves as a good reminder that the practice of microscopy involved a diverse variety of instruments throughout the century.

Microscopical publications flagged during the turn of the nineteenth century until Goring and Pritchard jumpstarted the pace with early essays and their short-lived 1829 serial *The Natural History of Several New, Popular, and Diverting Living Objects for the Microscope*, which led to a series of influential book collaborations. Pritchard claims in 1832 that their earlier microscopical texts were nearly the only works "of practical importance" on the subject to have been published in a half century.[6] Even after the new achromatic compound microscopes became available in the early 1830s, there was a lag of some years before microscopy gained ground as a standalone subject in British publishing. In the meantime, working with both older and newer technologies, naturalists renewed debates over histology, embryology, and cell theory, linking microscopical research into grand debates about the most primal elements of matter, the basic functions of living beings, and the relations between species.[7]

The **1840s** helped to establish new roles for the microscope in Victorian culture, perhaps prompted by a remarkable series of microscopical events in the late 1830s. European researchers had reported important new work: Christian Ehrenberg's discovery of fossilized infusoria and foraminifera in geological formations, and Theodor Schwann's formulation, with Matthias Jakob Schleiden, of a unified cell theory. Adding to the buzz, Louis Daguerre and Henry Fox-Talbot exhibited photomicrographs among their earliest efforts of photography. Another important technical development occurred in 1845 when Friedrich Adolph

Nobert started making his "test plates," using a diamond to cut fine lines on glass slides; finer lines closer together would be more difficult to resolve (see distinctly). Goring had suggested the use of test objects as early as 1827, but Nobert's plates provided standard objects to enable more accurate testing of microscopes and objectives. He produced more difficult challenges every few years once earlier sets had been resolved, sparking competition among microscope manufacturers and enthusiasts alike.

In England, the allure of the newly reliable instruments was amplified by a flurry of publications and the growth of natural history societies and soirées. The Microscopical Society of London (MSL) was founded in 1839, dedicated to improving the science of the microscope; one of its first actions was to regularize the size of microscopical slides.[8] It was at this point that Ross made the extravagant claim for microscopy quoted in the epigraph above. The microscope expanded its reach in Britain during this period. In 1842, the surgeon James Paget suggested that medical men should train in "microscopic investigation" to aid their work in "human anatomy and pathology."[9] Some followed his advice, traveling to Paris to study clinical medicine (George Eliot's fictional surgeon Lydgate is an early adopter of this approach), particularly the work in localized pathology that the microscope facilitated; or enrolling in extramural courses in microscopical diagnosis, which began to be offered for medical students about this time. Other practitioners resisted the new approaches. But medical microscopy—with other elements of clinical medicine, such as autopsies, chemical tests, new tools like the stethoscope and sphygmograph, and formalized training and record-keeping—would, over the following decades, help change nineteenth-century medicine from a bedside practice to a scientific profession, and transform the microscope from a *mere toy* to an icon of doctors' new, elevated status.

By mid-century, an 1846 reviewer of four microscopical books in the *Westminster and Foreign Quarterly Review* was decidedly bullish on Victorian microscopy. The author, George Thomas Fisher, Jr., dismisses the eighteenth-century microscope as "a mere philosophical toy," utterly "useless" for productive work. He firmly sets the *mere toy* trope in the past and sets up the new compound microscope, in contrast, as "one of the most beautiful and perfect instruments" available—in fact, "the most important instrument ever bestowed by art upon the investigator of nature." He points to its value in chemistry, pharmacy, physiology, pathology, medicine, the law, geology, paleontology, botany, and zoology.[10] This optimism may reflect Fisher's personal interest in this work: a laboratory assistant and devotee of new technologies, that year he also published a brief guide to microscopical practice, published by an optician's shop. This kind of text—a microscopy guidebook to be sold alongside the instrument—shows such firms catering to a lay public as technologies like microscopy and photography were more widely adopted.[11]

As the **1850s** began, then, Victorians were using the microscope in a wider variety of settings. Among the most active promoters of the instrument was the British physician Arthur Hill Hassall, who published a series of influential books on microscopic anatomy (1846) and on the use of the microscope to detect pollution in London water (1850) and food adulteration (1855). Buyers of microscopes and microscopical texts enjoyed broader choices in both instruments and texts, benefiting from an array of compound microscopes, which had been exhibited at the Crystal Palace, and a wave of popularizing guidebooks. In 1855, acting on a suggestion by William Carpenter, the School of Arts sponsored a medal for an affordable microscope; this prompted microscope-makers to produce a range of microscopes, which would eventually range from the inexpensive but reliable "school" or "student's" instruments to the magnificent Powell & Lealand No. 1 (1878), shown in Fig. 2.1.

A range of practical microscopical manuals, starting with John Quekett's 1848 *Practical Treatise on the Microscope*, also started to appear on the market, encouraged by new printing technologies and the repeal of the taxes on knowledge. In tabulating the 389 microscopy-related titles (books and periodicals) published in Britain 1800–1900, I find a marked rise beginning in 1839 (see Fig. AIV.1 in Appendix IV). The early part of the century yields only single-digit totals per decade, rising to 10 books published in the 1820s and around 20 in the 1830s, and a jump to about 40 in the 1840s, doubling to nearly 80 books published in the 1850s. After this the totals subside to almost 70 in the 1860s, almost 60 in the 1870s, and just over 50 in the 1880s, then rise again slightly to over 50 in the 1890s. When graphed with a five-year moving average to smooth out the disjunctions between decades, the trend rises from about two books per year in the 1830s to a peak of ten books per year in 1861, then leveling off at about six books per year until the mid-1880s and five per year after that (see Fig. AIV.2 in Appendix IV).[12] In both cases we see a pronounced increase of interest in the middle decades of the century (1840–1860), subsiding to a new normal at a level below the peak but still far above the starting point. The decreases after 1860 likely reflect the joint trends of a decline in hobbyist interest and the relabeling of microscopical investigations under reference to other scientific specialties. What these data capture is not so much the use of the microscope during the nineteenth century but the life cycle of *microscopy* as a field category.

This mid-century surge of books and articles strove to formalize microscopical practice and recruit new practitioners, helping fuel the growth of microscopical communities. Working microscopists increasingly turned to print publications in their efforts to regularize their practice.[13] Many of the most influential nineteenth-century manuals on microscopy first appeared in the 1850s, including guidebooks by Lionel Beale (1854, 1857), Jabez Hogg (1854), and William Carpenter (1856), as well as the dictionary published by J. W. Griffith and Arthur Henfrey (1856).

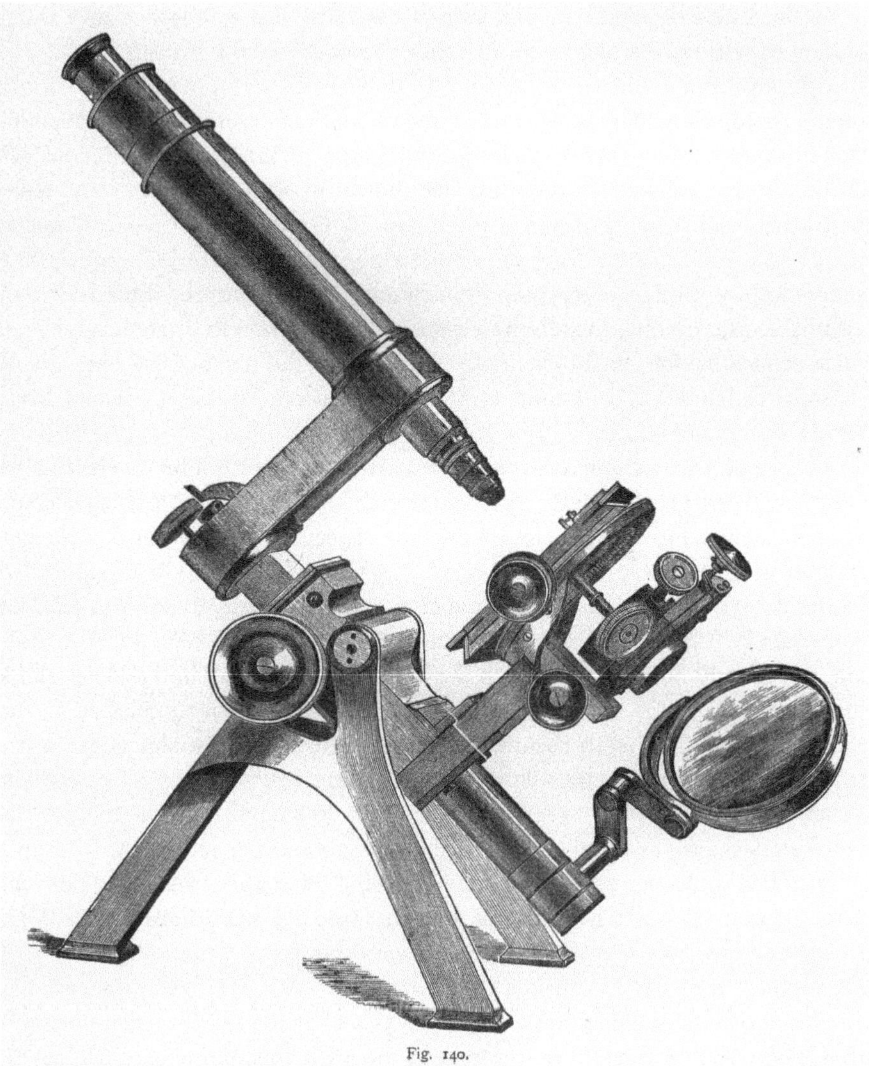

Fig. 2.1 Powell & Lealand No. 1 Stand.
Source: Henri Heurck, *The Microscope: Its Construction and Management.* London: Crosby, Lockwood & Son, 1893, p. 191.

This increased rate of microscopical book publication continued through the end of the century. While Goring, in an early article, had half-jokingly declined to be a "despot" or "Sir Oracle" in prescribing particular approaches to the practice, mid-century writers were not shy in doing so.[14] These books were ostensibly addressed to novices, but they served as comprehensive guides to practice and became standard references even for working microscopists, who had already mastered many skills. Other books, like Louisa Lane Clark's *The Microscope* (1858) and Mary Ward's *A World of Wonders Revealed by the Microscope* (1859), were directed to broader audiences in a drive to recruit more new microscopists from the general public. Even elite institutions like the Royal Society began to minimize the influence of aristocratic patrons and facilitate greater involvement by active researchers. The so-called "Christian philosopher" Thomas Dick, known for his blend of popular theology and science, had long been publishing books endorsing the microscope, but at mid-century his work was joined by a crowd of other instructive and amusing guides to microscopy, including books by Gideon Algernon Mantell (1846), George Thomas Fisher, Jr. (1846), Agnes Catlow (1851), Dionysius Lardner (1856), W. H. Olley (1857), Louisa Lane Clarke (1858, 1863), Mary Ward (1859, 1864), G. H. Lewes (1858), Philip Henry Gosse (1859), Edwin Lankester (1859), and J. G. Wood (1861). This partial list does not capture analogous developments in periodical publishing on the microscope nor public and private exhibits and lectures on the topic, but it does indicate the surging interest in microscopy in and around the 1850s, a trend that continued for decades.

Publishers of microscopical books were concentrated in metropolitan areas, such as London, Dublin, and Edinburgh. These cities hosted medical schools influenced by Continental models, giving rise to microscopical classes and scientific publishers.[15] Microscopical writing was distributed more widely, though, by provincial booksellers and circulating libraries, and by periodicals for scientific or general audiences.[16] The ready availability of microscopical guidebooks spurred public interest in the instrument; groups of enthusiasts in Birmingham, Manchester, and other locales followed the example of Bristol (where a Microscopical Society had been founded in 1843) and formed microscopical clubs.

Microscopy gained ground as a scientific tool during this period, although a plethora of conflicting observations and some high-profile errors, with the inherent difficulty of the work, meant that it was still not universally trusted. Some of the most influential guidebooks of this period emphasize questions of method, in an effort to regularize microscopical practice and secure more reliable observations. In 1855 Rudolf Virchow built on Schleiden and Schwann's cell theory, and on the work of British anatomist John Goodsir, to articulate the precept *omnis cellula e cellula* ("every cell stems from another cell"). Investigators narrowed their focus again, from localized to cellular pathology. It was a period of lively innovation in technology also; J. J. Plumer, writing in 1865 on the choice of a microscope, identified eight important technical advances from the 1850s, from the dark-ground illuminator to "Mr. Wenham's binocular arrangement."[17] Frederick Scott Archer's 1851 development of the complex but sensitive wet collodion photographic process encouraged experiments with photomicrography, although this mode of delineation would not be used much in publishing until the end of the century. In 1857 the MSL designated a universal screw size, allowing operators to mix and match components from different manufacturers. The standard screw helped develop microscopes as modular instruments and enabled better comparative testing of fittings like objectives. And new synthetic dyes developed at this time would greatly facilitate staining techniques. This approach to slide preparation helped define the structures within organic specimens, especially at high magnifications, and supported work in new specialties such as histology and, eventually, bacteriology.

During the **1860s**, microscopists were intimately concerned in the entwined questions of vitalism and various germ theories of disease, debates that were especially vigorous in this decade and the next. (Vitalists opposed the materialist doctrine of abiogenesis, commonly termed spontaneous generation.)[18] Microscopical practice in the 1860s followed out and accelerated trends originating in the 1850s, especially in staining, cell theory, photography, and the formation of provincial microscopical clubs. Physiology and experimental biology were buoyed by this surge in microscopical work. In London, the mycologist and editor of *Science-Gossip* Mordecai Cooke, following a suggestion published in the journal, helped found the Quekett Microscopical Club in 1865 (memorializing John Quekett, who had died in 1861). In 1866, under the leadership of James Glaisher, the MSL was awarded a Royal Charter converting their organization into the Royal Microscopical Society (RMS), thus implicitly recognizing microscopy, even with its emphasis on lower-status studies like physiology or histology, as a peer of inorganic sciences such as physics and astronomy. And in 1869 the American J. J. Woodward, an army surgeon and pathologist, demonstrated his skill in both microscopy and photography by producing a photomicrograph of Nobert's "19th band," a feat that was not bettered until electron microscopes were developed in the twentieth century. Even through the 1860s and into the 1870s, though, as Michael Worboys has shown, many practicing physicians, surgeons, veterinarians, even investigators doubted the reliability of microscopical science, skeptical of claims based on observations that were so difficult to achieve or replicate and that were subject to such interpretive dispute.[19]

In the **1870s**, microscopical communities were further shaken by the outbreak of theoretical controversies over the limits of the (optical) microscope itself. Amici had started experimenting with immersion lenses—first oil, then water—in the 1840s, encouraging further exploration of the technology, though they were not much taken up at first. In 1871, however, the American optician Robert Tolles exhibited an impressive homogeneous immersion with balsam, sparking the so-called Battle of the Glasses (also known as the Aperture Wars) on the question of maximum angular aperture. Thus while, as John Harley Warner notes, "before the 1850s microscopical enthusiasm comparable to England was almost entirely absent in America," the work of Tolles (and before him Charles Spencer) meant that by the 1870s the most powerful American objectives were outracing English ones.[20] The aperture wars gave a new turn to longstanding disputes over low powers versus high powers, workers versus diatomaniacs, and British versus

American or Continental technology, and it spurred an ugly chain of confrontations in print. At the same time, the vitalism debates also erupted in blunt *ad hominem* attacks. These ructions strained ideals of gentlemanly science. Microscopical optimism was also complicated in this period by Ernst Abbe's and Hermann von Helmholtz's theories on the limits of optical microscopy, first translated into English in 1874. Incredibly, the finest bands Nobert carved, on his 1873 test plate, were beyond the reach of the optical (light) microscope.[21]

In response to concerns about the physical limits of optical microscopy, investigators were encouraged to take up grueling longitudinal studies of the life-history of cells and organisms, like those of Dallinger and the physician John James Drysdale. William Carpenter had suggested such a course in his 1856 book *The Microscope and Its Revelations*, where he urges readers to "concentrate their attention upon some particular species...and work-out its entire history with patience and determination"; but such sustained observations were rarely reported at the time.[22] (25).

Ironically, given all the talk about microscopical limits, microscopists enjoyed significant improvements in working resolution with the introduction of the homogeneous oil immersion lenses that Abbe produced with Zeiss in 1878. These innovations, combined with improvements in photomicrography, supported the development of bacteriology and modern germ theory. Gerard L'E Turner has graphed the improvements in microscope objectives and correlated this trend with the ability to resolve ever-smaller objects.[23] Victorians' growing enthusiasm for and use of the microscope helped support the increasing specialization and professionalization of science. At the same time, the founding of the Postal Microscopical Society in 1873 (again through correspondence in *Science-Gossip*) made it possible for any microscopist within reach of the postal service to practice slide-making, observing, note-taking, and drawing with a congenial set of colleagues.

In medicine, some British physicians and surgeons had resisted laboratory medicine and microscopy long after James Paget, Robert Lyons, Lionel Beale, John Hughes Bennett, and others started advocating for its use in the 1840s and 1850s. However, the 1870s served as a watershed moment when more medical schools started incorporating laboratory training as a formal element of a medical degree. By the 1880s, licensing examinations covered the laboratory work of physiology and pathology, and medical textbooks discussed bacteriology—although use of the microscope for diagnosis was still challenging and uncertain for most practitioners, and doctors could not outsource such work to consulting laboratories until the 1890s.[24] The late 1870s and 1880s saw the development of tropical medicine as a specialty, amidst growing interest in parasitology and bacteriology; by the 1890s, clinical laboratories were established in a swath of British colonies—a process that, as Pratik Chakrabarti has argued—"embodied a political and moral regime" resonant with British imperialist ambitions.[25] The status of the microscope at this point is evident in that it joins an array of scientific apparatus and practices—including also animal experimentation and vaccination—exported from Britain to "improve" and control tropical lands and their peoples. *Writing Embodiment in Victorian Microscopy* focuses on mid-Victorian microscopy, during a period of tremendous growth and optimism for the role of *microscopist*, after the achromatic compound microscope was developed but largely before these significant milestones signaled a wider acceptance of the microscope as a scientific tool.

Most of my analysis here centers on texts about work with the compound microscope after Lister's and Tulley's advances, but use of the simple microscope persisted throughout the nineteenth century. Whether a dissecting microscope, a bedside microscope for urinalysis and other quick medical tests, a botanical glass for fieldwork, a pocket microscope used by customs officers to examine imported goods, or even a penny microscope made of a pill-box and drop of Canada balsam, versions of the simple microscope continued to make up a good portion of nineteenth-century microscopical observation. Victorian writers tend to refer to such instruments generally as "microscopes" or specifically as particular types of microscopes, rather than differentiating them under the category of simple microscopes. Most of the texts examined in this book, however, concern the compound microscope.

The microscope can be seen as the protagonist in a compelling success story—it became a valuable tool for a range of Victorian sciences—but as scientists began to organize their work in the distinct modern disciplines we recognize today, the user base for microscopy fractured into discrete professional and popular communities. Organizations centered on general microscopy lost momentum toward the end of the century, as hobbyism waned and investigators were drawn off to field-specific societies and journals.[26] The editor of the journal *The Microscope*, trying in 1893 to update a directory of American microscopical societies, noted a "remarkable decrease in societies during the past few years": fifteen out of seventy were "reported dead."[27] Even as microscopy became ever more important to modern science and culture, fewer observers would identify themselves as primarily a *microscopist.*

Victorian microscopy emerges as part of what Isobel Armstrong calls "an avidly scopic culture—a culture of looking" with "acute awareness of an intensely mediated culture."[28] Victorians experienced new ways of viewing the world on both a very large scale, through the "numerical method" (statistics), and a very small one, through microscopy. The microscope was joined by a wealth of new optical apparatus such as the kaleidoscope, stereoscope, zoetrope, ophthalmoscope, and of course camera. Contemporary writers also (as I will show) frame microscopy as a technology of travel or exploration, aligning it with the telescope, panorama, telegraph, railway, aquarium, hydrogen balloon, and tunneling shield, other contemporary technologies that shrank large distances to a manageable size or brought the viewer into or through unnavigable spaces.

Over the course of the century, the microscope became a valuable tool for scientific work, including geology and (with polarization) chemistry; indeed, one review enthused that "in the hands of an Owen and a Mantell the microscope becomes an instrument of magic power."[29] However, it was most influential in the life sciences. Hogg argues in *The Microscope* (1854) that "to point out in detail the discoveries made through [the microscope], as regards physiology, would be to give a history of modern biological science; for there is no department in this study which is not more or less grounded upon the facts and teachings of the microscope."[30] And although the microscope was useful in examining the structure of tissues, it was especially associated with minute plants, animals, and bacteria. As Hogg explains, he "ventured to dwell somewhat longer" in writing on "the very minute portion of creation . . . in the belief that that department is more especially the province of the microscopist than many others."[31] I focus on the study of microscopic organisms in part because they feature in such a wide range of texts.

The Telescope and the Microscope

Microscopical communities were not unique in developing a culture of wonder and skepticism; such an approach also emerged among, for example, geologists and astronomers, as Victorians grappled with expanded concepts of deep time and deep space. Indeed, microscopists borrowed some important tools from astronomy, such as test objects and the notion of a *personal equation* to address variations in individual observations. I am interested in examining how microscopists negotiated these approaches in ways specific to microscopy—for example, developing an aesthetic of delicacy and minuteness, or innovating ways to share a microscopical image that was inherently experienced in a solitary and private manner. Microscopists' self-definition helps us identify the most salient elements distinguishing Victorian microscopy from similar practices and fields of study.

The *telescope and microscope* were often paired rhetorically and compared in practical terms, since both of these instruments featured lenses arranged in a tube to enable a human operator to view previously unknown structures of nature, huge or minute. Both became iconic modern technologies and revealed *new worlds* to the observer. (The photographic camera boasts a similar structure and related aims but is usually framed as an adjunct to the microscope or vice versa, not as its functional counterpart.) As the physiologist Thomas Laycock said in 1860, "the microscope has done that for the world of organisation which the telescope has done for the universe at large."[32]

However, eighteenth-century telescopes enjoyed significant advantages over microscopes. While, like microscopes, they were sold as toys, the larger size of their lenses made it easier to produce reliable instruments capable of supporting scientific work. In 1758 the optician John Dolland patented a telescope lens combining crown and flint glass to combat spherical and chromatic aberration, spurring the manufacture of achromatic refracting telescopes in a range of powers.[33] In contrast, as noted above, the compound microscope's smaller objectives were more finicky to work with, and it was not reliably corrected until the 1830s.

Additionally, the telescope was glorified as transporting the viewer into deep celestial splendors, whereas the microscope had been used to traverse ever-smaller increments of space and to inspect (among other things) dirt, parasites, spermatic fluid, and urine. The Scottish poet Thomas Campbell drew upon this longstanding sense of the inferiority of the microscope in an 1821 lecture critiquing social satire. With its tendency to exaggerate our "follies" and "fluttering vanities," he wrote, ridicule is not like "the magnifying telescope, that makes us acquainted with the glories of heaven" but rather "the microscope, that amuses us with the plumage and panoply of the half-visible tribes of creation."[34] Campbell draws on common perceptions of microscopy in dismissing it here as obviously unserious, superficial, and vain, and preoccupied with the trivial and low. The telescope (it is implied) impels us to greater things, the microscope to petty ones. Both channel affective flows, but one offers the sublime; the other, disgust.

That said, the comparison between microscopy and telescopy shifted over the course of the nineteenth century as the status of the microscope rose and as well-made parlor instruments allowed more middle-class observers to become working microscopists; in contrast, astronomical investigation required large, expensive telescopes. The Scottish clergyman Thomas Chalmers, in his influential book *A Series of Discourses on the Christian Revelation, viewed in connection with the Modern Astronomy* (1817), made this comparison a cornerstone of his arguments. Thomas Dick, in his book of natural theology, *The Telescope and the Microscope* (1851), and William Whewell, in *On the Plurality of Worlds* (1853), followed suit. The two instruments were frequently paired in a dialectic defining them as complementary (and competing) tools with which humans could perceive both extremes of the great hierarchy of the created universe.[35] As Chalmers says, "The one led me to see a system in every star. The other leads me to see a world in every atom."[36] Whewell quotes Chalmers and adds, "Infinity in one direction was balanced by infinity in the other."[37] This framing stages both instruments as comparable, set off in harmonious balance one with the other. The idea of an equitable partnership of telescope and microscope persisted through the century; Arabella Buckley (longtime secretary to the geologist Charles Lyell), in her 1890 children's book *Through Magic Glasses*, describes a "magician" (schoolmaster) and his apparatus: "If in the stillness of night the telescope was his most cherished servant and familiar friend, the microscope by day opened out to him the fairy-land of nature."[38] In this scene, despite the teacher's greater comfort with the familiar telescope, the powers of microscopy are judged to be comparable.

The nineteenth-century partnering of telescope and microscope required reframing the microscope's focus on littleness as a benefit, not a shortcoming.

Chalmers praised the microscope in this way, posing it as an antidote to the religious doubts sparked by the telescope. While "the one [the telescope] is constantly widening the circle of [God's] territory," recontextualizing Earth as just one of many worlds, "the other [the microscope] is as constantly filling up its separate portions, with all that is rich, and various, and exquisite." That is, "the one told me of the insignificance of the world I tread upon," but "the other redeems it from all its insignificance" by testifying to God's care, even to the vastness of infinitesimal "worlds teeming with life... numberless as are the glories of the firmament," minute and hidden from human sight but all thriving as witnesses to God's creative powers.[39] This comparison appears over and over again, often reusing Chalmers's phrases without much attention to the theological arguments that frame them in the original—especially when reprinted in texts for general or youthful audiences.[40] Tropes of wonder abound in all these pairings: *revelations* of *new worlds*, *vastness*, *wonder everywhere*, and the like (see Part II).

Despite Chalmers' defense of the microscope, the conventional yoking of these instruments sparked nineteenth-century microscopists into a kind of competition with astronomers. The ubiquity of this comparison may have aided the microscope early in the century, when its status was low, by framing it as a partner or complement to the telescope, which already enjoyed considerable respect as a philosophical instrument. However, early-nineteenth-century microscopy did not necessarily gain by the comparison. Even after the improvements to the compound microscope sparked interest in the instrument, the telescope was associated with the high-status sciences of astronomy, physics, and mathematics, while the light microscope served a broader range of sciences but those generally of lesser status: the life sciences and medicine, geology, chemistry, and optics. Thus, even though many microscopists were adept in the use of the telescope, a rivalry persisted and microscopists remained in a defensive stance. For example, Quekett opens his influential 1848 *Practical Treatise on the Microscope* by arguing that recent technological improvements "have given to this instrument an importance second only to that of the Telescope," transforming "what might once have been an amusing instrument or a toy, into one of the most powerful auxiliaries that can be employed in scientific research."[41] While he makes great claims here for the microscope, he also ranks it below the telescope and acknowledges its history as a much more trivial instrument.

While Victorian microscopists admit the supposedly diminished status of the eighteenth-century microscope, they deflect that critique by claiming that, as a philosophical instrument, the nineteenth-century microscope was comparable or even superior to the telescope. For example, in 1829 Goring remarks that "[n]one are apt to treat microscopists with more contempt than some astronomers, and even mere star-gazers." He responds to this remembered insult by asserting the importance of the microscope, saying, "To the telescope certainly belongs the inspection of the *great* and *sublime* works of the creation; to the microscope belongs the *petites* and *beautiful* ones: if the former shews us the world above, the latter exhibits the world beneath us; telescopic objects are numbered, those of the microscope inexhaustible."[42] Here the microscope and telescope are as balanced as the measured periods in Goring's prose.

After the 1830s improvements to microscopical technology, some commentators go further: as early as 1846, the geologist Gideon Mantell compares microscopical to telescopical visions and deems the microscope "that noblest instrument of natural philosophy."[43] John Brocklesby, an American naturalist, extends the list of microscopical excellencies in 1851, saying that the microscope is of "superior importance" compared to the telescope because it shows animate nature.[44] Similarly, the physician Henry Maudsley writes in 1868 that "The exquisite minuteness and consummate delicacy of the operations going on in the most intimate recesses of nature are even more striking and wonderful than the vastness and grandeur with which the astronomer is concerned."[45] Brewster threads the needle in the *North British Review* in 1856, writing, "Second only to the telescope, though in many respects superior to it, the microscope transcends all other instruments in the scientific value as well as in the social interest of its results."[46] Brewster's careful, hedging language here cannot hide the reversal that promotes the microscope from "second" to "superior" and even transcendent.

Statements on the superiority of the microscope gathered momentum at mid-century, as manufacturers raced to outdo each other's offerings and communities formed promoting higher-quality microscopical investigation. Frank Crisp, a solicitor and tireless booster of the microscope who surveyed the field in 1878 for the RMS, contends that "our text-books give the uninstructed reader the impression that the microscope stands in point of principle on no higher level than the telescope. This is far too low an estimate, and it would be truer to say that the microscope stands to the telescope at as great a relative distance as the chronometer stands to the sun-dial."[47] Crisp's choice of the term "chronometer" rather than the homely "clock" or "watch" indicates the status he hopes to attribute to the microscope as a precision instrument suitable for research and accords with his longtime efforts to secure the status of microscopy as a science.

Goring also offered a humorous and pragmatic defense of microscopy that highlighted not its value for experiencing sublime mysteries but rather what was most convenient and amusing about it. He asserts, "I must here positively chant the vast superiority of a microscope, as a *hobby-horse*, over a telescope, at least when the latter is used merely *as a star-gazing instrument*," and he argues,

> Let us suppose you have, at a great expense, procured a very fine telescope. How many hours in the year will you be able to use it?... How many colds and quinzies are you to get by exposure to the night air, when you might be much better employed in your warm bed? How many disappointments are you to endure from finding the wind suddenly shift...? How many scoldings and curtain

> lectures from your wife? After all, when you have seen all the double stars, clusters, nebulae, within reach of your instrument...what remains but to dispatch your charmer up to the garret, as of no further use?[48]

Goring paints here a sad scene of the disappointed, shivering, sickly, and henpecked stargazer, who should be in his "warm bed," where he could be "better employed," presumably with the wife who had been displaced by the "charmer" of a telescope. (There may have been some truth to this: Gosse's wife attributed his death to having caught a chill while adjusting a telescope on a cold winter's night.)[49]

In contrast to the frustrated and ineffective stargazer, Goring suggests the delights of microscopy, which are always available and inexhaustible. In an 1827 article he had pointed out that microscopists, unlike astronomers, can work with artificial light to "compensate for any loss of rays by transmission through several surfaces."[50] In 1838 he points to other significant advantages over the telescope: "A microscope...is available at all times and seasons" and "[t]here is no end to the number of objects it may investigate." Indeed, he says, "In the solitude of the most dismal prison, you would only need a lamp, a microscope, and some vegetable infusions, to furnish you with an inexhaustible fund of amusement, totally independent of other external objects."[51] Although in his zeal he chooses an extreme and unlikely example of microscopical amusement, Goring counters the discomforts of astronomy by imagining how much more practical and pleasant is the microscopist's cozy workspace.

Victorian writers of all sorts use a common trope that posits another difference where perhaps none exists. Unlike the telescope, which allows us to "see far away," the microscope does more than simply allow us to "see more closely"; it reveals what is not just tiny but *invisible*. For example, Hogg evokes this concept in remarking, "Leeuwenhoek and Swammerdam, Lyonet and Ellis, examined the invisible forms of nature."[52] Thus, Martin Willis notes, while astronomers worried that the great distances traversed by the telescope distorted their observations, "microscopists were concerned to know whether the objects they saw actually existed."[53] Of course, far-distant stars also may be invisible to the naked eye, and observers may wonder if they exist at all. The trope of *seeing the invisible* appears often in Victorian writing on microscopy, though, lending a magical air to the microscope that should obviously give it the palm in any contest. Over the course of the century, other devices were invented that could reveal invisible forces or objects—the spectroscope, the polariscope, the Roentgen ray—but the trope of *seeing the invisible* still regularly appeared in microscopical writing.

The framing of telescope and microscope as obvious counterparts persists even in scholarship today. Some writers emphasize similarities between the instruments: Armstrong argues that both the telescope and microscope introduce a fundamental instability to vision—that "the *norms* of the lens were the action of bending light, ungroundedness, ideality and dissociation from matter"—and

Anna Henchman remarks that telescopes and microscopes both "radically change the appearance of what one observes" and "introduce a disconnect between the world as it is ordinarily seen and the way a scientist sees it."[54] In these respects, telescopes and microscopes share much with optical tools or amusements from the period such as the camera obscura, phantasmagoria, kaleidoscope, stereoscope, zoetrope, and panorama. All these play with human visual expectations in order to present reality under a new guise.

Other scholars emphasize the differences in their function, finding new reasons to give the palm to the microscope. The pairing of telescope and microscope even in these critical texts speaks to how powerfully and persistently these two technologies have been counterposed: two competing magnifiers, like mighty genii, offering opposite visions and vying for dominance in the Victorian visual ecology. Roberta McGrath, for example, suggests that the microscope "offered the opportunity for manipulation, while the telescope served only to make the observer aware of his trifling insignificance within a vast universe he could not hope to control."[55] Indeed, Victorian astronomers were at the mercy of local conditions—not just cold nights as in Goring's anecdote above, but also haze, light, or vibration interfering with the image. While Victorian microscopists did sometimes struggle with light or vibration at the observation site, they could much more easily control the conditions of observation.

Just as important, astronomers generally cannot probe or change what they study, whereas microscopists necessarily manipulate both instrument and object.[56] Astronomers' relations with heavenly objects must of necessity be one-sided; nothing they do can change the stars whose light finally reaches the questing eye. It is perhaps for this reason that Thomas Hardy's novel *Two on a Tower* (1882) sets a troubled love-affair against the indifference of the universe. The distance in scale can make the microscopist's role seem to be a watcher only, like an astronomer, but there are crucial differences.

The peculiarity of the microscopist's role comes to the fore in a humorous essay published in the *Pall Mall Gazette* in 1894 by H. G. Wells, "Through the Microscope," whose peevish commentator mounts a series of complaints about the instrument—fretting, for example, that the microscopist, like a "Peeping Tom," peers in at "innumerable things" who "are being watched by a creature of whose presence they do not dream": the observer "sees them, and they do not see him." This passage portrays the *microscopist as a voyeur* who is "too incredibly vast and strange to come, save as an overwhelming catastrophe, into [the] lives" of these unknowing beings. But that interpolated phrase, "save as an overwhelming catastrophe," recognizes that the watcher is not, actually, just a disembodied eye but rather a material presence; the microscopist's actions often affect the objects of study—inevitably provoking physical and other responses.[57]

In fact, setting aside the passive viewings at soirées and exhibitions, every microscopical observation begins with some manipulation of the object, multiplying the skills and work required in microscopical practice. As the great historian of microscopy Gerard L'S. Turner quipped, the microscope is "unlike its twin brother the telescope because objects for the telescope are obvious."[58] In order to access and visualize an object of study, microscopists shift the orientation of the object or illumination; introduce stimuli to observe a living specimen's reaction; kill, dissect, section, or stain the object; and mount it for repeated observation at their leisure. In fact, the tropes of *multiple views* and *many-sided comprehension* urged operators to investigate objects in as many different ways as possible. A set of skills grew up around those processes that became as much a part of working microscopy as the manipulation of the apparatus itself. This did not necessarily make microscopy easier. The physical interactions of the operator especially with a living object form a unique relation, complicating any observation. These interactions also prompt anthropomorphism and critiques about microscopical vivisection, as I will show in later chapters; even inanimate objects can present a kind of resistance to the operator's efforts.

I would note two other important differences between telescope and microscope. First, astronomers share a set of objects to be observed. They might use different telescopes or watch on different nights or under different conditions, but the ultimate objects of study are the same. But this is not necessarily the case in microscopy. True, at lectures or soirées, observers might take turns looking through the same instrument at the same object, and in performances using projection microscopy or magic lantern slides, audiences would all gaze at the same image. But in most cases, observers prepared and viewed different individual objects; even commercially marketed test slides were not identical. This made it much more difficult to secure agreement about the true structure of any difficult object. By and large, microscopists could not consolidate their work in a "collective ledger," as Lord Rosse's project did in its investigation of nebulae.[59]

Finally, where the telescope looks across a great *distance*, the microscope traverses *scale*. True, the microscopical observer sees only what is within easy reach, but within the microscope's brazen tube everyday space gains a new dimension that opens out of the present place and moment, a dimension available everywhere and yet existing seemingly nowhere, invisible to us. The microscope *creates depth out of flatness* by showing pockets of complexity within what looks simple, even empty to us ordinarily. The observer thus performs a virtual penetration into hidden layers of space. As Gaston Bachelard says, "The miniscule, a narrow gate, opens up an entire world.... Miniature is one of the refuges of greatness."[60] Indeed, the microscope's infinitely dividing gaze reveals objects as, unbelievably, "bigger on the inside." Susan Stewart similarly argues that "the miniature...threatens [an] infinity of description without hierarchization...whose profound interiority is . . . always unrecoverable."[61] Bachelard and Stewart suggest that the microscope can yield both smallness and largeness—no telescope necessary. (Indeed, the *vastness of the very small* was an important trope for Victorian microscopists.) While the telescope challenged Victorians with views that unsettled both mind and faith, the microscope destabilized vision and space like no other technology.

Writing Embodiment in Victorian Microscopy: Beautiful Mechanism. Meegan Kennedy, Oxford University Press.
 DOI: 10.1093/9780198940623.003.0002

3

Who Is the Microscopist?

> Microscopical inquiry may be undertaken by persons in almost any position... [yet] I fear it must be confessed that the observers are comparatively few.
>
> Lionel Beale, *How to Work with the Microscope*, 1857

Introduction

The term *microscopist* in British texts implicitly refers to a white man of the middle or upper classes: someone with the wherewithal to pursue microscopy as a hobby or vocation. The tradition of gentlemanly Anglican investigators ("parson-naturalists") remained strong, since microscopy posed a relatively lower bar to entry than sciences like physics or chemistry. Microscopical observers could contribute to scientific knowledge without the space or expense of a dedicated laboratory. Vicars and curates could readily find material on "rambles," suitably house the apparatus in a study or drawing room, and conveniently carry out observations during non-work hours. As a result, British microscopy developed a rich, quirky ecology ostensibly open to dedicated observers from any background. Many working microscopists, parsons or otherwise, had to confine their practice to leisure time yet still contributed to or confirmed microscopical knowledge, especially in fields deriving from natural history.

When microscopical evangelism began in earnest in the 1840s and 1850s, guidebooks cast a wide net to solicit workers, claiming that even a hobbyist with an indifferent instrument could contribute to microscopical science. From 1855, William Benjamin Carpenter and others encouraged the manufacture of more low-cost, accurate student microscopes. Classified advertisements in *Science-Gossip* and other periodicals allowed cash-strapped novices to find used copies of classic guidebooks. As late as 1886, C. T. Hudson (who had studied *Rotifera* for thirty years) declared that rotifers

> are to be found almost everywhere; they cost nothing; they require neither expensive lenses nor an elaborate apparatus; they tempt us to explore the country, and to take pleasant walks; they are beautiful themselves;... they suggest all kinds of difficult questions on life and being. Moreover there is happily still a great store of scientific ignorance concerning them, thus leaving an ample field for fresh discoveries.

For a modest outlay of time and money, then, readers can enrich their lives, contribute to scientific knowledge of "difficult questions on life and being," and enjoy "the contemplation of Him, whose almighty hand is as visible in an atom of this animated dust, as it is in the myriad sparkles of the starlit heavens."[1] Nineteenth-century microscopy offered a close encounter with the sublime that you could fit on a parlor table.

Despite the missionary rhetoric and attempts to make the field more accessible, it would be easy to overstate the ease of entry for would-be microscopists. Even the modest expense of a student's microscope might be too much for some—and advanced work demanded much more than just a basic apparatus and some ditch water. To become an adept required considerable time, training, and funding for books and equipment (multiple objectives, tools, slides, fixatives, stains, and the like), a stable, well-lighted workspace conducive to focused work and undisturbed storage of materials, facility in reading well enough to manage scientific terminology, and a community of interlocutors. John Hughes Bennett, a physician and promoter of medical microscopy, warns, "There are few studies where, owing to the sources of error inherent in the best microscopes, the manual dexterity required, and the difficulties to be overcome, a teacher is more necessary."[2] These inherent demands limited the ability of many women and working-class naturalists to participate in microscopical communities as equals. Historically, most working microscopists during this time were men; most elite spaces of microscopical science remained difficult of access for women and working-class microscopists throughout much of the century. Local microscopical communities were more likely to welcome such participants.[3]

That said, women were thought to have a particularly delicate touch, suitable for needlework, piano playing, or microscopy. The physician Lionel Beale even wrote in 1880 that microscopy "is a department in which ladies are likely to excel."[4] Women had already earned a reputation for skill in scientific illustration, and it is not unusual to find a male author thanking some woman (often a family member) for this. Some women did observe or even work as microscopists, especially in the later years of the century, whether as an investigator (alone or with a spouse), as a delineator or engraver, or as a student or hobbyist; but they were excluded from most scientific societies, especially those of higher standing, for much of the century. The Royal Microscopical Society (RMS) did not admit women as Fellows until 1884, and even these were not permitted to attend meetings until 1909; the Quekett Microscopical Club (QMC) held out until 1918.[5]

Most women and working-class Victorians probably experienced microscopy as viewers at soirées or exhibitions rather than as active observers. Others perhaps tried out the instrument but did not pursue it in any substantive way. The vast majority of these are lost to history. One suggestive exception appears in R. H. Moore's presidential address for the Postal Microscopical Society (PMS), published in the Society's journal in 1896. The PMS was among the more inclusive of societies, in part because its work was conducted asynchronously via boxes of slides circulated in round-robin mailings.[6] Moore comments that "in scientific societies ladies and children are usually rigidly excluded," while at home "the husband and father" is "either missed or dismissed from the drawing room to indulge in" his studies, limiting other family members' exposure to this work. To this discussion Alfred Allen, the editor, adds a footnote taken from one of the Society's circulating notebooks:

> After an evening spent in examining one of our boxes, [the PMS member] left his microscope with the ¼-inch objective attached, and sundry slides on his study-table, and was much annoyed in the morning to find one of the slides smashed and the ¼-inch objective totally destroyed. On enquiry he learned that his maidservant had been doing her microscopy before breakfast![7]

Although the anecdote is meant to be a humorous reflection on the bumbling efforts of someone obviously unsuited to observation, the passage also points to the curiosity and determined effort of at least one would-be microscopist, who likely screwed the objective right down onto the glass in an effort to focus the image (a common error among novices). This anecdote suggests the yearning and frustration that many women and servants must have felt in hearing about the master's important and fascinating "microscopical work" while barred from entering that marvelous world. The existence of noteworthy exceptions such as Mary Somerville (a systematizer and popularizer) and Frances Elizabeth Hoggan (an original investigator and popularizer) does not contravene the fact that many Britons, exposed to microscopical views through popular exhibitions, lectures, or articles, must have felt curiosity to further explore minute forms but faced obstacles in taking up any substantive study, for reasons both ideological and practical.

The print record of British microscopical communities is thus limited by an unmistakable class, gender, racial, and ethnic homogeneity. And as the microscope played a formative role in the studies of tropical diseases such as malaria and sleeping sickness, it helped to underwrite the spread of British imperialism. Scholars have begun identifying and tracing the contributions that historically marginalized (racialized and/or working-class) collectors and naturalists made to the Victorian natural sciences.[8] Work like this is valuable in its ability to correct, expand, and enrich our understanding of Victorian microscopy. Although working microscopy could make considerable demands upon time and purse, many other, even fleeting, encounters with microscopy were open to a wider range of Victorians and helped shape what it meant in Victorian culture generally.

In referring to anonymous or generic microscopists, I have tried to avoid gendered terms as much as possible so as not to further erase the presence of microscopists gendered as women. The primary texts I use often refer to a generic microscopist as "he." Where I quote such passages in this book, this locution signals the exclusion of women from many real and imagined spaces at this time.

Microscopical Roles

In response to a growing body of work on the types and sites of scientific work, historians are finding it more productive to study roles and practices, which are situated but mobile across time and place, rather than identities, a term which suggests stasis. Elite practitioners have garnered most of the scholarly attention, due to their dominance in the print record and importance in technological developments, but there were many who participated in microscopy apart from these few. Jutta Schickore argues that test objects helped differentiate between three types of microscopist: "the microscope enthusiast, the (mere) practitioner, and the (real) man of science."[9] These roles are perhaps most prominent within nineteenth-century microscopy, but Victorians developed an even more expansive hierarchy of such roles. This system, like Martin Rudwick's concentric circles representing "graduated zones of 'ascribed competence'" or Ludwik Fleck's exoteric/esoteric circles, describes a spectrum of personal experience, commitment, and competence.[10] The tropes of embodiment examined in *Writing Embodiment in Victorian Microscopy* helped define these roles and their behaviors.

I offer a catalogue of roles here not as an exhaustive list nor a rigid structure of identities but to support scholarly conversations about Victorian relations with the microscope and accurate descriptions of distinct microscopical communities. Victorian microscopists often spoke explicitly about discrete microscopical roles and formed communities around some of them, but individuals (like texts) were understood and indeed encouraged to move between roles. That said, ideological and logistical constraints limited movement into elite circles and the rhetoric around working-class contributions to microscopical science far outstripped the reality. The main roles are as follows:

1. *Original investigator* or *investigator*: one who conducts original investigation and creates new knowledge with the microscope. *Investigator* is an actors' category; Victorians used it to refer to any researcher who observes regularly and rigorously, often with a specialist focus (foraminifera, mosses, springtails), and whose observations build toward causal and theoretical explanation of the findings.[11] Such work includes experiments and analytic observations (such as ascertaining the relation between species); investigators might focus on accumulating facts or might propose theories supported by experimental facts—although, as Savile Bradbury points out, genteel British microscopists were more resistant to theorization than their Continental counterparts.[12] The term *investigator* might describe a university botanist working on cell theory, a physician examining the differences between human and animal blood, a clergyman identifying species of

algae, a lawyer dissecting muscle, a banker studying optics. The *investigator*, a common reference in nineteenth-century science, thus cuts across the supposed division of professional and amateur. The term *amateur* is unstable in this period. It is sometimes used to refer to an investigator whose microscopical work is not his vocation and at other times used to criticize the casual or faddish observer.[13]

The community of original investigators is often voiced in this book by publications of the Microscopical Society of London (MSL)/Royal Microscopical Society (RMS). Both elite microscopical societies and more informal ones serve as what Steven Shapin and Simon Schaffer identify (in the seventeenth-century context) as a *disciplined space*: dedicated to making knowledge as a "collective enterprise," although "not everybody could come in; not everybody's testimony was of equal worth; not everybody was equally able to influence the institutional consensus" (78). Such spaces could be defined with considerable flexibility even in the Victorian era, as shown by the remarks of the eminent meteorologist James Glaisher in opening his presidential address to the MSL in 1866. He acknowledges that the Society's "rules require an Annual Address from your President… which should speak with authority of the advance in Microscopical Science," but he cannot do so: "my usual avocations are not microscopical, and they are so engrossing I have no time to devote to these researches."[14] While it might seem that Glaisher's demurrals should instantly disqualify him for the post, this performance of doubt—his careful delineation of the bounds of his certainty—shows that he understood the importance of skepticism to the investigative community. However, as I will show, even investigators writing to investigators could use tropes of wonder, if carefully handled.

2. *Proficient* or *adept*: a skilled observer who does not produce original or theoretical work but works steadily as a student of the microscope, building focused expertise through systemic observations, perhaps collecting, identifying, and studying specimens of local species. These microscopists are known not by what they produce (since they do not usually conduct original investigation) but by their regular, skilled practice, where they follow the lead of others' observations and replicate or support others' work rather than produce new knowledge. Those with deep local knowledge could provide valuable guidance to investigators.

An example of this type of observer might be the physician Edward Dowson, who was memorialized in the *Journal of the Royal Microscopical Society* in 1880. The obituary remarked that Dowson

> belonged to a class which in England is happily a large one, and comprises thoughtful and highly intelligent persons who, from various circumstances, are prevented from adding much to the science of their time. Doing little in the way of writing books or memoirs, and perhaps little known beyond the circle of their immediate acquaintance, such men nevertheless often possess extensive information and sound judgment, and exert considerable influence upon

> scientific thought and progress.... [M]any are themselves thorough scientific workers and good observers, and always take a deep interest in the communications and researches of others.[15]

The MSL/RMS included some such members, but one would more typically find such a person to be a "Quekett club-man" (some of whom were also Fellows of the MSL) or member of a provincial natural history society. Although there were many microscopists of this description, I have not been able to identify a widely used contemporary term for them, perhaps because advice to experienced microscopists often addresses all working microscopists, investigators as well as non-investigators (see below). When they publish on microscopy, it would not usually be to communicate original work but rather to contribute confirmatory observations of others' theories or to compile or popularize accepted knowledge. Much of the print culture of Victorian microscopy was written and circulated by such writers. Their evangelical zeal suggests the term *enthusiast* but this term was also used for particularly energetic investigators. Victorian investigators in the latter decades of the century also disparaged so-called "diatomaniacs" for their enthusiastic application of skill and energy to a narrow, fanatical pursuit of "beads" or obsessive comparison of wide-apertured lenses rather than contributing to meaningful scientific progress. To attain expertise and skill, they had to adopt the practices described by skeptical tropes, but tropes of wonder can appear frequently in their writing—and in critiques of their work.

The term *proficient* is suggested by John Quekett in 1848: referring to the ivory sliders sold with George Adams's eighteenth-century microscopes, he remarks that the observer who "could exhibit these objects well, was considered a proficient in the art" of microscopy. These sliders offered a selection of objects ranging in difficulty such as "insects, wood, feathers, butterfly wings, hair, plants, minerals, and shells"; the microscopist who could reliably exhibit them had demonstrated core microscopical competencies.[16] Another potential term for a disciplined microscopist would be *adept*; Albert McCalla comments in 1883 that "it is not true that every eye can see what the trained adept at the Microscope can easily discover."[17] However, in Victorian microscopical writing both terms (*proficient* and *adept*) are occasionally used to refer generally to all skilled microscopists, including both investigators and non-investigators. The late-century term *expert* did not often refer to microscopists as we might use it today; it appears primarily in discussions of forensic microscopy.

3. *Working microscopist* or *worker*: any skilled observer who "works at the microscope" in a committed, regular, and focused practice; this includes both investigators (professional and amateur) and non-investigating proficients.[18] In 1876 the diatomist Frederic Kitton suggests the term *microscopic worker* to replace the offensive word *microscopist* (a hybrid of Greek and Latin).[19] Many, like the American zoologist J. S. Kingsley in 1884, simply used *worker* to refer to researchers

("those who are actually increasing the sum of human knowledge"), but the term was also used more broadly.[20] E. C. Bousfield uses the term in a general sense in 1887 when commenting that "A microscopist—that is to say, a microscopic *worker*—has been compelled to learn how to bring out the points of the objects he meets with, and how to get 'full duty' from his objectives and illuminating apparatus"; and Jabez Hogg, in a late edition (1898), even refers to "student workers."[21] The editors of the *Quarterly Journal of Microscopical Science* refer to this class as *advanced observers* in saying that they expect to "relieve the graver and more strictly scientific matter of the Journal" by incorporating also "lighter contributions, such as will be found useful to the beginner, not uninteresting to the advanced observer, and of interest perhaps to the general reader."[22]

Working microscopists would enjoy considerable (deep and broad) skill and experience, usually with particular areas of heightened expertise and interest. They were likely to belong to clubs or societies formally dedicated to natural history or microscopical practice. Texts addressed to them (books, periodicals, lectures) assume a reader already at least somewhat disciplined who enjoys a general knowledge of microscopy.[23] Such texts participate in theoretical debates, extend disciplinary boundary formation, and help construct microscopy as a field and practice. Passages of articulated practice written for workers may skip steps or details that were considered general knowledge for workers. It might seem that the presence of romantic (wonder-filled) language in such texts would endanger their authority and deter readers, but as this book will show, wonder—if properly handled—was an important component of all microscopical work.

4. *Instrumental observer*: someone who uses the microscope regularly for directed observation but primarily as a contribution to other work or for a restricted range of tasks. While some who used microscopes in their profession became working microscopists, many would use the microscope in a much more limited way. Instrumental observers (not a Victorian term) have gained microscopical skills, but only within a restricted set of practices. They use the microscope as an auxiliary to complete job-related or other tasks, as with Rudwick's "accredited" geologists, but are not particularly interested in expanding their expertise beyond those basic tasks and are not likely to publish on the topic.[24] Such observers might include a physician examining patient samples, a cook testing food for adulteration, a customs officer evaluating imported goods, and a botanical collector checking a specimen. An analogy might be to a naval captain using a telescope to peer at the stars or an enemy ship; he cares more about his immediate aim than the tool used to achieve it. Writing about the instrumental use of the microscope tends to use tropes of skepticism more than wonder, since the microscope was not a focus of interest.

5. *Student*: someone engaged in repeated study and observations in order to gain skill in using the microscope; a *novice* would be early in this endeavor. The term *student* would apply to diverse groups: medical students enrolled in an

external course, schoolteachers being trained in laboratory work by Thomas Henry Huxley in Kensington, or a group of children being instructed in natural history and natural theology by their vicar. Their skills might be much rougher than those of an instrumental observer (within a circumscribed set of tasks) but their relation to the instrument should be more committed and curious. Those addressing students typically offered a heady blend of wonder and skepticism, since they hoped to both inspire and direct serious study.

6. *Recreative observer* or *hobbyist*: someone who uses the microscope as an "instructive and amusing" pastime rather than committing to it as a study or using it regularly in the course of work. *Hobbyist* was not a term widely used in this context at the time but accurately captures the spirit of this person's essentially untutored and casual interest. They might acquire experience in specific microscopical tasks and knowledge about areas of particular interest, but they would lack advanced skill and knowledge in microscopy generally. Hogg mildly tells novices that "the difference of the philosopher and the common observer" (worker and hobbyist) is that "the one devotes the energies of his mind to the elucidation of truths, the other merely enjoys the fruits collected by skilled industry."[25] Children using a penny microscope or other ad hoc magnifier might also be considered as hobbyists due to their active manipulation of the lens and its objects.

Texts addressed to hobbyists (in print or lecture format) do not largely discuss or contribute to theoretical debates. Passages of articulated practice addressed to hobbyists may describe material and physical steps and actions in excruciating detail, assuming no previous knowledge in the reader. The hobbyist primarily seeks microscopical wonders and perhaps to learn about the natural world, but not to achieve expertise as an operator.

C. R. Goring, writing early in the period, condemns the ignorant hobbyist as a "half-taught booby," but the growing popularity of microscopy may be gauged by the fact that by the 1850s the hobbyist was often welcomed into the fold.[26] W. L. Notcutt, in an 1859 *Handbook of the Microscope*, promotes the microscope as "a perennial source of delight and instruction" requiring only a "tolerable instrument…obtained for a few shillings" and no significant investment of time or attention. He explains, "It makes no large demand on the time of those who use it, for it may be taken up at any moment, and be as quickly laid aside again. It requires no previous course of preparatory study to use and enjoy it."[27] Edmund Saul Dixon, a priest and regular contributor to Charles Dickens's periodical *Household Words*, in 1857 celebrates the microscope's ability to provide "an inexhaustible treasury of amusement to crowds of amateurs who…are content to admire beauty and variety of design, even when they cannot penetrate to final causes." Dixon paints sentimental vignettes depicting the microscope as a welcome entertainment "to the invalid or lame person confined to the house, to the worn man of business whose soul is weary of affairs, to the lonely dweller in a country residence where little or only uncongenial society is to be had." Microscopy, he says, requires only "a few plants and minerals from the nearest hedge or stone-heap, a box of the commonest insects, a half-score of wide-mouthed bottles containing water-weeds."[28] However, against this mid-century encouragement, Carpenter excludes the hobbyist from his imagined audience, addressing his guidebook instead to those "who desire to gain something more than a mere *sight* of the objects."[29]

Writers used the term *dilettante* to critique the hobbyist who doesn't know how to use the instrument, but they also used that term to refer to the diatomaniac-enthusiast who misapplies expert knowledge. In 1890 E. M. Nelson, an accomplished microscopist and future President of the RMS, joked to the QMC that "the Microscope of the dilettanti is usually a perfect instrument of its kind, fully supplied with apparatus, the greater part of which is absolutely useless"; he later admits to having been "a member of the latter class" in the past.[30]

7. *Viewer*: one who experiences occasional, usually social, exposure to microscopical images via small or large, private or public observations. A viewer observes or even studies the image but does not select the specimen or manipulate the apparatus to bring the object into view (apart from perhaps adjusting the fine focus). A viewer might be a medical student examining a patient sample in a portable microscope with its mechanism locked; a supervisor being shown an adulterated substance discovered by an inspector; a jury presented with microscopic evidence of forgery or murder; or a visitor to a private or public entertainment. Such entertainments might include a soirée, lecture, or oxyhydrogen microscope show; a lantern slide show that incorporates microscopical slides; a moving picture show with a short film that parodies microscopy; or a microscope offered to the public in a museum or in Trafalgar Square. Tropes of wonder or skepticism would be used differently in each of these contexts to achieve desired aims. As contemporary reports of oxyhydrogen shows demonstrate, the viewer may not enjoy much agency regarding the selection and presentation of the microscopical image but the experience may still be moving and memorable.

8. *Virtual viewer*: a *reader* of print texts about microscopy, experiencing the microscopical image only vicariously, through its written description or printed reproduction in illustrations, photographs, and cartoons of and about microscopy. Microscopy was often virtually accessed in this way via a magazine (a periodical miscellany) or newspaper but could also appear in private letters. The virtual viewer may also encounter the microscope mentioned in passing or invoked as metaphor, in texts that are not primarily about microscopy. Virtual viewers' relatively distanced relation to microscopical practice differentiates their reading from that of working microscopists, students, and hobbyists, who carefully study, mentally imagine, and even physically replicate the microscopical observations recorded in print.[31] Victorian writers did not overtly recognize this role, but many texts implicitly frame the reader as a virtual viewer of the microscopical image.

Virtual viewers might access the actual microscopical image through print reproduction, especially later in the century, allowing them to (as it were) look over the shoulder of actual observers. This form of observation would have contributed largely to the wider cultural understanding of microscopy, just as someone reading about the new technologies of photography would form and share opinions about it despite never having handled a camera.

Of the microscopical roles listed above, the first six are all *observers* (manipulators of the microscope who *select*, *produce*, and *manipulate* microscopical views as well as *experience* and *respond to* those views), while the last two are *viewers* (who *experience* and *respond to* the microscopic world but do not in this instance directly *produce* or *manipulate* it). This distinction of observers and viewers may seem to replicate the problematic division of *producers* and *consumers*, but it differentiates the roles, not the persons acting in these roles. Geoffrey Belknap calls out the "shared" or "communal practices" of "observing, talking, writing, drawing, engraving, and printing" that were produced through the "collaboration and community debate" of naturalists.[32] While Belknap writes about how a scientific practice could be shared by different persons in different roles, it is likewise the case that a single participant could engage in multiple practices or roles.

Persons often remained mobile within the system. Those who worked at the microscope also acted as lecturer, listener, author, viewer, or reader in other contexts. For example, some working microscopists oscillated between the RMS and the more sociable, playful QMC, although they remained active in both forums as well as in other scientific societies and clubs. The movement between roles could work in either direction: any observer who worked as an investigator by day might also operate an oxyhydrogen lantern microscope at an evening soirée; and a casual viewer of that microscopical show, given the necessary funds and opportunity, might be encouraged to take up more serious study. Individual microscopists thus did not sit quietly in one role but moved between roles, either over time (student to investigator, viewer to hobbyist, investigator to hobbyist) or dependent on context (laboratory, classroom, parish hall, parlor). Furthermore, one's self-definition might not be universally shared; a diatom seeker, for example, might be accepted as a working microscopist by some, dismissed as a hobbyist by others.

Even having achieved knowledge and skill in a particular role, individuals demonstrated great variation in how they acted upon or expressed it. In particular, working microscopists as different as Philip Henry Gosse and Thomas Henry Huxley also acted as popularizers.[33] They sometimes wrote, sometimes lectured or demonstrated; sometimes addressed elite audiences, sometimes broad ones; sometimes spoke on obscure points of structure or function, sometimes on familiar microscopical conventions. Gosse acknowledges the intersection of his roles as working microscopist and as popularizer by referring to his own research in his popular books. In *Evenings at the Microscope*, for instance, describing the threads of the Scarlet-fringed Anemone, he says, "As this was first described... by myself, I will take the liberty of citing some of my observations on the matter." At another point, he casually remarks, "You may like to hear what Réamur has to say" about Saw-flies, "especially as I have nothing of my own on the subject"—which implies that he cites his own work preferentially.[34] These links connecting the discrete roles and cultures of particular communities enabled a broader, shared culture of microscopy.

This mobility between microscopical roles was encouraged. Many Victorian popularizers of science hoped to motivate readers to join the *army of microscopical investigators*, an approach that, Bernard Lightman argues, continues through the 1880s.[35] Beale's lectures to medical students asked them to become not just vocational observers but also investigators; in book form his *How to Work with the Microscope* presents this message to a much broader audience. Carpenter's *Revelations* expects novices to aspire to contribute to microscopical knowledge and ultimately "benefit science."[36] Such invitations appeared even in the most affordable tracts: the Anglican clergyman J. G. Wood assures the readers of *Common Objects of the Microscope* (cheap edition sold at 1s.) that

> if any one with an observant mind were to set himself to work determinately merely at the study of the commonest weed or the most familiar insect, he would, in the course of some years' patient labour, produce a work that would be most valuable to science, and enrol the name of the investigator among the most honoured sons of knowledge.[37]

While working-class readers might in practice struggle to secure the time, training, funding, space, light, quiet, literacy, and community needed to become a microscopical adept, texts addressed to audiences including such readers (such as Wood's guidebook or the *English Mechanic and World of Science*) commonly endorse the upward mobility of individuals within and between microscopical roles, proposing the role of student or even investigator as an aspirational goal.

Such movement inevitably implies changes in social class status as well. Wood specifies that "It is not the wealthiest, but the acutest and most patient observer who makes the most discoveries, for a workman is not made, nor even known by his tools." Indeed, "a good observer will discover with a common pocket magnifier many a secret of nature which has escaped the notice of a whole array of *dilettanti* microscopists, in spite of all their expensive and accurate instruments." When Wood remarks that "upon a pinch, a very respectable microscope may be extemporized out of a strip of card, wood, or metal, and a little water," he implicitly promises the very poorest Victorians that a bourgeois practice is within their reach.[38]

Those who remained within the role of viewer or virtual viewer did not, of course, gain much if any skill or knowledge in microscopy. Relatively detailed and serious accounts of microscopy did appear in periodicals for general readers, especially in ambitious reviews such as Charles Appleton's *The Academy* (1869–1902), which summarized, reviewed, and quoted long passages from science periodicals including the *Monthly Microscopical Journal*. Of course, microscopical performances and articles in many general-audience periodicals often offered only a series of marvelous or shocking facts and anecdotes, relying heavily on tropes of wonder. These texts, unlike manuals for beginners, did not call upon readers to manipulate the instrument or mentally follow along with reported procedures. Even when they do offer instruction, texts addressed to viewers do not assume much underlying knowledge and they mention theory rarely. They can still, however, tell us much about how scientific work was positioned within Victorian culture. For example, discussions of microscopical error and skepticism do appear, albeit sparingly, in texts like these that typically feature more romantic content.

Although many viewers may have had only casual or infrequent exposure to microscopy, they still helped shaped microscopical culture. Audiences at soirées or mass microscopical events, through their expressed responses, shaped the experience and even affected the operator's actions—as is evident in reports on oxyhydrogen shows—and readers' responses similarly shaped authorial choices.[39] As Jonathan Topham explains, "scientific authors respond both proactively and reactively to readers."[40] Thus it would overstate the case to describe microscopical viewers as passive. Even those Victorians who accessed microscopy solely through the printed page or through performance would have communicated their intellectual or affective responses to others, thus helping to construct cultural expectations around microscopy. And most printed texts on microscopy for general audiences, unlike many performances, did ask some participation of their readers, often encouraging them to take up serious study with the instrument—to become observers or even workers.

Some viewers' remarks promoted notions of microscopy as a dangerous tool or sensational gimmick. However, overall, the involvement of ostensibly passive viewers helped working microscopists accomplish their goals:

- ensuring that members of the general public know something about microscopy (either as part of a natural theology, as support for another field or theory, or as an element in a new concept of general secular education, reflecting the changes brought about by the Education Act of 1870);
- recruiting more active microscopists;
- building a broad base of public funding support for microscopical work at all levels;
- enhancing the role, status, and visibility of scientists as professionals;
- engaging readers in the natural world;
- honing readers' aesthetic sense;

- enticing working-class readers into self-improving, inoffensive hobbies;
- attracting and retaining purchasers, readers, and viewers of print microscopy.

Alex Csiszar discusses the debates in the nineteenth century over the "vision of science as public knowledge" and the growing importance of public opinion as "a legitimating resource" for scientific work.[41] Although he defines science itself as the site of this "public and democratic" debate, Victorians increasingly recognized that scientific fields and practitioners might gain power and status from public support. Microscopical hobbyists and recreational viewers and readers helped to construct just such a base of support for working microscopists.

Writing Embodiment in Victorian Microscopy: Beautiful Mechanism. Meegan Kennedy, Oxford University Press.
 DOI: 10.1093/9780198940623.003.0003

4
Microscopy and Its Audiences

> My object will be to unite the provinces of the Guide Book and the Panorama;... making my remarks useful to those who are already in possession of a microscope, while [also seeking to]... attract those readers who, unversed in microscopic marvels, might possibly feel repelled by a complete and lengthened treatise.
>
> Mary Ward, *Microscope Teachings*, 1864

Introduction

I was first attracted to Victorian microscopy because of the fascinating range of genres it employs. While this is doubtless also true of other sciences and other national literatures, the corpus I examine surveys a period of remarkable change in microscopical culture through a particularly rich and eclectic blend of texts. I was trained and work as a literary scholar, so this book centers around texts by nineteenth-century British microscopists: how these texts were written, published, circulated, and read, and what studying them can tell us about how British microscopy was practiced, taught, shared, and understood.

I work here primarily with published texts, surveying the print record of Victorian microscopy to tease out how published depictions of microscopical practice help circulate cultural narratives about the instrument and its work. Hannah Landecker has questioned "the assumption often found in ethnographic and historical scholarship that interviews and archives are somehow closer to reality than [the] published papers" that historians view as having been "processed for public consumption."[1] In fact, as she implies, there are different ways to define historical "reality"; and scholars value diverse types of sources precisely because they tell different stories about the past. The print record chronicles Victorians' efforts to circulate microscopical work across time and space; I study these texts to tease out how microscopists' self-presentation helps form communities with distinct approaches to microscopical observation, within a shared culture of microscopy as a difficult but rewarding practice. James Secord has argued that, during the first half of the nineteenth century, publication was only one means of building a scientific reputation, and not a dominant one; taking part in informal conversations and formal meetings was sufficient to establish a reputation.[2] With a few exceptions (notably the work of C. R. Goring and Andrew Pritchard in the

1830s), I focus on a later period, from the 1840s to the 1880s, when published writing of various kinds was an important element in defining microscopical communities and individual roles within those communities. I am interested in how published texts find an audience; how writers design their texts to direct the practice of microscopy from afar; and how published materials in this period help non-microscopical readers construct an idea of the practice, aesthetic, and ethics of "the microscopist."

Mindful that published texts are not necessarily authoritative or static ones, I read revisions, reviews, and responses to understand how debates shuttle from one periodical or book to another, how readers take (and sometimes mis-take) writers, and how textual material is revised and reframed for new audiences and new purposes. I orient some chapters as surveys of the literature organized by concept, reading examples drawn from across a spectrum of microscopical sources, and anchor others in illustrative case studies. Print publication had an important role in shaping the developing genres, practices, roles, and communities of Victorian microscopy.

Nick Jardine has argued that "routinely authored works—instrument handbooks, instruction manuals, observatory and laboratory protocols, etc." are "crucial primary sources" for "the specificities of the past practices of the sciences" and that they "often [play] central roles in the formation and stabilisation of questions and discoveries in the sciences."[3] Most of the texts I examine here are not "routinely authored" in that they are written by figures well known in Victorian microscopical communities, but they do circulate a formalized or idealized practice within these communities. Of course, manuscript sources such as notebooks, journals, and letters would additionally illustrate microscopy in historical, not imagined, practice. However, I am here concerned primarily with how, in a period marked by an unprecedented increase in literacy and in print and visual culture, published texts focused on microscopy helped recruit new Victorian observers, develop their microscopical practice, and build both a widely shared culture and many locally differentiated communities centered on the instrument and its minute world.

Nineteenth-century microscopy was an international endeavor. However, "British microscopy" was a concept very much in discussion, especially among elite observers who, like Charles Babbage, decried the longtime lack of public research support in England. Research microscopy had gained a foothold on the Continent earlier than in England; workers from the German lands made notable strides in histology, optics, and bacteriology and published beautiful illustrated books—but British microscopes and objectives were for much of the century considered among the finest in the world, and British workers contributed materially to efforts in photomicrography, histology, polarized light microscopy, petrography, germ theory, and life-history studies, as well as relevant technologies like printing and staining. They also spoke out boldly in professional disputes over optics, vitalism, germ theory, diatoms, and more. The texts I examine here are anything but dry scientific prose; they pulse with the pride, loyalty, anxieties, fears, fascinations, affections, and ambitions of British microscopists.

In an effort to limn the national literature of British microscopy and its effects on microscopical practice, this book draws primarily on British texts, especially those (textbooks, manuals, and journals) that centrally address microscopical work. I do not, for example, focus on important contemporary work on germ theory or bacteriology but include such discussions as they appear in publishing contexts centered on microscopy as a discrete discipline. I include some non-British texts that were republished, summarized, or translated for British readers. Some microscopists read German, but the most widely referenced texts had been translated and reframed.[4] Microscopists reading periodicals like the *Monthly Microscopical Journal* would have encountered references to and quotations from these authors even if they lacked access to these texts themselves.[5] The importance of the Continental (especially German) context is evident in the *Quarterly Journal of Microscopical Science*, which began publication in 1853 featuring not just the *Transactions* of the Microscopical Society of London (MSL) but also a section of "Translations." Translations appeared elsewhere also, as when William James Hickie, a Cambridge linguist of Greek and German, excerpted Julius Schumann's research on the diatomaceae in the *Monthly Microscopical Journal* in 1875.[6] The American sources I have quoted are generally those found reviewed, quoted, or reprinted in British periodicals.

Most of the texts cited in this book are authored by working microscopists and address microscopy as a primary topic, although I have quoted some texts on more general topics. The analysis draws most heavily from books that achieved canonical status in their genre; these were likely the most widely distributed and influential readings of Victorian microscopy. Indeed, the historian John Lidwell-Durnin notes, "[William] Carpenter was, with the exception of Robert Chambers's anonymous *Vestiges*, arguably Churchill's best-selling author in the 1840s and 1850s."[7] Some of these books (by Jabez Hogg, Carpenter, Lionel Beale, and Louisa Lane Clarke) achieved many editions over decades of use. Since multiple editions do not necessarily indicate bestselling status, I also consider periodical reviews, the frequency with which books are cited in other general guidebooks, and the availability of such books in libraries.[8]

Some Victorian authors list the most useful texts in the field, outlining a contemporary consensus on the microscopical canon wherein the same names recur in each list. Arthur Hassall, in *Adulterations Detected* (1857), sends readers to John Quekett, Hogg, and Beale.[9] Among writers for general readers or beginners, the Anglican clergyman Edmund Saul Dixon, writing for Charles Dickens's weekly *Household Words* (1856), primarily recommends Hogg, Carpenter, and Dionysius Lardner; Louisa Lane Clarke (1858) suggests Quekett, Hogg, Beale, and Carpenter; Mary Ward (1864) endorses Quekett, Hogg, and Carpenter; and the "Microscope" entry in the 1874 *Chambers's Encyclopedia* ("a dictionary of universal knowledge for the people") advises readers to consult Quekett, Hogg, Beale, and Carpenter.[10] W. T. Suffolk, in a series of 1869 lectures on microscopical manipulation addressed to a more knowledgeable audience, the members of the Quekett Microscopical Club (QMC), recommends Arthur Hill Hassall, Beale, and Carpenter.[11] Essay reviews canvass a somewhat broader array of texts.[12] To assess influence by another measure, Quekett's work was translated (into German, 1854); and the earliest American editions of Victorian microscopy manuals reprinted Carpenter (1856), Clarke (1858), Hogg (1859), and Philip Henry Gosse (1860).[13]

Library catalogues show which Victorian books on microscopy would have been available to borrowers. Any middle-class reader with a circulating library subscription would have enjoyed access to a selection of books on microscopy for all levels of expertise, and thus to various models of microscopical embodiment. For example, Mudie's Select Library, an influential circulating library, in 1857 offered access to standard texts by Carpenter, Gosse, Hassall, and Hogg, as well as popular books by Agnes Catlow and Gideon Algernon Mantell, and Edwin Lankester's translation of Matthias Jakob Schleiden's treatise on botany, which included an influential chapter on microscopical procedure.[14] These holdings provided a spectrum of guidance on microscopical topics, from serious texts for would-be investigators to chatty introductions for recreative naturalists. (Mudie's also carried natural history texts that mention but do not focus on the microscope.) By 1865 Mudie's had dropped Mantell but listed additional books by Gosse and Lankester, adding also popular books by Henry James Slack, a geologist, and George Henry Lewes, a literary critic. In the 1865 catalogue Mudie's had increased its total holdings of Catlow, Gosse, Lankester, and Lewes, reflecting the surge in scientific publishing for general audiences at mid-century; they had also dropped six books by Carpenter (retaining *The Microscope and Its Revelations*), which may suggest a shift in their collection toward more accessible works on science.[15]

Whereas Mudie's offers a glimpse at the kinds of books likely to be available to a middle-class novice or hobbyist, the library catalogue for the QMC (1915) represents the interests of Queketters, who could be working microscopists, enthusiasts, or earnest students.[16] The published QMC library catalogue, looking back over nearly a century of active work, features a wide representation of microscopical books for a range of audiences, with a number of specialized treatises and journals but also books for novices and for broad or youthful audiences. The QMC maintained multiple copies or editions of books by Beale, Carpenter, Clarke, Goring and Pritchard, J. W. Griffith and Arthur Henfrey, Hogg, Quekett, Slack, Henri van Heurck, and Joseph Wythe, as well as a handful of late-century texts. These multiple-copy titles, representing a wide range of expertise from novice to investigator, likely denote patterns of particularly high usage.

The catalogue of the MSL library (1859) served an elite community of British microscopists.[17] This catalogue represents only the first few decades of Victorian microscopy. It does not indicate multiple copies but lists an extensive collection including multiple books or editions by Beale, Carpenter, Pritchard, Quekett, and others, and single copies of many other treatises. Notably, it includes a number of books for general readers as well (Clarke, Gosse, Lankester, Lardner, Mantell, and Notcutt, as well as *Chambers's Papers for the People*). These books were largely retained throughout the century, as shown in the 1929 catalogue.[18] These catalogues suggest that Beale, Carpenter, Hogg, Pritchard, and Quekett were considered standard works for students and aspiring workers, that Clarke, Gosse, Lankester, and Lardner provided an introduction to microscopy for more general audiences, and that most of these were widely available from either general or specialized circulating libraries.

I hope that this book will encourage more critical attention to the fascinating and influential corpus of microscopical writing, including a wealth of lesser-known microscopical texts. Of the authors noted above, Carpenter has probably attracted the most substantive scholarly attention, but scholarship on him rarely addresses his role in the canon-formation of Victorian microscopy. Among the most frequently cited books above, Carpenter's *Revelations* reached the most editions and was being actively revised and reissued over the longest period—from 1856 to 1901, if we include the posthumous seventh and eighth editions edited by W. H. Dallinger. The book quickly reached canonical status. J. J. Plumer, writing "A few words on the choice of a Microscope" for aspiring workers in 1865, comments that "Dr. Carpenter's great work...is, in fact, a complete Cyclopaedia of itself"; and J. E. Barnard, reflecting on the centenary of the Royal Microscopical Society (RMS) in 1941, remarks that *The Microscope and its Revelations* was "a book that will still be found in the hands of most microscopists at the present day."[19]

However, the book was not universally praised. Reviewing the first edition in the *North British Review*, David Brewster commended Carpenter's introductory remarks on the "educational value and uses of the microscope," but he criticized Carpenter's handling of optics, microscopical history, and polarization; disparaged the book's "unseemly shape" and "plethoric bulk"; and dismissed much of the volume as "pages of compilation." He concluded his remarks on Carpenter by charging that when "everybody writes upon everything to gratify a popular taste," our "intellectual and physical food is equally adulterated." He called for books not by compilers (like Carpenter, he implies) but by microscopists known primarily for their research, "men whose powers of research are not inferior to their powers of instruction."[20] Despite Brewster's harsh words at its birth, "Carpenter on the microscope" became a classic of Victorian science and deserves further study if only for its longevity and seeming ubiquity.

I also work with microscopical periodicals representing distinct strata of microscopical practitioners, such as the *Monthly Microscopical Journal* and the weekly *English Mechanic*, with its 1870s stream of microscopical articles. This book does not attempt a comprehensive survey of any one society or periodical. For practical reasons, I draw most heavily on the periodicals associated with the MSL/RMS. This body of microscopists claimed the status of an authority and arbiter, standardizing the size of microscope slides in their first year, and later instituting a "universal screw." I do refer to other periodicals, such as the *Journal of the Quekett Microscopical Club* (*JQMC*), *Quarterly Journal of Microscopical Science* (*QJMS*), and *Science-Gossip*, which provided important alternate channels for microscopical discourse. Additionally, journals associated with the many regional and local microscopical associations and the Postal Microscopical Society (PMS) represent the "army" of working microscopists powering enthusiasm for the instrument.[21] However, the periodicals of the MSL/RMS were particularly influential due to their association with the Society. Comments in RMS journals during the aperture wars shows that Society members felt that they represented British microscopy to the world. The *English Mechanic* provides a useful contrast to the MSL/RMS journals since it enjoyed a diverse readership ranging from correspondents with considerable expertise to working-class readers with little time or money. It reprinted articles from the various microscopical societies, but the influence was bidirectional. For example, the *Journal of the Royal Microscopical Society* (*JRMS*) reprinted much of a talk by E. M. Nelson that had been reported in the *English Mechanic*.[22]

I also examine texts that were less influential for working microscopists but likely to have been widely read, such as articles recirculated from the Chambers brothers' many popular publications. Victorians enjoyed a surge in print culture, including scientific print culture, from the second quarter of the nineteenth century. This was made possible by cheaper printing processes and, especially after 1840, the lifting of taxes and restrictions on printed materials. We can thus gauge the popularity of nineteenth-century microscopy not only by its most-purchased and longest-lived texts but also by the multiplication and reach of books and periodicals with smaller print runs or a shorter period in the public eye. My analysis thus also incorporates pamphlets, ephemera, advertisements, and other texts with low or very local circulation, especially texts that, like William Henry Olley's *Wonders of the Microscope*, speak to a moment of technological or cultural change.[23]

Microscopical culture developed in many other social and print spaces as well. Other societies practiced and discussed microscopy not as a subject in itself but as a tool assisting in their work, including the Royal Society and the British Association for the Advancement of Science, medical organizations and botanical and geological societies, and local and regional natural history societies around the world.[24] The print record of these organizations dedicated to science, medicine, or natural history, while beyond the scope of this book, would yield useful insights about how microscopical culture was constructed in scientific communities centered on other interests. Alex Csiszar points to the "extremely diverse" channels of scientific communication in the nineteenth century, including not only books and scientific journals of various types (original articles, memoirs, proceedings, and transactions) and commercial journals (printing digests, excerpts, reprints, book reviews, and correspondence) but also encyclopedia articles, "scientific meetings, newspapers, catalogs, historical treatises, [personal] correspondence, abstracts, patents, pamphlets, offprints, and index cards." As he notes, popular commercial periodicals ranging from *The Times* and *The Athenaeum* to *The Illustrated London News* and *Punch* "were substantive sources of scientific news" also, just as science-focused commercial journals typically presented a "hybrid" of original work with "educational or entertaining" pieces.[25] *Writing Embodiment in Victorian Microscopy* includes examples from many such sources, but not an exhaustive survey of them.

Other participants in microscopy include the many religious, civic, scholarly, philanthropic, and social organizations that hosted entertainments featuring projection microscopy or microscopical lantern slides. The richness and variety of the textual and visual archive of microscopy calls out for more in-depth study. The scope of this book limits my ability to engage in extended case studies of individual texts, periodicals, communities of practice, or forms of microscopical practice; but it is my hope that by canvassing the landscape of nineteenth-century microscopy, this book will foster future, more targeted work.

While much of this book examines conversations among elite microscopists, it is important to recognize the influence of everyday microscopists—those working at home or in school in ordinary settings. These often participated in communities organized around education and encouragement of microscopical culture, such as the QMC, PMS, and provincial natural history societies and inexpensive periodicals such as *Science-Gossip*. Everyday microscopists voice their experiences through such club magazines or periodical correspondence, and I have tried to include these where possible.

Many of the texts quoted in this book recur and overlap in new publication contexts. Book reviews commonly quoted large swaths of the book under review, and many periodical articles were recirculated and reprinted in other periodicals, as was conventional at the time whether in scientific journals or family magazines. The work of original investigators circulated and was debated in these broader forums. Sometimes—as with Henry Edward Fripp's translations of Ernst Abbé and Hermann von Helmholtz, first published in the *Proceedings of the Bristol Naturalists' Society*—articles from provincial or foreign societies' periodicals or the *JQMC* were reprinted or excerpted in periodicals such as the *JRMS*. This book offers a view onto the varied microscopical landscape in Victorian Britain, but much work remains to be done.

"Popular Microscopy"

The late twentieth century saw a historiographic turn toward the study of so-called "popular science," productively complicating the notion of a professional/popular divide in science. In an influential 1994 article, Roger Cooter and Stephen Pumfrey argue that historians of science have tended to replicate a separatist, elitist model of science by presenting a *diffusionist model* where scientific knowledge filters down from expert researchers to an undiscerning public.[26] This view had influential nineteenth-century adherents.[27] Cooter and Pumfrey propose a more dispersed model based on collaboration. Audiences might influence or repurpose elite science, in an "essentially dialectical" process better described by terms like "reciprocity," "translation and enrolment," "grafting, appropriation, and transformation."[28] Some scholars reject the term *popular science* as too vague or condescending and offer *commercial science* instead to refer to the paid work of writing and lecturing on scientific topics, often for general audiences, as a way for impecunious men of science to eke out a living.[29]

The terms *professional* and *amateur* have also been critiqued for fostering an unproductive and misleading distinction. For instance, the work of original investigators in botany relied upon a complex network of hobbyists, collectors, and scientists, paid and unpaid. The gentlemanly amateur often enjoyed greater status than a paid professional, although these meanings shifted over the course of the century.[30] Recognizing the complex landscape of nineteenth-century science and how much it changed over the course of the century, historians have come to reject any simple bimodal distinction between "high" and "low," elite and popular, science producers and science consumers, in order to explore a more nuanced range of roles and relations. They have examined the contributions and culture of non-elite scientific workers, from radical anatomists and artisan naturalists to the far-flung collectors who provided specimens from all over the world for well-known investigators such as Joseph Hooker and Charles Darwin.

The shift away from a diffusionist model has also shaped print culture studies. Scholars working on Victorian science are attending more to the differentiation of scientific, literary, and popular audiences that emerged in early nineteenth-century commercial science periodicals.[31] Ralph O'Connor and others distinguish between a "popularizing" mode of scientific writing *for* the people, where a text is consumed by a more-or-less passive audience, and science *by* the people, where a text encourages, often records, and sometimes disciplines the interactive practice of science at all levels of participation.[32] This is a valuable distinction; but many microscopical texts—especially instructional ones—shift between these modes, both lecturing and encouraging the reader. And while texts overtly courted particular audiences and claimed specific aims, once published they found other audiences and caused other effects. James Secord, for instance, notes that the inexpensive popular science books that publishers developed for working-class readers reached a largely middle-class audience.[33] Conversely, the artisan naturalists discussed by Anne Secord and portrayed in Elizabeth Gaskell's novel *Mary Barton* (1848) often pooled their funds to access scientific texts.[34] As Jonathan Topham notes, "Printed material is well known for getting into the 'wrong hands.'"[35] That is, texts, like individuals, could shift between or bridge over distinct audiences. Topham identifies a "multifaceted and uneven contest for meaning" that occurs as scientific texts circulate through disparate communities of readers.[36] He draws attention to a cycling-back effect, where work addressed to broader audiences also shapes the work of more specialized or expert readers. He argues that "the history of science popularization" should be read as "part of a wider history of knowledge in transit" (using James Secord's influential term).[37] This applies to microscopy as well.

While scholars have been careful to nuance the distinctions between various Victorian periodical and book genres, these genres also shared important approaches to their readership. For example, the members of various microscopical societies were expected to contribute their observations to fill out meetings and journals and to foster a sense of print community; commercial popular science periodicals such as *Science-Gossip* and the *English Mechanic* likewise encouraged readers to submit their own experiences in scientific and technical work. Similarly, introductory guidebooks expected readers to follow step-by-step instructions, as did technical manuals for working microscopists; the detailed description of practice in research reports implicitly promotes this expectation as well.

Individual authors often shifted between different textual approaches as directed by circumstances and context, whether writing a book for young novices, an article in a scientific society journal, an article in a family magazine, or a public lecture. Some prominent Victorian scientific naturalists were also popularizers. T. H. Huxley, for instance, produced many reviews, lectures, articles, and books for a range of popular audiences; John Tyndall completed a hugely successful American lecture tour and published his lectures in *Household Words*. Similarly, many popularizers—such as Mordecai Cooke, Gosse, and Lewes—were also original investigators or serious workers, although with differing access to elite communities of science. It is convenient and often appropriate to differentiate communities along textual lines—for example, differentiating between the work of the RMS and QMC—but the memberships of such groups overlapped, sometimes significantly.

It is useful, then, to consider not only the expected audience for any text but also its modes of address, which usually relate to the expressed or implicit goals of the text. In this book I consider four modes of address: research, instruction, community-building, and recreation. *Research* texts communicate, archive, or critique original investigation or observation. They are primarily addressed to working microscopists, especially other investigators. Articles in scientific periodicals usually present themselves as primarily research texts. Such texts are very likely to include minute description and articulated practice. They model a skeptical stance, but they do exhibit wonder, usually in the Introduction or Conclusion or in fleeting references.

Texts or passages of *instruction* generally summarize research and emphasize articulated practice. They can be addressed to diverse audiences, from working microscopists to students to rank novices. Instructive texts often display both wonder and skepticism. The mid-century manuals for working microscopists (Quekett, Hogg, Carpenter, Beale) as well as reference texts (encyclopedia articles and textbooks) present their role as primarily instruction. However, other texts adopting this mode of address include introductory guidebooks for children or novices, such as Clarke's 1858 *The Microscope*, which opens with a section "On the use of the microscope."[38] Other texts, like Catlow's fervid *Drops of Water* (1851) and Eliza Lynn Linton's playful "Under the Microscope" in *Household Words* (1859), provide copious information about the microscopical world but little practical instruction.

Some texts enable *community-building*. In texts addressed to those who already practice microscopy, such texts might be letters (published or private), presidential addresses, book reviews, the published Transactions of society meetings, the brief queries and requests from readers published in *Science-Gossip* or the *English Mechanic*, and the monthly packets circulated among members of the PMS. When texts framed as research or instruction refer to other scientific authorities, this also serves as a form of community-building. These bolster the sociality of microscopical practice, working to solidify and direct microscopical communities. Other, evangelizing texts (such as Catlow's and Linton's) attempt to build community by promoting the microscope to lay audiences, working to attract students or to boost cultural respect for the work. Introductions to guidebooks and manuals (for example, Hogg and Beale) often reference evangelism as a goal of these books, but much of the microscopical writing in general-circulation periodicals suggests this aim as well.

Finally, many microscopical texts present a form of *recreation*. Some texts, such as oxyhydrogen microscope shows or soirées, very much emphasize the amusing or entertaining aspects of microscopy, especially its wonders. Mary Ward ostensibly designed her *Microscope Teachings* (1864) to serve as both "Guide Book" (instructive text) and "Panorama" (recreative one), as noted in my epigraph above. Recreative texts are often advertised as *instructive and amusing*, borrowing this trope from the Horatian concept of *utile et dulce*; but while they generally do impart information about the microscopical world, most recreative texts do not stage much actual instruction on microscopical practice.

These modes of address often overlap within any particular text. For example, the comic songs performed at Quekett Microscopical Club dinners excel in both community-building and amusement.[39] In another example, in an 1857 article on

"Microscopic Preparations" written for *Household Words*, Dixon adopts an amusing voice—the article is full of clever phrases and arresting analogies—but he also presents extended passages of articulated practice purportedly intended as instruction. It's likely that most readers experienced such passages as performative instruction only. Most readers did not actually compare objectives of different magnification or execute the multi-step, multi-day process of cleaning, sorting, and mounting diatomaceae. The effect of this article was likely to be evangelical more than instructive, demonstrating the kind of work microscopists do, for an audience that is not engaged in microscopical study but might enjoy reading about the excitingly complex apparatus and disciplined practice of the adept. Lewes, in the Preface to his *Physiology of Common Life* (1859–60), frames his book as research ("much original matter" resulting from "very considerable research"), instruction (for "the Student"), and community-building (for "the general reader...unacquainted with anatomy and physiology").[40] His adoption of the unusual practice of reporting original investigation in a text for general audiences (serialized in *Blackwood's Edinburgh Magazine*) reflected and probably reinforced his unstable position within microscopical circles.

Authors move through various modes of address and textual strategies as needed even within one article or book. Most texts feature multiple modes of address in different proportions depending on author, publication, and audience. These distinctions can be understood as a spectrum of approaches to microscopical writing. Authorial agency or personal style appears in how an author employs these distinct approaches to meet specific audiences and aims. Microscopical genres such as the scientific article, the popular lecture, the introductory guidebook, the ramble, the gossip, and the soirée are built up from the pattern formed by many writers' decisions, collated with publishers' expectations and readers' engagement with these texts. Gradually adapting existing or new genres to microscopical contexts in this way, Victorian observers assembled a stable of recognized forms understood to achieve common aims.

Microscopy books provide ample evidence of the wide range of "popular" writing. Rather than easy categorization, they demand assessment across a range of factors: author, aims, audiences, writing style, mode of publication, and the like. Aileen Fyfe notes that a text "for 'the people'" had to consider not just cost but also "literary accessibility" if it was to reach working-class readers.[41] While many texts do seem to strive for a simple, direct prose style, many books addressed to a broad audience can be considerably denser and more difficult to read than their counterparts in twenty-first-century popular culture. An instructive example appears in *The Microscope: Its History, Construction, and Applications*, published in 1854 by the ophthalmic surgeon and photographer Jabez Hogg, who would later (1872) become the first President of the Medical Microscopical Society. Hogg's Preface explains that he had written *The Microscope* to provide a "concise, yet sufficiently comprehensive *popular* account" of microscopy to "the great mass of the general public."[42] He repeatedly emphasizes not only the great extent of microscopic knowledge but also his hopes to provide a brief and "popular" account of it. For example, the book attempts to "point out and elucidate, at once in a practical manner and in a popular style, the vast fund of utility and amusement which the Microscope affords... as fully as the restrictions of a limited space, and the nature of a succinct summary, would permit." Hogg voices a desire to provide "concise," "limited," and "succinct" guidance on microscopy, but the book is laden with turgid prose—even the first edition of his text reaches nearly five hundred pages. However, he presents this hefty volume as a popular guidebook, fulfilling "the humble, but... not useless task, of setting up a finger-post... to direct the inquirer into the wider road."[43]

Readers seemed to accept this designation; a review in the *Natural History Review* commented, "It is a work for the *million*; and we gladly hail it, as we do all *such* works tending to direct their too often misspent energies into a wholesome direction." The reviewer praised the book's plates but critiqued Hogg's lack of accuracy and originality compared to Quekett, adding, "We regret that it is, in many places, disfigured by what, in charity, we must call misprints," since no surgeon could be "guilty" of allowing them.[44] A reviewer for the *New Quarterly Review* less charitably remarked, "Mr. Hogg is no naturalist," and criticized the book as "a wretched bungle" displaying "not simply... ignorance and incompetence" but also "great carelessness." The reviewer charged that the book is "so attractively illustrated, and so cheaply presented to the public, that it is obtaining an extent of circulation which is doing harm to science."[45] Indeed, despite its flaws, the book sold reasonably well (the first three editions rapidly sold five thousand copies each), and by 1898, having swelled to some seven hundred pages, it reached a fifteenth edition.[46]

The evolution of Hogg's text into a tome is echoed in the journey of Carpenter's *Revelations*. Carpenter explicitly framed his book as a guide for novices, saying, "the Author has constantly endeavoured to meet the wants of those, who come to the study of the minute forms of Animal and Vegetable life with little or no previous scientific preparation, but who desire to gain something more than a mere *sight* of the objects." He reaches well beyond introductory material, however; Carpenter imagined his reader as a future investigator and hoped to provide, "in a moderate compass, that information... which is most essential to the working Microscopist" and which "might prepare him to benefit science."[47] The book overflowed its originally envisioned "moderate compass," reaching 778 pages and canonical status. A review of the 1862 third edition exclaims, "it must not be imagined... that it bears any resemblance to that very unsatisfactory kind of literature which is usually described as 'popular science.' It is an admirable concentration of substantial learning and profound research."[48] After many decades and new editions, the 1901 eighth edition, edited and expanded by W. H. Dallinger and E. M. Nelson, offered nearly 1,200 pages of technical instruction, excessive for most beginners or hobbyists. Hogg's and Carpenter's reference texts, weighty in every sense, were offered to the general reading public alongside lighter reading such as Gosse's enthusiastic rambles and Dixon's lively meanderings. They demonstrate the variety and range of so-called "popular microscopy."

While individual texts self-described as "popular" often attempt to combine amusement and instruction, they can balance these imperatives in very different ways: one might provide recreation for general audiences, while another might tutor students of the microscope. Even when addressing similar audiences, different authors adopt different rhetorical strategies and prose styles. Quekett, for example, in his *Practical Treatise on the Microscope* (1848), says that "in order to render the matters...clear and intelligible to the general reader, as many technicalities as possible have been avoided, and the simplest language made use of; which will account for the plainness of style and composition."[49] Hogg proposes a similar tactic, albeit in less apologetic, and more richly periodic, prose. Writing on "*Polygastric Animalculae*," at the time a class of infusoria, Hogg intones,

> As our object is to give a popular account of the wonders disclosed by the microscope, we will not closely follow the path of the industrious and learned investigators of this branch of science; and hence will avoid as much as possible the adoption of those terms that so alarm the uninitiated; culling, as we pass, only those objects of most interest, likely to create a desire for further knowledge, whereby there will be found a future pleasure in overcoming the obstacles by which the learned render so mysterious and repugnant to the popular mind the stores of their gathered wealth.[50]

Hogg here declares (although his tortuous syntax does not perhaps fulfill) his commitment to a brisk, "popular," and inviting mode of address. He repeatedly reiterates this stance, identifying vocabulary as a particular hurdle for novices. In discussing the *Pennatulidae*, Hogg remarks,

> Naturalists call them *Sea-pens*; and if they had in general applied such simple, appropriate terms to the objects of nature, they would have removed a bar to the study of science, and overcome a repugnance commonly felt upon entering the enchanting field of knowledge. The crabbed, far-fetched, and learned cognomens usually give an air of mystery and difficulty where none ought to exist. The wonderful discoveries in nature ought to be a book open to all, and like herself, simple and unadorned.

Hogg here emphasizes his readers' anticipated needs in order to excuse his use of common names rather than the binomial nomenclature and other scientific terms used in classifying and describing living objects. He contends that technical language deters would-be students. Thus "the young and plainly educated are driven to idle, irregular reading" instead of study, he explains, "when they might have adorned a science, and brought honour to their country."[51] Hogg's declared (if not achieved) simplicity is, it seems, chosen in service to Science, Morality, and the Nation. Despite this extended apologia, the book does occasionally use scientific nomenclature as well as the skeptical tropes that guide practice.

Carpenter, however, adopts both a different approach and a different tone, unapologetically using scientific terms in *The Microscope and Its Revelations*. Given the inexperience of his anticipated audience, he remarks that some readers "may perhaps object to the general tone of his work as too highly-pitched, and may think that he might have rendered his descriptions simpler by employing fewer scientific terms." However, he rejects this suggestion, arguing that "the use of scientific terms cannot be dispensed with, since there are no others in which the facts can be expressed." Further, he says, he "has made a point of explaining these, in the places where they are first introduced," and so he "cannot think that any of his readers need find much difficulty in apprehending their meaning."[52] Indeed, the wisdom of Carpenter's strategy is suggested in Anne Pratt's *Chapters on the Common Things of the Sea-Side* (1850), where her pronouncement that "the use of scientific terms would here be out of place" results in multiple instances where she mentions a particularly intriguing object only to skip over it because it is "too difficult of description without the aid of scientific terms."[53] In comparing Carpenter's scientific terminology with Hogg's reiterated pledge of simplicity, Hogg's writhing prose seems more likely to trip up unsteady readers.

Just as "popular" books cannot be readily pigeonholed, "popular" audiences also resist easy characterization. While some microscopical texts—association journals, for example—tracked closely with a particular community, others—such as those by Carpenter and Beale—served multiple communities. Other texts, like the *English Mechanic*, support beginners, who sometimes wrote in for advice, while also hosting more advanced discussions, such as the 1870s articles penned by "Essex." Fyfe notes that some children learned from books meant for adults, and some adults, perhaps especially those with lower literacy levels, studied books marketed to children.[54]

Furthermore, individual readers might participate in multiple imagined audiences, even ones that seem quite distinct. Martin Rudwick's classic diagram of intersecting "zones" of 1830s geology suggests a useful analogy.[55] The exact contours of these overlapping audiences may be difficult to retrieve from existing documents, but we can begin to sketch out a landscape of reading practices by analyzing citations over a range of texts, tracing some of these border crossings in the print record. Even a short, low-status, and/or ephemeral text can provide useful clues. For example, a reader of *Science-Gossip* in 1888 advertises "WANTED, recent editions of 'Carpenter on Microscope,' Beale's 'How to work with Microscope,' and Lankester's 'Half-hours with Microscope,' will exchange numbers of SCIENCE-GOSSIP."[56] While Carpenter's, Beale's, and Lankester's guidebooks all supposedly address themselves to students of microscopy, the first two had built a reputation as serious textbooks and standard texts; they were often cited by investigators or other authors. The physician Lionel Beale's 1880 fifth edition of *How to Work with the Microscope* was a bulky volume of well over five hundred pages, originally instructing medical students (whom he hoped would become professional users of the instrument) on the finer points of manipulation ("work"). Carpenter's 1881 sixth edition of *The Microscope and Its Revelations* was a behemoth coming in at nearly nine hundred pages. These volumes were regularly updated with the latest research and procedures. It is reasonable to distinguish these from the lighter reading suggested by the physician Edwin Lankester's 1873 third edition of *Half-hours with the Microscope: A Popular Guide to the Use of the Microscope as a Means of Amusement and Instruction*, which presented a series of objects using what we might call soirée format, in a mere 130 pages. Yet this correspondent grouped them together as desired texts, much as *Science-Gossip* itself brought together diverse approaches to microscopy. (Lankester's text was plentifully illustrated by the renowned Tuffen West, which may have formed part of its appeal to this ambitious reader). Beale himself refers his readers to "the many excellent books" covering the construction of the microscope, including not just books by fellow investigators Quekett, Hermann Schacht, Carpenter, Hogg, Adolphe Hannover, and John Hughes Bennett, but also "Lardner on the Microscope," a reference to Dionysius Lardner's articles on microscopy for his serialized, eclectic encyclopedia *The Museum of Science and Art* (1854–6).[57] In the 1865 third edition of *How to Work with the Microscope*, Beale added a two-page bibliography of "Works on the Microscope, &c., Useful to the Student," in which textbooks such as Albert von Kölliker's *Manual of Human Microscopic Anatomy* (in translation) and Heinrich Frey's *Das Mikroskop und die Mikroskopische Technik* rub shoulders with Slack's *Marvels of Pond Life*, Lewes's *Sea-Side Studies* (cited as "Lewis"), and the Rev. J. G. Wood's *Common Objects of the Microscope*.[58]

Lewes's *Sea-Side Studies*, addressed to a much broader audience than Carpenter or Beale, also demonstrates the heterogeneity of a supposedly "popular" voice. First published serially in *Blackwood's Edinburgh Magazine* (an influential monthly costing half-a-crown per issue), these chapters blithely combine a spritely, humorous travelogue with references to "Nordmann's *Mikrographische Beiträge zur Naturgeschichte d. Wirbellosen Thiere* 1832" and "a *Mémoire* by Van Beneden, in the *Annales des Sciences*, 1851." Lewes even quotes the anatomist Charles-Philippe Robin in the original French.[59] Lewes's *Studies in Animal Life* (1860), which first appeared as articles in a competitor to *Blackwood's*, the shilling monthly *Cornhill Magazine*, offers a similarly cosmopolitan range of references.[60] Similarly, Lardner's encyclopedia, marketed as a popular text, discusses complex technical matters of microscopical optics. David Brewster acknowledges this dual approach, praising Lardner's "Treatise on the Microscope" in 1856 as an "elegant volume... distinguished by a happy combination of theoretical and practical knowledge." In fact, Brewster deemed it "the most valuable contribution that has ever been made to the scientific instruction of every class in society."[61]

The 1915 catalogue of the QMC library showcases this interpolation of nominally distinct audiences. The QMC was founded in 1865 upon the suggestion of a letter in *Science-Gossip* suggesting that London could use an "amateur" alternative to the MSL. The Club encouraged and trained novice microscopists through convivial events (classes, rambles, and gossips), seeking to circulate microscopical ardor and discipline through a broader class of observers, but it was largely run by experienced men of science. Thus, as shown earlier, both specialist texts and general-audience books appeared on the QMC shelves. Two of the latter—*My Microscope and Some Objects from My Cabinet: A Simple Introduction to the Study of "The Infinitely Little"* (1887) and *The Microscope and How to Use It* (1893), which was marketed to "the youngest beginner," "the merest tyro"—might be included because they are "by a Quekett Club-Man" (T. Charters White, a surgeon and avid photomicrographer).[62] It was not unusual for member-authors to present a copy of their books to the MSL or QMC library. But the library carried books by other popularizers, too, including a number of guidebooks published by the Religious Tract Society, such as Henry Scherren's *Ponds and Rock Pools* (1894).[63] Some popularizers, like Scherren (a zoologist), were established scientific or medical observers, but many were not.

The movement of microscopical writing across multiple audiences was made possible in part because nineteenth-century print culture offered Victorians many modes of reading apart from the model where a reader purchases a text, takes it home, and reads it cover to cover in solitary splendor. Readers frequently borrowed or shared books, periodicals, and newspapers; many audiences encountered a book in periodical form or by hearing it read aloud. In addition, while the well-recognized cycle linking author, publisher, printer, shipper, bookseller, and reader anchors Robert Darnton's "communications circuit" of published materials, where readerly responses track back to shape future editions of the text, this does not capture the lateral movement of textual passages from their origin text into other texts and even other genres.[64]

The work of the French structuralist Gerard Genette on paratexts and intertexts has sparked productive work on this intermediation of texts by scholars working in the history of the book, periodical studies, and media studies. Intertextuality is particularly relevant to the Victorians, who were barraged by an explosion of new media, and nineteenth-century publishers engaged in a lively recirculation of texts. Periodicals and printed lectures frequently recycled textual material quoted or plagiarized from elsewhere in order to fulfill the need for what we now call "content." Pithy quotations were collected into anthologies; choice anecdotes made the rounds in diverse formats; long passages were commonly quoted in books and articles or excerpted in reviews; and whole periodicals (such as the *British and Foreign Medical Review*) were founded on the premise of digesting or reprinting material from other sources.[65] Thus, while some Victorians read authors' words in the original book or article, many readers—perhaps most—encountered them in other settings.[66] Working-class audiences might hear an investigator's article quoted in a public lecture; the investigator, reading a family magazine, might encounter quoted material from the lecturer's introductory book on the microscope.

Any particular observation or experiment might convey one meaning in one context and a very different meaning in another one.[67] A reader even of the original might understand something very different from what the author intended, and the movement of textual passages into other print and performance frameworks only increases the likelihood of this. For instance, an author like Gosse invokes microscopical wonder as an expression of faith and solicits the reader's assent in his statement. Yet Gosse's wonder-filled prose, encountered either in the original book or in one of the many other sources that quote him, might act on a reader in different ways: as an advertisement for the book, an attraction in itself, an inducement to take up microscopy, or a subject for debate. Similarly, Beale exhorts readers to observe skeptically so that the community may progress toward greater knowledge, but a reader might take his words as an attack on lackadaisical colleagues, or a warning away from difficult microscopical practice, or an assertion of scientific authority, or a challenge for the curious.

Our understanding of scientific genres, roles, and communities must be flexible enough to account for texts drawing from and recirculating in multiple genres, addressing audiences that shift and overlap. O'Connor remarks, "popular science worked, not as a one-way process of knowledge transfer, but as a heterogeneous network of cultural exchanges and feedback loops between different social groups."[68] We find a similar pattern in microscopy as within science writ large: not a clear-cut opposition between "research" and "popular" microscopy, or "science" and "non-science," but rather a diversified spectrum of practices populating a multitude of ecological niches or communities. O'Connor, noting that "nonspecialist literary history" often conceives of a text as belonging to a single discrete genre, describes textual genres instead as "a set of norms or conventions" and "interpretive tools rather than prescriptive categories." He sees any individual text as "float[ing] in a constellation of different genres" with "affiliations of varying proximity to various genres."[69] Similarly, the microscopical texts, genres, and communities I examine here demonstrate multiple affinities with one another. Sally Shuttleworth and Berris Charnley have called attention to "a web of interactions between different printed forms—the transactions of learned societies, commercial publications, popular and technical and specialist titles, or indeed encyclopaedias—and the spoken word and performance of science, in popular venues as much as at learned societies."[70] A visual representation of Victorian microscopical communities and texts would demonstrate a similar significant overlap with bidirectional interrelation. Rather than a periodic table of elements or phylogenetic tree, it would look more like a network diagram, with clusters focalizing areas of particular energy, innovation, or creativity.

Given that texts of different genres, addressing different audiences for different aims, can turn to similar textual strategies, and that even within a single genre authors can disagree about the best approach to reach their readers, of what use are generic terms like "popular" and "handbook"? O'Connor calls upon scholars to judiciously preserve "popular science" as what he calls an "umbrella-term."[71] Similarly, Bernard Lightman retains the terms *popular science* and *popularizers*—terms that were used in the Victorian period—while foregrounding the complicated history of such terms and pushing back against any implications of diffusionism or simplification.[72] In this book I do occasionally use the term "popular" as a widely recognized general category for cultural artifacts: a placeholder that allows us to set up more fine-grained analysis and that enables translation and exchange with cognate disciplines. However, each text is of course shaped by differences in author, publication, audience, and historical moment. That is especially true regarding tropes of wonder; their use varies considerably from one setting to another.

Writing Embodiment in Victorian Microscopy: Beautiful Mechanism. Meegan Kennedy, Oxford University Press.
 DOI: 10.1093/9780198940623.003.0004

5
Physiological Psychology and Victorian Microscopy

Seeing, as Schleiden very justly observes, is a difficult art; seeing with the microscope is yet more difficult.

Hermann Schacht, *The Microscope in its Special Application to Vegetable Anatomy*, 1853

Introduction

Nineteenth-century microscopy became important to many sciences, but it had a particularly close and reciprocal relation to physiology. Simple and compound microscopes, along with graphing techniques and technologies like the kymograph, galvanometer, and camera that could visualize physical events in new ways, gave investigators new ways to probe the body, and new interest in not just biological structures but also systems: organs, muscles, nerves, and the like. From Charles Darwin's theories about how external stressors provoke species adaptation, to Claude Bernard's concept of *le milieu intérieur* (the internal environment), to Herbert Spencer's claims for society as itself a kind of organism, nineteenth-century investigators increasingly explored the interactions—the patterns of response, reflex, and environment—that shape biological life. As microscopy gained acceptance and eventually stature as a scientific instrument, it helped underwrite research into biological systems and functions. The nineteenth-century study of physiology was particularly strong on the Continent. Investigators such as Johannes Müller, Hermann von Helmholtz, and Claude Bernard, with British workers like Charles Bell and David Brewster, argued for more research into the physical bodily functions underlying human perception, cognition, emotion, and muscular action. They emphasized that sensory perceptions are subjective, vulnerable to interference, prone to error, and contingent on circumstance, and that even thought is embodied.

Influenced by this new work in physiology, some researchers argued that not just biology but also psychology is grounded in the physical structures of the body. The philosopher John Stuart Mill (*A System of Logic*, 1843), argued for a material basis of mind and empirical approaches to psychology, although he advocated for primarily psychological rather than physiological enquiry. Other investigators did study questions of human thought and feeling using the tools of Victorian physiology, including the microscope and microtome (which could cut very thin, precise sections of a specimen). These *physiological psychologists* at mid-century included William Benjamin Carpenter, Alexander Bain, Herbert Spencer, Thomas Laycock, Henry Holland, George Henry Lewes, and Henry Maudsley. They published titles like *Chapters on Mental Physiology* (Holland, 1852), *Mind and Brain* (Laycock, 1860), and *The Physiology and Pathology of the Mind* (Maudsley, 1867).[1] Despite their commitment to physiology, these were not deterministic, naïve materialists. Countering the premise that the mind is static and siloed into discrete pre-existing "faculties," they argued instead that mind (perception, thought, feeling) arises and functions within embodied physical and social systems—that it is distributed, relational, fluid, and dynamic. They emphasized that functions of mind, including unconscious functions and automatic, reflex actions, can be trained, but that these functions are also a form of physical labor that cause and are degraded by bodily fatigue.

This research, much of it based on experimental investigation, challenged the grounds of empirical science itself. Although the work of the physiological psychologists was not universally accepted, it shifted the ideas of influential investigators like John Tyndall, who (as Bernard Lightman argues) concluded that a naïve "ontological materialism was contradicted by the most recent research in physiological psychology."[2] These conclusions spurred researchers to develop methodologies protecting the validity of empirical work. As Part III of this book will show, such matters were especially relevant to microscopists, given the unique demands of their practice. They saw operator, instrument, and object as connected in a nervous network shaped by deliberate action, unconscious or reflex function, and passive or active physical resistance. Microscopical investigators had to reckon with how the material grounds of perception, thought, and feeling interacted with their own body and the bodies of instrument and object to shape their work.

In this chapter, I examine the tenets of physiological psychology and their relation with the microscopical tropes and practices examined in this book. Despite their frustrations with microanatomy—it was difficult to pin down how nervous functions were related to the minute structures visualized in brain and nerve tissue—many physiological psychologists used the microscope in their research, which was reported and reviewed in general-interest as well as scientific periodicals. In return, the tenets of physiological psychology supported Victorian microscopical practices, suggesting how body, mind, and apparatus work together: how they might foster affective energies and how they might err.

Indeed, the research published by physiological psychologists provides a remarkably consistent theoretical foundation for the ideas governing everyday Victorian microscopical practice. Even setting aside those who worked in both fields, microscopists and physiological psychologists shared many assumptions about the management and circulation of emotion, as is evident in how both groups write about wonder and the formation of communities as systems of affective exchange. (As this chapter will show, their ideas in some ways prefigure work by contemporary affect theorists.) At the same time, research by Carpenter and others on perception, cognition, and neuromuscular function chimes with microscopists' concerns about these matters and their efforts to develop skeptical methodologies to safeguard their work. I am not arguing that all Victorian microscopists accepted the tenets of physiological psychology, but that this integrated approach to mind and body aligns with important aspects of Victorian microscopical practice.

This chapter focuses on the work of Carpenter, Lewes, Spencer, and Bain, mostly during the mid-century decades. Spencer and Bain were primarily philosophers and rarely discuss microscopy except to cite studies. However, I include them here because they were well informed in physiology and served as important early advocates for physiological psychology.[3] Carpenter and Lewes wrote multiple books and articles on microscopy as well as physiological psychology.[4] All four of these writers produced work that was foundational in the field. Spencer's *Study of Psychology* (1855) and Bain's *The Senses and the Intellect* (1855) and (especially) *The Emotions and the Will* (1859) were influential early texts. Carpenter's first physiology treatise was *Principles of General and Comparative Physiology* (1839); over time he focused more on mental and emotional functions in *Principles of Human Physiology* (1842) and especially *Principles of Mental Physiology* (1874). Carpenter's physiology monographs enjoyed a publishing career similar to that of *The Microscope and Its Revelations* (1856), reaching multiple English and American editions. Lewes's most sustained writing on microscopy as a practice appears in *Sea-Side Studies* (1858), a book of popular natural history addressed to a general educated audience, written as Lewes was himself still more an enthusiast than a worker in microscopy, but pressing for recognition. His contributions to physiological psychology mostly derive from the treatise *Problems of Life and Mind*, published in three parts (1874–9) and finding a good reception after his intensive work on neurophysiology in the 1860s and 1870s.[5] Each of these writers had particular interests. Bain influentially argued for the embodiment and contingency of mind (among other things); Carpenter, the unconscious functions of mind; Spencer, the evolution of the individual and social mind; Lewes, the contingent, language-bound, and social elements of knowledge. As a group, they insisted on the physical, multifactorial nature of mental actions.

These writers were known and discussed in educated circles. As Roger Smith shows, "[t]his new science was much in evidence in the periodicals."[6] Their books were reviewed in elite quarterly and shilling monthly periodicals as well as medical and scientific journals—though not always positively.[7] These books were available to interested readers not just in booksellers but also circulating libraries. In 1865, Mudie's Library did not carry Carpenter's physiology books, but it offered

Bain's and Spencer's volumes and Lewes's *Physiology of Common Life*.[8] Lewis's Medical and Scientific Library, a specialized library in London that Royal Microscopical Society (RMS) members could use, offered all of the books I discuss here, including six different physiology titles by Carpenter.[9] Despite the stigma associated with materialism at mid-century—its associations with atheism and radical politics—the tenets of physiological psychology, which insisted on the physical, material elements of mind, still received wide circulation in print.

I am not claiming that Victorian microscopists regularly cited this research when extolling microscopical revelations or arguing for specific methodological practices. Indeed, Spencer and Bain's philosophical treatises were not generally cited whether in research texts or guidebooks. But Carpenter's and Lewes's work on nervous function was recognized in discussions on the subject; Carpenter's treatises on physiology were particularly well regarded, kept available in the Royal Microscopical Society library and cited.[10] Carpenter's massive manual *The Microscope and Its Revelations*, probably the most widely read and influential mid-century microscopical text, was certainly informed by his study of physiology. When he published it in 1856, he had been writing about physiology for seventeen years and his *Principles of Human Physiology* was in its fifth edition. Even apart from the obvious relevance to microscopy of the physical construction of the eye and its resulting vulnerability to error, elements of Carpenter's physiological psychology do appear in *Revelations*, shaping his advice on operating the instrument. For example, we can recognize his interest in mental training when he advises that the brain can be taught to ignore the image on the retina of the eye not placed at the eyepiece; we see his interest in embodied perception when he warns that the stance of the body vis-à-vis the microscope can affect the distinctness of vision; and we note his suggestion of unconscious thought when he cautions readers about the role played by cognition (expectation) in our visual acuity, as when the "retinal pictures" representing cells in animal tissue were overlooked by physiologists before Schwann.[11]

Overall, microscopy and physiological psychology form two strands of nineteenth-century thought that run largely parallel to each other and sometimes overlap. The physiological psychologists understood thought and feeling as fundamentally embodied experiences, an approach that is also implicit in the attractions and protocols that fill the microscopical corpus. Microscopists developed a pragmatic understanding of their day-to-day work as an embodied practice that calls for explicit direction in describing affective flows and skeptical methodologies; while physiological psychologists strove to theorize perception, cognition, emotion, and action as functions of physical embodiment, turning to the microscope as both an experimental tool and a source of real-world case studies.

While all four writers in this chapter cite microscopical research, they also often use microscopical practice to provide examples illustrating their ideas. For instance, Carpenter cites microscopical vision in discussing (for example) focused visual attention, false perceptions, minute hand tremors, and unconscious cerebration. Discussing false perceptions, he remarks, "It has long been known that when a *seal* is looked-at through a Microscope, it will appear sometimes projecting as a cameo, sometimes excavated as an intaglio"; on unconscious movements, he notes, "[T]he unconscious rhythmical motion of the hand...will not surprise any one who knows how difficult it is to prevent the tremors of a Telescope or a Microscope, by the most careful construction of its supporting frame-work."[12] It is not surprising to find microscopy featuring in their arguments. Even apart from these writers' own microscopical experience, the difficulty of microscopical work, which pressed eye, hand, and brain to their limits, served as a good test case for physiological psychology.

Especially relevant to Victorian microscopy, and to the tropes I examine in this book, are these core propositions derived from the work of the physiological psychologists, which I summarize here in simplified form:

1. Perception, thinking, and feeling are **entwined**. Thoughts can affect emotions and vice versa.
2. Perception, thinking, and feeling are **embodied**.
3. Perception, thinking, and feeling are **distributed** across the body.
4. Perception, thinking, and feeling are **dynamic and mobile phenomena**; they emerge through a flow of **nervous "currents"** that circulate within the body.
5. Perception, thinking, and feeling also **circulate** between people.
6. Perception, thinking, and feeling can be **trained**.
7. Perception, thinking, and feeling are forms of **labor** that can cause and are affected by physical **fatigue**.
8. We are also entwined with **things**.

These tenets, drawn from the study of bodies and minds, correspond in many ways with Victorian microscopists' depiction of their work.

Perception, Thinking, and Feeling Are Entwined

While faculty psychologists and phrenologists sought to parse distinct "faculties" of mental life, the physiological psychologists argued instead for the interrelation of reasoning, volition, emotion, and sensation. This also served to expand associationist psychologists' focus on ideas more than feelings. Spencer, for instance, argues in *Principles of Psychology* (1855) that Feeling is inextricable from thought (Memory, Reason); they are simply "different sides of the same psychical phenomenon." Some states of mind, in fact, such as that "produced by seeing a beautiful statue," are "clearly both cognitive and emotive." He reflects that "it is impossible to [see such a sight] without a greater or less feeling of pleasure—without some emotion." Great art is not needed to elicit such a response. He points to "the gratification given on looking at a simple curve—an ellipse or parabola," for example, or "the perception of even one colour, when of great purity or brilliance." Indeed, he insists, "[N]o act of cognition can be absolutely free from emotion... [and] no emotion can be absolutely free from cognition." In sum, "[A]ll cognition implies emotion, and all emotion implies cognition." In fact, he contends, any "conscious action... must exhibit all the essential elements of a conscious action: that is—must simultaneously exhibit Memory, Reason, Feeling, and Will."[13] Spencer's argument about the entwinement of perception, thinking, and feeling applies to microscopical beauty even more than to "a beautiful statue" or a parabola, given that Victorians observing microscopical objects often (as this book shows) report that the image evokes responses "both cognitive and emotive."

Emotion Is Embodied

It may be obvious that sensory perception occurs through bodily events, but, with the advent of physiology, Victorians became more likely to discuss affective feeling (emotion) in such terms as well. Alexander Bain influentially argues in *The Emotions and the Will* (1859) that "no emotion, however tranquil, is possible without a full participation of the physical system."[14] He explains,

> Almost all the members of the body are brought into action, in displaying the stronger degrees of emotion; the arms gesticulate in many modes, the limbs are thrown out and retracted, the trunk and head are agitated in many ways, the features are especially acted on, the voice is stimulated, the muscles of respiration are affected; in short, it would seem as if no movement were left dormant in the round of our various manifested feelings.[15]

More generally, Bain proposes that "[t]he pleasure or pain, that is, the mental antecedent of a voluntary act, is embodied in the nervous and other organs, and rises and falls with their physical condition."[16] The effect of two decades' research into the physiology of emotion is evident in Charles Darwin's *The Expression of the Emotions in Man and Animals* (1872), which everywhere asserts the joint action of emotion and muscle movement—not just in regard to the muscles of the face but also the voice, the limbs, the whole being. In Victorian microscopy, where (as Part III of this book will show) a trembling hand, laboring pulse, or even heavy breathing can endanger the integrity of the image, the embodied nature of emotion means that it can both derive from and disturb one's microscopical practice.

The link between body and emotion is so strong that it works in reverse as well: physiological events can actually beget the mental and emotional ones that ordinarily would rouse that effect. Bain argues that "if the entire physical condition accompanying any feeling can be aroused from whatever cause, the feeling itself will necessarily co-exist." Indeed, "[c]ould we reinstate all the outward gesticulation, the play of feature, the vocal tones, the altered secretions, the nerve-currents instigated in a burst of grief, we should have the very emotion itself enacted in our mental experience."[17] This statement serves as a reminder that, for Bain, thought, emotion, and action do not direct but rather emerge out of particular physical configurations.

Thinking Is Embodied

Physiological psychologists argued that thinking is an embodied act, not just given the brain's physical housing within the body but also in its reliance upon the conditions throughout the body. Their emphasis on the relations between mind and its material environment was an important revision to associationist psychology.[18] For example, Bain argues in *The Emotions and the Will* that "we are at liberty to say mind is a source of power; but we must then mean by mind, the consciousness in conjunction with the whole body." In a footnote, he emphasizes, "Our consciousness in this life is an *embodied* consciousness."[19]

In *Principles of Psychology*, Spencer uses an unusual metaphor that insists on the materiality of sensory impressions; he suggests that "what is really given in a visual impression" is a kind of touch. He explains, "The retina, as examined microscopically, presents...a tesselated [sic] pavement made up of minute rods packed side by side, with their ends exposed so as to form its surface," where "each of these rods is supplied by a separate nerve" and "capable of independent stimulation." He proposes an unsettling analogy: "Imagine that an immense number of fingers could be packed side by side, so that their ends made a flat surface; and that each of them had a separate nervous connection with the same sensorium." Within this image of fingertips packed together side-by-side, retinal images then correspond to physical "things laid upon that surface." Each finger "touches some particular part of one of the images; and conveys to the sensorium the feeling produced by the touch."[20] Spencer's comparison of the retina to fingers here ironically draws on microscopical investigation to represent vision, including microscopical vision, as a form of physical touch.

More boldly, Carpenter argues that material conditions determine thought as well as sense. He emphasizes in *Principles of Human Physiology* in 1842 that mental functions rest upon physical ones. He explains that "the *psychical* operations of Man"—which include "intellectual, emotional, or instinctive" actions—are "dependent...upon the nutritive apparatus" and "the due supply of his bodily wants." In fact, the "organic life" of humans "appears entirely subordinate to [the] animal functions" like breath, pulse, and digestion.[21] Much later, in his *Principles of Mental Physiology* (1874) Carpenter describes the human body as "a Material Instrument; whose function it is to bridge over the hiatus between the individual Consciousness and the External World, and thus to bring them into mutual communication."[22] He reassures readers that he does not see the human subject "as a mere puppet" of material forces, but he repeatedly returns to "the direct and immediate relation between Mental and Corporeal agency." He goes so far as to say that "the operations of the Mind are in a great degree determined...by Material conditions." It is for this reason that "very slight departures from the health of the Body exert a most powerful influence upon our *intellectual*, and still more upon our *emotional* condition."[23] These comments align well with Victorian microscopists' concerns (detailed in Part III) about working when agitated or too soon after a heavy meal.

Bain similarly argues that bodily action actually *precedes* the will to that action—that "the spontaneity of muscular action" is "an essential prelude to voluntary control." Echoing Carpenter's 1842 comments about oxygen and nutrition as a prime cause, Bain proposes that our "organic energy," derived from air and food, "is the general and fundamental fact; consciousness is the occasional and accessory fact attached to it." Without food and air, "the whole system is run down like a steam-engine with its fire burnt out." This point is so important that he repeats it: "The transformations of the food and tissue are the *sine quá non*, the consciousness is the accidental part." And again, "the physical energy is the indispensable condition, and the consciousness the casual."[24] Bain's strong, simple phrasing insists that mind—the central fact of so much previous philosophy and psychology—is not just embodied but contingent, belated, and even "accidental" in relation to the one fundamental need for food and air. Nineteenth-century microscopists do not challenge these ideas, but they are more concerned that digestion might prevent rather than enable good observation; similarly, the frequent recommendation to work near a window is more about accessing good-quality light than fresh air.[25]

Lewes, of course, had centered *The Physical Basis of Mind* (1877) around the claim of embodied mind, and he continued to argue for it in the strongest terms. In *The Study of Psychology* (1879) he says, for instance, "the Mind [is] an expression of organic and social conditions" and "every mental fact is at once a state of Feeling and a state of the Organism."[26] But he also recognizes how unlikely this may seem to physiologists studying nervous structure. He muses,

> Who that had ever looked upon the pulpy mass of brain substance, and the nervous cords connecting it with the organs, could resist the shock of incredulity on hearing that all he knew of passion, intellect, and will was nothing more than molecular change in this pulpy mass? Who that had ever seen a nerve-cell

> could be patient on being told that Thought was a property of such cells, as Gravitation was a property of Matter?[27]

Lewes here turns his considerable skill as a writer toward articulating the *frisson* of unease one might feel in siting consciousness within the body. Rather than backing off the assertion of embodied mind, though, he uses this moment to double down on it. Indeed, this moment of apparent reflection actually ventriloquizes anti-materialist arguments in order to critique them; these questions represent a "revolt of Sentiment" that (wrongly) damns empirical science as "moral degradation." Only a few lines later, Lewes insists that "mental phenomena are functions of the organism" and can only be understood by studying the physical whole.[28] Working microscopists echoed claims like these—that it is necessary to scrutinize microscopical structures in order to understand biological functions, including neural and cognitive function—when arguing for the importance of the instrument to scientific progress.

Perception, Thinking, and Feeling Are Distributed across the Body

Physiological psychologists argued not just that Mind is embodied but that it is distributed: it emerges from phenomena involving the whole body. Thus, Bain argues in *The Senses and the Intellect* that "the organ of mind is not the brain by itself; it is the brain, nerves, muscles, and organs of sense." He exhorts readers to "discard for ever the notion of the *sensorium commune*, the cerebral closet, as a central seat of mind, or receptable of sensation and imagery." He says, "it is doubtful if even thought, reminiscence, or the emotions of the past and absent, could be sustained without the more distant communications between the brain and the rest of the body—the organs of sense and of movement."[29] Lewes also addresses this topic in *The Physiology of Common Life*, serialized in the *Cornhill* and published in book form in 1859–60 to generally good reviews. He notes that "intelligence" resides "mainly, if not wholly" in the brain, but he argues that "that the Brain is only *one* organ of the Mind, and not by any means the exclusive centre of Consciousness." While "Intelligence and Emotion are mainly, if not wholly, dependent on the Brain," other forms of consciousness ("Sensation and Volition") extend throughout the body.[30]

Later, in his 1877 *Physical Basis of Mind* (the second part of *Problems of Life and Mind*, written for a specialist audience), Lewes expands upon this idea, saying, "Abundant Experiment has shown that the spinal cord, apart from the encephalon, is capable of acting as a sensorial and volitional centre." He clarifies that "[t]he neural process is one continuous excitation along the whole line of ingoing nerve, centre, and outgoing nerve" and insists that "there is no Consciousness... unless the whole organism is involved."[31] As scholars like Elfed Huw Price have noted, Lewes was an outspoken advocate for what we now term embodied cognition, a concept that may underlie the nineteenth-century notion that microscopical observers should stand and move about from time to time in order to "invigorate" their vision and cognition.[32]

Perception, Thinking, and Feeling Are Dynamic and Mobile Phenomena

The physiological psychologists viewed perception, thinking, and feeling as dynamic and mobile phenomena that emerge through a flow of nervous "currents" circulating within the body. Bain's work is suffused with metaphors of forces flowing through "a system of connexion."[33] In *The Senses and the Intellect* he frequently refers to "currents" of nerve action (referencing both electrical and fluid metaphors) and figures nervous events as a "wave" washing through the body.[34] This central notion of *flow* describes the action of consciousness generally. Bain explains "the Nerve Force and the course of Power in the Brain" as "of a *current* nature."[35] This "current character of the nerve force" means, he argues, that we should not see the "brain [alone] as the organ of mind." Indeed, "no currents, no mind."[36] Later, in *the Emotions and the Will*, he explains emotion in such terms as well, saying, "In the shock of an impression of sense [or] emotion proper, there is at first a cerebral participation, and a diffusive agency, extending to the active extremities of the body." More simply, "[e]verything new that strikes us through any one of the senses, must be assumed to prompt [an] emotional wave" that diffuses throughout the body.[37]

Lewes emphasizes that this current is infinitely variable, carried along "a vast network of pathways, all equally capable of conducting an excitation, but not all equally and at all moments open." Furthermore, "*each stimulation leaves behind it a tremor which does not immediately subside*." This is a dynamic, everchanging system, "its elements being combined, recombined, and resolved under infinite variations of stimulation." He says, "Picture to yourself [a] sentient organism incessantly stimulated from without and from within, and adjusting itself in response to such stimulations." This stimulation forms consciousness, which he likens to "a mass of stationary waves," which are yet "fluctuating, because of the incessant arrival of fresh waves, incoming and reflected." (It is precisely this assault of incoming and interior stimuli that the microscopist must learn to manage and control.) For this reason Lewes pushes back against the trope of the *body as machine*, saying "machines have fixed and calculated mechanisms; whereas organisms are variable and to a great extent incalculable mechanisms."[38]

The transmutation of *flow*, *currents*, and *waves* from analogies to working theory did concern some critics, since the resemblances were only partial.[39] However, the trope of nervous transmission as *currents and waves* became conventional within physiology, associationist psychology, and even literature, revising the longstanding discourse on mental states as static entities and promoting interest in exploring how mind, body, and environment work together as mutable elements in a complex dynamic system.[40]

In fact, Bain originates the phrase "stream of consciousness," discussing "the concurrence of Sensations in one common stream of consciousness,—in the same cerebral highway."[41] Both Bain and Lewes referred to a stream of consciousness long before William James took up the notion.[42] Victorian microscopists were well aware of the currents of nerve-action given that research on the phenomenon required the use of the microscope, and research on the topic regularly appeared in texts for working microscopists. As Part III of this book will show, guidebooks' emphasis on maintaining a calm, well-balanced stance while observing reflects not just the need to prevent inadvertent movements of the apparatus but also the need to ground embodied vision in neuromuscular systems that are not unduly agitated or fatigued. This idea—that both thinking and feeling emerge from the circulation of nervous currents throughout the body—undergirds Victorian precepts relating to both microscopical skepticism (where the observer trains in specific practices in order to secure the authority of microscopical vision) and wonder (where the observer expresses and invites preferred forms of affective enchainment with the instrument and object).

Perception, Thinking, and Feeling Circulate between People

This model of mind as embodied and linked to a system of currents within the body undergirds Victorian models of emotional transmission between persons: if emotion diffuses in waves throughout the body, it may also diffuse *across* bodies. Among the physiological psychologists Bain and Lewes were particularly fascinated by this phenomenon (Spencer became more interested in questions of inheritance and evolutionary biology).[43] As Rick Rylance says in describing Lewes's work, "the *whole* organic and mental entity is to be regarded as in immediate, continual, integrative, and enabling contact with the particular environmental medium in which it functions," a process that applies "whether this medium is (for example) human society, evolutionary stock, or general biological conditions."[44] The same model that studies the flows entwining thought with feeling and the body's other organic processes, also posits our currents of thought and feeling as integrally, even physically, circulating with similar processes in other people.

In *The Emotions and the Will*, Bain interests himself in the phenomena of what he calls "Sympathy and Imitation"—that is, "the tendency of one individual to fall in with the emotional or active states of others," especially if we echo their physical behavior as well (since, as noted above, Bain believes that physical states can trigger the mental or emotional ones usually associated with them). He clarifies that "we *sympathize* with grief, anger, or astonishment; we *imitate* the handicraft, or the behaviour, the elocution, or the language, of one that we consider a model. Imitation...is more frequently voluntary on our part." That "susceptibility" is "greater in some men than in others to the outspoken feelings of their fellows, by which they yield more promptly to sympathetic influences."[45]

This description of the communication of nervous energies across bodies is a natural expansion of Bain's work in *The Senses and the Intellect*. There, he argues that experience attained at secondhand—via memory or imagination—is embodied just as the original thought, emotion, or sensation would be. Although a memory will be less intense than the original experience, he says, "a sensation enduring after the fact is essentially the same as...during the fact; the same organs are occupied, the same current action goes on." He asserts, "*The renewed feeling occupies the very same parts and in the same manner as the original feeling.*" Memory thus reinvigorates our earlier nervous energies, offering a physical reprise of prior experience, where "the train of feeling is re-instated on the same parts as first vibrated to the original stimulus."[46] He goes further: the imagination, he says, can rouse nervous flow in us de novo even if we did not experience the described event in person. Bain explains, "The imagination of visible objects is a process of seeing; the musician's imagination is hearing; the phantasies of the cook and the gourmand tickle the palate; the fear of a whipping actually makes the skin to tingle." Here "language that might otherwise be deemed figurative becomes literal."[47] Bain insists here that our physical, emotional, and cognitive experiences co-create each other. This dynamic relation applies both within an individual body and throughout a group. Bain's work here anticipates what contemporary theorist Karen Barad calls the *material-discursive*, which is the proposition that physical and cultural phenomena are inevitably entangled and co-produce each other.[48]

This theoretical alignment across bodies can be put to practical use; Bain suggests that a skillful teacher will communicate not just facts but also appropriate emotions to pupils, thus enlivening difficult scientific subjects for them. In *The Emotions and the Will* he explains, "we love to come under the influence of one, that wields with ease, the matter, and the emotions of scientific truth, and we may in fact have the greater enjoyment of the two."[49] This passage supports the persistent framing of popular science as *instructive and amusing*, even as it insists that the study of "scientific truth" is not factual and cognitive only but brings its own charge of "emotions."

Despite the pedagogical value of emotional transmission, Bain also figures such transfers as forms of involuntary physical consumption from external sources: the body "catches fire" or is taken over by "outward contagion."[50] The trope of *emotional contagion* for affective alignment continues even today. But Bain's use of this term reveals his concerns that humans are vulnerable to emotional influence and that speakers may attempt to direct audience responses. For instance,

Bain explains that "we are most ready to assume through outward contagion the states belonging to our own individuality" but wonder is "peculiarly catching from the boldness and energy of the expression belonging to it, and from the easy susceptibility of most natures to the state."[51] For Bain, unchecked wonder, terror, and other extreme emotions can overrun reason, and we should strive to curb our susceptibility to them.

Despite his concerns about these strong, "peculiarly catching" emotions, Bain sees the transmission of emotion as natural, even desirable. In 1879 Lewes, too, generally endorses "that sympathetic inward movement which may be called mental contagion."[52] Despite the potentially unsettling nature of the *contagion* metaphor, both Lewes and Bain turn to poetic language to evoke the power of shared feeling. Bain draws on biblical language as he argues, "To rejoice with them that rejoice, and weep with them that weep, to be carried away by the enthusiasm of a throng, to conform to the society that we live among, and to imbibe the beliefs of our generation, are a part of the human constitution." Indeed, he sees this "chiming in with the mental states of others" as an important "pleasure and interest of life."[53] Lewes echoes this phrasing, saying, "I see you reacting as I myself react; I hear you speaking as I myself speak, reasoning as I reason, loving as I love, trembling as I tremble."[54] In both instances emotion aroused in one person is readily carried by words and images to inflame the body, mind, and spirit of another.

Notably, while the immediacy of vision and hearing contribute to the phenomenon of emotional transmission generally, language also allows imagination to ignite others' "train of feeling." The words exchanged between friends carry a charge beyond mere meaning: they elicit a physical mimicry in us, sparking an experiential train of events in mind and body. Lewes argues in *Study of Psychology* that Mind and Language are experienced in a way "at once individual and social," helping to constitute both the subject and the community. He remarks, "What [a man's] tribe speaks he repeats; but he does not simply echo their words, he rethinks them." (This claim echoes Bain's idea that remembered or imagined emotions trigger the same bodily events that originally accompanied that emotion.) Without language, Lewes says, this alchemy cannot occur: a man "cannot think [his tribe's] thoughts as long as his experiences refuse to be condensed in their symbols." Thus "Language belongs essentially to the community" and to its shared affects.[55] For Lewes, as Rylance points out, "cultural transmission, largely through language, plays a part equally significant to that of biology" in forming both persons and communities.[56] Victorian microscopists were urged to achieve "exactness of description" and often debated the value and meaning of specific words and phrases, such as *microscopical*, *microscopist*, *penetrating power*, *granule*, and *reagents*. This emphasis on clarity of language suggests that many of them, like Lewes, recognized the role of language in sharing experiences, training observers, and anchoring communities.[57]

As suggested above, contemporary affect theorists echo this interest in the physical and cultural foundation of emotional experience and the alignment of affect within groups. They have, over the past twenty years, elaborated a lively discussion on emotion as shaped by and shaping human sociality. This discussion builds on the interests and concerns of Victorian physiological psychologists, and of Victorian microscopists.

William Reddy reaches insights similar to Lewes, emphasizing the importance of linguistic expressions (*emotives*) in establishing an *emotional regime*—a "set of normative emotions and the official rituals, practices, and emotives that express and inculcate them."[58] Barbara Rosenwein offers a less overtly political model, which I adapt here, of *emotional communities* characterized by distinct "systems of feeling." Here, implicit scripts demonstrate "the modes of emotional expression that [communities] expect, encourage, tolerate, and deplore."[59] The tropes I identify in this book function much like these scripts in attempting to channel the circulation of affect and fix it in favored forms; Victorian microscopists developed emotional communities around shared wonder and curiosity.

While Lewes and these theorists emphasize language as an important channel for circulating affect, Teresa Brennan (like Bain) examines the embodiment of affect instead, seeing it as a physical transmission across bodies. She notes that this "transmission of affect"—which neurologists term *entrainment*—is a "process whereby one person's or one group's nervous and hormonal systems are brought into alignment with another's."[60] Bain, Darwin, and their contemporaries study the role of vision and nervous currents in thus diffusing emotion, a notion that also powers tropes like *"oh law" objects* or the *body as machine*. Brennan draws upon twentieth-century science to broaden this mechanism, identifying not just an "electric entrainment" forged by the vibrations of touch and sound but also a "chemical entrainment" communicated by pheromones and our sense of smell.[61]

Other scholars see subjects, emotions, and social bonds as more contingent and provisional than these models suggest. Brian Massumi prefers *affective attunement*, a phrase borrowed from the psychologist Daniel Stern. The notion of *attunement* recognizes that different persons may respond to the same stimulus in different ways; they are directed toward the same event but not exactly aligned. Massumi explains, "Say there are a number of bodies indexed to the same cut, primed to the same cue, shocked in concert. What happens is a collective event. It's distributed across those bodies." Attunement thus acknowledges both difference and similarity. He says, "However different their eventual actions, all will have unfolded from the same suspense. They will have been attuned—differentially—to the same interruptive commotion." Unlike models of imitation or contagion, he points out, attunement can comprehend "difference in unison, and concertation in difference." His Spinozan approach emphasizes the role of affect in connecting us to others, but Massumi emphasizes that these experiences also call out each participant in specific responses. These "collective individuations" echo

Lewes's model, where affect and language both connect us in "tribes" and individuate us.[62]

Sara Ahmed also criticizes the model of *emotional contagion* because, she says, it "risks transforming emotion into a property, as something that one has, and can then pass on, as if what passes on is the same thing." Ahmed also points, as Massumi does, to the possibility of disparate responses to a shared stimulus. She says, "Shared feelings are at stake, and seem to surround us, like a thickness in the air, or an atmosphere. But these feelings not only heighten tension, they are also in tension. Emotions in their very intensity involve miscommunication, such that even when we feel we have the same feeling, we don't necessarily have the same relationship to the feeling."[63] Ahmed and Massumi thus complicate the model proposed by Victorian physiological psychologists. While Bain does acknowledge that individuals may be more or less vulnerable to specific feelings, he sees emotions as tightly bound to certain physical configurations. Microscopists, however, recognized the potential for varied response. All who viewed the microscopical image were thought to share in the affective energies excited by it. But, just as individual observers enjoy different levels of visual acuity (requiring the development of a *personal equation* to allow comparison of their reports), not all viewers would experience those affective energies in the same way. Tropes of wonder attempt to redirect any wayward emotions.

Some scholars of affect theory expand upon the physiological psychologists' notion of consciousness as dynamic and mobile, seeing emotions not so much as static properties exchanged by stable subjects, but as affective practices. Notably, Monique Scheer uses Bourdieu's concept of *habitus* to discuss emotions as action and process. She describes "emotional practices" as "manipulations of body and mind to evoke feelings where there are none, to focus diffuse arousals and give them an intelligible shape, or to change or remove emotions already there." She emphasizes "the mutual [and historically specific] embeddedness of minds, bodies, and social relations," addressing emotional transmission by noting that "these practices are very often distributed, that is, carried out together with other people, artifacts, aesthetic arrangements, and technologies."[64]

Ahmed similarly argues that "emotions should not be regarded as psychological states, but as social and cultural practices." In this light, she revises Rosenwein's "emotional communities," suggesting instead a model of "emotional collectives" since (as Bain and Lewes point out) the circulation of emotion actually helps *constitute* a community. In these "affective economies," Ahmed argues, "feelings do not reside in subjects or objects, but are produced as effects of circulation." She presents a "model of sociality of emotions" where "it is through emotions, or how we respond to objects and others, that surfaces or boundaries are made." That is, emotions in circulation "produce the very surfaces and boundaries that allow the individual and the social to be delineated as if they are objects." Individuals and communities, registering impressions from these emotional contacts, are formed as "effects of [this] circulation." However, Ahmed argues, particular objects or sites become "sticky, or saturated with affects, as sites of personal and social tension." In fact, "[I]t is [these sticky] objects of emotion that circulate, rather than emotion as such." Despite this accretion of affect on particular objects, she cautions, "Affect does not reside in an object or sign, but is an effect of the circulation between objects and signs," where "the more signs circulate, the more affective they become." Adapting the "wave" metaphor for the movement of emotion, Ahmed traces a "'rippling' effect" where emotions are produced and move in various ways: "sideways (through 'sticky' associations between signs, figures and objects) as well as forwards and backwards" into past or future.[65] Emotion here figures as a form of energy that ideas and objects accumulate in transit.

Ahmed goes beyond the physiological psychologists in describing affect in this way, but her model usefully explains the overdetermined meanings of the microscope in Victorian culture, where microscopes are especially sticky objects in their association with conflicting powerful emotions like wonder, astonishment, disgust, fear, and the sublime, and powerful ideas like faith, science, vision, and progress. Their role as sites of human–machine intercourse, at a time when Victorian society was strained by debates over the role of industry and the machine, only increases that affectual charge.

Nicholas Dames has argued that Bain and Lewes—who were also, of course, literary critics—understood the novel to be "a machine for the production of affect."[66] While Victorians describe novel-reading as an addictive, consuming experience, the microscope offers a similar experience in that the observer plunges into another world. Tropes of wonder like *microscopical fairylands* and *the vastness of the very small*, shared in books, periodical articles, lectures, soirées, and the like, promoted the arousal of affective waves and their circulation between author, microscope, object, and readers. This book argues that Victorian tropes of microscopy attempt to channel and organize roused currents (Bain's term) of affective energy into favored emotions such as wonder or admiration and thus to help constitute microscopical communities.

Perception, Thinking, and Feeling Can Be Trained

The physiological psychologists argued that even our unconscious or automatic nervous processes can be trained. For them, the deep grooves worn by mental habit represent an opportunity to direct and harness reflex functions toward external goals.[67] As I will show in later chapters on the trained eye, mind, and body (Part III), Victorian microscopists similarly recognized how unconscious function affected their observations and sought to train particular habits of mind and body that would support their scientific aims. With an implicit nod to the interlinkage of body and mind, and of unconscious and conscious thought,

Victorian microscopists discussed how the repetition of particular movements during one's practice facilitated particular physical and mental habits of observation. As Victorians gained interest in microscopy during the 1850s and 1860s, Lionel Beale and others stressed the importance of habit in building a strong microscopical practice. Carpenter, for example, remarks that "the experienced Microscopist...constantly keeps his finger on the milled-head of the 'fine movement'" when making a first pass through the object.[68] Assuming the habitual crouch over the table microscope, feeling the familiar ridges of the milled head (the focal adjustment) under their fingers—these physical sensations would place the worker in readiness and focus his attention on the process of observation. Given a history of prior "revelations" at the microscope, and a strong association between these physical sensations, this stance might also incline the observer to new experiences of wonder.

Perception, Thinking, and Feeling Are Forms of Labor that Can Cause (and Are Affected by) Physical Fatigue

In the dual context of the physiological psychologists' model of *embodied mind* and mid-century discussions of energy networks, mental fatigue is understood as a form of physical exhaustion. Anson Rabinbach discusses the cultural effects when, in the 1840s, the law of the conservation of energy—that "there was a single, indestructible, and infinitely transformable energy basic to all nature"—was first stated.[69] In this model, Rabinbach says, "[t]he body, the steam engine, and the cosmos were thus connected by a single and unbroken chain of energy" and by the concept of a fungible "labor power" or work-energy (*Arbeitskraft*, to use the term introduced by Hermann von Helmholtz in 1854).[70] Mark Paterson describes how this "transferability of energy" created networks linking different forms of labor: "An ox with its plough, a windmill grinding wheat, or a worker in a factory were situated in a continual network of energy transfer."[71] So, too, to cite a common phrase, the operator "works with" his microscope and object.

Microscopists working within this system participate in a system of circulating energy, drawing from and contributing to it in partnership with their apparatus and objects of study. Indeed, in *The Microscope and Its Revelations* (1856), published only two years after Helmholtz's introduction of *Arbeitskraft*, Carpenter positions the microscope as an element of the modern industrial world in just this way. He writes that "there is a large quantity of valuable *Microscope-power* at present running to waste in this country,—being applied in such desultory observations as are of no service whatever to science, and of very little to the mind of the observer."[72] The phrase "Microscope-power," though not widely used, here introduces and frames what became a canonical microscopical guidebook for half a century. The term construes microscopical observation in terms of the "labor power" circulating within networks, like the "water-power" and "steam-power" driving England's factories. This factory model of microscopical labor proposed by Carpenter requires efficiency; it should not be left "running to waste." Training and habit are seen as valuable tools in this context also; Carpenter suggests that, to "direct this power to more systematic labours," microscopists should adopt more methodical habits of observation, perhaps with his own book as guide and corrective force.

The principle of the conservation of energy was soon countered by that of entropy, and Victorians increasingly became concerned about the penalty every laborer suffers: fatigue. Rabinbach locates the "discovery of fatigue" after 1870, noting, "Before 1860 almost no medical or scientific studies of fatigue are recorded."[73] That is, he says, "The first signs of a change in the perception of work appears in the medical literature of the late 1870s, which began to consider overwork, over-exertion, and fatigue in the taxonomy of modern disorders." Before that, fatigue "did not yet appear as a medical term, nor did it receive significant attention." He also comments that "fatigue did not play as large a role in English thought" generally.[74] As I will show in Part III, however, physical fatigue appears as a concern for microscopists as early as the 1840s.

Physical fatigue was one concern; mental fatigue was another. Cognitive work is often associated with physical inaction, as in contemporary concerns about over-studious youth. But physiological psychologists insisted that thinking entails a form of physical labor, drawing parallels between muscular and mental fatigue. As early as the 1846 (third) edition of *Human Physiology*, Carpenter explains that "the activity of the Nervous system is immediately dependent, not upon a process of development or nutrition, but upon one of disintegration or destruction"; he points to "the sense of fatigue which is experienced, when the Mind has been long acting through its instrument, the Brain." However, Carpenter's interest is more in human health than the bodily efficiency that later experts would emphasize. This physiological fact, he says, has potentially dangerous consequences if ignored; he warns, "Where a prolonged exertion of the mind has been made, and the natural tendency to sleep has been habitually resisted by a strong effort of the will, injurious results are sure to follow. The bodily health breaks down; and too frequently the mind itself is permanently enfeebled." He even offers a diagnostic test doctors can use with patients; an "excessive phosphatic discharge" in the urine demonstrates this kind of "undue waste of nervous matter."[75] To counter this problem, in the 1853 (fourth) edition of *Principles of Human Physiology* he recommends a system of regularity and habit to minimize the fatigues of cognitive work, rhapsodizing over "the immense saving of exertion which arises from the formation of *methodical habits* of mental labour."[76] Carpenter's added advice for "methodical habits of labour" in 1853 chimes well with Beale's proposed training regimen for would-be microscopists in 1857.

Among the physiological psychologists writing in the 1850s, Bain discusses fatigue in terms that resonate most with later concerns. He dedicates considerable effort in *The Senses and the Intellect* to explaining the mechanism of "nervous fatigue and exhaustion." He proposes that "the conducting power of nerve fibre is a wasting operation" that "draws upon the vitality of the fibre," necessitating periods of "rest and a copious supply of nourishment" to recuperate. In fact, he cites this fatigue to challenge the metaphor of *nervous system functions like a telegraph*: unlike wires, he says, "the nerve takes fire from a slight stimulus like a train of gunpowder, and is wasted by the current that it propagates."[77]

Bain addresses fatigue again in *The Emotions and the Will*. There he argues that when the mind attends to "one engrossing outlet," it restricts others, in a process that he calls "the calling in of the stray movements or wanderings of the faculties" or "a shutting up" of unnecessary functions, thus achieving "a real economy of the mental energy" and "perform[ing] an important service to the overworked machine."[78] Bain warns of nervous fatigue but notes that some types of mental labor are more likely to cause it than others. "Abstract thinking and reflection task the brain severely," he says. On the other hand, "the dealing with actual objects of sense, the employment of the hand, the ear, the eye, in their respective spheres, is comparatively quieting and anæsthetic, and can be continued much longer before incurring cerebral exhaustion." Thus "the mental philosopher is more prone to the pains of excessive consciousness than the astronomer, the chemist, or the zoologist, who mingle much more largely the activity of the senses, and the body generally, with their studious life."[79] Microscopical work, of course, involves the easier work of the hand and eye but requires exhausting feats of abstract thinking.

The physiological psychologists' view of physical and mental work makes a clear break with earlier perceptions of fatigue. In his Bridgewater Treatise on *The Hand* (1833), Charles Bell portrays fatigue as a luxuriant, pleasurable experience. He says that any great activity "is followed by weariness and a desire for rest: ... [wherein] there is diffused through every part of the frame after fatigue and whilst the active powers are sinking into repose, a feeling almost voluptuous."[80] Fatigue is not framed in these texts as a threat to public health or social progress, as it would later be, but it is described as a natural mechanical and cognitive limit to human work in terms that presage later discourse. If, as Rabinbach says, fatigue "emerged at the threshold of the body's economy of energies [as] the corporal horizon of a mechanical universe,"[81] it did so in Victorian microscopical writing well before the general societal and scientific "turn" in the 1870s.

We Are also Entwined with Things

Just as the physiological psychologists saw mind as part of an embodied system, they also understood it as (as Vanessa Ryan says) in "dynamic, bidirectional relationship" with its environment, as part of a "continuous, learned, and adaptive interplay" between mental, physical, and social functions.[82] In their efforts to direct practice, Victorian microscopists recognized the power of community in molding individual action and insisted on the interaction between mind, body, and other bodies. For example, Spencer identifies the entwined embodiment of the microscopist and apparatus directly; he proposes that we recognize how the microscope, among other tools, works not just with the body but as an extension of the body. He recognizes the many prostheses we use in everyday life, arguing that "all observing instruments, all weights, measures, scales, micrometers, verniers, microscopes, thermometers, barometers, &c., are artificial extensions of the senses; and all levers, screws, hammers, wedges, wheels, lathes, &c., are artificial extensions of the limbs." He points out that these artificial aids mirror and extend physical systems in the body, saying, "The magnifying glass adds but another lense [sic] to the lenses existing in the eye. The crowbar is but one more lever attached to the series of levers forming the arm and hand." Spencer here describes microscopists as both interacting with the instrument and becoming one with it. His human–machine hybrid anticipates the arguments of later scholars like Donna Haraway (1991) who proposed a model of tool-using humans as cyborgs. Spencer goes on to suggest that humans and these tools, like the Sciences and the Arts, are bound by relations of "reciprocity" and "mutual dependence."[83]

Although technically speaking, the microscope serves as a visual prosthesis, the trope of the *new sense* speaks to observers' experience of the instrument as an integral part of their body. For example, in *Sea-Side Studies* G. H. Lewes writes, "The microscope is not the mere extension of a faculty, it is a new sense."[84] Lewes's phrasing takes for granted that the instrument extends the human body. He takes issue only with the nature of that extension. For Lewes, the microscope does not aid or extend existing power; rather it transforms him by adding a new kind of power. Despite this trope's allusion to a wonderful new power, microscopists quoted in society journals use the phrase to call for more patient labor. Thomas Henry Hennah, a photographer who was serving as President of the Brighton and Sussex Natural History Society, cautions in 1870 that the microscope can be used either as "a new sense, or a mere toy."[85] Similarly, the evolutionary biologist Edward Bagnell Poulton warns in 1888, "The process by which the student first learns to see with the Microscope is almost like the education of a new sense-organ suddenly conferred upon a mature organism.... A long course of discipline is necessary" to master microscopical observation.[86]

This integration of human and microscopical bodies reaches an apogee when, as physician Forbes Winslow reported in his 1852 Lettsomian Lectures, a man wrote in his will that he wanted his body after death to be "converted into fiddle strings," "sublimed into smelling salts," and "vitrified into lenses for optical purposes." A court determined that this man was eccentric but not insane; he simply wished to be "converted to purposes useful to mankind."[87] These objects—violin strings, optical lenses, smelling salts—all blend material and transcendent elements, body and spirit. Unlike this imaginative fellow, Victorian microscopists remain physically separable from their apparatus; but, embracing their instruments, like him they join a dynamic system distributing force and feeling, and making meaning.

Indeed, the trope of the *body* of the microscope reflects an implicit analogy between the microscope and the human form, which interact during observation. This common trope invokes a reciprocal address of microscopical and human bodies, figuring them as partners bending toward one another, lenses and limbs meeting in mutual labor. The observer hunched over his instrument, reaching out to adjust the screw, embraces the body of the microscope as a lover might. As Part III of this book shows, microscopical guidebooks often laid out preferred stances or movements in this intricate dance. But "touch" in this arrangement is not solely in the power of the operator; as Pablo Maurette notes, "To touch is to be touched. To feel something is also to feel oneself." The "simultaneity" of touch means that it is bidirectional: the bodies of both operator and instrument touch and are touched.[88] Stephanie Downes, Sally Holloway, and Sarah Randles argue that objects are agential and participate in affective economies in that they "have a capacity to 'act' on people's feelings." They see them as "actors which do not emote; which produce and transmit feeling, but do not, themselves, feel."[89] The texts I read in this book demonstrate how powerfully Victorian microscopists saw themselves as physically and emotionally "touched" by the instruments and objects they worked with.

In this book I generally describe the observer, apparatus, and object as though these were all separate roles, linked in a single system. By distinguishing between these roles, however, I am making what Karen Barad would call an "agential cut" between subject and object, articulating a distinction that is contingent and provisional.[90] Microscopical organisms like infusoria registered their own "observations" of the microscopical setup by responding to the actions of human operators—a reciprocity that observers acknowledged. Indeed, John Ellor Taylor, in *Playtime Naturalist* (1889), invokes an intentional gaze directed back at the eye of the observer. He says of rotifers, "Most of them swam so rapidly that they came and went before you got the note of admiration out of your mouth! Then an individual would boldly crawl or slowly swim right up to the inner surface of the glass trough, as if it were looking through one end of the microscope at you, whilst you were gazing through the other end at it."[91] Here the observer is also observed; the studying subject is also the object of study. Not all commentators adopted Taylor's whimsical anthropomorphism, but (as I show in Chapter 13) even in describing inanimate objects, Victorian observers did regularly note their physical responses or implicit material resistance to the operator's actions.

The observer becomes the object or recipient of attention in other ways, too: both humans and microscopical apparatus were themselves commonly studied under the lens, as in the examination of scratched slides or the practice of examining bits of one's own body.[92] Louisa Lane Clarke suggests, for instance, "snipping a papilla from your own tongue with a sharp pair of curved scissors."[93] All participants in the microscopical system may thus be observers and may also be objects, observed. Despite the overtly visual aims of the instrument and its users, Victorian microscopical texts, echoing the insights of physiological psychology, insist on the material elements that give rise to these visions. They insist on the physical grounds of observation, on the overlapping and contingent bodily processes of thinking and feeling, and on the entangled materiality of observer, instrument, and object.

Writing Embodiment in Victorian Microscopy: Beautiful Mechanism. Meegan Kennedy, Oxford University Press.
 DOI: 10.1093/9780198940623.003.0005

PART II
BEAUTY AND WONDER

6
Beauty and Wonder

My magic glass has taught me that there is not even a living speck which does not open out into something marvellous and beautiful.

Arabella Buckley, *Through Magic Glasses*, 1890

You want a *microscope*. You are burning to penetrate the arcana of Nature and to see those things which to most mortals are only seen through the eye of Faith, to mingle in the world of unknown infinite and infinitessimal [sic] whose forms quaint or grotesque move about in their contracted spheres of action with hopes and fears.

Francis West Lewis, letter, 1859

Introduction

As the physician Francis West Lewis humorously suggests in the epigraph above, the gleaming, desirable microscope promised nineteenth-century observers a tantalizing glimpse at nature's hidden mysteries—specifically an infinite invisible space, often troped as a *world*, and its minute but curious and lively occupants. Lewis mocks these familiar tropes of microscopical wonder but borrows their rhetorical power anyway. Indeed, wonder anchors many foundational texts in the history of microscopy. Leeuwenhoek, in a 1674 letter to Henry Oldenburg, Secretary of the Royal Society, describes animalcules in pond water, reporting that "[t]he motion of most of these animalcules in the water was so swift, and so various upwards, downwards, and round about that 'twas wonderful to see: and I judged that some of these little creatures were above a thousand times smaller than the smallest ones I have ever yet seen."[1] Hooke, in a 1677 letter to the Royal Society confirming some of Leeuwenhoek's observations, comments, "much to my wonder I discovered vast multitudes of those exceeding small creatures, which Mr. Leeuwenhoeck [sic] had described...some of these so exceeding small, that millions of millions might be contained in one drop of water. I was very much surprized at this so wonderful a spectacle."[2] Dispute over wonder also dates to those early years: Leeuwenhoek complained to Hooke, "I...oft-times hear it said that I do but tell fairy tales about the little animals."[3] Such tensions persisted into the Victorian era.

Of course, microscopy and its settings underwent considerable change between these sixteenth-century observers and Victorian practitioners. Nineteenth-century microscopy emerged from two traditions: *natural history* (the empirical study of the natural world, including what we now call botany, geology, and zoology, by observations in and out of the field) and *natural philosophy* (the experimental and theoretical study of physics, mathematics, astronomy, and, in the case of the microscope, optics). Victorian microscopists, while retaining a strong natural history tradition, helped develop an investigative practice increasingly shaped by experiment and theory, by a secular empiricism, and by newly formal standards of scientific writing, now differentiated from the more informal, literary, and eventually popular style of natural history writing.[4] Victorian texts in both natural history and science use tropes of beauty and wonder to channel and direct observers' affective responses.

Microscopists writing for general audiences or children, as in other genres of natural history, often use tropes of wonder (such as *the object as jewel* or *diatom as Gothic rose window*). These tropes stage particular kinds of visual pleasure for the reader, overtly seeking to recruit and entrain novice observers into favored modes of microscopical observation such as admiration or awe. Spectacular tropes prompting shock or horror, such as the *phantasmagoria*, appear relatively rarely in writing on table microscopy, but they were a staple of projection microscopy shows. Beauty serves as perhaps the most versatile trope of wonder, the most likely to be found in secular empiricist texts despite its long association with natural theology. Texts for working microscopists generally avoid the extended passages of fanciful language that draw the eye in popular texts. They most often use tropes of *beauty* or *analogy* to identify objects or to praise a particular apparatus, preparation, manipulation, or theory; but such texts do sometimes reference more exuberant tropes of wonder as well.

Writers call upon other forms of wonder most often in moments calling for heightened rhetoric—ceremonial addresses, or paratextual sites such as introductions and conclusions, where it is appropriate to call on the larger meanings of microscopical work.[5] The generic expectations of particular textual forms were generally well known; at least one society President delivering his annual Address overtly considered whether his text would fulfill his listeners' expectations.[6] Overall, the print record of Victorian microscopy suggests that microscopists at all levels of expertise valued both theistic and non-theistic tropes of wonder as an important means of directing microscopical observation.

Disenchantment

> "Analysis" of thistle-top, I say; not "dissection," nor microscopic poring into.
>
> John Ruskin, *Praeterita*, 1887

The nineteenth-century expansion of empirical investigation sparked ongoing debates about the appropriate way to approach the natural world. Some posed science and its new technologies as a cure for superstition; others voiced fears about a crisis of faith and a general secularization and loss of wonder. Visual perception was often at issue, and microscopical technology was sometimes explicitly caught up in the tumult.

At the outset of the Victorian period, Walter Scott published *Letters on Demonology and Witchcraft* (1830), a book that found long-lasting market success in explaining various supernatural anecdotes. The volume emerged naturally from Scott's antiquarian interests; he had long collected old Scottish legends, ballads, and the like, and reused them in his literary work. For some tales, the explanation derives from physiological issues, where "the external organs" of the viewer "may... become so much deranged, as to make false representations to the mind"; glitches in visual perception are "peculiarly apt" to be misread as real.[7] Other tales are explainable upon optical principles. For instance, three singing ghosts prove to be a "reflection thrown forward" from an adjoining room "by the effect of a concave mirror"; a "gleamy figure" bending earthward toward an astounded viewer turns out to be an old woman gathering her cabbages nearby, whose lantern "threw up the refracted reflection of her form" onto a church window.[8] Scott's curious but skeptical analysis is nowhere more realist than in this substitution of the elderly cabbage-gatherer for the glowing, angelic vision. *Letters* suggests that the "scrutiny" of empirical analysis is inherently demystifying, but it also frames this process as congruent with Christian efforts to "dispel the clouds of superstition."[9]

Scott's book sparked a number of follow-up efforts in the *explained supernatural*, including David Brewster's *Letters on Natural Magic* (1832), part of publisher John Murray's "Family Library." Brewster frequently mentions microscopy in this volume, such as the optical illusion in which elevations appear to be depressions, which he explains at length.[10] These visual deceptions are countered with experiments in microscopical vision, so that microscopy acts as both a cause and cure of illusion. In this book, Brewster aims not to dispel magical effects but to assign them to physical causes ("natural magic"). He waxes lyrical in praising the microscope's unique form of magic: its "power of bringing the remotest objects within the very grasp of the observer, and of swelling into gigantic magnitude the almost invisible bodies of the material world."[11] His romantic phrasing here shows how tropes of wonder can serve an ultimately skeptical project.

Such efforts in demystification rapidly invited pushback from an influential Victorian sage, Thomas Carlyle. He famously warned, in *Heroes and Hero-Worship* (1840), of the limits of empiricism and its misguided conceptions of knowledge. Nature, he argues, is "beautiful, awful, unspeakable." He asks, "This green flowery rock-built earth" and all its elements, sea and sky, "what *is* it? Ay, what? At bottom we do not yet know; we can never know at all." Implicitly referencing the vastness of microscopical and astronomical space and the limits of human instrumentation and cognition, he argues that Science looks a puny and deficient enterprise next to "the great deep sacred infinitude of Nescience, whither we can never penetrate, on which all science swims as a mere superficial film."[12] He here urges readers not to use empirical methods when wonder and awe are more appropriate responses to Creation.

However, Carlyle was not necessarily opposed to microscopy itself. In 1827 he translated the words of the German Romantic writer Jean Paul (Richter) that one route to happiness is for people to "take a compound microscope, and with it to discover, and convince themselves, that their drop of Burgundy is properly a Red Sea, that butterfly-dust is peacock-feathers, mouldiness a flowery-field, and sand a heap of jewels." This metaphor parrots the tropes of the *common wonder* and *the repulsive revealed to be beautiful* (which I examine in Chapter 9) to convey the moral that "we ought to value little joys more than great ones." Even allowing for Jean Paul's mordant humor and for the self-deception or self-discipline required of this philosophy ("convince themselves"), microscopical vision here represents an interest in and contentment with "little good-haps" that, the narrator says, is "more lasting" than social status and material pleasures.[13] In this case the microscope does not demystify but remystifies the world for the viewer.

Concerns over a disenchanted world persisted or even intensified at the other end of the Victorian period, after half a century's growth in empirical investigations. John Ruskin (another sage) was known for his critique of Victorian technologies from the factory to the bicycle, although he endorsed photography. In 1887 he goes beyond Carlyle to explicitly oppose all forms of optical magnification, promoting instead a holistic natural vision. He writes,

> Flowers, like everything else that is lovely in the visible world, are only to be seen rightly with the eyes which the God who made them gave us; and neither with microscope nor spectacles. These have their uses for the curious and the aged...but the use of instruments for exaggerating the powers of sight necessarily deprives us of the best pleasures of sight. A flower is to be watched as it grows, in its association with the earth, the air, and the dew; its leaves are to be seen as they expand in sunshine; its colors, as they embroider the field, or illumine the forest. Dissect or magnify them, and all you discover or learn at last will be that oaks, roses, and daisies, are all made of fibres and bubbles; and these, again, of charcoal and water; but, for all their peeping and probing, nobody knows how.[14]

Ruskin urges readers to seek detail with their own God-given sight rather than seeking to improve or supplement that sight with microscope or spectacles. Magnification, he argues, can disclose the mechanical workings but not the mysterious essence of living forms. Microscopy cannot fathom the world's mystery and indeed prevents us from seeing the "best pleasures" in nature.

In a famous 1917 lecture, Max Weber extends this claim, arguing that science actually evacuates wonder from the world—that increased rationalization over "millennia" of human life had "disenchanted" the world, leaving only a sterile set of problems to be dully worked out. He laments that "there are no mysterious incalculable forces that come into play [in the world], but rather that one can, in principle, master all things by calculation." He complains that even the "spheres of the irrational" are being "put under the lens" of empirical analysis in a misguided attempt to know the divine.[15] Microscopy ("the lens") here stands in for the "rationalism and intellectualism of science" that Weber so opposed. In both Ruskin and Weber, microscopy's power to move us—to delight, inspire, enchant, or terrify the observer—seems to have dropped out of the discourse.

Scholars have challenged Weber's dystopian vision of the modern era. George Levine argues that "Weber is right to recognize the power of 'rationalization' to demoralize," but that effect "is far from universal." In fact, he asserts, even in the nineteenth century, "Enchantment, of a sort, follows positively, quite naturally, from intense engagement with the entirely secular."[16] Steven Shapin similarly contends that "the 'scientization' of Western culture advertised by Weber, Veblen, and others did not accurately describe late nineteenth- and early twentieth-century realities: the world may well have been disenchanted for a few but not for the many."[17] It is also significant that Weber constructed his depiction of scientific rationalization as a contrast to the knowledge of "an American Indian or a Hottentot," or simply "the savage."[18] That is, he constructs this critique on a flawed (indeed, racist) stack of oppositions—enchanted/disenchanted, savage/rational—which collapse the variety and nuance of different kinds of culture, reason, technology, and knowledge into a simple binary.

Many Victorian microscopical texts—even those addressed to broad audiences—do suggest that the microscope would free the modern world from the grip of old superstitions. However, it is difficult to find any complaint about disenchantment here. Indeed, writers endorsing modernization and embracing a process of disenchantment often also use tropes of wonder. They suggest that the microscope's defining power does not eliminate such pleasures but rather enhances and enables them.

In this regard, Victorian microscopists formulate a theory of observation that chimes with theorist Jane Bennett's claim, "Enchantment as a mood requires a cultivated form of perception, a discerning and meticulous attentiveness to the singular specificity of things."[19] This kind of attentiveness to particulars appears in, for example, an 1850 article in *Chambers's Papers for the People*—an inexpensive periodical for working-class readers—explaining the phenomenon of "Natural Paper" or "Natural Flannel." The author describes the specimen with care, noting that "the outside resembled fine paper in texture, or more nearly soft-dressed glove-leather; it had a glistening surface, and was smooth to the touch, and as tough in texture as ordinary unsized paper." The correction from "fine paper" to "more nearly soft-dressed glove-leather" indicates this concern for accurate representation.

Similar substances found in nature had puzzled the curious for over a century. But, using an apt metaphor for investigating this bundled "paper," the narrator says that "the microscope unfolded its structure" to show it as "a compact belt of confervae" (algae) combined with "a number of siliceous infusoria." The narrator moralizes, "Vulgar superstition in a past age would have, and in all probability did attach a supernatural character to all these productions." He asks, "How many a fairy tale does science resolve into hard facts and ungilded expressions of truth" instead of superstition?[20] The microscope here is presented as a scourge of "vulgar superstition," stripping the gilt off fairy tales to reveal the "hard facts" underneath for a readership likely presumed to be particularly gullible. Despite this explicitly anti-romance stance, however, the narrator still frames the microscope in romantic terms, whether in the revelatory framing of the "unfolded" facts or the naïve opposition between superstition and truth. The language in such discussions thus often re-mystifies both the object and the instrument; the Victorian microscope serves here as both a wonder-dispersing device and a wonder-dispensing one.

This pattern is especially notable when the narrator cites a historical tale from sixteenth-century Switzerland, when "blood-spots, or *signacula*" appeared in a widow's house. She "suddenly beheld a frightful spectacle—blood springing from the earth all around her!" The widow "rushed in alarm into the cottage—but oh horrible! Blood is flowing everywhere—from the wainscot and from the stones! It falls in a stream from a basin on the shelf, and even the child's cradle overflows with it." The widow runs for help, "crying murder! Murder!" But the apparition spreads throughout the village, "blood bubbling up in a pond, blood flowing from the loft, blood covering all the walls of the house! Blood, blood—everywhere blood!" After retailing this extremely romantic tale with its Gothic overtones, the narrator turns to a more restrained, realist tone, commenting wryly, "It is very evident that there is much in this account which is overdrawn." Here "the microscope, among other instruments of science, comes acceptably to our aid" to show that the supposed blood was likely caused by "a little monad (*Monas prodigiosa*)."

Despite the performative rejection of romance, the microscope's revelatory power is compellingly staged in this anecdote—the striking shift in tone itself suggesting that the microscope is no ordinary instrument, with its power to transform the blood-stained terrors of the past into dry scientific nomenclature and modern facts. There is more than a hint, in this dramatic staging, of the microscope as *beautiful mechanism* (a trope I explore in Chapter 8). Indeed, much of the article relies on romantic language to make its claims, and almost every page describes some microscopical image or power as "wonderful." Whereas the narrator claims that the microscope "has contributed its share to the dethronement of fiction and the erection of fact," the title of the article—"The Microscope and its Marvels"—again insists on the fantastical more than the factual elements of microscopy.[21] The narrator muses, "It is always an unsatisfactory course to set down as false certain wonderful phenomena reported by history" but luckily does not have to do so now that the microscope "admits the phenomenon, and explains the circumstances." The microscope in this example explains away the bloody horror, but it also preserves and enshrines it, using tropes of wonder to preserve but redirect the affective flows presumably aroused in readers by the sensational tale.

It was not uncommon to find tropes of wonder even in the research papers that might most effectively disenchant the natural world. Many Victorian investigators found both wonder and deep feeling in microscopical investigation. Some of them, describing diatoms (unicellular algae) and other test objects, were working at the very limits of microscopical power and technique. They were among the most skilled and experienced observers in the world, examining the objects that working microscopists relied on to assess the quality of their lenses and thus secure their work—yet their writing frequently expresses a sense of awe and even spiritual purpose in what they see and infer. As I will show, wonder persists, albeit sometimes under restrictive conditions, through much of the nineteenth century. Indeed, Barbara Stafford argues, "The typically modern, 'Enlightened' association of technology with secularization tends to overlook its historical role in materializing the sacred."[22] While Stafford speaks of eighteenth-century technologies, her insight applies in the Victorian period as well, where the microscope's ability to access hidden natural beauties allowed it to recruit new observers and gain moral stature.

One of the most canonical Victorian microscopy texts acknowledges the importance of these wonders in its title. William Benjamin Carpenter, a physician and an acclaimed physiologist and zoologist, was well respected in the scientific community. He was affiliated with the popular lectures of the Royal Institution and helped build the fledgling University College London in its early decades. Scholars have examined his interest in the physiology of mind and in debunking spiritualists. However, they have largely overlooked his influence in Victorian microscopy, which can hardly be overstated. Carpenter's 778-page *The Microscope and Its Revelations* (1856) became a classic text in the field, reaching eight editions by 1901.[23] Carpenter proposed this work as a guide for "the working Microscopist" so that the observer might "benefit science, whilst expanding and refreshing his own mind."[24] The book was frequently recommended as a comprehensive introduction to microscopic work. A reviewer in the *Monthly Microscopical Journal* (*MMJ*) of the Royal Microscopical Society in 1873 called it an "excellent treatise... assuredly, as regards the wants of the general student... the very best and most abundantly illustrated work on the subject, that any language possesses."[25] It is one of the few general works to be cited regularly in the *MMJ*.

Carpenter's use of the trope of *revelations* was typical of Victorian texts across multiple genres. Such terms were common in microscope text titles and in natural history texts as a whole, especially in religious, popular, and children's texts.[26] In "The Microscope and Its Marvels," the instrument is called the "this wonder-working apparatus" and the "all-revealing microscope."[27] Authors most frequently used the phrase "wonders of the microscope" or the term "wonderful" to call attention to objects of study. "Marvels" and "marvelous" also appeared regularly, or terms evoking revelation ("revelation" or "reveals"), suggesting the magical agency often attributed to the microscope and evoking its support of natural theology. Jabez Hogg, for example, on a single page of his 1854 manual *The Microscope*, writes not only that "the microscope... *reveals* to us" the structure of nerves, but also that "the microscope *tells* us" of their composition, "the microscope has *shown* us" the network of capillaries, "the microscope *teaches* us" of the relation between glands, and "the microscope *discovered* [the] real anatomy [of the skin] and showed us" all its components.[28] In similar phrases, the microscope commonly *divulges, discloses, unfolds, unravels*, or *unveils* natural truths, implicitly referencing the trope of *Science peering into Nature's hidden secrets*. As the author of an 1847 essay review wrote, his discussion of "the wondrous revelations" of the microscope would "show how the eye of man is here opened to penetrate the most profound, and formerly unsuspected, secrets of nature."[29]

The frequent appearance of extravagant language of wonder and revelation in titles points to the importance of this concept in microscope texts (books, articles, and lectures) and within the ecology of natural history generally. Scholars examining wonder in microscopy have tended to emphasize its role in popular rather than scientific texts. The work of naturalists (in natural history) and popularizers was indeed different from a developing "scientific" practice understood to be gaining dominance and status. During the second half of the century, scientific naturalists including T. H. Huxley, John Tyndall, and Herbert Spencer promoted a model of dispassionate, professional science sited in laboratories and grounded in empirical method. However, Victorian scientific communities demonstrated considerable range and overlap. Individual observers could and did belong to or read in multiple communities, and conversations ranged between them.

Given this overlap, some elements of microscopical practice and writing appear across contexts, while others remain anchored in a particular setting. The environment for microscopical work can vary widely, from field, to school, to laboratory.[30] Those writing in natural history could adopt a more affective, narrative style, while scientific naturalists increasingly sought to maintain a disinterested tone using plain speech and precise, quantified language. Specific tropes might appear across decades and in diverse contexts but function somewhat differently in each temporal and generic setting.

Examining the history of the scientific article, Alan Gross, Joseph Harmon, and Michael Reidy find a depersonalizing trend emerging in the nineteenth century: more abstraction, less personal narrative and action, less poetic expression, more precise quantification, and a more systematized, formal approach. They also identify an increased willingness to explain and theorize from the facts.[31] Alec Csiszar traces how nineteenth-century scientific journals borrow and develop "practices of validation and judgment," such as the peer-review process, to enforce expectations for research and writing. By the twentieth century, scientific journal articles had become an authoritative voice of "modern public-oriented scientific expertise."[32]

Lorraine Daston and Peter Galison tell a slightly different version of this story in their study of nineteenth-century scientific atlases. They identify an ideal of mechanical objectivity that became a goal of scientific observation and reporting during the nineteenth century, creating "a newly disciplined scientific self bound to a highly restrained way of seeing."[33] The "regulative ideal" of mechanical objectivity directed how scientists saw and how they recorded what they saw—or rather, what they should *not* do when seeing and recording the world.[34] They explain, for example, that the observer should resist "the temptations of aesthetics, the lure of seductive theories, the desire to schematize, beautify, simplify."[35] More broadly, the observer should restrain judgment, inference, correction, any intervention; check any impulse to frame or contextualize the information; suppress personal feeling and interest, indeed any subjective access to the material.

At the same time that they strove to achieve a mechanical objectivity, however, both naturalists and scientific naturalists developed discipline-specific practices that refuse, or at least revise, the asceticism of mechanical objectivity. Martin Willis has disputed the suggestion that mechanical objectivity dominated Victorian science. He argues, "Imagination . . . neither disappeared nor was devalued, but continued throughout the nineteenth century to give impetus to the making of knowledge."[36] I have similarly found that the undoubted influence of mechanical objectivity did not suppress the affective experience of wonder in texts written by and for investigators. Wonder was not, then, restricted to popular, pedagogical, or theological settings, but it served different ends in different communities. While naturalists posed admiration and pleasure as primary aims, scientific naturalists like Tyndall and George Henry Lewes instead promoted the judicious exercise of speculation, pairing observation with hypothesis and experiment. They developed a model of science as (in Jonathan Smith's words) "an imaginative, speculative, creative enterprise" that brought its own form of pleasure.[37]

While it is important to recognize the differences between distinct microscopical settings and groups, their commonalities remain significant. Most Victorian microscopists urge readers to observe closely. Many recommend carefully recording what is seen, by word and image. And many also write about how wonder entangled the observer with the microscope and with the observed.

Victorian microscopists reference a broad landscape of microscopical wonder, drawing upon elements of the beautiful, the sublime, and romance. Tropes of wonder act as textual traces of these observers' affective arousal through the microscope; the generally romantic language of these tropes helps entrain readers in similar affective patterns. The writers I examine very often use tropes of wonder, especially the beautiful and the sublime, in the service of natural theology and the discourse of design. Microscopists expressed a swooning sense of the sublime inspired by the beauty, intricacy, and infinitesimal size of the objects of the microscope; the incomprehensible numbers of microscopic organisms that must then exist; and the apparent vastness and infinite regress of microscopic space. Many of these writers explained their experience of the sublime by referring to the narrative of natural theology, which can be understood as either a romantic or (in its links to natural history) realist genre. Not just microscopic objects but microscopes themselves could be depicted in aesthetic terms; they were exquisite *objects d'art* that reveal unexpected beauty once you peer inside.

Victorian writers sometimes exhibit another form of wonder when they describe a charming, fantastical, grotesque, or phantasmagorical "fairy-land of science," to quote Arabella Buckley's book title. These romantic fantasies, most often found in texts for broad audiences, draw upon fanciful or Gothic traditions to portray microscopic objects in the enticing language of traditional literary romance. They suggest that the instrument gives observers entrée to a picturesque fairyland, exotic land of Orientalist myth, or Gothic space of trickery, horror, and the grotesque. Each of these discourses of wonder signals a distinct approach in the microscopical observer.

The art historian James Elkins has suggested that critics eschew "words like 'awe,' 'wonder,' and 'the sublime'—words that have been said without thought so many times that they are like blank coins, rubbed by thousands of fingers until they are nothing but thin blank disks."[38] But these terms are historically appropriate for nineteenth-century visual experience, and writers of the time use them to describe their experience of microscopy. While such terms were at times used in the service of an argument directly referencing natural theology, they can also serve to stress the power and glory of the natural world or even of the microscope itself. This power is not simple nor does it remain unchallenged, as shown in the chapters that follow. But it is an integral element across various nineteenth-century cultures of microscopy.

Wonder

> A working man came, half shyly and half pleased,...to look through the instrument at some striking object. He gazed attentively for a moment, and then exclaimed, in considerable surprise, "It is beautiful—but, *is it true*?"
>
> Mary Ward, *Microscopic Teachings*, 1864

Wonder was itself the object of scientific study by Victorian investigators like Alexander Bain. With other physiological psychologists, he studied the physical aspects of human cognition and feeling, a topic that I will address in Part III. In *The Emotions and the Will* (1859), Bain defines wonder as a "shock" sparked by discontinuity and difference, by "anything that very much surpasses or deviates from our habitual experience," any "rupture of [the] accustomed continuity of events." Wonder is a physical experience, expressed through "the rounding and protrusion of the mouth," "elevation of the upper eyelid" and eyebrow, "the fixing of the eyeball," and other muscular actions. It "arises only from what is novel, rare, or surpassing" in our interactions with the world, such as "[n]ew scenes, new events, new acquisitions in knowledge, new discoveries." These "impart a racy stimulus to the mind." He states that "The *feeling* of wonder or astonishment... is rather a coarse pleasure, like mere intensity of sensation, or violent bodily exercise," characterized by "vivid sensation and muscular manifestation."[39] Bain warns that wonder is "one of the corrupting emotions";[40] but his overall position toward wonder is actually somewhat ambivalent.

Bain is primarily concerned that emotions like wonder, terror, love, and self-pride can "bar out the impressions of reality." He argues that this "intellectual blindness" allows "falsehood, mistakes, confusion, and fatality" to "ramify in every province of affairs and every relation of human life." He poses emotion as a threat to empirical work and reprises Bacon's warning against the "'idola'—idols, or false deities"—that mislead human seekers after truth. In science, he explains, "the eye, cold and indifferent, should see nothing that cannot always be seen the same, and the mind should keep record in like manner." Unfortunately, he says, this dispassionate approach is "not natural to the human intellect." As a result, "[t]he narrations of matter of fact are constantly perverted by" wonder.[41] His tone expresses frustration and disappointment; such corrupting emotions obstruct scientific progress.

Bain argues that investigators should preserve the integrity of their research by training the mind to restrain any excessive surge of emotion. This equanimity can be acquired with practice; he says, "science and extended study naturally bring a man more or less to the position of 'nil admirari.'" Eventually, the observer can govern his interactions with the world: "[e]ven the unexplained phenomena can be looked at with composure by the philosophic mind." Ultimately, he claims, "[t]he true intellectual nature" is "that which takes on... the abiding impression of difference (and resemblance) with the least emotional shock."[42]

But Bain's suspicions of wonder are not absolute. Despite his concerns about wonder in science, and his advice that investigators should learn to moderate their emotions, he condones the arousal of wonder and excitement in some contexts; he even celebrates the wonders inherent in scientific work. He quotes Carlyle's concern about the "decay of wonder" but argues that science has its wonders, too. Bain remarks, "The discovery of uniform laws makes wonder to cease in one way by showing that nothing in nature is singular or exceptional; but even out of this very circumstance the scientific expositor will often find out new matter of surprise.... Indeed, the 'wonders of science' have in our day come into rivalry with the marvels of fiction."[43] He also recognizes that botanists and other scientific investigators share with the huntsman the pleasures of the chase, such as "bright points of victorious achievement," "intervals of suspense and emotional stillness, like brilliant stars over the azure spaces," "danger and difficulty," and "fiery interest."[44] Such language invites and endorses wonder as much as pleasure. Later, Bain notes that "the study of the objects of the outer world may engross the very highest pitch of mental energy. A man contemplating the sun, moon, and stars, the varieties of the earth's surface, the vegetable and animal life,—may rise to the utmost excitement that human nature is capable of."[45] Bain's examples and approving language here frame scientific work as a sublime endeavor, pushing the human frame (body and mind) to its full capacity. Such statements authorize the thrilling arousal that can be generated by intensive study of the natural world. The dedicated investigator, then, can enjoy wonder and excitement as the legitimate prize gained from exploring the unknown.

Beauty and Intricacy

> The wonders disclosed by microscopic observation have opened to us a new world, in which we discover, with astonishment, the extremes of minuteness and complexity of structure united.
>
> John Herschel, *A Preliminary Discourse on the Study of Natural Philosophy*, 1830

The beautiful was the most common and versatile trope of microscopical wonder throughout the nineteenth century.[46] Even after florid expressions of romance or the sublime became rare in scientific articles and treatises, beauty continued to appear regularly. Long associated with natural theology, the trope of beauty can solicit admiration and awe for God's creative work, but it can also express and arouse admiration for a secular world, in the same way that, Levine argues, Darwin channels the reader's wonder toward nature instead of God.[47] Victorian microscopists use the term *beautiful* to encourage admiration of nineteenth-century technology and the deft or intelligent work of elite microscopists. "Difficult" natural objects such as diatoms were valued not just for their own beautiful forms but also for providing the opportunity to recognize and appreciate human skill. The physician Lionel Beale, advocating for medical microscopy in 1854, praises one investigator's "beautiful dissections of the nervous system" and another's "beautiful and highly important investigations" on the spinal cord; he offers instruction on how to produce a "most beautiful section" of human skin "showing all the structures, including the sweat-ducts."[48] Such passages not only instruct but also inspire students and working microscopists.

The term *beautiful* could thus describe microscopic objects themselves, microscopical apparatus, theories about microscopy, or the skill displayed by a microscopist in preparing or resolving the object. This is one important way that wonder remains valid within the changing scientific vernacular, by proposing science itself as an object of wonder; but *beauty* remains a viable approach to the natural object as well. The versatility of *beauty* leads writers of research reports, which over the course of the nineteenth century develop a restricted rhetorical range, to reference beauty more than other tropes of wonder. Indeed, in research settings over the course of the century, beauty largely displaces overtly faith-based or frankly romantic tropes of wonder such as *microscopical fairylands* and the *Oriental tale.*

Generally, *beautiful* describes the objects of observation themselves, in a usage found across microscopical genres. It draws on a notion of beauty derived from the work of Edmund Burke, in *A Philosophical Inquiry into the Origins of Our Feelings of the Sublime and Beautiful* (1757), and Immanuel Kant, in *The Critique of Judgment* (1790). Both approaches were available to nineteenth-century writers. Romantic poets such as Wordsworth and Coleridge popularized an aesthetic of the sublime drawn from these theorists, and William Whewell was a Kantian.[49] Burke, an empiricist interested in the psychological effects of vision, emphasizes material structure or markings of delicacy, symmetry, and regularity, and smooth, gradual variation.[50] Kant, distinguishing himself from both empiricists and rationalists, specifies that there is no intrinsic, independent property of beauty in an object, but that our sense of beauty results from the work of our judgment in coordinating our sensible appreciation of a material object with an a priori idea. He argues that such aesthetic judgments must be disinterested, universal, and necessary (for example, this belief can be communicated), and that the beautiful is "purposive" without an immediate purpose (it appears "as if" designed for our judgment), demonstrating pure form and composition.[51] Kant grounds the discernment of beauty in a cognitive act by the innate faculty of judgment, a position very different from the materialism of the physiological psychologists examined in Part I. But his definition of beauty aligns well with microscopical culture in an important respect: the writers I examine in this book, and Burke and Kant, all generally agree that beauty brings pleasure through the effect on us of sensible form. Some Victorian microscopists—especially those closely aligned with natural theology—describe beauty as though it inheres in an object apart from human perception. However, perhaps due to the importance of sensory discipline in microscope culture, many other observers, writing for a range of audiences, are fascinated by the role of human sensibility in materializing beauty, discussing beauty as a pleasure effected by and through human observation.

Microscopic visions revise philosophical ideals of beauty in important ways. Burke warns against strong or glaring color, and Kant dismisses color as merely charming, a distraction from pure beauty.[52] In general microscopy, color is associated with the errors of chromatic aberration or misfocused lenses; in contrast, the image of a perfectly adjusted superb lens might be celebrated for its water-clear transparency. But expert microscopists employed a number of techniques that brought out spectacular color in the image, and they sometimes describe these colorful appearances in the most admiring terms, even in texts addressed to other investigators. For example, the physician George West Royston-Pigott rhapsodizes about the "colour test" (flashes of deep color) that heralds the "beading" he saw in the scales of the *Podura plumbea* (springtail).[53] The use of injected stains, dark field illumination, or polarization also provided occasion for writers to exclaim over the beauty of brilliant colors. The *MMJ* reported, for instance, that Thomas Cushing presented to the Croydon Microscopical Club "a series of three selenite plates superposed on each other, and revolving at different speeds, capable of giving remarkably beautiful combinations of colour in conjunction with the polariscope, arranged by Mr. Becker."[54] Color inspires some of the most effusive language about microscopical beauty.

Just as important, microscopists also diverge from philosophical notions of beauty in that they rarely praise objects displaying the qualities of simplicity and smoothness, while they frequently admire *beauty as intricacy*. Microscopic objects are especially likely to be admired for their delicate, intricate structure. The beauty and structural intricacy of *Podura* and diatoms—valued as test objects for their intricate structure—especially provoke tropes of wonder. Diatoms, for instance, were frequently described as "exquisitely delicate and beautiful forms" or the like.[55] Lynn Merrill argues that intricacy was increasingly valued in Victorian art and literature due to microscopists' new awareness of, and interest in, detail; indeed, Ariane Varela Braga traces the influence of intricate microscopical forms in Victorian art and design.[56] The minuteness of such structures doubtless encourages microscopical communities to elevate an *aesthetic of the detail*. This aesthetic helps define nineteenth-century domains as diverse as the ordinary, the picturesque, and the decadent. Naomi Schor argues that forms associated with the detail tend to be gendered as feminine.[57] Indeed, the trope *beauty as intricacy* aligns well with her claims. But the surge of microscopical instruction at mid-century demonstrates a collective effort to elevate the detail as a signal of skeptical practice, traditionally marked as a masculine domain. Just as in literary texts the detail can function as ornamental or realist, detail anchors most microscopical texts, from theologies of nature to technical manuals, and it can function there within tropes of either wonder or skepticism.

Analogy, Metaphor, Hybridity

The unfamiliar beauties seen through the microscope often invite comparison with the otherworldly or transcendent. To describe these strange microscopical forms—beautiful or not—writers had long turned to *analogy* and *metaphor*, a practice that continued throughout the nineteenth century.[58] Microscopical beauties often inspire comparison to an opulent world filled with lovely things that one might recognize but not, perhaps, possess: precious objects like jewels, crystal, and pearls, and sumptuous materials like lace, silver, gold, velvet, and silk. The trope of *microscopical richness* speaks to not only the profusion but also the variety and apparent opulence of microscopical life. The popularizer Mary Somerville draws on this trope when she describes the antennae of the *Sapphirina fulgens* (sea sapphires) as "silky plumes."[59] It appears also when Louisa Lane Clarke urges her young readers to study oil cells on plants, exclaiming, "Cells containing oil are beautiful objects ... raised upon delicate stalks, often brightly coloured, or glittering diamond-like in the sunshine." She elaborates, "Sometimes the oil cells are sessile, in golden spots upon the back of a black-currant leaf; or white and silvery in the recesses of a Sage leaf, a leaf of Rue, or Hop, or Mulberry." Clarke here uses color, intricacy, and richness (diamonds, gold, silver) to venerate these little objects hidden in the most ordinary places.[60] Forty years later, the naturalist Edward A. Butler, in his 1886 book *Pond Life: Insects*, similarly effuses over the "beautiful china-mark" moth, which "when viewed by reflected light with a low power of the compound microscope" reveals "satiny scales" along the wings and "velvet pile" lining the antennae.[61] The unlovely fact that Clarke's oil cells "are called glandular hairs," or that Butler's beautiful moth is now considered a pest, does not detract from the beauty and apparent (if not actual) rarity of these objects under the microscope.[62]

The clergyman J. G. Wood, in his long-lived guidebook *Common Objects of the Microscope* (1861), doubles down on this convention of using metaphors of rarity and luxury to describe the microscopical world. Wood's little microscopical guidebook, with its stunning illustrations, earned long-lived success in the British and American marketplace. In it, he regularly compared microscopical objects to rich, gorgeous materials. Describing objects illuminated by a condenser, Wood says, "when [the light] is refracted from hairs, feathers, or scales ... the whole field of the microscope seems to be filled with resplendent gems, flashing with a radiance that is almost painful in its intensity." Of a lavender leaf, he notes, "the globular perfume-cells may be seen scattered plentifully about; gleaming like pearls through the hair branches under which they repose." Similarly, "While the diatoms are alive, swimming through the water, their beautiful markings are clearly distinct, glittering as if the form were spun from crystalline glass."[63] (Since diatom frustules are formed of silica, these intricate shells are indeed "spun from crystalline glass.") Wood's continued references to the field of vision of an implied observer, his emphasis on what "seems" to occur in the microscopic field—what "may be seen"—and his practice of comparing microscopical structures to tangible objects such as jewels show that he values human perception and understanding as integral to the effect of these visions, even as the metaphors' inherent tension between like and unlike things (cells and pearls, for instance) sets up these analogies to powerfully suggest sensory pleasures. Such passages blend instruction, recreation, and community building; they entice more diligent students into microscopical practice by showcasing the delightful rewards on offer.

Victorian microscopists sometimes overindulged in such comparisons. For instance, the clergyman Edmund Saul Dixon, writing in *Household Words*, was evidently not content with conventional tropes of beauty, like jewels and pearls. Instead, in a frenzy of metaphor, he piles up a motley jumble of comparisons drawn from an array of settings. In just a single page he compares microscopic objects to a whale, a bunch of grapes, a "curdled fluid," gossamer cloth, the spokes of a wheel, a "simple iron wire," pearl necklaces, rosaries, gelatin, "a pencil of hairs like the stamens of flowers," an ox carcass, "young palmtrees," a wedge of cake, a hand made up of "broad miller's thumbs," a corkscrew, a "centenarian oak," a dodo, and a turkey-cock.[64] He heaps metaphor upon metaphor in a feverish collection of oddly specific images. He describes pollen, for example, not just with the lovely image of "Chinese carved ivory balls" but also as "grains of wheat, magnified into semi-transparent weavers' shuttles," and "cannon-balls covered with spikes."[65] Dixon usually does not offer these metaphors as explanations, apart from his discussions of a diatom frustule as a "shield" and the Navicula as a boat, which do suggest function; he usually proffers metaphors as simple descriptions.[66] Unlike the proponents of the "cell-state" theory that Andrew Reynolds discusses, Dixon argues only that spiderwebs *look* like pearl necklaces once magnified, not that they *work* like pearl necklaces also. His comparison of the notch in a butterfly scale to a "wedge of cake" is not offered as a theoretical statement and does not provide any transferable insight about the structure or function of the scale.

However, Dixon's use of metaphor does argue for specific qualities of microscopic vision. Most obviously, his metaphors foreground the marvelousness of microscopic images, which often strain credulity. Even the basic descriptive function of metaphor proves inadequate to the unusual physical appearance of objects magnified under the microscope. Each of his comparisons invokes a particular form or body, but many of them combine disparate objects in nightmarish *hybrid forms* such as the hand made of miller's thumbs. He draws on a long history of describing "animalcules" in such terms, depicting the microscopical world as so surreal that any single metaphor cannot encapsulate its grotesqueries.[67] To explain these strange forms to lay readers, Dixon crams together increasingly unlikely, detailed analogies such as "bits of half-melted sugar-candy, with which a small quantity of horse-hair has become entangled."[68] He jolts the reader from one metaphor to another much as the microscopic image does when the operator changes the magnification. Elkins has termed such far-reaching metaphors a "visual desperation" prompted by extraordinary, unfamiliar images.[69] This fantastical language, desperate though it is, builds a detailed and accurate mental picture of the microscopical image. Dixon's emphasis on texture in these comparisons (spiky, sticky) goes further, encouraging readers to grasp these strange things cognitively by imagining a richly material, tactile encounter, wherein the image becomes itself an almost tangible object, as though the observer could reach out and grasp it. The sensory overload of Dixon's writing here aligns well with the setting of this article. He does not offer any practical instruction to readers but emphasizes the recreative potential of the microscope and the plenitude and creativity of the microscopical world.

The trope of *hybrid form* anchored Victorian disputes over how to classify various tiny lifeforms. The enthusiasm for classifying and reclassifying the objects of the natural world was common enough among Victorian naturalists that it frustrated some.[70] Microscopy provided a wealth of occasions to dispute what its objects of study were, at the most basic level: plants or animals. George Thomas Fisher, describing minute freshwater infusoria, warns novices about "how closely the lower forms of vegetables and animals resemble each other." Indeed, "the question…'Do all these Infusoria belong to the animal or vegetable kingdom?'" is "difficult of solution." (Today most of them are considered protists, neither plants nor animals.) Fisher's footnote on the subject emphasizes the ever-changing form and activity of such creatures. He quotes at length from a recent treatise on algae comparing young infusoria with balloons, spinning tops, and "the fleetest race-horse" before they fall "as though dead." The passage concludes, "Strange transition, from the roving life of the animal to the fixed existence of the plant!"[71] The unsettled form of the infusoria here (balloons, tops, horses) seems to underscore the uncertainty over their identity as plants or animals: plants aren't supposed to move or prey on other creatures. Conversely, Carpenter notes in 1856 that the Hydra (freshwater polyp) can "propagate itself by buds like a plant."[72] The hybridity of many of the organisms now classified as protists thus encompassed both form and action. This confounding set of forms and behaviors led to protracted debates over such definitions.

Hybrid form similarly appeared in mid-century arguments over vitalism and the boundary between animate and inanimate matter—debates in which British microscopical skill and knowledge were leveraged on all sides. Carpenter and Lionel Beale, among the most influential Victorian writers of microscopical manuals, played important roles in these discussions: Carpenter's 1862 *Introduction to the Study of the Foraminifera* triggered disputes over whether the foraminifera (unicellular protists) were animate, and Beale was a staunch defender of the theory of a distinct vital force differentiating living from non-living beings.[73] Observers sparred over whether particular forms, like the foraminifera or "germs" (a term that could refer to a number of different things), were animate, inanimate, or something in between.[74] The established tradition of *hybrid forms* in microscopy—the bewildering panoply of structures found under the lens—supported the notion that, at the microscopical level, the usual rules of form and life might not apply, and borders might be blurred or broken. The boundary between life and not-life looks less solid when investigators are already describing tiny, virtually invisible beings that look like an unlikely mashup of a zoo, a jewelry box, and a rubbish heap.

The microscope was pivotal in nineteenth-century research helping to determine the answers to questions about what separates animals and plants, life and not-life. It enabled Robert Brown's discovery in 1827 that tiny pollen particles suspended in water swarm about randomly (Brownian motion), proving that *independent movement* does not necessarily indicate *life*. It was also central to the experiments conducted by Pasteur and Tyndall that helped eventually disprove the vitalist hypothesis, and to the development of a science of bacteriology later in the century. Nineteenth-century microscopy probably contributed as much to answering as to asking these questions about the boundaries of form and life.

While effusive, hybridizing comparisons were more common in popularizing texts like Dixon's, analogy and metaphor can appear in texts addressing very different audiences. James Secord has noted how Victorian science books for adults—even some influential treatises—shared "some of the same rhetorical techniques" used in children's science books, since "introducing new scientific concepts involved radical changes of perspective, often best explained by analogy, metaphor and anthropomorphism."[75] The popular writer Anne Pratt thus uses a string of similes to describe the *Diatomaceae* (unicellular algae) for her young readers in 1850, saying that "Some…seem like strings of beads; some like tubes of glass filled with bright colouring matter; others composed of networks finer than the finest cobwebs; some like little bunches of slender hairs; others branched, like minute trees of emerald green or sparkling brown."[76] But the physician John Hughes Bennett also explicitly recommends description by metaphor, although his similes are relatively pedestrian. In a textbook on clinical medicine, he suggests that metaphor is a useful tool for instruction and research, advising readers that "[m]inute structures seen under the microscope may…be likened to the shape of well known objects, such as that of a pear, balloon, kidney, heart, &c. &c."[77] This endorsement of figurative language has a long pedigree, as shown in Chapter 1, as does the trope of *metaphor as a lens*. The metaphorical language commonly used to describe microscopical objects provides a textual magnifier to supplement the material one.

The intricate, mesmerizing beauty of diatoms inspired many recurring literary tropes. Their elaborate patterning inspired comparisons to exquisitely crafted objects such as a *spiderweb*, the engraved Guilloché patterns machined onto an *engine-turned watch*, and a stained-glass window, specifically the *rose window* traditionally found above the west wall of a Gothic cathedral. For example, Louisa Lane Clarke repeatedly describes various diatoms as "beautiful" or "lovely," explaining to beginning microscopists that the "wonderful delicacy of [the] tracery" and "perfection of…form" in *Arachnoidiscus* is "like a gothic window."[78] Carpenter describes *Arachnoidiscus japonicus* as "[a]nother very beautiful discoidal diatom" with "beautiful markings" on its outer shell like "a spider's web," and an inner structure like "a circular Gothic window."[79] In 1868 he added a spectacular fold-out engraving based on Richard Beck's delineation of this diatom and its seaweed environment under a ¼-inch objective[80] (Fig. 6.1). In my

Fig. 6.1 *Arachnoidiscus japonicus.*

Source: William Carpenter, *The Microscope and Its Revelations*. London: John Churchill, 1868, plate X (facing p. 300).

copy of this edition, the plate has been detached, perhaps from frequent use or for ease of reference. Despite the lines that have worn through the paper as readers repeatedly unfolded and folded the paper over time, the image remains striking, emphasizing the intricate structures that demarcate this species and inspire these tropes. Such depictions persisted throughout the century. T. Charters White (the "Quekett Club-man"), in 1887 describes *Aulacodiscus orientalis* as "a little disc, sculptured and fretted with marvelous workmanship, with ornamentation more elaborate and more beautiful than that of any rose window in the finest continental cathedral." Marveling that "this very individual may possibly have tenanted the earth about the time of Abraham," he even gives this ecclesiastical aesthetic an actual link to biblical history.[81]

Jane Bennett argues that "enchantment...seems to hang out [on] the border between humans and animals and between organisms and machines, as well as those places where one confronts a perplexing, almost overwhelming, degree of complexity."[82] Microscopical texts frequently figure diatoms in such terms: a living artifact, always defined by the complicated patterning on each frustule, revealing exquisite beauty at unimaginably small scale. It is no wonder that so many were captivated by these forms.

Sometimes a writer will admire a beautiful image even if that image might not be a reliable representation of form. In an 1874 article that the *MMJ* reprinted from *The American Naturalist*, the American G. W. Morehouse discourses on the beauty of diatoms, even though he is aware that their structure was then "unstable and contested ground" among expert microscopists.[83] The 1870s brought many disagreements over whether particular diatoms were composed of patterned lines or series of holes, hollows, or hills. Morehouse comments of *Hyalodiscus subtilis* Bailey, "On this beautiful little shell the 'engine rulings' [engraved patterns] are readily seen with almost any illumination." *Surirella gemma* is a "beautiful form" and he notes "no trouble...in making the little beauty *seem* to 'wear beads'" (show off its hills). He concludes, "Here as elsewhere nature, without waste of material, combines utility with beauty."[84] As his italicized "*seem*" shows, Morehouse could admire the lines or hills shown in the image even in this investigative article—and even if they might not represent the true structure of those objects.

Beautiful metaphors travelled widely between different publishing contexts, but—as with wonder generally—those writing for working microscopists were likely to use more care in framing their references to beauty. It is perhaps this cautious use of the beautiful that allows it to persist in some working microscopists' texts quite late in the century. The marine biologist William Saville Kent dedicated his *Manual of the Infusoria* (1880–2) to T. H. Huxley, thus signaling not just that the book offered research and instruction more than simple recreation but that Saville Kent identified his own work with the secular, rationalist, and empirical principles of Huxley's scientific naturalism.[85] Saville Kent also indicates the seriousness of his text by emphasizing beauty more than awe, amazement, or whimsy, and by using strongly emotive references only rarely, preferring the milder term "delicate" instead. In one passage, however, he describes the *Flagellata* as "exquisitely beautiful representatives of this, excepting to the initiated, practically invisible world."[86] Here the interpolation of "practically" helps hedge the romantic force of "exquisitely beautiful" and the wonder-filled trope of *seeing the invisible*. In another outlier example, Saville Kent remarks that "the *Vorticellae* or 'Bell-Animalcules' include, perhaps, the most familiar and graceful representatives of the entire class of the Ciliata; their "wineglass- or bell-shaped bodies and accompanying elastic, thread-like pedicles" provide "a spectacle of unparalleled beauty and activity."[87] He does not honor many objects with such praise, and these infrequent, exuberant references to beauty convey additional force because they are used so carefully.

Just as writers may use tropes of beauty with more or less restraint depending on context, they also may use it for different purposes. Texts for workers, for instance, are more likely to mention beauty when helping readers identify particular species. Carpenter—writing for "working microscopists" but adapting his prose to those with "little or no" experience—generally refrains from exaggerated aesthetic judgments and most often turns to beauty and analogy when identifying specific diatoms.[88] The "very beautiful" *Hyalodiscus subtilis* features radiating markings "very much like those of an engine-turned watch," while "very beautiful" *Arachnoidiscus* is "so named from the resemblance which the beautiful markings on its disk cause it to bear to a spider's web." He explains that the outer layer of *Arachnoidiscus* "has upon it the peculiar spider's-web-like markings"; the inner layer "bears a very strong resemblance to that of a circular Gothic window."[89] John Quekett, in his influential 1848 treatise on the microscope, frequently uses the term "beautiful" but distills metaphors of beauty to the minimum necessary for identification; for example, he mentions "the beautiful engine-turned discs (Coscinodisci)" without elaborating on the resemblance.[90] Saville Kent's phrasing also demonstrates this instructive function. Rather than describing the bell-animalcules fancifully *as* wineglasses, he says they have "wineglass- or bell-shaped bodies," which frames these analogies as overtly descriptive and instructive.

In contrast, the popularizers Clarke and White use tropes of wonder such as the rose-window metaphor to emphasize the otherworldly beauty of the diatoms as their central feature. White offers only a brief review of the diatomaceae, while Clarke describes at length what to look for in these exquisite forms. But both Clarke and White foreground beauty not just as an aid to memory but also as a means to recruit readers into microscopical practice: they both suggest that the beautiful diatoms provide a reason to study microscopy.

For the more serious writers, in contrast, beauty is simply one among many attributes worthy of remark. (The exception here might be Royston-Pigott, who insists upon the beauty of the "beading" revealed in *Podura* scales by his Aplanatic

Searcher; but he exceeds the conventions of gentlemanly science in multiple ways.) Carpenter and Quekett mention beauty and the watch analogy almost in passing, as an aid to identification. Elsewhere, Carpenter places beauty in a broader context, recommending that the "scientific Microscopist" should study the "structure of flowers" to gratify "not merely…his love of knowledge but also…his taste for the beautiful." The description of the geranium that follows accordingly notes its "most beautiful variety of vivid coloration" and the "velvety appearance" of its petals.[91] Unlike Clarke and White, Carpenter extensively discusses the structure of diatoms, important research questions and reputable investigators, and methods of further investigation. He cites research supporting his claims and details the process of separating the layers of the diatom. Importantly, despite these different approaches to discussing diatoms, however, these authors all repeatedly mention their beauty.

The naturalist Philip Henry Gosse offers a striking example of how publication context may affect these decisions. He uses tropes of wonder to describe rotifers differently in two very different texts: *Evenings at the Microscope*, published in 1859 to provide recreation and some basic instruction for a general audience, and an 1887 research article in the *Journal of the Royal Microscopical Society* (*JRMS*). In *Evenings* he applauds the beauty, personality, and obliging transparency of the little "wheel-bearers" (*Rotifera*), saying,

> Their numerous and varied forms, often of remarkable symmetry and elegance, their swiftly-revolving wheels, their vigorous and sprightly motions, their curious habits and instincts, their complete organization, and the ease and correctness with which this is discerned through their tissues, which have the transparent brilliance of the purest crystal—all combine to impart a charm to the Wheel-bearers, which makes the observer hail their appearance in his drops of water with pleasure, and linger over them with unwearied delight.[92]

The keywords here are "pleasure" and "delight"—Gosse here remarks on the observer's affective engagement with what is "remarkable" and "curious" about the rotifers. He was an active member of the nonconformist Plymouth Brethren, whose beliefs centered on biblical literalism and the Second Coming, and Jonathan Smith argues that this faith underlies the general optimism that is visible in his work. "For Gosse," Smith explains, "nature is full of almost unlimited joy and happiness, untainted by competition or suffering," providing a glimpse into something like "the unfallen Garden of Eden" or our existence after Christ's return.[93]

With the tropes *life in a drop of water* and *the object as crystal*, then, Gosse directs readers toward admiration and awe rather than shock or disquiet at these peculiar forms. He frequently models the observer's enjoyment like this, luxuriating in description: he invites his readers to "take a peep at this beautiful *Brachionus pala*," with its "cup of elegant form" and its *lorica* (shell) "of the most glossy transparency," through a pocket lens, which "reveals its beautiful form." The cilia he describes as "two perfect crowns…which we can compare with nothing else than the crown-wheel of a watch, if allowed to run down," adding, "Is this not a charming sight?…The beauty and wonderfulness of these ciliary wheels are so striking…that for awhile we see nothing else: we cannot take our eye off from them."[94] This popularizing text affords Gosse ample leeway for fanciful aesthetic description, which he uses to overtly solicit the reader's affective engagement.

Gosse uses wonder more judiciously when he presents newly discovered species of *Rotifera* to his scientific colleagues. Here he details each rotifer with a crisply scientific voice: "*Nommata Theodora.* Near to *N. aurita, naias,* and *potamis*; from all which it differs in that the eye is small, and quite frontal, while the slender straight foot is protusile to an immense length, or wholly retractile. Length, when fully extended, about 1/60 in. Lacustrine." He uses specialist terminology and quantified measurement, with occasional conversational qualifiers (quite, immense). However, Gosse supplements these empirical reports with familiar tropes of wonder to help identify each species. He remarks upon the beauty of *Nommata Theodora*, "[a] noble form, of great elegance, and of glassy clearness." *Furcularia molaris* boasts an eye "of ruby brilliance" and is "an elegant and attractive little species." *Furcularia sphaerica* has "a pleasing little form" with "a minute ruby eye." The tropes of *intricate beauty* and the *object as jewel* recur here. Gosse dips into whimsy as well: *Furcularia sterea* "was not much addicted to swimming, but crept vivaciously about the vegetation, grubbing and browsing." *Dispinthera capsa* "suggests the odd notion of a creature carrying its great clumsy head in a bandbox."[95]

While some articles in this relatively late volume of the *JRMS*, such as Henry B. Brady's "Synopsis of the British Recent Foraminifera," eschew tropes of wonder, others join Gosse in employing these devices. The physician Alfred C. Stokes frames his "Notices of new American Fresh-water Infusoria" with wonder. He introduces "these minute but charming forms of animal life" by describing an observer interacting with them mechanically, cognitively, and affectively: the "student…has only to sweep his collecting vessel among the water-weeks, or gently scrape the soft ooze," to be "amply rewarded" with infusoria whose "beauty shall stir his aesthetic nature" and "fill him with wonder." The infusoria shift grotesquely and apparently infinitely from shape to shape; different species are described as "remarkably plastic," "curious," even "indescribable," with "incessant alterations" and "endless changes."[96] Stokes's work shows how, as late as the 1880s, tropes of beauty and wonder continued to feature in a range of contexts. Tropes of beauty and wonder do appear less often in articles reporting on experimental work, more often in research articles that primarily describe or distinguish particular species—surely to aid readers in identifying and remembering those species, but also showing how the work of species identification often retained an affiliation with older natural history traditions.

In sum, then, the texts addressed to elite audiences do not avoid the language of beauty, but they acknowledge beauty in the course of addressing other aspects of the microscopical object. Where popular natural history texts, hoping to entice not just readers but also would-be microscopists, often make beauty an object in itself, research-related texts addressing an audience of fellow investigators do not eschew beauty altogether but rather propose it as one object among many. They may evoke wonder in passing, most likely by dropping a single word (such as "beautiful" or "exquisite") into their prose.

Whether this adjective describes the structure of a diatom or the skill of its photographer, it acknowledges the aesthetic and affective dimensions of microscopical work. Beale, whose praise of "beautiful" sections I quoted above, explicitly marks out these dimensions in a later series of lectures for medical men. He begins by recommending a particular staining technique for its beauty. "It is not easy," he says, "to find any object which will rival in beauty...tissues which have been freely injected with vermilion or chromate of lead." However, he continues, "from such preparations we learn but little save the general arrangement" of the circulatory vessels. Indeed, he explains, when studying tissues prepared with this technique "we cannot but admire the beautiful arrangement of the vessels, but at the same time we are not led to consider" function, process, or relations of structure, especially since "essential structures" in these tissues "are rendered quite invisible by this mode of injection."[97] Beauty is not irrelevant to Beale's text; he references it frequently, repeatedly drawing our attention to the beauty of stained objects in particular. However, he wants readers to know both that vermilion staining produces beautiful results and that this technique may destroy the opportunity for a valuable physiological observation. But this is not a warning against beauty. Rather, he evidently considers that microscopical work can engage observers in multiple ways. Beale suggests that microscopists may enjoy admiration and pleasure in observation—they may even seek out beautiful forms—as long as their practice does not obstruct knowledge formation.

Writing Embodiment in Victorian Microscopy: Beautiful Mechanism. Meegan Kennedy, Oxford University Press.
 DOI: 10.1093/9780198940623.003.0006

7
The Sublime
The Microscopical Infinite

> The microscopist bends in astonished awe before the infinitude of God's works in the uncountable varieties and exquisite beauty of the minute Diatoms.
>
> Louisa Lane Clarke, *The Microscope*, 1858

> The small becomes the dreadful and immense!
>
> Robert Browning, "Mr. Sludge, 'The Medium,'" 1864

Introduction

By its very name (which pairs "small" and "look at"), the microscope yields access to what is minute. According to Edmund Burke, smallness of form is associated with the beautiful, while greatness or magnitude engenders the feeling of the sublime.[1] But surprisingly, despite the fact that microscopy is by definition about smallness, nineteenth-century microscopists consistently also associate it with immensity, expressing their experience of *vastness* (their amazement at the infinite scope of microscopic space) and *mystery* (their bewilderment at the unknowable extent and variety of that space). These latter two invoke the mode of the *sublime*.[2] The sublime, especially popular in the Romantic era, taps the immensity and power of the natural world, and the visceral and intellectual responses of human observers. Philosophers and scholars have differed over the relationship of wonder to the sublime.[3] Practically speaking, in nineteenth-century microscopical writing the sublime often appears as one form of wonder, appearing in company with other types of wonder and working in concert with them.

Nineteenth-century microscopists liberally drew upon but also revised this discourse of the sublime as set out by Burke and Immanuel Kant. Burke writes that viewing the sublime brings us both pleasure and terror, triggered by a realization of our insignificance and powerlessness in relation to the grand scale and infinitude of nature.[4] Kant similarly relates the sublime to "what is *absolutely large*," what "displays magnitude and might," to those natural "appearances whose intuition [sensible apprehension] carries with it the idea of their infinity." He adds the caveat, "all we are entitled to say is that [an] object is suitable for exhibiting a sublimity that can be found in the mind." Thus the feelings we attribute to sublimity do not originate in the object as inherent properties of it but rather express our response to it: "the expansion of the imagination" while viewing the object.[5] The sublime complexly blends pleasure and displeasure; Kant says we are flooded "by *amazement* bordering on terror, by horror and a sacred thrill."[6] We seek it out in order to subject ourselves to an affective, visceral force.

Burke and Kant do present somewhat different approaches to the sublime. Burke's sublime makes terror more central to the experience than does Kant's; and Kant, unlike Burke, finds human reason to be ultimately triumphant in identifying, for example, infinite space. Our ability to judge something sublime "proves," he says, "that the mind has a power surpassing any standard of sense" ("supersensible") (§25, 26). Burke and Kant agree, though, on linking the sublime to vastness and on the psychological force of both beauty and the sublime. It names what happens when nature's immensity shakes us from our usual standpoint.[7]

Nineteenth-century microscopists often echo Kant's argument that we experience the sublime when confronted with an "unboundedness" that our imagination cannot grasp in sensible (empirical) terms even while our reason is able to recognize its immense "totality."[8] That is, when we view the infinitely large or infinitely small, our imaginations boggle in attempting to realize these extreme scales of existence, even though we can reason about the concept of infinity. Our sense of space and scale collapses. In microscopy these tensions are heightened by the contradictions of the microscopical sublime, where the solidly human macroscopic world seems small and insignificant in light of what is even smaller: the vast world of minuteness that surrounds and infuses us. We can access a kind of transcendence through this experience, in which we are both diminished and, paradoxically, exalted. Such incomprehensible scales present us with our inadequacy and insignificance (indeed, our nothingness) even as we are free to know this rationally, by glimpsing a god's-eye view of our relation to the whole.[9]

The notion of a microscopical sublime follows earlier thinkers such as Blaise Pascal, who, considering a mite, finds the "immensity of nature, within the compass of this abbreviation of an atom." Here the *microscopical world* in fact reveals itself to be *multiple worlds* or, as Pascal says, an "infinity of worlds" that are "as astonishing in their littleness as [the planets] in their magnitude," with humans "suspended... between these two abysses" of the starry firmament and the infinity of microscopical worlds. He comments, "Who ever shall thus consider himself, will be frightened at himself."[10] Although it is the vastness of some natural object or phenomenon that inspires this experience, the sublime is not so much about the object as it is about its effect on us, in particular on how we understand our relation to the world; we are overwhelmed but also transformed by our brush with it, both annihilated and exhilarated. It is a phenomenon not ignored by

Victorian microscopists, who are as concerned with the internal experience of microscopical vision as with the external objects they study. They frequently invoke tropes of the sublime, especially the *dizzying vista of infinite minuteness* or the *vastness of the very small*, which can be described as formlessness or boundlessness or as great height or depth: microscopists look down vertiginously into endless worlds "below" them. Allied with these are *mystery* and *magnificence* (Burke notes a "great profusion" of "splendid" things).

The tension between the infinity or vastness of nature and the limited capacity of a human observer famously animates Victorian writing on astronomy. Such texts often describe the observer as inadequate in the face of the boundless distances displayed through the telescope; Anna Henchman has persuasively shown how astronomy contributed to "the Victorian preoccupation with extreme shifts in perspective."[11] But the microscope unsettles even telescopic distinctions in scale by revealing what is at once both unimaginably tiny and unimaginably vast: the world of minute life. To be more exact, the microscope *suggests* this vastness, in its gradual discoveries (so often typed as "revelations") of progressively smaller organisms and in their imagined analogy to the infinite starscape.[12] This difference often animates the conventional comparison of telescope to microscope. For example, in *The Microscope and Its Revelations*, his renowned 1856 manual, William Benjamin Carpenter explains,

> The universe which the Microscope brings under our ken, seems as unbounded in its limit as that whose remotest depths the Telescope still vainly attempts to fathom. Wonders as great are disclosed in a speck of whose minuteness the mind can scarcely form any distinct conception, as in the most mysterious of those nebulae whose incalculable distance baffles our hopes of attaining a more intimate knowledge of their constitution.[13]

He links this trope of *infinite minuteness* to that of *life in a drop of water*, comparing "the unmeasurable depths of space" to "those countless multitudes of living beings" in a water-drop, and arguing, "There is something the extreme of minuteness, which is no less wonderful, might it not almost be said, no less majestic?—than the extreme of vastness."[14] In these passages, where Carpenter prioritizes community-building over instruction, he uses first-person plural ("our hopes") to figure the sublime as a communal experience of transcendence. He invites the reader to share his awe, explicitly calling out "wonders," what is "mysterious," "wonderful," and "majestic," and imagining the observer who responds to these images with "wonder and admiration." Carpenter refers to an "unbounded" or "extreme . . . vastness" that the astronomer "vainly attempts to fathom"; of which the mind can "scarcely form any distinct conception"; in which "the mind loses itself"—a clear reference to the sublime. The failure of imagination itself generates wonder.

This conventional comparison of the astronomic and microscopical sublime persisted throughout the century, with, if anything, more emphasis on the observer's visceral response to the microscopical image. In 1891, an unsigned essay on "The Horror of Astronomy" in the moderate British weekly *The Spectator* discusses the horror experienced by lay stargazers as a "physical experience of the intellect," a "disturbing element" so extreme as to provoke "an actual sense of pain." While the writer does not use the term "sublime," the references are unmistakable. The article continues, "[M]any people experience as great a horror at the discoveries of the microscope as of the telescope. There is something intellectually horrible in the thought of the infinite multiplication of life in a drop of dirty water. If each drop contains a universe of its own, how appalling is the sum-total of vitality in the pond." (However, the writer suggests, astronomy and microscopy "have only got to be properly known and understood to lose all their horror," since "reason will soon bring...a corrective.")[15] This article omits any mention of beauty or wonder in astronomy or microscopy, but it extensively speculates about the physical disturbance ("pain") felt by the observer of infinite vastness, perhaps prompted by contemporary interest in the dynamic relation between nerves, brain, and mind. The essay is instructive only in the most general philosophical manner, and its somewhat Gothic emphasis on horror suggests that this discourse on the sublime is more to entertain readers than to create a community, as Carpenter seeks to do.

However, the *Spectator* essay leans heavily on the Kantian suggestion that these visions of the sublime can yield to the power and audacity of human reason. Carpenter also admires human reason, praising microscopists' "general doctrines...[on] the laws of Organization and the nature of Vital Action."[16] In fact, he declares that the microscopical sublime shows "Microscopic enquiry" to be "especially fitted to correct" a flaw in our philosophical approach to the world. It will help to shift us away from "the estimation of things by their merely sensuous or material greatness," and toward a focus on "their value in extending our ideas and elevating our aspirations." After all, he says, "all *size* is but relative, and...*mass* has nothing to do with real grandeur."[17] Thus the microscopical sublime, according to Carpenter, refines and exalts human reason by unshackling it from a too-close reliance on material objects and sensory experience (in Kant, this would refer to the "intuitions" of imagination), freeing and "elevating" it instead in the realm of the supersensible. Not all microscopy texts include this idea of the triumph of reason, but most of them foreground the elements common to both Burke and Kant: viewing the vastness of nature can evoke an exhilaration born of both pleasure and terror.

The Microscopical Sublime

However, microscopists also revise these inherited versions of the sublime. For example, some texts show how even reason falters when confronted with the microscopical sublime. Popularizers especially used the innate tension in the trope of *counting the uncountable* to promote microscopy to lay readers. This approach combines a rational, quantified assessment of the scope and scale of the microscopic world with a romantic shudder at the inconceivably minute myriads of the microscopic world, yielding both pleasure and horror. Victorians witnessed a remarkable series of expansions of the world's known species, leaping from the 12,000 species named by Linnaeus to approximately 250,000 species by the end of the century. While earlier microscopists did know of the minute objects that invisibly surround us, nineteenth-century workers came to recognize a much wider variety of these, and to boggle at the sheer multitudes and increasing minuteness of the biological organisms (or their remains) that fill the world. The diversity of the known creation seemed inconceivably to expand without limit even as the size of these organisms seemed to recede into the infinitely small. Victorians recoiled not only at the vastness of the very small but at the invisible, inescapable presence of these organisms massed in and around us, in food, water, earth, and air.

A writer might measure the dimensions of such miniscule objects in "diameters" or in numbers per inch or per Paris inch, but these objects are at the same time frequently described as so small as to be "infinitely minute" or even *invisible*. With language combining measurement and the immeasurable, microscopists can start from an empirical base and build to an expression of the uncountable sublime, dramatizing how calculation ultimately fails. By attempting to measure the volume of microscopical life, they can point simultaneously to both the minuteness and the magnitude—the infinitesimal size and incalculable number—of microscopical forms.

Mid-century popularizing texts frequently draw on investigators' publications to make this point. The "Christian philosopher" Thomas Dick, writing in 1851 for the Religious Tract Society (RTS), comments,

> The minuteness of some animalcules is almost inconceivable. Under the most powerful microscope, they appear only as moving points, of which . . . eight hundred thousand millions are calculated to occupy a cubic inch of the water in which they are found. Superior instruments, in refinement and accuracy, would probably disclose creatures far inferior in size. Let it not be supposed that such statements are extravagant and baseless. These are facts, which have been repeatedly observed; and, startling as the statement may be, it is proved that a drop of water, tenanted by the smallest animalcules, termed monads, . . . contain a number equal to that of the whole human population of the globe.[18]

Dick here offers "facts" and numbers (800,000,000,000 in a cubic inch of water, or approximately 1,200,000,000 in a drop) drawn from the work of Christian Ehrenberg, alongside what is "startling," even "extravagant." His gesture toward measurement paradoxically just underscores the uncountability of these multitudes—a strategy that makes sense here given that Dick's ultimate aim is not so much instruction in microscopical science, as microscopical support for natural theology.

While Dick does provide some quantification, other microscopical writers bring up the idea of counting, only to reject it altogether. Authors will gesture toward an actual count, then fall back on terms like *myriads* or *uncountable*. For instance, the Irish phycologist William Henry Harvey in his popular book on *The Sea-Side* (1849) used this trope to describe the vast submarine strata formed by the glassy remains of *Diatomaceae* (unicellular algae). He says, "We cannot count them by millions simply, but by hundreds of thousands of millions. Indeed it is futile to speak of numbers in relation to things so uncountable.... The world is a vast catacomb of *Diatomaceae*."[19] Harvey here first attempts to count the diatoms ("millions" or "hundreds of thousands of millions"), then re-characterizes them as "uncountable."

This rhetorical two-step appears also in more ambitious texts. Jabez Hogg, a surgeon who wrote one of the first successful mid-century guides to microscopy, also uses this paradoxical trope to entice new observers, both quantifying the magnitude of microscopical existence and declaring it uncountable. Hogg also wrote for microscopical novices, but unlike Dick and the *Chambers'* author, he was a recognized microscopist with considerable experience; his guidebook was a massive volume and became a valued resource for microscopical work. He opens the second half of his exhaustive 1854 manual by portraying microscopical life as both tiny and vast, counted and unknowable. Also drawing on Ehrenberg's work, he reports the calculation that the "mass of animal remains" yearly deposited in the mud "at the harbour of Wismar, Mecklenburg-Schwerin" is "22,885 cubic feet in bulk, and weighing forty tons." At this figure Hogg pauses to intone, "How vast, how utterly incomprehensible, then, must be the number of once living beings whose remains have in the lapse of time accumulated!" He demonstrates here the overlap between the microscopical and the geological sublime, the effect deriving in part from the notion of uncountable past eons—what Ralph O'Connor calls the "thrill of deep time"—augmenting the unsettling vastness of the very small.[20] Again, in discussing the monads Hogg first works out their size and mass:

> Vainly does man, with all the subtlety of his mind, endeavour to conceive an object the twenty-four thousandth of an inch in length; yet this is the size of the monad.... If each be allowed three times its bulk to move about in, a cubic inch of water will then contain 800,000 millions of these organised beings. Nay, some philosophers say this family of Infusoria are seen only the one forty-thousandth of an inch in length.

But after this barrage of calculation, he concludes, "What incalculable numbers of animaculae must swarm the waters of creation!"[21] Over and over again, in popularizing texts or in textbooks, writers count microscopic organisms while declaring them uncountable.

Microscopical visions of the sublime may emphasize the vastness or infinite aspects of what is seen but often pair this with exclamations about other forms of *greatness* as well. One frequent concept is the *mystery* or *magnificence* of the sight. As Peter Mark Roget asks in the fifth Bridgewater Treatise (1834), "What sublime views of the magnificence of creation have been disclosed by the microscope, in the world of infinite minuteness, peopled by countless multitudes of atomic beings which animate almost every fluid in nature?"[22] This dizzying statement manages to juxtapose many tropes of greatness and minuteness, including a reference to atoms (then a term for the tiniest, indivisible particles of matter).

Cell theory, which envisages immense numbers of minute cells organized in accordance with a great natural law, proved a compelling example of majesty in the microscopic world. The surgeon Ernst Abraham Hart, writing a largely recreative article for Charles Dickens's family periodical *Household Words* in 1857, draws heavily on tropes of magnificence and mystery. He acknowledges that "the colossal grandeur of the Alps" or the "Pyramids among the master-pieces of art" display a "magnitude" that commands "unmingled reverence and awe," but (like Carpenter) he argues for a "higher standard of value than [great] size," since our "[s]ense bows before the majesty of sublime proportion" as well. In particular, he says, Victorian technology has proven the "profound axiom—tota natura in minimis—in smallest things is nature greatest." (This trope of *Nature is greatest in the smallest things* slightly revises a phrase from Pliny the Younger that might also be rendered *Natura in minima maxima.*)

Comparing cell theory to the masses of tiny infusoria that can build up immense geological strata, Hart turns to example after example of the "strange secret of...structure," this "one secret that is the key" to all form, the "hidden unity amidst all variety." He exclaims,

> We see now, by the aid of the microscope, that greatness has no existence but as composed of infinite littleness. Who that bowed before the oak could have thought the lord of the forest to be a compound mass of many millions of independent organisms...? Who that looked upon the mountain chains of western Asia, or the white cliffs of Dover, could surmise that they were the handiwork of infusorial animalcules, whose shells make up the mass in numbers of thirty millions to a cubic inch? These are the revelations of the microscope. Gifted with this new power, the naturalist has traversed the material universe as though armed with a magician's wand.

Hart nods to theologies of nature, posing the "secret" of cell theory as the "one word which all creation perpetually utters; a witness to the one source whence all derives." But his encomium seems equally directed at the microscope itself, the magic wand that frames this passage in the article and his own passage into the sublime mystery of natural form. The microscope's ability to show cell structures, he suggests, counters the nightmare posed by Earth's infinitely proliferating life-forms. The lens reveals instead a hidden continuity of structure that binds complex organisms together by clustering countless numbers of tiny cells. That is, after shuddering at "the contemplation of endless dissimilarity, and apparently inharmonious difference" of natural objects, he turns with relief to the microscope, which "resolves this complexity, and bridges easily this chasm" between "the noble form of man...and the green mould that clothes his tomb."[23] By joining with the microscope, the observer can experience both a swooning admiration for the world's "infinite littleness" and a rational analysis that can plumb those vertiginous depths.

Victorians' microscopical sublime is triggered by the observer's visual experience of nature, the conventional locus of the sublime; but the infinitude generating that powerful experience cannot be experienced visually in the way that a thunderstorm, mountain range, volcano, waterfall, or tumultuous ocean might generate the natural sublime for Burke or Kant. What makes it different? Joseph Roach, discussing the visual style of eighteenth-century microscopy, argues, "The power of the microscopical sublime resides in the definitive articulation of its details. Its very tangibility resists allegorization."[24] For nineteenth-century observers, though, the microscopical sublime did not seem to depend on clarity or detail in the image. Microscopists do discuss clear or blurry imaging, but those conversations generally center on the realism of the image, a different issue. Instead, texts describing the microscopical sublime are more likely to pivot from describing a single specimen to elaborating on microscopic infinities of scale. Indeed, the power of the microscopical sublime rests in the suspicion that life and form extend on infinitely beyond what can be seen or known; unlike the natural sublime, it emerges where vision ends. Of course, the Kantian sublime also derives from a failure of vision in that it arises from the inability of imagination to fully comprehend sensible images. But the natural immensities that Kant associates with the mathematical sublime are evoked by something that *looks vast* to us.

In contrast, the vastness of the microscopical sublime is not *seen* but *inferred from what is not seen*. The physiologist Thomas Rymer Jones, in his Fullerian Lectures at the Royal Institution of Great Britain (1845), makes this process excruciatingly clear, urging his readers, "Exhaust the art of the optician, strain your eyes to the utmost, until the aching sense refuses to perceive the little quivering movement that indicates the presence of life, and you will find that you have not exhausted nature in the descending scale."[25] (Jones's "straining" rhetoric may emerge from his own experiences with sensory limitation; though a qualified surgeon, he left practice due to a hearing impairment.) The power of the microscopical sublime arises from a mental image more than a visual one. It is felt when the imagination struggles to extend the sense of littleness or of great numbers from an observed specimen into infinite microscopical space. The imagination boggles just as in the traditional sublime, but the leverage—the movement of transport—is based less on sensory evidence than on the analogy from what is seen to what is inferred.

We associate the sublime especially with Romantic-era gestures of transcendence, but this approach to the microscopical minute continued throughout the century, especially in popularizing works of natural history, as in the geologist Richard Kerr's 1895 tract on *Hidden Beauties of Nature*. Kerr's discussion of diatoms echoes Kant's description of the failure of imagination in the face of the supersensible sublime. Kerr says,

> The diatoms in the illustrations are about on the borderland of our vision. Although our eyes may see, when aided by microscopic power, the minutely complex but orderly system of lines and spaces on the surfaces of these flinty atoms, yet our keenest perceptions become exhausted in the effort to grasp a fixed idea of what we are actually beholding. We see, but we have a certain difficulty in believing. For a time we question the existence of such a possibility, and we finally conclude that we are on the threshold of an infinity in microscopy as great as that which confronts the devout astronomer when studying the realms of space.[26]

This text, published by the RTS, seems largely concerned with community-building—in this case, using microscopy instrumentally, as a means of drawing readers to religion. So Kerr emphasizes here that "we" humans share these limits in sight and belief, suggesting that these limits in imagining the sublime might be taken as an analogy for the "difficulty in believing" we may have with religion as well. Kerr chronicles the failures of human perception confronting the infinitely small, and he points to a kind of success of human understanding when "we" finally "conclude" (reason) that there are infinities we can never fully know.

Later, Kerr argues that the microscopical sublime is more difficult to comprehend than the astronomic sublime, since "it seems more possible to come face to face with infinity in the great outer space, than to form any conception of infinity as regards the minute side of creation." He concludes, "Few objects bring home to us more effectively the limits of imaginative power than do the diatoms."[27] The tension between "bring home to us" and "the limits of imaginative power" reveals this as a version of counting the uncountable: the microscopical sublime is about knowing, failing to know, and knowing that you fail to know.

Multiple Worlds

Microscopic worlds are not only "infinitely minute," according to nineteenth-century texts; they are also infinitely recursive, disclosing *worlds within worlds*, each animated by increasingly tiny beings. Nineteenth-century microscopists express a crushing sense of these infinitely retreating spheres of life. This vision of life often rouses a romantic rhetoric of excess. One early commentator, the radical Unitarian Jeremiah Joyce, exclaims in 1806, in another evangelical text, "Each drop of stagnant water contains a little world of animated beings…every plant, every flower, affords food for millions of creatures.…What prodigious shoals must there be over the whole earth, and in the immense expanse of the atmosphere!" Yet within these hidden caches of life are nested still others: "How many millions of still smaller insects and worms are crawling on the ground, or living beneath its surface!" Indeed, "[t]he microscope displays to us, in each object, a thousand others which escaped our knowledge. Yet, in every object discovered by this instrument, others yet remain unseen."[28] This vision provokes both wonder and an exhaustion of the imagination, signaled by the exclamation points on his rhetorical (because unanswerable) questions. For Joyce, the microscope was a valuable way to build Christian communities because of its ready access to these seemingly unanswerable puzzles.

George Henry Lewes, the novelist, literary critic, editor, and physiological investigator, was not, like Joyce, a believer. And, unlike Joyce, he wrote during the heady momentum of the mid-century surge in microscopy. But he offers a version of this vision in his popularizing text *Studies of Animal Life* (1860):

> Nature…fills the illimitable heavens with planetary and starry grandeurs, and the tiny atoms moving over the crust of earth she makes the homes of the infinitely little. Far as the mightiest telescope can reach, it detects worlds in clusters, like pebbles on the shore of infinitude; deep as the microscope can penetrate, it detects life within life, generation within generation, as if the very universe itself were not vast enough for the energies of life![29]

In the face of these recursive infinities of being, Lewes, like Joyce, turns to exclamation and rhetorical questions, even though he likely had very different aims for the microscopical community his text might build.

Lewes's image of "life within life" echoes the old recursive trope that would be, in 1872, humorously recomposed by the mathematician Augustus De Morgan:

> Great fleas have little fleas
> upon their backs to bite 'em,
> And little fleas have lesser fleas,
> and so ad infinitum.[30]

Lewes, however, examines not just the physical nesting of infinitely minute organisms but also their temporal nesting in the form of generations. He opens an aphid to find its children nestled within, and tiny grandchildren within those.

He imagines the reader's amazement: "In your astonishment you will ask, Where is this to end?" In dissecting microscopical beings, he suggests, we entangle ourselves in a physical and temporal chain of unending minute life, the scope of which overshadows and outlasts human endeavor. He muses, "The Apennines and Cordilleras, the chalk cliffs...these are the pyramids of by-gone generations of atomies. Ages ago these tiny architects secreted the tiny shells which were their palaces; from the ruins of these palaces we build our Parthenons, our St. Peters, and our Louvres."[31] Even our monuments are inscribed with (and within) these hidden generations of minute life.

These exclamations at infinitely nested microscopical elements recall a geometrical theory that was both resisted and ultimately recognized by the eighteenth-century empiricist philosopher David Hume, in his *Enquiry Concerning Human Understanding*. Hume, whose skeptical empiricism provoked Kant's idealism, writes, "No priestly *dogmas*, invented on purpose to tame and subdue the rebellious reason of mankind, ever shocked common sense more than the doctrine of the infinite divisibility of extension." In this doctrine, "A real quantity, infinitely less than any finite quantity, contain[s] quantities infinitely less than itself, and so on *in infinitum*." Hume comments, "[T]his is an edifice so bold and prodigious, that it is too weighty for any pretended demonstration to support, because it shocks the clearest and most natural principles of human reason." However, he continues, "But what renders the matter more extraordinary, is, that these seemingly absurd opinions are supported by a chain of reasoning, the clearest and most natural." Despite the logical grounds for this doctrine, then, "Reason here seems to be thrown into a kind of amazement and suspence" that shakes her confidence. She retreats into "diffidence of herself" and her grounds. He explains, "She sees a full light, which illuminates certain places; but that light borders upon the most profound darkness. And between these she is so dazzled and confounded, that she scarcely can pronounce with certainty and assurance concerning any one object."[32] Hume here evokes how the rational mind, "dazzled," revolts at being asked to comprehend this mathematical doctrine even as it grounds the claim in logic. He depicts a standoff between reason and the disbelief, confusion, uncertainty, and extreme skepticism that challenge "her" conclusions; he suggests later that we should be skeptical even of our skepticism.[33] Victorian microscopists explored this notion of a vertiginous, infinite smallness, grappling with the proper role of skepticism, materialism, and empirical investigation, as their rational enquiries ventured into the realms of the minute, beyond the invisibly tiny, to the unfathomably small.

The evidence of multiple "worlds" nestled within one another—a vastness of the very small—may have been especially relevant at mid-century, when mid-Victorians were expanding upon astronomical research, debating the existence of a "plurality of worlds" in the universe or, encountering evidence of extinct animals in the fossil record, supporting the possibility that species had emerged not in one week but throughout an unfathomable sequence of distinct ages—successive "worlds" in time, with generations springing from generations.[34]

There are other models for the notion of multiple spheres of existence, but they differ from microscopic vision in important ways. The Great Chain of Being provided a mental model for a hierarchized spectrum of creatures coexisting in our world and the heavens, with the higher beings enjoying more complexity of structure and a richer, more powerful, and more noble sphere of existence. But microscopic observation challenged this model. The German botanist Christian Gottfried Ehrenberg influentially (although wrongly) argued in *Die Infusionsthierchen als vollkommene Organismen* (1838) that even minute infusoria boast the same internal organs that humans do.[35]

The microscope fostered multiple-world theories by offering observers a way to imaginatively enter "distant" spaces and identify with their inhabitants even though these spaces and creatures could be accessed only through virtual images. The microscope was not the only modern technology to do so—the telescope being an obvious example—but it was unique in its affordances. For instance, although Ehrenberg's arguments were rapidly challenged, the movements of microscopic beings suggested that some of the supposedly lower creatures experience full emotional lives, even consciousness; the infusoria in particular were often anthropomorphized.[36]

Another model for this kind of imaginative exploration might be through contemporary fictional genres. Elaine Freedgood argues that the development of the realist novel, especially in the nineteenth century, "allows us to imagine that we can, in the privacy of our own minds, live in more than one world." She explains, "Criticism of the novel and the novel itself…gave us one hugely ruptured but continuous world in which we are, as imperial liberal subjects, always in more than one place at the same time, always inhabiting multiple domains in person or by proxy."[37] The microscope emerges at the same time as the modern Western novel—although it takes longer to become popular—and does just this and more, for its virtual worlds are offered not as realistic fictions but as part of the real.

The microscope also transforms our understanding of space and our place in it. Space becomes not singular, homogeneous, and extensive but multiple, differentiated, and recursive. This microscopic vision broaches the unsettling perspective that we ourselves may be only mites in a larger, enclosing world we know nothing of. The compound microscope thus opened up for nineteenth-century observers a dizzyingly recursive depth haunted by current and past life that is both nowhere (invisible to the eye at least) and everywhere around us, both empty (to our eyes) and filled with a dizzying plenitude. The microscopist is radically empowered to plumb this depth but is constantly brought up against human impotence and insignificance. This abyss swarms with life but threatens to extinguish our own agency. Faced with this existential crisis, Victorians turned to reason and imagination to supplement microscopical vision. Freedgood discusses a distinction between "outside" and "inside" the narrative—"we are left with the vertiginous feeling that we ourselves may not be at the outermost level of narrative"—but for her that distinction is both imagined and spurious.[38] In contrast, the Victorian microscope places us within a vision of infinitely recursive space and scale and tells us that is real.

The above examples of the microscopical infinite draw from a range of popularizing texts engaged mostly in instruction, community-building, and recreation. The language of infinitely recursive minuteness does appear in research publications, but more rarely. For example, the optician Andrew Pritchard, in his 1841 *History of Infusoria*, orients readers by quoting Ehrenberg's much-circulated summary: "experience shows an unfathomableness of organic creations, when attention is directed to the smallest space, as it does of stars, when reverting to the most immense."[39] Pritchard's and Ehrenberg's texts reference natural theology, as was not unusual at the time. But they direct their treatises to experienced microscopists and frame their work within a scientific system in which original investigation is published, exchanged, made use of, corrected, and updated—as in Pritchard's late addition of an appendix abstracting recently published work of Ehrenberg on infusorial chalk.[40] The microscopical sublime most often inflects research texts in ceremonial or paratextual moments such as this, where the author reviews the subject and asserts some larger meaning for the work. Such texts rarely grapple with the philosophical issues raised by the microscopical sublime; instead, their references to the limits of microscopical vision tend to frame the work as important and, like microscopical infinitude, never-ending.

The microscopical sublime shares some features with the *technological sublime*, a term coined by Leo Marx to describe the romantic nineteenth-century American trope about the relation of technology to the natural world. The technological sublime celebrates the awe-inspiring force of technologies such as dams, tunnels, and railways that remake the natural landscape for human use.[41] Victorian microscopists similarly articulate a recognition that technology can powerfully mediate between humans and the natural world, but with a crucial difference: the microscope may reveal nature's hidden secrets but it largely does not destroy them. While poorly managed observations can injure the viewer's vision or a particular specimen, properly conducted microscopical work renders the natural world even more terrible and potent in human eyes.

The Beautiful and the Sublime

Both Burke and Kant distinguish the beautiful from the sublime: the beautiful expressing what is harmonious in form (small, smooth, delicate, gradual, orderly, distinct, pleasing in design) and empirically graspable; and the sublime in its greatness, terrifying obscurity, and formlessness straining human sensibility.[42]

The "confounding" of these distinct ideas in fact impelled Burke to write his famous treatise, and he cautions against taking evidence of design ("dexterity" or "care") as sublime.[43] Kant goes further: while both the beautiful and the sublime for him reference an act of human judgment, the beautiful has its seat in material form (sensory experience) whereas the sublime does not. He explains, "The beautiful in nature concerns the form of the object," whereas he associates the sublime with "a formless object." While form in beauty "is compatible with charms and with an imagination at play" and promotes "*restful* contemplation," the sublime brings what he calls a "negative pleasure" where the mind is "*agitated*" in oscillation between attraction and repulsion.[44] Kant explains that "unboundedness" momentarily checks us—it is "contrapurposive for our power of judgment, incommensurate for our power of exhibition, and as it were violent to our imagination"—before releasing us to what Wordsworth characterizes in poetry as a "spontaneous overflow of powerful feeling." Further, Kant adds, "for the beautiful in nature we must seek a basis [a prompt] outside ourselves, but for the sublime a basis merely within ourselves and in the way of thinking that introduces sublimity into our presentation of nature."[45]

However, there are similarities between them: both the beautiful and the sublime place us into affective relationships against our other inclinations. Kant explains, "the beautiful prepares us for loving something, even nature, without interest [e.g. in a disinterested way]; the sublime, for esteeming it even against our interests (of sense)."[46] Both bring us out of narrow interests into a larger view. It is perhaps for this reason that they appear so frequently in ceremonial or paratextual moments, as I have noted above, where they can authorize the text against a backdrop of great purpose.

I have shown that Victorian texts on the microscope cite both the beautiful and the sublime. Despite the philosophical distinction between these concepts, some writers do refer to elements of the beautiful and the sublime together, linking form and formlessness, delicacy and greatness in the microscopic world. This appears most notably in discussions of diatoms, which powerfully combine elements of both the beautiful and the sublime. For example, Mary Ward, who wrote her *Microscope Teachings* in 1864 for readers "unversed in microscopic marvels," describes diatoms as "objects of extraordinary beauty."[47] She exclaims at their "exquisitely beautiful" shells and says of *Heliopelta*, "You will be at once astonished and delighted at the perfection of its structure, and the wonderful delicacy of its tracery." But at the same time that Ward speaks of the beauty of diatoms—their delicacy and smallness—she also speaks of their physical greatness and extremity: they "remain almost indestructible... testifying of their existence in past ages, as they occur fossilized in countless myriads in different parts of the world." Here the "tiny diatom" demonstrates greatness of extent in time ("past ages"), of number ("countless myriads"), and of location (all over the world).[48] The popularizer Mary Somerville offers a similar example in *On Molecular and*

Microscopic Science (1869), which synthesizes scientific theories for committed general readers. Somerville is not generally effusive in her style, but the case of diatoms rouses her into a more romantic discourse. She explains that diatoms "have deposited their silicious (sic) valves at death for countless generations, producing geological changes of enormous magnitude." She comments that the "Supreme Being" thus "creates generations of infinitesimal creatures, adorns them with exquisite beauty, and makes them His agents to form future continents."[49] Her references to extreme mass, extreme beauty, infinite minuteness, and uncountable generations all associate diatoms with both the sublime and the beautiful.

Citing a different trigger of the sublime, Kerr, in his 1895 *Hidden Beauties of Nature*, clearly notes the sense of both beauty and infinity roused in the observer by the *Diatomaceae*. He frames the diatom as "beautiful...complex, symmetrical," with a pattern "of exquisite beauty and...geometrical accuracy." Then, visually diving inside one of its perforations, he describes the "magnificent design" of this tiny interstice and exclaims, "There seems to be no end to minuteness."[50] While Ward and others point to the countless numbers of diatoms that constitute physically vast tranches of the earth, Kerr emphasizes how their intricate inner structure suggests the vertiginousness of infinitely nested scales of being—the *worlds within worlds* of infinite minuteness.

Such passages express wonder at and admiration for a beautiful object as well as awe at the notion that this exquisitely detailed object should also be infinitely small or that there should be infinitely many of them. The perspective shuttles back and forth between the close-up on diatoms' structural detail and the panoramic view of their incredible extent and, by implication, number. This oscillation is in a way analogous to the alternation Kant posits in the sublime between overwhelm and reason, except that diatoms represented obscurity more than rational control, since Victorians admitted their empirical knowledge of diatomaceae to be imperfect and incomplete. The shift back to exquisite detail simply underlines the incomprehensible magnitude and variety represented in diatom deposits. For such writers, the diatom's beauty does not counter but rather intensifies its sublimity.

This synergistic relationship with beauty forms an important difference between the microscopical sublime and the Burkean or Kantian sublime. We can account for it partly by considering the role of faith in the microscopical tradition. For instance, Somerville's description of the sublimity of diatoms emphasizes the work of "the Supreme Being." While not every Victorian microscopist was a person of faith, they all had to reckon with the dominance of natural theology in past microscopical writing and its continued influence on nineteenth-century microscopy. In contrast, Burke and Kant anchor their concepts of the sublime on a "vision of nature"; this experience is not explicitly tied to an experience of God. Burke, indeed, distinguishes the contrivance of natural theology from the innate pleasure or reverence inspired by the beautiful or sublime.[51] And Kant, in *Critique of Pure Reason* (1781) and *Critique of Teleological Judgment* (1790), argues that faith must rest on a moral judgment distinct from the arguments from reason and design. So Burke and Kant differentiate between religious expression and writing about the sublime. There is, however, a tradition of writing about the natural world that does combine exclamations of beauty and the sublime (narrowly conceived) and expressions of reverence for a Creator. That tradition is natural theology.

Victorian microscopical texts often pair the beautiful with the sublime because they write not just from a strictly Burkean or a Kantian aesthetic, concerned with distinguishing the effects of natural objects on human sensibility, but also from a natural history tradition strongly influenced by natural theology, concerned instead with these as signs of an omnipotent Creator, where exquisite beauty of detail amplifies the glory of a world sublimely infinite of extent. The microscope provides valuable support for this tradition, and this longstanding association inflects even those authors, like Lewes, who were not believers. Most microscope texts are not structured as explicit arguments for the existence of God—many, indeed, do not make explicit reference to a Creator anywhere. But the historical alignment of Victorian microscopy with natural history and natural theology provides a set of aesthetic and philosophical conventions that allow writers to pivot quickly from swooning at the horror of the infinite mystery to gasping at its elevating power.

Writing Embodiment in Victorian Microscopy: Beautiful Mechanism. Meegan Kennedy, Oxford University Press.
 DOI: 10.1093/9780198940623.003.0007

8
Natural Theology and the Beautiful Mechanism

Abroad into the fields! nor longer mope
In deep abstraction in your book-filled room
For the bright sun's dispelled a week of gloom!
Take with you Chaucer and your microscope;
For life's awake, and the wee flowrets bloom
Under the grass, lovely of rainbow hues.
Here's a gem! a gossamer steeped in dews,
Hanging like silver veil on budding broom.
God's work of wonder everywhere is seen—
In air, in water, and the world of green;
Germs feel His breath and start up into life.
Gaze through the microscope upon this leaf:
Another world's revealed, where it is rife,
Joyful and sad as in our own, but, oh, how brief!

James Howell, "To Wentworth Holworthy, Esq.," 1860

Introduction

When nineteenth-century microscopists expressed awe at visions of beauty or the sublime, as in Chapter 7 or the epigraph above, they sometimes made explicit reference to the tradition of *natural theology*. This tradition aimed to prove the existence of God as Creator by rational means (natural reason), studying nature for evidence of God's care for the world. Even before the improved compound microscopes of the 1830s, microscopy offered compelling support for such arguments. Although the tradition of natural theology has ancient roots, William Paley was an influential nineteenth-century exponent of this argument in his treatise *Natural Theology or Evidences of the Existence and Attributes of the Deity, Collected from the Appearances of Nature* (1802). But the tradition of natural theology fulfilled a number of theological and social functions beyond conversion. The scholar John Brooke lists several other aims: expressing awe, confirming faith, mediating between different faith traditions, validating the practice of scientific inquiry, even aligning scientific practice with a conservative, law-governed polity.[1] Most scholars agree that natural theology was a dominant discourse through the 1840s and that versions of it persisted to the end of the century, especially in texts published by religious entities or addressed to Christians, women, children, or the working classes.

Microscopists' statements of wonder throughout the nineteenth century reveal the influence of natural theology as a tradition, a genre, and a social function. While more common in early or non-specialist texts, natural theology shapes a wide range of texts including research papers written by investigators for their peers. Nineteenth-century microscopical authors used a discourse of design to showcase Christian faith without necessarily expecting to convert skeptical readers. In fact, as I will show, they used tropes of natural theology (such as *contrivance*) that address microscopical aims as much as theological ones.

Brooke influentially distinguishes between *natural theology* and *theologies of nature*.[2] *Natural theology*, narrowly defined, uses the wonders of the natural world as evidence in a logical argument for the existence of a Creator. The argument from nature is often counterposed to the argument from Revelation (the record of which is found in the Bible). One approach cites the Book of Nature; the other, the Holy Book. *Theologies of nature*, by contrast, take a broader and more diverse approach to natural wonders. They assume the reader is already a Christian; the text aims not to convert readers but to confirm, support, and strengthen an existing faith based on revelation. Brooke specifies that a theology of nature is "simply…[a] belief about God's relationship to the natural world.…There is no attempt to demonstrate anything, and the belief is almost certainly an article of faith."[3] Theologies of nature eschew elaborate argument, but they do often make reference to the argument from design, which finds proof of God's presence in finely fitted natural adaptations. The argument from design can appear in something as simple as the overdetermined term "contrivance." Jonathan Topham identifies in such references a *discourse of design*. He warns, "Passing references to the evidence of design were often clearly intended to stop far short of any form of inductive inference…[T]heir pervasiveness should not be taken as demonstrating the pervasiveness of natural theology." They rather reflect a "prior commitment to the truth of the Christian revelation."[4] Microscopical texts generally use the discourse of design to present these more capacious theologies of nature rather than a strict natural theology. These writers typically admire and study the apparent contrivance and design of the natural world rather than laboriously proving the existence of a Creator.

Natural theology even formed a literary genre; John Brooke and Geoffrey Cantor, in their Gifford Lectures, discuss it in this light.[5] They adapt George Campbell's *Philosophy of Rhetoric* (1776) to list rhetorical approaches common to the genre: beauty, vivacity, novelty, analogy, metaphor, and antithesis.[6] Many microscopical tropes rely on these approaches as well, perhaps even more than in writing about the telescope, whose objects are more ponderous. And in Brooke and Cantor's definition of the natural theology genre, texts use the argument from analogy (including the trope of the *watchmaker God*), the argument from design (emphasizing intricacy and contrivance), or the argument from beauty (privileging order, symmetry, and coherence). The microscope readily provides compelling evidence for these arguments, especially in privileging intricacy and delicacy in design.

Wonder persists in another genre that often uses the discourse of design, the *narrative of natural history*.[7] Many of the rhetorical approaches taken by the narrative of natural history overlap with those that Brooke identifies in natural theology texts: vivacity, beauty, sublime, and novelty. The most important features of the narrative of natural history as it appears in microscopy include a loose relation to natural theology through references to design, beauty, and the sublime; an emphasis on close-reading the object; and a tone of curiosity and wonder expressed with friendly familiarity, sometimes blended with scientific terminology.

Amy King identifies a related nineteenth-century prose genre, *reverent empiricism*, which she locates in nineteenth-century natural theology and natural history texts and in provincial realist novels. Microscopical wonder often shares the central qualities of reverent empiricism as King defines it—"the valorization of *commonness*...the prevalence of *detail*...and [the use of] *dilation* as a form of reverence."[8] Microscopical wonder, as I have shown in earlier chapters, often appears entwined with the theologies of nature that power reverent empiricism. But King associates reverent empiricism primarily with realist prose, whereas microscopical wonder appears in many *modes*, including references to frankly romantic genres such as the fairy tale, the scientific romance, and the Oriental tale. As with the sublime, these genres may involve disorienting, fantastical, or sensational affects that are generally foreign to reverent empiricism. Microscopists made effective use of a range of literary forms to defend and promote their work.

Nineteenth-century microscopists frequently repurpose the formal, argument-based structures of natural theology to advocate for their new science. For example, the physician C. R. Goring uses these structures in his 1829 "Exordium," an introduction to the microscope that opens a new publication on microscopical natural history that he was co-writing with optician Andrew Pritchard in an effort to encourage more microscopical work. The "Exordium" puffs the latest technology, urges the need for microscopical publications like theirs, jokes about the liveliness of microscopical objects, and concludes with a "Vindication of Microscopic Science and its Votaries" where Goring asks,

> Suppose some individual, greatly distinguished by his talents in ship-building, in making astronomical instruments, or steam-engines, &c., was also to evince a passion for making minute automata and watch-work, such as tarantula spiders, minute singing birds, musical seals, or even such curiosities as a coach drawn by

fleas, &c. would it shew good breeding, or good taste, to despise or ridicule his minute labours, while we admired his grander and more imposing works?

Goring clarifies that the "individual" mentioned in this rhetorical question refers, of course, to "the Supreme Being."[9] He cites here the trope of the *watchmaker God*, well known from Paley's and earlier work, which points to the intricate adaptations of the natural world as proof of a Creator.

However, Goring calls on natural theology here primarily to defend the microscope against claims that it was only good for examining insects and other lowly pests. He explains, "Men are perpetually wondering what can be the use of bugs, and fleas, and wasps, and such kind of vermin; and speak of them as absolute blots in the escutcheon of the Almighty. The use of these little insects is surely to teach man *a perpetual lesson of humility*." Goring argues that the microscope, by enabling us to study these delicate little structures, opens us to God's lesson of humility. He puts a humorous twist on this moral lesson by shifting to a mosquito's point of view, suggesting, "If a bug (I beg pardon—a *cimex lectularius*) could reason, it would probably suppose that man was of no other use in the creation than to prepare its habitation and supply its food; and would think that no better evidence need be adduced to prove the immense importance of a bug than that such bountiful provision had been made for it."[10] This whimsical story turns natural theology on its head by suggesting the weakness of the argument from nature, though Goring likely did not intend this effect. The shift of perspective works together with the microscope to further destabilize the human standpoint and entangle observer and object: the purpose of "man" is just to provide "food" for the bug, whose proboscis the lens reveals to be beautifully adapted to this purpose.[11] Goring writes here in the style of William Paley's argument, but he turns Paley's techniques toward different (indeed, conflicting) ends, exalting the man-made microscope even as he claims that it chastens human hubris.

Explicit arguments like this became less common later in the century, but theologies of nature persisted, often inviting the reader's admiration for microscopical wonder. They appeared throughout the century especially in texts addressed to women, churchgoers, or newly literate audiences such as working-class readers or children. The popularizer Rev. J. G. Wood, in *Common Objects of the Microscope* (1861), uses the argument from analogy and the argument from beauty when he adopts the trope of *jewels* to describe ordinary objects, such as the perfume-cells of a lavender leaf. And the photographer William Henry Olley uses beauty to solicit religious awe in his book *The Wonders of the Microscope* (1857). He notes of the "common Fly (*Musca domestica*)" that "though an insect familiar to every one, there are comparatively few who are acquainted with the exquisite and surprising manifestations of divine contrivance which the microscope discovers in all its organs."[12] Olley's use of the trope of contrivance marks microscopic beauty with the discourse of design. He rarely names the Creator directly, so the appearance of this discourse of design in his text shows how unremarkable such language was at the time. However, the effect is blunted when he describes almost every object as "beautiful" or "exquisite." For example, in the case of "the Honey Bee (*Apis mellifica*)," he remarks on its "complex and beautiful apparatus" such as the "exquisitely pointed, horny sheath" of its sting. Similarly, he exclaims how "remarkable for beauty and exquisite organization" is "the Tongue of the common Cricket (*Acheta domestica*)."[13] These terms lose force from such frequent repetition.

The theological trope of *revelations*, with its suggestions of an overwhelming, even life-changing visual and affective force, very often was adopted to microscopical aims as well. This trope splices the argument from Revelation (the Bible), accessed by reading, with the argument from the Book of Nature, accessed by peering through the microscope. John Ferguson, a nonconformist minister, explains in his book *The Microscope, Its Revelations in Science and Art* (1858) that "between the Work and the Word of the Author and Lord of all, there can be no opposition."[14] The most famous use of the term is, of course, in the title of William Benjamin Carpenter's classic manual for workers, *The Microscope and Its Revelations* (1856). A reviewer for a medical journal emphasizes the theological underpinnings of *revelation* and the microscope's claim to that power, in reviewing this book. The review says, "Our author [Carpenter] has chosen an intelligent and meaning word when he speaks of the *revelations* of the microscope. It is truly a revelation—an evangel of knowledge."[15] This passage frames the microscope as a literal "evangel"—a form or purveyor of the Gospel, whose purpose is to transform its readers. The trope of revelations cites the entangled embodiment of microscopical practice, but it reverses the conventional notion of the microscopical operator as agent acting upon a passive instrument and object. Instead, the microscopical object acts through the instrument to move a passive viewer. The object exhibits what is marvelous; the microscope spectacularly reveals it; and the observer is acted upon, both transfixed and elevated.

Popularizing texts regularly blend theologies of nature with the sublime, but this can be found in other settings also. The medical journal reviewer declares,

> If the [telescope] has opened the path through space to new firmaments, and drawn the veil which covers distant worlds, the [microscope] has given us a more exceeding astonishment in the worlds within worlds, the life within life, the endless lengthening and multiplication of the links of that unbroken chain which reaches from dust to Deity, which it reveals.[16]

Here the nod to "exceeding astonishment" and the tropes of *telescope and microscope*, *worlds within worlds*, and *revelations* invoke the sublime extremity of the Creation, best visible through the microscope. The passage rhetorically amplifies the power of the Almighty, but it also glorifies the instrument that showcases that power for a human viewer.

The trope *counting the uncountable* also readily pairs with theologies of nature. Louisa Lane Clarke, writing an introduction to microscopy for "young persons" in 1858, appends it to a discourse of beauty and design.[17] She explains that "the microscopist bends in astonished awe before the infinitude of God's works in the uncountable varieties and exquisite beauty of the minute Diatoms." She continues, "Wherefore are they thus highly wrought, and why in such abundance? Take up that slide of Bilin slate and know that in one single cubic inch 40,000 millions of these delicate forms are found!"[18] Emphasizing that a substance is available in abundance might ordinarily cheapen or debase it, and quantification is a central element in Victorian empirical science. But Clarke leverages the profusion of diatoms as a kind of richness, to heighten her wonder at the lavish beauty of the Creation and rouse that wonder in her readers. In her efforts to interest young readers and enlist them into microscopical practice, she explicitly depicts them interacting with the object and turns to second-person imperatives to secure them in a cognitive and affective relationship: *take up* that slide *and know* the extent of the enormity implied there. Her effort to number the diatoms is swamped in their plenty, leaving her with estimates ("40,000 millions") that should only intensify our awe. The discourse of design is ultimately not diminished but augmented by Clarke's unsuccessful attempt to count the uncountable.

While the above texts mostly just pique readers' interest about microscopy, others provide detailed instruction on microscopical work. These also pair tropes of the sublime with the discourse of design, although they tend to do so more fleetingly or in paratextual spaces (introductions, conclusions) where they reflect upon the text as a whole. For instance, Carpenter describes the "majesty" of "the extreme of minuteness" and simply argues that "in that Infinity alone can we seek for a Wisdom to design, or a Power to execute" natural laws.[19]

Two years earlier, Jabez Hogg, a surgeon and photographer, had published a similar mid-century manual, *The Microscope*. Hogg offers a provocative glimpse of changing views around such subjects; he discusses them at length in his "Concluding Remarks" and revises those remarks heavily between editions. In 1854 he emphasizes our place within a great chain of being and how the microscopic sublime confounds human observers and glorifies a Creator. He says (rephrasing the eighteenth-century microscopical writer Henry Baker without attribution), "The beginnings and endings, the exceeding greatness or the very littleness of things, are to us, in a certain sense, a source of perplexity and confusion."[20] He quotes a long passage from Addison in *The Spectator* on the subject—a passage also quoted by Baker—asserting,

> The understanding, indeed, opens an infinite space on every side of us; but the imagination, after a few faint efforts, is immediately at a stand, and finds itself swallowed up in the immensity of the void that surrounds it. Our reason can

pursue a particle of matter through an infinite variety of divisions; but the fancy soon loses sight of it, and feels in itself a kind of chasm.[21]

This passage evokes Kant in its emphasis on how the imagination collapses into "a kind of chasm" while reason forges on.

But Hogg reframes the Kantian sublime with the discourse of design, using both to support a renewed appreciation for human reason as providing a kind of microscopic vision. He quotes Thomas Chalmers, a dissenting clergyman and well-known proponent of natural theology, on the microscopical "regions of invisibles." Chalmers frames the nesting of infinite worlds under the explicit rubric of design, envisaging "a universe...so small as to elude all the powers of the microscope; but where the wonder-working God...can raise another mechanism of worlds, and fill and animate them" with glory.[22] Again quoting Baker without attribution, Hogg remarks that although "the sphere of [human] bodily action is comparatively narrow and restricted," the human "mind is untrammeled in its sphere, and can soar and penetrate" into remote regions "beyond the bounds of matter."[23] With a nod to the trope of *microscopy as a metaphor for mental scrutiny*, reason here acts as a kind of microscope that can "penetrate" into the invisible arcana of creation. The passage continues,

there remains an illimitable scope for our inquiries and discoveries, and every step...enlarge[s] our capacities, and give[s] us still more noble and just ideas of the power, wisdom, and goodness of God. This marvellous universe is so full of wonders, so teems with objects of latent beauty, that perhaps eternity alone will open up and develop sufficient opportunities to enable us to survey and admire and appreciate them all.[24]

Here discourses of design and the sublime combine to glorify both microscopical inquiry and reason as revelations of God's greatness.

Interestingly, Hogg chooses not to quote Baker on how microscopy "may possibly increase our Pride and Folly, and make us imagine ourselves of mighty Consequence" when really "we...sink to nothing" in the scale of God's Creation.[25] Hogg emphasizes instead a very Victorian view of microscopical infinity as an opportunity for human research: providing "an illimitable scope for" "our inquiries" and "discoveries." This declaration forms the last prose paragraph in the book, although Hogg follows it with an abridged stanza from the eighteenth-century poet James Thomson's *Seasons: Summer* that turns again to our limited vision and God's unlimited power. Hogg's 1854 conclusion thus pieces together a series of quoted and plagiarized passages to argue that the microscopist encounters infinite worlds that are both beautiful and sublime, that strain but do not shake our capacities, that urge us toward further discovery, and that ultimately remind us of "the power, wisdom, and goodness of God."

Two years later, in the 1856 edition, Hogg shifts toward a greater emphasis on human power and capability. He deletes many of the quoted remarks including the long passage from Chalmers, perhaps because he was concerned to keep the volume at about the same length.[26] However, he retains Baker's passage about the "illimitable scope" of the microscope (still unattributed), prefacing this with a page of remarks that significantly reframe it. In this new section he describes the current limits and anticipates the future powers of the "human intellect" and, by extension, the microscope. He says,

> The infinitely small, and the infinitely great…have for us no bounds—nothing that we can reach and measure. Infinity is every where around us.…The more powerful the instrument, the more astounding its revelations; until we marvel in what sized atom organic matter ceases; and our facts become stranger than fiction, and far beyond the imaginings of the most poetic brain.…[H]ow many regions will remain to be explored in that, as yet almost unknown country, which belongs to the microscope?[27]

The "powerful" microscope and its "astounding…revelations" here act as a kind of extension of ourselves. When Hogg says here that the "infinitely small…[has] for us no bounds," he is suggesting the powers more than the limits of observer and microscope. Although Hogg greatly reduced the number of pages devoted to these concluding remarks (from five to one-and-a-half pages), he retained many familiar tropes: the telescope and microscope; the microscopic sublime; the discourse of design; revelations; microscopical worlds. His revisions suggest how these tropes were being reframed at mid-century. While Hogg's later edition retains the major points of the earlier conclusion, his changes reduce the references to God and put new emphasis on human and microscopical powers.

Created or Constructed?

Theologies of nature often venerate natural objects by comparing them to human-made ones, devaluing human work in the comparison. Microscopists had expressed wonder for the delicacy of natural objects in contrast to the coarseness of human-made ones since Hooke influentially performed this unflattering comparison in his *Micrographia* in 1665.[28] This *opposition of "created" and "constructed" objects* appears when microscopical texts compare the delicate patterns of a butterfly's wing or flea's carapace to the ragged outline of a needle point or a printed punctuation mark. We find it in, for example, Jeremiah Joyce's *The Wonders of the Microscope*, a guidebook for children anonymously published in 1806, with much of its matter derived from earlier sources. Joyce exclaims at the "beauty and perfection" of the "works of God," in a passage that he (like Hogg) quotes from Baker without attribution. He says,

> If we examine the numberless species of insects that swim, creep, or fly around us, what proportion, exactness, uniformity, and symmetry shall we perceive in all their organs! What a profusion of colouring! Azure, green, and vermilion, gold, silver, pearls, rubies, and diamonds; fringe and embroidery on their bodies, wings, heads, and every other part! How high the finishing! How inimitable the polish we every where behold!... If we go farther and examine the animalcules, many of which the unassisted human eye could never discern... how amazingly curious must the internal structure of these creatures be! The heart, the stomach, the entrails, and the brain! How minute and fine the bones, joints, muscles, and tendons! How exquisitely delicate, beyond all conception, are the arteries, veins, and nerves![29]

Here we find the familiar conjunction of beauty, infinity, and amazement, with an emphasis on how these elements are designed and constructed. The tropes of *the uncountable* ("numberless species of insects") and *objects like jewels* add force to his exhortations. Human-made objects and actions appear as reference here in the "fringe and embroidery" on insect bodies and in their "polish" and "finish." The discourse of design pervades this passage, proposing a human process of construction as the imagined model for the divine one.

But the human-made world itself looks shoddy to Joyce in contrast to these insects. He explains (again plagiarizing Baker), "A microscope will prove the most boasted performances of art to be as ill-shaped, rugged, and uneven, as if they were hewn with an axe, or struck out with a mallet and chissel [sic]. It will show the inequality and imperfection of every part."[30] In sum, "The most perfect works of art betray a meanness, a poverty, an inability in the workman; but the works of nature plainly prove, that the hand which formed them was absolute master." Microscopic natural objects boast a quality of workmanship that human work cannot rival: "Every hair, feather, or scale, even of the meanest insect, appears rounded, polished, and finished to the highest pitch, and shews the abundant riches, munificence, and skill of its Maker."[31] Wonder here is used to criticize the efforts of human makers—with one exception. "This instrument," he says (again plagiarizing Baker), "has given to the human race a new sense, and displays to us wonders never thought of by former ages."[32] The only human-made product to escape this critique is the microscope itself.

Hogg highlights this conflict, which is inherent in the trope of *created vs. constructed objects,* since only with the amazing powers of the microscope—a human-made artifact—can we see the flaws of human-made artifacts. He introduces human-made objects as the supposed focus of our admiration, commenting, "Man boasts of the fineness and perfection of his handicraft; his manufacture will be so small that a magnifying-glass is required to observe it; he will proudly display a piece of lace of most fragile material." Here both lace and microscope feature as human achievements. Then he pivots sharply to the inadequacy of one of these objects: any observer who looks through the microscope at objects "invisible to his naked eye" will see "the perfection of Nature's works" and must recognize the relative coarseness of human works. Even as he disparages the lace, he champions the microscope for yielding access to "invisible" works of "perfection." This epiphany is a deeply affective as well as intellectual event. Hogg's imagined observer "feels abashed, and sinks, conscious of the futility of his attempts to rival the accuracy and completeness given by a guiding power to the most insignificant of its creations."[33] Ironically, this belittling statement appears as Hogg discusses the use of *Navicula* diatoms as test objects, a process that enabled microscopists to validate scientific observations and, just as exciting, exhibit their achievements in technology and skill. The instrument's ability to display the intricate patterns of diatoms testifies to the skill of both God, in making these tiny objects, and the microscopist, in resolving them under the lens.

This trope of *created or constructed* persists throughout the century; the comparison was especially likely to appear as an exercise in children's texts. Margaret Gatty in 1871 recommends that a young correspondent to *Aunt Judy's Magazine* compare the algae *Zygnema* or *Conferva* "with the very finest bit of thread your mother can find in her workbox; for under the microscope *that* is sure to look like a coarse, rugged rope, whereas the more you magnify the other, the more delicate its structure proves to be." The narrator cites "Honiton lace, and Irish 'cobweb mits'" as feats of handicraft, but warns, "not one of these will bear the test of the microscope, as the commonest objects in God's creation do. Into them you cannot look deep enough to find a point where order and beauty cease."[34] Gatty uses the comparison between thread and algae to reference both God's design and the microscopical infinite. (The comparison is often to delicate objects gendered as feminine, such as the thread or lace noted here, or a needle's point, although sometimes the comparison is to printed matter such as an engraving or a punctuation mark). The microscope once again is valorized here even as other artifacts are disparaged. While Gatty says "you cannot look deep enough" to find disorder, her point is not that "the test of the microscope" is limited or faulty but that God's creation is infinitely orderly, and that the microscope allows us to plumb these wonders to unimagined depths.

Beautiful Mechanism

The discourse of design sometimes draws on models of law-bound or specifically "mechanical" nature dating back to the seventeenth century.[35] In addition to citing a watchmaker God, Victorian microscopists often refer to the body (human or otherwise) as a *machine* or *mechanism*, another familiar concept from the discourse of design.[36] Philip Henry Gosse, a naturalist, popularizer, and evangelical church leader in the Plymouth Brethren, takes this trope to an unusual extreme, citing the "mechanism" evident in the many-toothed tongue of the periwinkle. "When he eats," says Gosse in his 1859 book *Evenings at the Microscope*, "he separates two little fleshy lips, and the glistening glass-like tongue is seen...rapidly running round like an endless band in some piece of machinery; only that the tooth-points, as they run by, remind one rather of a watch-wheel." Gosse's book was explicitly Christian, published with the Society for Promoting Christian Knowledge, so it is no surprise to find a link between mechanism and design here. However, as a member of the Plymouth Brethren, Gosse privileged the argument from Revelation and eschewed any formal argument from nature. Instead, natural wonders serve as signs activating the reader's existing faith. He exclaims, "It is wonderful to see,—perhaps not more wonderful than any other of God's great works, never more great than when minutely great."[37] Gosse promises to exhibit many curious things like this; he vows that the book would "open the path to the myriad wonders of creation, which...are made cognisable to sight by the aid of the microscope."[38] The grotesque example of the periwinkle tongue uses the discourse of design to praise what is "minutely great" and, implicitly, the microscope that affords us such visions.

The trope of mechanism is sometimes rendered as a specifically *beautiful mechanism*. For instance, Hogg points to the argument from mechanism in the "*Saw-fly*," remarking that "the instrument which the little insect uses for the purpose of cutting the leaf, is one of the most remarkable pieces of perfect mechanism."[39] A cluster of such references appears in mid-century texts. The popular theologian Thomas Dick describes fish scales by combining the trope of *beautiful mechanism* with that of *created or constructed*; he declares, "There is no work of art we can compare with their beautiful mechanism. The finest needle-work which has ever been wrought cannot for a moment be put in competition with the beautiful net-work, and interweavings, and divarications which appear in the contexture of these scales."[40] And Hogg enthuses that insects display "the most beautiful mechanism in their frame-work," noting in particular the "beautiful mechanism of...the tracheae of the silk-worm moth."[41] The trope of the beautiful mechanism aptly conveys the observer's sense of awe, inspired by the structures of the marvelously intricate natural world.

The trope of the *beautiful mechanism* does appear in research reports, although more rarely and without such emphasis on the writer's awe. For example, Joseph Bancroft Reade, a clergyman and founder of the Microscopical Society of London, in 1844 gave a presentation to the Society where he described the digestive system of the oyster, visible in fossilized specimens, as "so beautiful a contrivance" and a "beautiful mechanism."[42] When Reade died in 1870, during his tenure as President of the Royal Microscopical Society, he was eulogized in the *Monthly*

Microscopical Journal with specific reference to this "beautiful mechanism" of the oyster.[43] With its echoes of natural theology, the phrase provided Reade's obituarist a context-appropriate memento of his microscopical accomplishment.

The trope also often is used to refer to the microscope itself, a beautiful mechanism that yields visual access to other beautiful mechanisms. Contemporary discourse on the protracted efforts to produce an achromatic compound instrument emphasizes the unique status of microscopes, constructed of delicate components much like the objects they examine. Goring argues, "[i]t is perfectly well known that the difficulties in the construction of any piece of exquisite machinery are wonderfully increased and aggravated, if the artist is compelled to work upon an exceedingly small scale."[44] Andrew Ross, an optician, explains that constructing objectives requires "operations on portions of glass so small that they are themselves almost microscopic objects."[45] And William Carpenter, expanding upon this notion, comments,

> When it is considered...that in the highest powers now made, the largest of three pairs of lenses is very little larger than a pin's head...we can easily understand the difficulty of producing the required achromatic corrections....The aperture of one-sixteenth of an inch is not more than would be made by the prick of a small pin, and yet through that small hole the most perfectly distinct and beautiful images could be produced.[46]

Indeed, craftsmen had to use microscopes to work on these high-powered objectives. Yet the difficulties of construction were framed as proof of the instrument's value, playing into the microscope's signature aesthetic by privileging what is tiny, complex, exquisitely constructed, and beautiful.

Many Victorian microscopists were fascinated by innovations in the glittering, delicate materiality of the instrument and its supplementary apparatus; the instrument's splendid trappings enhanced the overall sense of *microscopical richness* inside and out. This technical interest was referred to, sometimes condescendingly, under the trope *brass and glass*. "Brass and glass men" celebrated magnificent artisanship in the precise machining and exquisite finish of fine microscopes; skilled manipulation of the instrument in practice; expert preparation of delicate objects; and superb performance of the technology in conjunction with some object being examined. Their position overlaps to a degree with the *diatomaniacs* who acquired powerful instruments in order to resolve spectacularly difficult, intricate test objects both natural and artificial.

Both types of observer serve as a nineteenth-century version of what Tita Chico calls the *Gimcrack*, from Thomas Shadwell's play *The Virtuoso* (1676). The Gimcrack, a satirical take on the *virtuosi* of the Royal Society, "fritters away his time and wealth on nonsensical experiments" and fancy gadgets.[47] By the nineteenth century, though, this satire had been somewhat revised. The *diatomaniac* and *brass and glass man* were mocked not for their dedication to science but for misapplying their skills, apparatus, and purse on impractical, speculative pursuits. The implicit moral of these tropes is that scientific endeavor is a social good, but only if applied to useful ends.

The tropes of *brass and glass* and *diatomaniacs* appear in Victorian texts to address a schism dividing committed microscopists, where workers devoted to technological advances disputed with those who valued greater access and practical utility instead. The "brass and glass" enthusiasts paid dearly for the most advanced, elaborate equipment and often did use it to its utmost, at least within a narrow range of tasks—but did they pursue better performance for the benefit of science, or simply due to their interest in pretty gadgets and petty competition? Goring defends microscopical craftsmanship when he argues in 1832 that "the art of working metals and object-glasses is one of extreme delicacy,—I know of no parallel to it; it ought perhaps to be called one of the fine arts, for it ultimately consists in producing, though indirectly, the most exact pictures of objects which the imagination can possibly conceive."[48] But the debate continued throughout the century, "*brass and glass* men" countering "*bug and slug* men" (ardent naturalists, identified with their lowly objects of study). An 1882 letter in the *English Mechanic* remarked, "The 'brass and glass' men, with heavy purses, have had a long innings, and it is hoped...that they have met with instruments sufficiently elaborate to satisfy their requirements." The letter-writer asks, "of what use is the brilliant lacquer" applied to "non-working or moving parts"? Costs could be substantially reduced by "confining the skill of the highly-trained workmen to the lenses," and housing the whole in a tube "like...old bronze" that would not, like elegant microscopes and other valuable objects, require storage under a "glass shade" to protect against dust.[49] More utilitarian design did appear in these later decades in response to concerns about practicality and access.

This admiration for the instrument cuts against the natural-theology trope contrasting exquisite natural objects to coarse human-made ones. However, shining brasswork, unlike human-made lace or needles, advertises the suitability of the microscope for investigating the precious, intricate, and enduring machinery of Creation. The gleaming exterior pays tribute to the minute contrivances visible within. The beauty of elite instruments elevates the practice of microscopy through analogy to beautiful nature, a refutation of the old critique that microscopists wasted their hours studying filthy vermin. Expertise similarly received its due. In his influential 1848 treatise, John Quekett frequently dubs some element of microscopical manipulation "beautiful." He acknowledges, for instance, Andrew Ross's "beautiful dividing-engine"; the "very beautiful manner" in which Wollaston's doublet shows test objects; Lieberkuhn's "beautiful injections" (staining of tissues); and the "beautiful sections" of wood produced by George Adams' innovative microtome of 1770.[50] Even in passing, such language recognizes and exalts what was often a delicate and tedious practice. Mastering these complex practices granted the eager microscopist visions of unseen structures within human and animal bodies. The observer who read the image as an instance of Creation could find wonder, reverence, and meaning where one might otherwise feel alienation or even disgust.

Writing Embodiment in Victorian Microscopy: Beautiful Mechanism. Meegan Kennedy, Oxford University Press.
 DOI: 10.1093/9780198940623.003.0008

9
Wonder Everywhere?

We find great things are made of little things,
And little things go lessening till at last
Comes God behind them. Talk of mountains now?
We talk of mould that heaps the mountain, mites
That throng the mould, and God that makes the mites.
The Name comes close behind a stomach-cyst.
Robert Browning, "Mr. Sludge, 'The Medium,'" 1864

Introduction

I have been arguing that Victorian microscopists turned to a handful of powerful literary forms to claim relevance and account for the range and diversity of their study. This chapter examines how they adapted the longstanding natural-theology trope of the *divine in the commonplace* to address contemporary debates around microscopical work. The trope of the *common wonder*, for example, can serve sacred or secular aims. Microscopical writers use such tropes to channel particular affective flows, charging lay observers and readers to move beyond their initial disinterest or even disgust at microscopical objects such as insects or dirt and into a relation of wonder instead. Microscopical specimens may be very ordinary little things—hair, blood, food, a flea, a spider, a drop of ditch water, a blot of mud, a scrape of mildew—but Victorians assured readers that they boast hidden marvels, and were it not for the microscope, those marvels would have remained unrecognized. By prompting certain affective responses to such objects and restraining others, these writers defend microscopy and its objects.

Victorian microscopical texts revise the paradoxical trope of the *divine in the commonplace* that Amy King identifies in nineteenth-century natural history, which expresses a "'reverence' for the commonplace as precious and truth-bearing."[1] That philosophy argues that the commonplace is actually beautiful; it finds a simpler, minimalist beauty in a breeze-ruffled tree, a sea-smoothed stone, a homely dandelion. Microscopy shares much with this approach. It begins as a practice of natural history and during the nineteenth century develops a practice of science as well. But even its naturalist texts, when they examine common things, usually make a somewhat different argument. Microscopists tend to argue that the commonplace contains exquisite wonders, not that the commonplace is itself valuable for its directness and simplicity. The microscopic aesthetic is not so much about redefining what *beautiful* or *wonderful* looks like but rather about where to find it; we can find it not just in the stars but also under our feet. The popularizer George Thomas Fisher exclaims, for example, that "multitudes" of infusoria can be found in "the chalk coating given to painted chambers, paper, or even glazed visiting cards."[2] Microscopists' cries of wonder usually refer to a familiar notion of beauty—elevating what is intricate, exquisite, and delicate—and validating the admiration it produces. The microscopic aesthetic revises not our definition of what beauty looks like but where we find it.

Microscopists thus dwell upon wonders that may be unexpected to the lay reader. For example, the American medical student James Jackson, Jr., who was abroad training in Paris (with Pierre Louis), London, and Edinburgh, wrote to his father in 1832 about a marvelous sight he observed:

> I dined with Dr. M. Hall [in London], and was indebted to him for one of the grandest spectacles that human eye can behold. I witnessed the circulation in the web of a frog, under a microscope of one hundred magnifying power. I know not whether you have ever seen this or not; if not, you have no conception of its beauty. With what ideas does it fill the mind! the whole animal world thus teeming with life and motion, and this motion in the most defined and regular vessels, and regulated by unvarying laws.... It is a most glorious sight, and most ennobling to see the blood thus actively pursuing its course.[3]

What elicits the younger Jackson's wonder in the clammy, hapless batrachian is not simply the intricate tracery of the blood vessels (that is, their form) but their lively yet harmonious and regular function—their mechanism as an expression of natural law. They speak to his sense of design and rouse his admiration for the impressive efficacy and even nobility of a coherent, created nature. While Jackson does not explicitly address the question of a deity, his vision embraces the aesthetic of precision associated with natural-theology tropes such as the *contrivance* and *beautiful mechanism*. The young American expresses his amazement at glimpsing, in the frog, an exquisitely machined, law-directed creation. In nineteenth-century medicine, those advocating for modern scientific approaches, like Jackson, were fostering an uneven shift away from artisanal medicine and the curious case history (with its interest in medical wonders) toward a clinical realism. But as Jackson's letter shows, even converts to scientific medicine may use expressions of wonder outlining an affective circulation that loops the observer into the spectacle of the frog's blood coursing through its veins.[4]

Nineteenth-century writing about the microscope in this way supplements the *divine in the commonplace* with the *common wonder*, a favorite trope of popularizers.[5] The *common wonder*, closely allied to the discourse of design, provided a powerful argument for microscopy: while some microscopists studied rare substances like deep-sea diatoms, the instrument most often allowed Victorian observers to find new meanings in ordinary things. In James Howell's microscopical sonnet, the epigraph for Chapter 8, the notion of the common wonder opens his final sestet, declaring that "God's work of wonder everywhere is seen—in air, in water, and the world of green." Robert Browning, in the epigraph above, cites the microscopical sublime to get to the same place—"God's work of wonder everywhere is seen." However, Browning (unlike Howell) destabilizes this sentiment by putting it in the mouth of a charlatan spiritualist, who cheapens religious faith with his quicksilver patter, and who debases the common wonder by finding it in mould, mites, even a stomach-cyst. Yet it is, indisputably, a wonder: "just a sac / That's mouth, heart, legs and belly at once, yet lives / And feels."[6] Both poets leverage the trope of the common wonder to dramatize the microscope's remarkable powers and to address its revelation of God's creative force. Nineteenth-century microscopists regularly refer to common wonders, even locating them (as Browning's speaker does) in what is repulsive. They do so not to disrupt the trope, as Browning does, but rather leverage the trope to rebut early slurs about the lowly objects of microscopical study—to insist upon the transformative power of microscopical practice in revealing the minute marvels hidden within common things. Hogg's paean to insects, quoted in Chapter 8, uses the common wonder in this way.

Diatoms as a Common Wonder

The quintessential common wonder is found in diatoms, unicellular algae notable for their intricate structure and difficulty of resolution. As T. Charters White (a "Quekett Club-Man") recalls in describing the diatom *Aulacodiscus orientalis*, "A somewhat irreverent friend . . . exclaimed 'What a waste of beautiful work, for the chances must have been a million to one that no one ever saw it!' "[7] In the discourse of design, natural beauty functions rhetorically as a way to provoke human admiration of God's powers. So the staging of these little beauties, difficult of access—underground or in bodies of water or aloft in the wind, and virtually invisible to the eye—seems puzzling indeed.

Lorraine Daston and Katharine Park note that early modern "wonders had to be rare, mysterious, and real"; and Philip Fisher similarly identifies wonder as one of the "aesthetics of rare experiences."[8] Diatoms qualify as *rare* in the sense of "exceptional"—for their extraordinarily fine and delicate structure—and perhaps also "uncommon"—for not many organisms can boast such a structure, and not many people can view it. But we cannot consider diatoms *rare* in the usual sense of "scarce." The geologist Richard Kerr, writing for a broad audience in *Hidden Beauties of Nature* (1895), notes the extensive deposits of diatomaceous earth "in every country of the world" and jokes, "The hoardings of our cities advertise in large characters myriads of diatoms for sale; only substituting for their correct name the words 'knife polish.' Many a man has more diatoms in his scullery than he can examine and mount as microscopic objects."[9] Kerr's wry observation paints diatoms as common not only in their geological abundance worldwide but also in their guise as a commodity—a workaday staple of the household and ubiquitous feature in the urban landscape.

If not technically *rare* meaning scarce, diatoms are at least *real*, although microscopists' access to them—via the image in the microscope—might not be. Diatoms were *mysterious* in that they prompted many debates about whether their appearance under the lens represented "true structure." In fact, they are so tiny and difficult to examine in the light microscope that they served as test objects for high powers. Overall, the extreme intricacy and minuteness of diatoms ultimately makes them a rarity in human experience even though (as noted in earlier chapters) they are not just as common as dirt—they *are* dirt. Margaret Gatty, writing for young readers in 1871, discusses diatoms in a passage typical of popular late-century microscopy texts. She describes diatoms as "aboriginal forms of creation . . . which come no one knows how, but never pass away, for they are indestructible." She stresses their ubiquity:

> They exist in every runnel of water that has stood for any time by the roadside, as well as in the great oceans of the world: their fossil remains are found in masses which you can dig into with a spade; they have been wafted from country by the winds; they are the last things that come up from deep-sea soundings.

But Gatty also highlights the "exquisite markings and patterns with which they are ornamented," saying that she "would need a volume to describe [their beauty]; indeed, many volumes would not avail to do it." She leans into the apparent contradiction diatoms pose for the observer, exclaiming, "And yet they are so delightfully common!"[10]

The Repulsive Revealed to Be Beautiful

While diatoms provide a standard example of the common wonder, microscopists often exclaim over even less promising objects. The physician C. R. Goring, an early promoter of microscopy, enthuses in 1827, "A human hair is a beautiful proof object [test object]; I repent that I have not given a figure of it." He presents this unpromising article as an object of mystery, noting that "Mr. W. Tulley has discovered a singular property in almost all hairs," the "greater resistance and different sensation" when drawn between the finger and thumb in one direction compared to the other. "The mystery is resolved by the microscope," which shows "teeth somewhat resembling those of a coarse round rasp" shingling the strand of hair.[11] Goring does not pause to comment on the apparent tension between the commonness of a stray hair, an object usually seen as debris or even a contaminant, and its status as a vehicle for mystery and even, in a sense, beauty ("a beautiful proof object"). Other writers point more directly to the fact that microscopic wonders underlie many objects that we would ordinarily see as rubbish or even find disgusting. An 1850 article in *Chambers's Papers for the People* on "The Microscope and Its Marvels" muses that "A morsel of ripe fruit, a little water spilt on a crumb of bread, a drop of stale ink, a neglected bottle of medicine, afford ample evidence of the activity of this teeming life-world around us."[12] The detritus of a busy household, an unappealing clutter of half-used objects, offers an unlikely but compelling window into the marvels of the microscopical world.

Microscopists' interest in such common things, an interest that extends even to dirt, rubbish, and parasites, countered a longstanding ambivalence toward the lowness and littleness of microscopic objects. Microscopists had long been scorned for interest in such things, their interests forming a stark contrast to the astronomer's study of the immortal stars. Hooke was mocked as "'a Sot that has spent two thousand pounds in Microscopes, to find out the nature of…mites in cheese."[13] And in the seventeenth century, Francis Bacon insisted in his twentieth axiom that natural philosophers must study not only what was grand and inspiring but also what was "mean or even filthy."[14] By the eighteenth century, this curiosity about the repulsive bits of nature shaped science and literature alike. Jonathan Swift's *Gulliver's Travels*, with its disgust at the magnification of human or insect parts, can be read as a fictional critique of eighteenth-century microscopy culture.[15]

These entrenched cultural narratives must have formed a perceived obstacle to widespread adoption of the new compound microscope, for the defense against them appears even in the penny press: the previously mentioned article on "The Microscope and Its Marvels" opens with a history of such critiques before launching into an extended argument for the microscope.[16] The use of the word "microscope" to describe works of social criticism no doubt reflects not only the common trope of *microscopy as scrutiny* but also this longstanding connection of microscopy with what is "low, grov'ling, mean" (as Catharine Quigley says in her 1819 poem *The Microscope; or Village Flies*).[17] Indeed, in *Under the Microscope* (1872), the poet Swinburne eventually abandons this "entomological" effort to study contemporary literary critics (whom he figures as fleas, bedbugs, and other vermin) for this reason; he scoffs, "The savours, the forms, the sounds, the contortions, of the singular living things which this science commands us to submit to examination, need a stouter stomach to cope with them than mine."[18]

This lingering association of the microscope with meanness in the object, disgust in the observer, may account for the resistance to early-nineteenth-century efforts to build a better compound microscope. Who would invest money or time into a pursuit that disgusts and demeans the user, that puts your body and mind into contact with what is "low, grov'ling, mean" with little apparent personal or societal gain? Goring reports in 1827 that his own attempts to improve the compound microscope were "ridiculed" as "frivolous and contemptible."[19] In 1829 he opened an ambitious serial publication on microscopy with an extended "vindication of microscopic science," where (like the *Chambers's* article two decades later) he first repeats, then repudiates, the critiques associating microscopy with what is trivial and nasty. He writes,

> Were [microscopes] applicable to no other purposes than the dissection of blackguard vermin, the observation of stinking ditch-water, or the amorous passions of ants and worms, I should, perhaps, for argument's sake, admit that they were but the tools of a puny, pitiful pedant, whose passions and amusements were of a trifling, if not of a degrading complexion: but I would ask whether, in the hands of men like Bauer, they are not applied to the development of the most curious, important, and interesting details of anatomy and physiology, which, without their assistance, could never have been known?[20]

Goring here negotiates a careful line indicative of the ambiguous status of the microscope at the time, just before improvements to its construction boosted its status as a philosophical and eventually scientific instrument. He scornfully rejects any unworthy objects of study—emphasizing his distance from such pursuits by damning them as "blackguard," "vermin," "stinking," "amorous" (with a smirk), and "degrading"—then pivots to a spirited defense of microscopic research as "curious, important, and interesting." He also rebuts any critique based only on littleness by referring minutiae to the greatness of God. He argues, "Surely he who is but a *work of God* may be allowed *to admire the works of his Creator*, without incurring derision or ridicule, even though they are minutiae." Indeed, he insists, "*Minutiae are not necessarily trifles.*" He concludes with some heat, "It is not only, in my opinion, unscientific, but even swinish and ridiculous, to contemn any thing merely on account of its minuteness."[21] With this outburst he deftly turns the tables on the critics by showing how they are themselves petty and contemptible.

The critique of microscopists' degrading obsession with vermin faded as the instrument gained status and adherents in the Victorian period and microscopy became more widely recognized as a valuable, though difficult, practice. While Goring acknowledges the unseemliness of many microscopical objects, a shift occurs in later authors, who often explicitly defend the study of such objects. They continue the tradition of inquiring into what is small or even "mean," but they show little disgust at such objects. Their descriptions redirect readers' attention and affective energies toward the minute size and novel form of these objects, modeling curiosity or admiration rather than disgust. The great microscopist John Quekett, for instance, assures his readers that insects are "the most numerous and, perhaps, the most beautiful class of objects for examination, as there is scarcely a part of the body of an insect that does not exhibit some remarkable structure."[22] While Quekett was writing for microscopical workers, the priest and popularizer Charles Kingsley similarly defends his attention to small objects that might seem trifling or mean, in his sea-shore book *Glaucus* (1855). Kingsley remarks that Nature is "not only 'maxima in minimis'—greatest in her least [works], but often 'pulcherrima in abditis'—fairest in her most hidden works."[23] Many nineteenth-century microscopists quote this old maxim, *Natura in minima maxima*, to justify their studies.[24]

Thomas Carlyle famously wrote that "Romance exists... in Reality alone."[25] But this relation between romance and reality is a complicated one. The scholar Jane Bennett notes that such revelations can be disruptive, even violent; for "[t]o be enchanted is to be struck and shaken by the extraordinary that lives amid the familiar and the everyday."[26] Charles Dickens's version of this insight acknowledges the discomfort involved in such experiences but he emphasizes instead the duty of the observer to seek out the extraordinary regardless of its costs: one aim of his periodical *Household Words* was "[t]o show to all, that in all familiar things, even in those which are repellent on the surface, there is Romance enough, if we will find it out."[27] Dickens was referring to "repellent" sociological realities, but his argument exactly fits microscopists' narrative about repellent objects as well. Sienne Ngai argues that "disgust is constituted by the vehement rejection or *exclusion* of its object"; that it is "directed toward the negation of" that object, "either by denying [it] or by subjecting [it] to epistemological skepticism."[28] Where (as Ngai comments) the feeling of disgust "strengthens and polices [the] boundary" between self and other, Victorian microscopists addressed their own, and their readers', disgust by pressing observers to approach even more closely to what repelled them.[29] In contrast to "the strange kind of sociability" formed by shared feelings of disgust, they ask observers to join a community looking so closely into the repulsive object that their disgust dissipates.[30] They reframe the uncomfortable response of disgust as curiosity and, eventually, as wonder, challenging the traditional belief that what is disgusting cannot be beautiful or marvelous.

Indeed, when they do acknowledge that most readers are repulsed by these objects, Victorian microscopists turn the association of microscopy with vermin and filth into a strength: they argue that the instrument displays the true beauty of objects that we mistakenly regard as repellant or even repulsive. This sentiment extended a general argument that studying natural history promoted the appreciation of forms that might not ordinarily attract our interest or sympathy. When Victorian microscopists appreciate objects that are not necessarily beautiful, valuable, or rare in everyday life but that become so in their transformation under the lens, they revise *the divine in the commonplace* into a powerful transformation narrative that we might call *the repulsive revealed to be wonderful*. Significantly, as the object is transmuted into a wonder, the observers are changed as well, their initial disgust translated into fascinated curiosity and awe. White, the "Quekett Club-Man," advises novices that "the repulsion generally felt at the sight of [a spider], or even at the mention of its name, is quite unearned by the insect." He muses, "We can imagine the feeling of horror overpowering [a] fly… when he views this fierce pair of eyes and dreadful body." In White's reframing of the spider, the microscope is proposed as a corrective to an uninformed observer's visceral response. It first transposes the reader into the fly's perspective (where a spider is dreadful but a fly obviously is not), then returns us to a recalibrated human perspective that can see what is wonderful about both creatures—for, White asserts, "both spider and fly" provide "a charming and interesting object for our cabinet."[31] Ultimately, White praises these creatures for their strength, resourcefulness, and elegance.

The louse does usually inspire disgust across a range of texts. For instance, Jabez Hogg, writing for would-be microscopical workers, praises the "beautiful little gem-like marks" on the diadem spider, the "surprising agility and…great strength" of the "curious" flea, even the "pretty" eggs of the bedbug. He pauses, however, on the louse, "this horrid parasite," commenting that "when [it is] examined under the microscope, our disgust of it is in no way diminished."[32] The botanist Mary Pirie, in a natural history textbook (1864), similarly lauds the flea as "extremely elegant in form…as if clad in a coat of mail" and defends the earwig against the "calumny" that it seeks the human ear, noting as well that "its four wings when expanded are like the finest lace-work."[33] But she is repelled by the louse, and not necessarily by its activities: "its external appearance, when examined through a microscope, is peculiarly disgusting."[34]

However, some microscopists could admire even the louse. W. L. Notcutt, a dissenting chemist (druggist), botanist, and teacher who collated a popular guide to microscopical objects in 1859, remarks of lice, "Notwithstanding the disgust which their vulgar name inspires, they will be found to present many objects of beauty and interest."[35] This rehabilitation of the louse despite its "vulgar name" is in keeping with the microscopical aesthetic generally. The ordinary objects listed earlier—hair, blood, food, flea, spider, ditch water, mud, mildew—are often considered dirt or even filth, but the *repulsive revealed as wonderful* allows microscopists to insist upon their hidden wonders. Observers may choose to study an ordinary or even disgusting object (fly wing, ocean-floor ooze, scummy water) in order to seek transformation through the microscopical image.

This notion animates microscopical observers across the century and across genres. The "Christian Philosopher" Thomas Dick explicitly references this trope in 1851 as he commends the microscope's role in natural theology. He says,

> The microscope…has unfolded to our view wonders unknown and unthought of in former ages. Three hundred years ago, who would have conceived it possible to distinguish myriads of living creatures in a single drop of water? Or, that

> blood could be distinctly seen circulating through veins and arteries, smaller than the finest hair? Or, that not only the exterior form, but even the internal structure of the viscera, and the motion of the interior fluids, should be rendered perceptible to the sight? Or, that numberless species of animated beings should be made visible to the eye, though so minute that a million of them are less than a grain of sand?[36]

Amidst this flurry of microscopical tropes, Dick lauds diatoms, blood circulation, and even the "fluids" of microscopic "viscera" as unexpected but common wonders, "unknown and unthought of" before modern microscopy. Given this context, when the 1850 *Chambers's* article on "The Microscope and Its Marvels" casually remarks, "Nothing, in fact, is more wonderful than the apparent omnipresence of fungus germs in the air,"[37] the term "wonderful" could readily suggest both "surprising" and "marvelous."

Authors speaking to an expert audience also find wonder and awe in the ordinary, although they usually omit noting any initial revulsion and are less likely to indulge in elaborate panegyrics. An 1878 "manualette" on section-cutting for working microscopists admires stained, sectioned tissues as "beautiful," "very beautiful," even "particularly beautiful." These marvels were slices of fat, ovary, skin, and stomach.[38] Such visceral objects—elements of our own unruly materiality—provide an effective but no less wonderful counterpoint to the elevated tropes of beauty, natural theology, and the sublime, and to the playful and sensational romantic tropes (fairylands, Oriental myths, phantasmagoria) that enliven nineteenth-century microscopical texts.

Writing Embodiment in Victorian Microscopy: Beautiful Mechanism. Meegan Kennedy, Oxford University Press.
 DOI: 10.1093/9780198940623.003.0009

10

Romances of Microscopy

Fairylands, Fantasies, and Phantasms

> [T]he microscope...unveils beauties among the minute and all but invisible tribes which people our earth, which equal, if they do not surpass, anything on which the unassisted eye gazes. It brings before us marvels so strange, and apparently incredible, that all the inventions of romance and fiction are rendered trivial and tame by the side of the facts which it presents to our senses.
>
> William Lowndes Notcutt, *A Handbook of the Microscope and Microscopic Objects*, 1859

Introduction

The elements of wonder discussed so far draw from ostensibly high-minded pursuits such as philosophy (beauty and the sublime) and theology (natural theology). But nineteenth-century scientific texts, including microscopy texts, display many forms of wonder other than the admiration and awe evoked by beauty, the sublime, or natural theology. Victorians also leveraged whimsical, exotic, and grotesque elements of the microscopical world as spectacle. These more playful or sensational elements are most likely to appear in entertaining books and articles for children or general audiences, in soirées for family friends, or in mass-visuality events such as the oxyhydrogen microscope show.[1] Observers metaphorically dally with fanciful fairylands, exotic lands of "Oriental" myth, and—particularly in discussing projection technology such as the solar or oxyhydrogen microscope—Gothic phantasmagoria, indebted to the sensational magic lantern displays of the early nineteenth century. These diverse examples share a floridly romantic bent. Spectacular tropes like this do appear in texts for working microscopists, but less frequently, and they are usually confined to informal, paratextual, or ceremonial settings.

The romances of the microscope often frame microscopical practice as a form of travel to and within a virtual landscape. Observers enter an immersive experience of exotic worlds and interact with the inhabitants, in a setting that exemplifies the reciprocal effects of observer upon object and vice versa. Passages using these tropes express and encourage affective flows between object and observer, by placing observers (and readers or viewers) actually within the image, where they are entertained, perturbed, and transformed. The trope of the *microscopical image as landscape* also powers some realist narratives, which model skeptical practice by portraying the investigator's intellectual progress as a virtual journey through the image; Part IV of this book will examine this process of *narrative investigation*. Both romantic and realist versions of the *image as landscape* solicit readers' interest with a promise that microscopical work will transport them into strange landscapes, much like the bold souls who were, at the time, exploring the oceans and mountains, Africa, the East, and the polar regions.

I'd like to disambiguate my work here from the lively tradition of scholarship on "Romantic science." The capital-R "Romantic" here identifies that work as focusing on *the Romantic period* in Britain and Europe and its general cultural approach. John Tresch defines *Romanticism* as it is traditionally understood: exploring nature and the self through "emotion, expression, aesthetics, and purposeful action." The Romantics grounded their work in organicism, emphasizing patterns "associated with the processes of life—growth and holism, circular causality, and the productive struggle and harmony between polarities."[2] Other historians of science also use "Romantic" to describe an understanding of nature emerging from the Romantic period and based on concepts of creative, organic unity and flow.

I use the term "romantic" instead as Lynn Merrill does, to describe the *literary genre of romance*, which emerged much earlier—in French and English medieval literature—and can appear in any literary or historical period. Merrill, in *The Romance of Victorian Natural History*, defines *romance* as a literary genre: "a *romance* may be thought of as an extravagant fiction, a pleasurable excursion into a lavish or fantastic realm…it may be said to be adventurous or exotic."[3] The small-r "romantic" is thus a term of genre, distinct from the capital-R "Romantic" denoting a historical period (although much of the work in that period does draw on small-r romantic elements). For example, Merrill examines Victorian naturalist texts, including texts on microscopy specifically, and discusses these as forms of romance.

To elaborate on Merrill's definition, *romance* is characterized by a rhetoric of excess and extremity and often describes exotic or extraordinary characters, settings, and events. Tresch's definition of Romantic science borrows from this literary and aesthetic sense of *romance*. He argues that "romantic machines" during the early years of the nineteenth century provided

> conversions, imaginative flights, and metamorphoses of the fantastic. They drew forth invisible powers, converted them, and put them to use. [They] involved the active participation of the observer and articulated a spontaneous, living, and constantly developing nature; [they] produced aesthetic effects and emotional states.…[They were] imbued with the aesthetics and the affects of the organic, the vital, and even the transcendent.[4]

While Tresch does not discuss microscopy in depth, this passage describes many of the tropes of microscopical wonder—tropes that persist in a range of microscopical settings well into the second half of the century. One could argue, in fact, that any Victorian microscopical text is at least somewhat romantic in that it describes an exotic, otherworldly setting and the beings therein; some texts emphasize these elements more than others. All the forms of wonder I examine in this chapter use exaggerated (that is, romantic) language and also stage outlandish scripts around the affect-charged encounter of observer, instrument, and object.

Fairylands, Fantasies, and Phantasms

Victorian descriptions of the object as a marvelous *fairyland* have attracted considerable scholarly attention, especially in the context of popularizing texts and their promise to be *instructive and amusing*.[5] Nineteenth-century authors writing for a popular audience or for children sometimes used fanciful language describing (as Arabella Buckley phrases it in 1878) "the fairyland of science."[6] The trope of fairyland, a version of the *microscopical image as landscape* and a whimsical, secular alternative to theologies of nature, invites readers into a memorable, enticing landscape of beauty and wonder, inhabited by diverting little beings.

Microscopy offers a natural opening for fairy-talk, given that it so often presents strange creatures to the gaze that can be as lively and willful as they are minute. The Irish-born writer Fitz-James O'Brien published a story, "The Diamond Lens," in *The Atlantic* in 1858 that exemplifies many of the tropes of fairyland.[7] This sensational tale combines elements of scientific romance, spiritualism, and a locked-room murder mystery, with a hint of slave-trading and a dose of Orientalism and anti-Semitism. The narrator, a votary of microscopy, raves about "a land of enchantments" under the lens; he claims to perceive in a raindrop "a universe of beings animated with all the passions common to physical life," or to see in mold-spots "enchanted gardens, filled with dells and avenues of the densest foliage," all bearing "strange fruits glittering with green and silver and gold."[8] The story centers on his use of a uniquely powerful diamond lens to gaze into an even tinier hidden fairyland within a water-drop, where he could "gaze upon a vast space" filled with cloudlike "fairy foliage" hung about with "fruits or flowers, pied with a thousand hues lustrous and ever-varying." Within the drop he discovered a lovely "female human shape" like a "Naiad" who enchanted him. Intoxicated by these visions, the narrator became obsessed with the beautiful creature (whom he named Animula), exclaiming, "I loved an animalcule!" He tortured himself with the thought that, because they existed on different physical scales, they could never communicate—although even one encounter would allow them to enjoy a "remote sympathy." But in fact, just as her movements provoked his fascination and delight, his actions did spark responses from her. The "suppleness of her motions sent a thrill of delight" through the narrator's body; then, when his "heart beat madly" she "turned her beautiful eyes" toward him, "suddenly started," listened, and flitted away into the "forest." Again, when he "lessened the lamp-light," he saw "an expression of pain flit across her face"; when he "flooded the stage...with a full stream of light," she "sprang forward" in delight. Finally, his inattention to the enclosing water-drop brought tragedy: "it had evaporated" in the light he had placed full on the slide, and as the life-giving moisture dissipated Animula "writhed, as if with some internal agony" before "shrivelling up" and "blackening" into death.[9] The narrator's initial pain at being physically separated from his loved one and at the impossibility of any reciprocal affection resolves into a spectacular finale in which her death throes spring from his actions, and observer and observed agonize together at her moment of death.

O'Brien's fantasy appeared at mid-century, but the romantic language of fairyland more often appears late in the century, when fairy-talk was more popular in the culture broadly.[10] For instance, John Badcock, a Fellow of the Royal Microscopical Society (RMS), offers in *Vignettes from Invisible Life* (1883) a charming anecdote in which even the "slime or filth" that accumulates "along the banks of our canals and ponds" wafts the observer to fairyland. He explains,

> You take up a portion of this apparently dirty slimy mucous matter—say on the point of your walking-stick—and your first impulse is to throw it away in disgust. Put it again into the water, and it is seen to spread itself out in a less repulsive appearance. Take up a small portion again, put it in a small glass bottle of water, and peer at it through a pocket lens; it begins to look attractive.

For this instructive and community-building text, Badcock uses the second person ("you take up") to direct readers' actions and places them within the scene, guiding them not only in what they see but also in what they feel (at first, "disgust"). Having ushered readers through a process of increasing attention and attraction to this "repulsive bit of slime," Badcock brings them to the microscope for his climax. Viewed "under a good microscope and with proper illumination," the lump of slime "is transformed into trees and flowers, and all in active and beautiful motion." One tree, with "wide-spreading branches," even has flowers, fruit ("like large oranges"), and "birds sitting on its branches." The birds are not "structurally like any known species of birds; but here one is, so to speak, in fairyland." But Badcock also discusses the species of "tree" (*Zoothamnium arbuscula*, a marine protozoan) and "birds" (rotifers), alternating between fancy and instruction. He had prefaced this parable with the warning, "things are not always what they seem here," but it is equally true that slime may really be a marvelous fairyland and that the fairyland is really slime.[11]

Badcock's lovely fairyland is fragile, however; as with O'Brien's Animula, it folds up in response to the observer's actions. Suddenly, the tree comes alive, even becomes sentient and responsive to the observer's actions—it "looks alarmed." Badcock suggests that the observer's emotional agitation has communicated a physical and affective thrill back to the object and spurs it to a remarkable response. Badcock explains, "You have, perhaps, touched the table or the microscope, and so produced a slight vibration which has thus reached the highly sensitive group before you; and down it sweeps, branches, flowers, fruit, birds, trunk, and all, in tiny contraction."[12] (The caution about vibration is a common trope of instruction in skeptical practice.) This evident entanglement of observer, instrument, and object provides a vivid example of Victorian microscopy as a fully embodied practice. Badcock uses the tropes of *the repulsive revealed to be wonderful* and *fairyland* here not only to transport readers to this fanciful scene but also to underscore how observer, microscope, and object are caught up in a dynamic system where each changes the other through optical, mechanical, cognitive, and affective links. The trope of fairyland, with its premise that humans can spy on or capture fairies and fairies can enchant humans, emphasizes the possibility for interaction within such scenes.

Badcock is writing for a broad audience, but elite microscopists addressing fellow investigators can also build fanciful word-pictures for their audience. Unsurprisingly, however, they employ frankly romantic prose less often and usually in particular spaces; those settings are also more likely to stage any extended references to beauty and the sublime. An informal setting for microscopical romance might be found in personal correspondence, as in the Francis West Lewis letter that opens Part II of this book. Romantic discourse is also more likely to appear in formal settings such as a ceremonial lecture, or the (paratextual) introduction or conclusion to a lecture or essay, which typically invite readers into the text and set the argument in a larger context. The appearance of romantic discourse in these framing sections accords with the generally more catholic range of styles in these sections.[13]

Speakers were well aware of the generic expectations for presidential lectures at the time. Over the course of the century, lectures became more focused on a particular area of expertise, but the framing sections still permitted more romantic discourse.[14] We should not dismiss these moments of overt eloquence as boilerplate oratory. It was often in presidential addresses, for instance, that leaders of scientific societies could set out their vision for the organization, the terms on which they proposed to work, and their priorities for the development of their science.[15] These are not throwaway moments but points of inflection for microscopical communities.

In a striking example of the presidential address as a romantic space soliciting affective engagement with the microscope, the schoolmaster C. T. Hudson, in his 1890 President's Address to the RMS, draws upon flagrantly romantic language to describe an exotic world within the microscope. He exclaims that "the Microscope dazzles us with the sight of a world whose inhabitants and their doings surpass the wildest flights of nightmare or fairy tale." He continues, "[T]he living creature...chains us to the magic tube. For there we see that the dream of worlds peopled with unimagined forms of life, with sentient beings whose ways are a mystery, and whose thoughts we cannot even guess at, is a reality that lies at our very feet."[16] Such anthropomorphic language might seem more suitable to a popularizing text. The microscope here is figured as a *magic tube* offering access to alien worlds and strange but sentient beings who enchant observers into intense attention to, and attraction to, the image—much as in "The Diamond Lens." Hudson's address employs tropes about *microscopical worlds, fairylands, mystery*, and *magic* to model an affective entanglement ("chains") with the object (the "living creature") and instrument (the "magic tube"). Hudson uses the first-person plural ("we see") to draw listeners as if physically into the fantastical world that "lies at our very feet."

Hudson's field of research specialization—*Rotifera* (also known as the "Wheel animalcule")—provides an especially fertile ground for such fancies. These strange-looking, tiny aquatic creatures, with the apparently spinning corona atop their bodies, had inspired many fanciful passages of nineteenth-century microscopical prose. And with long experience in the classroom, Hudson was known as an entertaining lecturer; his obituary in the *Journal of the Royal Microscopical Society* remarked that "his addresses were charming in style," and he was noted for using large transparencies that mimicked the striking dark-ground illumination of the microscope to further enchant audiences.[17] But even apart from what might be expected of the whimsical rotifers and their "charming" interlocutor, it is no surprise that these fantastical descriptions appear in a presidential address. This genre served as an important tool of community-building; its ritual form licenses the speaker to share microscopical wonders as one adept to another.

Other tropes also place the reader as an actor within a virtual storyspace, as an explorer of romantic scenes or a voyeur peering into them. The naturalist and popularizer Philip Henry Gosse, in *A Naturalist's Rambles on the Devonshire Coast* (1853), offers a singular blend of romance and science in describing *nature's laboratory* for general readers who were likely novices. This inventive scene was likely inspired by the longstanding trope of scientific work as *peering into nature's secrets*. The popularizer George Thomas Fisher, for example, in 1846 proposes that, in microscopy, "the eye of man is here opened to penetrate the most profound...secrets of nature."[18] Gosse revises this to figure Nature as a woman scientist working at a mysterious laboratory accessed through the peep-hole of the microscope. He notes that microscopical observing means "to take a stolen peep into the *Adyta* of nature's mysteries" and perceive "LIFE, carrying on its more secret and recondite functions." Sketching out an imaginary scene, he invites the reader to "fancy that we are slyly peeping through the keyhole of Madam Nature's door, her laboratory door, where she is actually at work, concocting and fashioning those marvellous forms which constitute the world of living beings around us." Nature's hidden laboratory is here a magical space inhabited by mysterious beings, doubtless inspired in part by nineteenth-century scientific romances such as Mary Wollstonecraft Shelley's *Frankenstein* (1818). The microscopist is figured as a *voyeur* stealing an illicit "peep" in on this scene. Gosse's use of the first-person plural ("we are slyly peeping") overtly places him and readers physically within this thrilling scene, crouched at the keyhole and peering in together. The scene hints at the sensuous pleasures of microscopy even as it argues for microscopical observation as a recreative good, drawing lay readers into a book designed for instruction and community-building. Gosse even comments on the "peculiar pleasure" achieved by "the reflecting and curious," promising his lay readers that the diligent student of microscopy reaps an exciting reward.[19]

If writers did not want to waft their readers to fairyland or a laboratory, they could turn to the romantic literary genre known as the *Oriental tale*, which figures microscopical observation as a portal to an enchanting Eastern land of myth and magic. The Oriental tale was launched into popularity in England in 1704–17, when Antoine Galland published (in French) the first translation of the *Thousand and One Nights* (*Mille et une nuit*), a collection of Arabic folk tales dating back to the eighth century. Gosse, in the Preface to *Evenings at the Microscope* (1859), invokes the Oriental tale in describing the powers of the microscope: "Like the work of some mighty genie of Oriental fable, the brazen tube is the key which unlocks a world of wonder and beauty before invisible, which one who has once gazed upon it can never forget, and never cease to admire."[20] The analogy persisted throughout the century; in George Blackham's 1882 President's Address to the American Society of Microscopists, he praises "the splendid instruments of to-day, the magic tubes that open to us treasures far richer than those of Ali Baba's cave."[21] The Oriental tale allows writers to refigure the unsettling proliferation of microscopical life to a form of wealth and abundance instead.

The Oriental tale also allows writers to frame microscopy as a form of scientific magic. They figure the brass body of the microscope as a *magic tube*—as in Hudson's and Blackham's flights of fancy—or a *magician's wand*—as when the medical journalist Ernst Abraham Hart writes in *Household Words* that "the revelations of the microscope" had "allowed the naturalist to traverse the material universe as though armed with a magician's wand."[22] Although these writers are describing scientific advances, the Oriental tale provides a romantic setting that celebrates microscopical vision as a kind of magic. These romantic analogies to an Orientalized land of myth or fable, as with references to fairyland, tap into longstanding cultural myths, with all their power to engage readers' affections. Such exoticizing descriptions paint the microscopist as enjoying magical agency and access to rare visions at the same time that these visions compel his awe and admiration.

The frankly romantic tropes of wonder discussed above can manifest in diverse ways, but they consistently emphasize what is most extraordinary or otherworldly about microscopy. They often appear in company with more versatile, more broadly applicable tropes, such as the sublime or natural theology, that also power literary or theological texts. The affinity between these genres is evident in the ease with which W. H. Harvey, in his popular *Sea-Side Book* (1849), quotes Edmund Spenser—"So fertile be the flouds in generation, So huge their numbers, and so numberlesse their nation"—as an epigraph for the book in an echo of the microscopical sublime. The common ground becomes even more clear in this passage from John Henry Newman, also quoted by Harvey:

> It is, indeed, a very overpowering thought, when we get to fix our minds on it, that we…hold intercourse with…creatures who are…as mysterious as if they were the fabulous, unearthly beings, more powerful than man, yet his slaves, which Eastern superstitions have invented…. [S]ay whether the presence of such countless multitudes, so various in their natures, so strange and wild in their shapes…is not as mysterious as anything which Scripture says about the angels?[23]

Here Newman, writing in support of theologies of nature, draws on tropes of the *sublime* and the *Oriental tale*, in language as relevant to microscopy as to theology.

Finally, while the above magical scenes rely largely upon fancy to solicit reader engagement, the Gothic trope of the *phantasmagoria* gripped observers with dramatic spectacle and visual trickery. Descriptions of popular exhibitions, especially projection microscope shows, often describe such spectacles in terms reminiscent of the eighteenth- and early-nineteenth-century display known as the phantasmagoria. This famous illusion used magic lanterns (not microscopes) to project images of supernatural beings onto smoke or screens with a frightening animation. The microscopical trope of the phantasmagoria does not often reference this historical tradition specifically but it portrays the microscopical image as a similarly dreamlike—or nightmarish—fantasy of the grotesque, filled with *hybrid forms*.

The optician Charles Gould in 1835 describes an oxyhydrogen exhibition (one of the first) in such terms, describing not just "beautiful scenes displayed of inanimate and animate objects [that] baffle all description" but more unsettling scenes as well, such as a humongous fly-eye-egg: "The eye of a fly is magnified to six feet, each of the lens [sic] as large as an egg." The technology also stages insect-lions and insect-devils, whirling about in a frenzy of violence:

> Here are generally seen the ditiscus or water-lion, the hydropholus or water-devil, destroying the smaller class of insects…and when this is accomplished they devour one another; in a few minutes a scene of carnage takes place,

> producing one of the most imposing and magnificent microscopic effects that the eye can witness or imagination conceive.[24]

Gould's promotion of his oxyhydrogen show is itself hybrid in that it draws upon many types of wonder, from enticing to disturbing; his audience should experience beauty, mystery, and spectacle, with elements of grotesquerie (the fly's many eyes like a cluster of eggs) and horror (the voracious projected animals writhing hugely on the screen). His phrasing promises great "effect" of "scene" that will "strike the spectator with astonishment." Spectacles like this commonly embellish texts (especially advertisements) describing projection microscopy through the end of the century.

The language of spectacle can appear in texts for working microscopists when writers describe the kinds of sights—animated animalcules in a drop of water, or blood cells in circulation—that played well in exhibitions. For example, John Quekett describes the circulation of a frog as "one of the most beautiful and marvelous spectacles which can possibly be witnessed."[25] However, research texts are less likely to dwell upon what is sensational or spectacular about the image, and the phantasmagoria proper—involving Gothic grotesquerie as well as spectacle—is rare. Hudson does cite the phantasmagoria tradition briefly in his 1890 President's Address to the RMS when he invokes "the wildest flights of nightmare or fairy tale," discussed above. The observers he imagines, who are "chained" to the "magic tube," perhaps experience some of the Gothic atmosphere of the phantasmagoria.

Life in a Water-Drop

As may be suggested by its appearance in multiple examples above, viewing *life in a water-drop* was a staple exercise in Victorian microscopy, one that pairs well with *fairylands* and other whimsical tropes as well as grotesque or horrifying tropes like the *phantasmagoria*.[26] It enjoys a similar versatility in its print settings. The *water-drop* helps drive instructional texts like Jabez Hogg's massive *The Microscope* (1854), popularizing books and articles, and recreative settings such as projection microscope shows. Agnes Catlow, for instance, organized a playful book, *Drops of Water: Their Marvelous and Beautiful Inhabitants Displayed by the Microscope* (1851), around the water-drop, even numbering her chapters "Drop I" and so on. She explicitly invokes *microscopical worlds* as she asks readers, "[P]ass with me through a wonderful brazen tunnel, with crystal doors at the entrance." Within, "your minds are bewildered with the variety of new beings and forms you behold" and the strange "vegetation you see around you."[27] Catlow's use of the second person ("you behold") places the reader within the fantastical scene she describes within a water-drop and guides them to an appropriate affective response of wonder.

Working microscopists could dive into fantastical scenes within a water-drop as well. Hudson again offers a striking example of this. In his 1889 presidential address to the RMS, he elaborates on "that wonderful world which lies in a drop of water." He muses that to "gaze into" this world is

> to see transparent living mechanism at work, and to gain some idea of its modes of action; to watch a tiny speck that can sail through the prick of a needle's point; to see its crystal armour flashing with ever-varying tint, its head glorious through the halo of its quivering cilia; to see it gliding through the emerald stems, hunting for its food, snatching at its prey, fleeing from its enemy, chasing its mate (the fiercest of our passions blazing in an invisible speck); to see it whirling in a mad dance to the sound of its own music, the music of its happiness, the exquisite happiness of living—can any one, who has once enjoyed this sight, ever turn from it to mere books and drawings, without the sense that he has left all fairyland behind him?[28]

Hudson here uses tropes of the *mechanism*, *jewels*, the *invisible*, and *fairyland* to prepare readers to view each tiny organism (in this case, a rotifer) as an enchanting little being capable of thought and feeling very like our own. Hudson points to the observer's affective entrainment with microscopical life by modeling his own sympathetic identification with this little rotifer. He tracks its movements, envisions its agency and intent, and finally declares his knowledge of its desire ("the fiercest of our passions") and its "exquisite happiness of living." While sympathy and identification are not themselves forms of wonder, they can powerfully support wonder, as they do here in recreating the observer's affective entanglement with the busy rotifer. But Hudson turns to them only after a long disquisition on the geographical dispersion of *Rotifera*. That is, his scientific argument in this presidential address unfolds in a textual space largely free of wonder. He values the beauty, mystery, and charm of microscopical lands, but he confines expressions of wonder to paratextual space, using them to celebrate and inspire, but not elaborate, a line of research.

The frequent use of the *water-drop* trope likely derives from the ready availability of water for experimentation, the importance of water in daily life, the often dramatic effect of its squirming "animalcules," and the emotional force of its suggestion that audience members were imbibing swarms of creatures with every sip. The trope thus served purposes far beyond the dreamy discourse of fairyland. It developed a synergistic relation with the periodic water-panics in nineteenth-century London, each lending strength to the other.[29] An 1833 article titled "Microscopic View of a Drop of Water" in the abolitionist periodical *The Tourist* paired a startling engraving of a water-drop with an explication of "the various species of…*animalcules*" visible there; and an advertisement for an 1837 oxyhydrogen show in London promises that "LIVING OBJECTS ARE SHOWN in separate drops of water, from different sources, viz., the well, the cistern, ponds, ditches, &c having living insects, animalculae, &c peculiar to each water."[30] So microscopic "animalculae" in water could serve as an example of the *common wonder* as in Catlow—or as something more ominous.

The ubiquity of water in daily life lent these revelations a certain piquancy and ensured a fascinated readership on the topic. Water-drops became a particular concern during the sanitation debates at mid-century, famously appearing in comic form in *Punch* and with instructive whimsy in *Household Words*.[31] The comic Charles Mathews worked up a "bit" where a temperance society becomes alarmed by the "animalculae" seen in a drop of water. The society members exclaim, "what do you think of having fifty such things as that stumping about in your epigastric region?" and the like; they imbibe bowl after bowl of alcoholic punch because the animalcules "do not live in punch as well as in water."[32] The joke is that punch is safer than water because microscopic creatures are both in the water and, horribly, in us.

As Mathews' joke suggests, the *water-drop* trope reminds readers and listeners of their physical contact with animalcules from the inside and out, inviting revulsion and horror as well as wonder, curiosity, and amusement. This trope is one of the few places where microscopical writers foment disgust in the audience. The shocking contents of a water-drop usefully demonstrate the power of the lens and suggest its value in evaluating whether drinking water is more or less infused with microscopic life. Water can flow, pool, and saturate its surroundings. This fluid energy makes the *water-drop* trope a compelling way to express and encourage an affective flow between these tiny denizens of the world and those who observe them.

The water-drop could thus simultaneously provoke both uneasiness and delight. Catlow, above, describes the contents of a water-drop in the most romantic terms, but she also acknowledges the uncomfortable truth about water: it is likely, she explains later, that "water may still teem with life even beyond the reach of the highest powers of our glasses, and we may never be able to say with perfect truth of any drop of water, that it is free from animal or vegetable life."[33] Catlow cheerily locates the reader's body both within the drop and outside it, as a potential imbiber of it.

David Brewster, writing on "Life in a Drop of Water" for the religious periodical *Good Words* in 1864, goes even further, expanding beyond sanitation to scientific theories and asserting that tiny animalcules inescapably surround us, not just in water but also in soil and air. In this article, he repeatedly puts quotation marks around the phrase "life in a drop of water," indicating its prevalence in Victorian popular culture. He exclaims at one of the "wonderful revelations" of the microscope: that "there is life in a drop of water—living beings born in millions, and living and dying in a dwelling so small."[34] He links this revelation to Pasteur's research on germ theory, which he cites repeatedly. Brewster dismisses the spectacular effects achieved by popular showmen who add sensational larvae and insects to the drops they display, but, drawing on a versatile trope that I will call *invisible life surrounds us*, he warns that "in our food and in our drink, we cannot escape from the atoms of poison which we breathe, and the legions of swarming, crawling, and whirling life which are ever at work within us, and without us, and around us."[35] Although Brewster does not place observers directly within the water-drop, he argues that human and microscopical life are inextricably entangled in overlapping, dynamic zones of risk and vitality. He tells readers that these strange bits of "life" suffuse the air, water, and earth around us—they surround and infuse our bodies—yet frames this phenomenon with wonder.

The spectacular claims of the *drop of water* and *invisible life surrounds us* tropes were especially well suited to popularizing texts, but they did appear in texts for working microscopists as well. As Hogg insists in his massive 1854 guidebook for novices and working microscopists, "water" is everywhere "animated," even to the tiniest droplets that waft on the air. He explains,

> [T]here is no condition of water, from the most fetid to the most crystal, but teems with animated nature. Thus the stagnant pond, the limpid stream, the bubbling spring, the extensive watershed, the majestic river, the broad lake, the mighty ocean, the deep mine, the unstable bog, the torrid zone, the arctic regions, the forest-leaves, the lowly herb, the lovely flowers, the living plants, the aerial ocean, the guano of Shabac, the skin and interior of animals, as well as that of haughty man,—are all invested with living animalculae; many having even parasites existing on them more minute than their own invisible bodies.[36]

Hogg's remarkable catalogue here, taken from his first chapter, engages readers in an imaginary progress through many kinds of waters and types of earth to "the aerial ocean" of the atmosphere, ultimately involving even the readers' own bodies, "all invested" (perilously close to "infested"!) "with living animalculae" inside and out. Opening his guidebook by locating the trope of the *water-drop* everywhere on earth—even in apparently non-watery substances—helps him make the case that microscopy is needed for readers to truly know the extent of the natural world.

This idea that *invisible life surrounds us* does appear in early Victorian texts (often paired with the *drop of water*) but it surges in the later decades of the century. The knowledge that microscopic beings inescapably surround us was not new, but it became more broadly known and relevant to society in this period, with the new popularity of inexpensive microscopes, with mass microscopy shows that exaggerated this phenomenon for show, and with contemporary debates on vitalism, sanitation, parasitology, and especially germ theory and the growth of bacteriology. Victorians had to learn to accommodate the fact that we are always in contact with "invisible" beings in our food, water, soil, and air.

References suggest that *invisible life surrounds us* would have been general knowledge for educated people by the latter decades of the century. It features, for example, in an 1878 *Punch* cartoon by Charles Keene. Here an agitated woman struggles to leave an oxyhydrogen show, saying to her husband, "Gude save us a', man! What wad come o' us if thae awfu-like brutes was to brek oot o' the watter!!"[37] Although the cartoon primarily mocks her naïveté in taking projected representation for reality—in thinking the "brutes" live on the screen and could swim into the room to attack her—a secondary joke is that she doesn't know they are already in fact all around her.

Like the *water-drop*, the trope of *invisible life surrounds us* could elicit disgust, fear, and horror, even as it suggests that microscopical science may help combat these unseen armies.[38] Deborah Jean Warner and Marina Benjamin argue that the work of Koch and Pasteur helped cause the decline of popular microscopy; lay observers had become squeamish about the disease-carrying organisms that lay hidden in the microscopical realm.[39] The advent of bacteriology likely had some effect on popular microscopy, but oxyhydrogen shows had been showcasing microscopical horrors since the 1830s, and solar microscopes since before then, drawing crowds avid for a glimpse of animalcules. More important, as these examples show, horror and wonder often coincide in the *water-drop* and *invisible life surrounds us* tropes. Despite the alarming possibilities disclosed by microscopical vision, Catlow, Brewster, and Hogg still frame the drop of water as a marvel well worth seeing.

Anne Secord has argued that science exhibitors felt that provoking "involuntary emotional responses" in their audiences was "necessary for stimulating engagement" with the topic, but that such affective flows also ignited a "kind of horror" in bourgeois viewers at their own loss of bodily control.[40] Apart from fostering a horrified fascination that would engross reluctant audiences in these scientific performances, the trope of *invisible life surrounds us* generally eschews the transformation from disgust to curiosity and wonder that is found in the *repulsive revealed to be wonderful*. Instead, it sets up viewers to be more receptive to the author's proposed solution, whether deeper faith (as Brewster urges) or more microscopic research, better housekeeping, or public sanitation. Accordingly, this trope is often paired with more reassuring narratives about microscopical life: objects as jewels, the watchmaker God, the common wonder, and the like.

Hudson's presidential address uses romantic language to describe life in a water-drop, but investigative texts generally are more restrained. They are less likely to use the water-drop trope, even when they are focused on water quality; they stress instead the procedures used to collect, verify, and address the enveloping throng. Arthur Hill Hassall, a physician interested in adulteration, in 1850 published *A Microscopic Examination of the Water Supplied to the Inhabitants of London and the Suburban Districts*, one of a number of such reports over the latter half of the century.[41] Hassall examines "Thames water from Lambeth" and explains, "A few drops examined with the microscope displayed the presence of minute *Infusoria*, *Vibriones*, filaments of *Fungi*, globular colourless bodies resembling ova, and granular and grumous masses, consisting of both dead and living organic matter: it was to the presence of all these that the water owed its colour and opacity."[42] Hassall refers to the contents using passive voice, scientific nomenclature, and precise Latinate descriptors ("granular and grumous masses"). Such a description could not be more different in tone from the whimsical fairyland drops or even Hogg's expansive litany of the waters of the world; the only lovely thing in Hassall's document is the microscope itself, a "beautiful instrument" (a form of *beautiful mechanism*). He argues that the microscope affords greater specificity in public health investigations than other methods. In earlier chemical analyses, for example, "we find, under the heading 'Organic Matter,' the word *traces*." Under the microscope, however, "it will become apparent that these traces are not inconsiderable in amount, that they are complex in organization, endowed with life, and in many cases possessed of active powers of locomotion."[43] It is striking to see Hassall's avoidance of the water-drop trope here. Microscopical texts on sanitation, in fact, despite their overriding concern with the contents of drops of water, rarely use the term, perhaps because of its romantic associations.

Romance Mystified or Demystified?

The romantic tropes described above all require, for their effect, a considerable suspension of disbelief—a willingness to be at least momentarily taken in by these outlandish fictions in the service of some greater microscopical truth. The original phantasmagoria (magic lantern) shows gained force from the complex tension between the deceit and revelation of the apparatus and operator, and the absorbed belief and distanced skepticism of the viewer. In historical phantasmagoria shows, the spectacular visual effect and the rational, mechanical explanation for it are held in balance, neither one negating or necessarily undermining the other, much as with the *explained supernatural* discussed in Chapter 6. This productive tension allows the phantasmagoria to serve as a useful analogue of other screen practices and of cultural forms of mystification. As Susan Zieger explains, "A dynamic alternation between knowingness and absorption structures all such illusionistic culture involving screens."[44] The historical phantasmagoria shows often demonstrated how (some of) these frightening images were made, but critics have seen them as inherently mystifying both as capitalist ventures and as visual spectacles. They have been used by Walter Benjamin, Theodor Adorno, and other modern critics as a figure for the shuttling of real and unreal in modern commodity culture.

Literary critics, following on from Benjamin and Adorno, have also tended to be more interested in calling out the mystification of the phantasmagoria than in its countering movement of demystification. Terry Castle notes, for example, that while these spectacular displays purportedly demystify their illusions, they also immerse viewers in the deceptive operations of vision and technology. And the phantasmagoria do more than just project spectral lightshows; they also reframe illusion itself by exposing the "ghosts of the mind," hallucinated phenomena that undermine the integrity of thought and perception.[45] Jonathan Crary points out that David Brewster developed the kaleidoscope and stereoscope specifically to demystify—by demonstrating—the creative operation of vision. The users of these optical toys and tools experienced the illusion even as they were instructed on its provenance; they were "simultaneously the magician and the deceived." However, he discusses how materialist critics find post-1850 visual culture—in fact, modern capitalism in general—to be "phantasmagoric" not just in producing spectral illusions but perhaps especially in its "effacement or mystification of [its]operation."[46] Although these critics argue that historical phantasmagoria shows both reveal and obscure the workings of their technology, they often emphasize its work of deception, implying that illusion dominates the relation of viewer and apparatus. This collapse of action into illusion powers the symbolic meaning of the phantasmagoria in describing Victorian commodity culture.

Some critics apply these concerns about Victorian visual culture to microscopic vision specifically, again emphasizing the deceptive quality of the microscopical image. Verity Hunt, for example, argues that scientific magic lantern shows (including projection microscopy) produced wonder in a middle ground between outright superstition and extreme skepticism.[47] Isobel Armstrong sees microscopic vision as hallucinatory.[48] Roberta McGrath argues that as viewers "bask[ed] in the fantastic glow" of microscopic vision, "the means of production, mechanical instrumentation and human intervention were all rendered invisible…kept hidden from the non-scientific, wide-eyed public."[49] More recently, Martin Willis defines the phantasmagoria as borne of "a complex interplay of science, technology and the imagination." He argues that the microscope is phantasmagorical in that its technology involves the viewer in an imaginary relation to the real, hiding its true action (its labor) from the consumer.[50] These concerns can apply to performance microscopy: oxyhydrogen operators gesture at the mechanics of practice, but largely as a fig leaf to authorize their spectacular visions.[51] This sleight-of-hand where ostensible instruction slips into deception helps explain how oxyhydrogen microscope shows gripped their viewers.

This symbolic reading has power and reach, but I use the trope of the phantasmagoria more narrowly here. I define the microscopical trope of the phantasmagoria as the use of magnification technologies to engage viewers in Gothic (spectral, grotesque) illusion. Such references do appear in table microscopy, but they are common in projection microscopy, perhaps due to the historical use of the magic lantern to produce phantasmagoria; the projection microscope uses a technology similar to the magic lantern and trades in a similarly mystified visual spectacle. When discussing projection microscopy, Victorians often mention techniques of projection, illusion, and Gothic performance. (However, the water-drop illusion of the projection microscope is illusory only in size; it represents real beings actually present in the room, though visual access to them is mediated through the microscope.) Microscopical texts that evoke the phantasmagoria are often explicitly Gothic: they present mysterious animated beings, images with agency that "should not" have agency and that threaten viewers' own agency.

Victorian table microscopy differs in important ways from the historical phantasmagoria. Tabletop instruments project their images not onto an external screen but directly into viewers' eyes. And the wonder expressed by table microscopists, while it can be Gothic or spectral, often takes other forms, as I have shown above. Additionally, Willis comments that microscopical technology is mystified because it is "unseen (at the moment of viewing)."[52] This was certainly true for observers at the more spectacular public exhibitions. The lecturer might "instruct" the audience about the apparatus but often in terms designed to stun the tyro into silent admiration—for example, by emphasizing the many thousands of diameters of magnification achieved. The working situation for table microscopists was somewhat more complex. While the eye looking down the eyepiece sees only the glowing image surrounded by the black interior of the tube, microscopists were urged to work with the other eye open.[53] This eye would, of course, register the gleaming brass of the instrument and its appurtenances, although a viewer could learn to ignore these images. Further, microscopical operators were trained to be insistently aware of the setting of their apparatus (fine focus, illumination, etc.) during observation and to manipulate it while considering how optical laws affect the system and noting how the image was changed by the adjustment.[54]

Most important, while projection microscopy shows often emphasize Gothic illusion over practical mechanism, table microscopy texts balance illusion and practice with much more of a bias toward practice. Working microscopists, whether addressing neophyte or expert audiences, do present microscopical images as wonderful revelations, using tropes of romantic illusion and wonder. However, they also demonstrate at length how these are crafted through technique (skeptical practice). Their work to demystify the apparatus more than balances out their attempts to foster wonder. Such authors incessantly rehearse the optics and mechanism of the microscope and how to manipulate the apparatus to produce the desired effects. The apparatus, optics, and manipulative technology in table microscopy are anything but hidden. They are a subject of endless exposition, anxiety, and pride, discussed at length and in detail. While Victorian microscopy can be mystified—certainly it was framed as wonderful, rare, and mysterious—it was often demystified at the same time: the instrument, the object, and the observation were relentlessly directed and explained.

It is hard to exaggerate how much attention Victorians gave to explaining the mechanics of microscopical observation and attempting to direct readers' own practice with the microscope. For instance, novices were likely to be disoriented by the use of too-high powers at the outset of an observation. The popularizer J. G. Wood advises, "In examining the objects enumerated, we may be permitted to recommend the novice always to commence the examination with the lowest power of his microscope, and then, where necessary, to proceed with the higher powers. It is well never to commence the examination of an object with a power higher than one inch."[55] And Hogg remarks, "The author has minutely described the use and manipulations of the Microscope, so as to render its management, he ventures to hope, intelligible to all."[56] The effect of this "minutely described" practice can be mind-numbing to the modern reader but reflects Victorian microscopists' determination to explicate microscopical work and thus direct the practice of their readers.

Such passages of relentless explanation frequently coexist in Victorian texts with tropes of romantic wonder like those discussed in this chapter. In 1858 Louisa Lane Clarke, for example, walks young novices through their first encounter with the table instrument (Chapter 17 of this book will explore this form, which I call *articulated practice*).[57] Early in her first chapter, she instructs them,

> Having inclined the microscope at such an angle that you can look through it conveniently, screw on the adapter, and remove the two object glasses D and E; next turn the mirror G about, so as to reflect the light up into the body of the instrument, and illuminate the field of view equally; then place an object on the stage F, and by turning the milled heads H raise or depress the microscope, until the distance between the lower lens and the object on the stage is about one inch. Now look through the instrument, and turn the milled heads H backwards and forwards until the most distinct view of the object is obtained: by adding the other lenses D and E, the magnifying power is increased and the focus shortened, as already explained. The slow motion I is used for the accurate adjustment of the focus, when the high powers are employed; it offers the advantage of much greater delicacy and exactness than is attainable by the use of the milled heads H.[58]

Here Clarke explains, step by step and with reference to an engraving of a microscope, each basic action that is necessary to observe through the instrument. The effect of this passage will differ depending on the position of the reader. It might function as mystification to a passive reader, due to the sheer effect of so many technical details. However, as I show in Part III of this book, many guidebooks were written with the expectation that the reader would consult the book with the microscope at hand. An active reader—sitting at a table with microscope to one side and book to the other, as was recommended—would work their way through these instructions, which serve as a script to direct the reader's movements in preparing the object and manipulating the apparatus.

The book was very likely to reach such readers. Clarke explains that the "object [of her book] was solely to accompany the Object-box and explain its contents; the idea for the book was suggested to her by an optician whose customers, buying a microscope and some slides, would "perplex" him with questions.[59] Clarke's little book thus serves as a resource for perplexed opticians and novice microscopists alike. And that concentrated passage of instruction leads readers into what she calls, just a few pages on, the "glorious edifice" of a plant, with its "tesselated pavement" of the flower cells and "romantic caves, or microscopic grottos" in the leaf. Clarke blends these instances of the trope of *landscape* seamlessly with references to a famed botanical investigator (Franz Unger) and to scientific terms (stomata) and theories (cell theory).[60] Clarke's use of *the image as landscape* and *Oriental tale* ("romantic caves, or microscopic grottos") solicits wonder from novice observers even as she guides their fumbling efforts with the unfamiliar apparatus and language of the microscope. Her intermingling of wonder and skepticism here—her address to both affective interactions and optical, cognitive, and mechanical ones—exemplifies how, even with the most flamboyant romantic tropes, these contrasting approaches can work together to express and direct Victorian observers' embodied entanglement with the microscope and the object.

Anthropomorphism

The romances and fantasias detailed above exemplify better than any other genre how Victorian microscopical writers may anthropomorphize the tiny creatures they study. The pages of nineteenth-century British microscopy sometimes seem saturated with *anthropomorphism* (the ascription of human attributes to non-human beings), perhaps due to its strong roots in the natural history tradition. Some of the earliest nineteenth-century texts on the subject engage in this practice, perhaps as a means of attracting readers' curiosity. When C. R. Goring writes in 1830, in one of the earliest and most influential microscopical texts of nineteenth-century Britain, he not only compares them to "a menagerie of living wild beasts," but also notes their "thousand diverting pranks and humours, quite peculiar to themselves."[61] Goring's phrasing invites readers into the microscopical world with both a sensational emphasis on these objects' bestial "ferocity" and a tantalizing hint of their human-like and individualized "pranks and humours." Such anthropomorphic analogies continued throughout the century, especially in (but not limited to) texts addressed to novices or broad audiences. Writers perhaps hoped that this curious notion—that humans and microscopical objects think and feel in similar ways—could help with community-building on the human scale as well.

Anthropomorphism became a more contested strategy across the century. It was criticized as a romantic fancy, too mired in ignorant or childlike sentimentality to be suitable for scientific settings; but Victorian investigators did not universally condemn it. George Henry Lewes and F. Max Müller, for example, cautioned against it; but Charles Darwin and George Romanes argued that such analogies serve as a useful recognition of traits shared by human and non-human animals—that they point to a common ancestry.[62] Anthropomorphism originally referred to the misattribution of human qualities to the gods; Lewes was likely the first to suggest that this could apply to our descriptions of animals as well. He was also among the first to flag it as inappropriate in scientific work.

However, Lewes did so, ironically, in writing for a broad audience in *Sea-side Studies* (1858), a popularizing series of essays on marine life first published in *Blackwood's Edinburgh Magazine*, one of several texts driving the mid-1850s fad for beachcombing. Discussing the behavior of the sea-anemone, Lewes lectures his readers that "as we are just now looking with scientific seriousness at our animals, we will discard all anthropomorphic interpretations, such as point to 'alarm.'"[63] (He would surely have rejected Badcock's description of the fairy tree folding up in "alarm" after the observer's disturbance of it, perhaps attributing this to an instinctive response instead.) Similarly, Lewes breezily dismisses the notion that molluscs see through their eyes, saying, "we speak with large latitude of anthropomorphism when we speak of the 'vision' of these animals." And in describing the razor clam (which he calls "Razor-fish"), he notes that "the Solen can have only slender pretensions to mental vigour of any kind," and continues, "Indeed, we are incessantly at fault in our tendency to anthropomorphise, a tendency which causes us to interpret the actions of animals according to the analogies of human nature."[64] This tendency derives from overweighting the apparent similarities between animal actions and human ones, ignoring significant differences in structure that point to concomitant differences in function. Thus, he reminds readers, many actions are instinctual or involuntary; "there may [even] be Sensibility without nerves, and Contractility without muscles."[65] Lewes argues that writers should not just refrain from depicting animals as possessed of human manners but even resist attributing will, reason, or emotion to them.

As others have noted, Lewes lays out this restriction in the same pages where he also insistently anthropomorphizes the creatures he studies. That reprimand about molluscs' vision follows five pages in which he repeatedly refers to the Solen as "he" and depicts him in human terms as one who may "suddenly change his mind," perhaps "poking up his inquisitive person" or "defiantly throwing [his] head at our feet."[66] (This last phrase, referring to the ready detachment of portions of the body, suggests both the clam's willful resistance to collectors and his loss or lack of mental capacity.) Even the reference to the clam's "slender pretensions to mental vigour" demonstrates the drily mocking tone Lewes often uses in such writings to romanticize marine creatures as an assortment of quirky personalities.

Both Lewes's decree and its ironic self-violation have been well canvassed by scholars. But what is not as well appreciated is that he extends this prohibition to microscopic creatures, who are even more different from humans in both structure and action—and that he anthropomorphizes them as well. Elaborating on the reasons to resist anthropomorphism, he explains,

> We do not, except in loose latitude of phrase, speak of the *legs* of an animalcule, meaning its organs of Progression... nor should we, otherwise than in easy speech, talk of the Digestion of a Polype, meaning thereby its Nutrition. The *purpose* of a leg, progression, is fulfilled by the cilia, which move the animalcule; the *purpose* of Digestion, preparation of food, is performed by the cavity of the Polype; but the specific organs named legs and alimentary canal, and the specific functions of those organs (Walking and Digestion), are in both cases absent.[67]

It should be noted that Lewes's prohibition on anthropomorphism is not absolute, as is evident here in his throwaway phrase "otherwise than in easy speech." He suggests, that is, that context dictates whether or not anthropomorphism is an acceptable discourse. Within a single text, then—in the pages of *Sea-side Studies*, for example—a writer may veer into and out of stages of argument where fanciful analogies to human forms and behaviors are more or less constrained.

Lewes claims to restrict such fancies to "the higher organisms," saying of "the lower organisms" (molluscs, for example) that "although they have their manners and their morals, [they] are too far removed from us to be intelligible."[68] But even as he disavows anthropomorphism with "the lower organisms" in this way, he engages in it; his popular writing offers many examples in discussing infusoria and other microscopical objects. For example, he conducted an experiment on a fragment of filament from a sea-anemone, which he describes in terms that repeatedly blur the boundary between human, animal, and (possibly) dead matter. He says, "I was examining one of those white filaments which certain *Actiniae* copiously throw out when disturbed," when he noticed something very odd:

> [T]his filament... although it had long been removed from the animal, was twisting and twirling itself like a worm in an unhappy state of mind, and moving across the stage with motions it was impossible to distinguish from voluntary motions. I then crushed it into many minute fragments; but long afterwards I observed some of these moving about like so many animalcules. Another day I observed what seemed a tiny white annelid crawling at the bottom of a vase; on securing it, I found it was one of the Actinia's filaments.[69]

In this passage Lewes conflates a long-severed bit of sea-anemone, a worm, an "animalcule," and a human (given the anthropomorphic implications of the object's "unhappy state of mind"). The observation also blurs the distinction between what is animate and inanimate.

Anthropomorphism like this may persist in microscopical texts because it functions somewhat like James Elkins' concept of *visual desperation* (see Chapter 6). Microscopical beings are so strange and alienating that perplexed and unsettled observers seek analogies not just to familiar forms but also to familiar behaviors, in order to make sense of what they perceive under the lens. By presenting non-human beings as if human, anthropomorphism encourages the observer to form or perceive an affective bond with the object—and thus with microscopical practice as well.

While anthropomorphic narratives are not uncommon in Victorian microscopical writing, I must stress that they are but one element of observers' entangled embodiment with their objects of study. Such narratives are characteristic of but not necessary to Victorian microscopists' understanding of the dynamic interrelations between observer, apparatus, and object. Significant forms of response exist even outside the anthropomorphic realms of will, reason, emotion, and other quintessentially human behaviors. After the botanist Robert Brown's 1827 work demonstrating that tiny inanimate particles will move about randomly when in suspension (Brownian motion), microscopists had learned to be cautious about attributing life to inanimate objects; they likewise were urged to take care in presuming will or even agency when observing changes in microscopical objects. However, they do not reject the concept of responsiveness overall. They generally accept the notion that the microscopist's actions in macroscopic space materially affect and change what is observed in microscopic space, whether simply through optical and physical changes or in cognitive and affective ones as well; prompting changes in their own engagement and practice in return.

Lewes, as a worker in physiological psychology, a popularizer of science, and a self-designated spokesman for an empirical outlook, was an outspoken skeptic on anthropomorphism, as shown above; yet the very culture of experiment that he endorses relies upon this interpenetration of macroscopic and microscopic actions by operator and object. He warns readers against mistaking involuntary for voluntary actions, and against imagining *will* where none properly exists; but he repeatedly describes how organisms respond to environmental stimuli, including the actions of a microscopical operator. Recalling an experiment he conducted to assess the sting of a tiny Hydra (a freshwater cnidarian), he reports,

> I removed a Hydra from the phial, in a little water, and placing it on a slip of glass, allowed it to settle and expand there for two hours, when I added several water-fleas (*Cypridae*) to the little pond.... At length one was caught, and held

> for some seconds; it then fell to the bottom, and remained motionless for at least two minutes, after which it started up, and was off as if its course had never been arrested. This certainly had very much the appearance of a case of slight paralysis; the animal seemed arrested by some benumbing influence, which for two minutes rendered it powerless; at the expiration of that time it seemed to have sufficiently recovered itself to swim away.... With a needle I gently arrested one of these water-fleas; it suddenly sank motionless, remained thus for more than a minute, and then darted off again. Thrice I repeated this act, and each time with similar result.[70]

Here the operator intervenes in the scene by four times "arresting" the water-flea as the Hydra does, "each time with similar result." It is not in question that the tiny creature acts in response to him, just as it reacts to the Hydra's touch. The water-flea was not paralyzed in either case, but Lewes leaves open the possibility that its pseudo-paralysis could be an involuntary or instinctive action.

However, eager to make the case for his hypothesis, Lewes concludes by turning to an unmistakably anthropomorphic analogy, where the very stillness of the little creature represents a form of human-like activity. He says, "And does not the reader at once recognise in this sudden motionlessness of the animal a very familiar phenomenon? The spider, the crab, the oniscus [pillbug], and very many animals 'sham dead,' as schoolboys know, when danger threatens; these water-fleas 'sham dead' when the Polype or the needle touches them."[71] Lewes's placement of quotation marks around "sham dead" likely signal his conviction that the polyp is incapable of *real* shamming—deliberate, playful pretense—as a human might do. Lewes himself has been shamming insofar as his pricking with the needle imitates the Hydra's sting. The value of his analogy between the water-flea and the spider, crab, pillbug, and polyp is evident; those present what he implies (via the scare quotes) is an involuntary physical response (a mock shamming), deceiving us twice in our tendency, first, to think they are dead; then that they possess the mental capacity to feign death. His use of an anthropomorphic reference here thus sets up an anti-anthropomorphic argument. Of course, this analogy to "shamming dead" likely calls out childhood memories in many readers, setting up a compelling anthropomorphic story (a slyly deceptive animal, a gullible observer) that may take root.

In sum, the nineteenth-century debates over anthropomorphism extended to discussions of microscopical creatures, which seem to present quirks and whims not unlike our own. But whether or not Victorian observers attribute will or reason or emotion or even consciousness to the living beings under the lens, writers of the period do not question that those beings predictably and materially respond to the microscopist's interventions with them or their environment. Some actions may provoke dramatic consequences; others, not so much. J. G. Wood, for example, humorously notes in *Common Objects of the*

Microscope (1861) that infusoria "may be dried over and over again, without seeming to care anything about the matter."[72] But regarding the collection of infusoria, Wood counsels his readers that the best technique is "placing the inverted bottle delicately near the intended victim, and then quietly turning it over without alarming the easily frightened creature." This instruction implies some anthropomorphic concern for the wellbeing of the little creatures, but more likely it simply supports a very practical aim, preventing them from scattering when disturbed.[73] Indeed, Victorian microscopists regularly record how even inanimate substances react to physical interventions like light, cutting, reagents, staining, or pressure. In many cases the worker then responds in turn (as Lewes and Wood do), adjusting the instrument or intervention, and forming a chain of interlocking actions between operator and object, executed through the instrument and apparatus of microscopy.

Writing Embodiment in Victorian Microscopy: Beautiful Mechanism. Meegan Kennedy, Oxford University Press.
 DOI: 10.1093/9780198940623.003.0010

11
The Limits, Pleasures, and Persistence of Wonder

> Nature is a free domain; and the profound conceptions and enjoyments she awakens within us can only be vividly delineated by thought clothed in exalted forms of speech, worthy of bearing witness to the majesty and greatness of the creation.
>
> Alexander von Humboldt, *Kosmos*, 1845

Introduction

Tropes of microscopical beauty and wonder would seem to run counter to the tradition of "plain speech" set forth as an investigative ideal by the Royal Society in the seventeenth century. Thomas Sprat, in his 1667 *History of the Royal Society*, notes that the members of the Royal Society are committed to disciplining "the manner of their *Discourse*." He warns that a florid style amplifies not reason but "its Slaves, *the Passions*." He deplores the "specious *Tropes* and *Figures*" blotting English knowledge and vows to reveal "the shallowness of all these seeming Mysteries…this vicious abundance of *Phrase*, this trick of *Metaphors*, this volubility of *Tongue*." And he praises the Royal Society for vowing "to reject all the amplifications, digressions, and swellings of style: to return back to the primitive purity, and shortness, when men deliver'd so many *things*, almost in an equal number of *words*"—adopting a "close, naked, natural way of speaking…bringing all things as near the Mathematical plainness, as they can: and preferring the language of Artizans, Countrymen, and Merchants, before that, of Wits, or Scholars." Crucially, "the *Experimenter*" relies on empirical means and "indevors rather to know, than to admire; and looks upon *admiration*, not as the end, but the imperfection of our *knowledge*."[1] Despite Sprat's own often-admiring prose, he clearly wishes to set a limit on the use of wonder in experimental philosophy.

In a similar vein, the eighteenth-century empiricist philosopher David Hume, in his *Enquiry concerning Human Understanding*, found wonder inappropriate in the pursuit of knowledge. He explains, "A correct *Judgment*…confines itself to common life, and to such subjects as fall under daily practice and experience; leaving the more sublime topics to the embellishment of poets and orators, or to the arts of priests and politicians."[2] Hume thus discourages florid content or style for philosophers committed to empirical work. It was during this period, as Lorraine Daston and Katharine Park argue, that "marvels…become vulgar" (that is, common, demotic).[3]

These trends are generally agreed to have accelerated during the nineteenth century, perhaps urged on by William Whewell's insistence in 1840 that "Facts, for the purposes of material science, must involve Conceptions of the Intellect only, and not Emotions." (This dictum did not prevent him from openly admiring the "exquisitely beautiful mechanism of the eye," presumably because his philosophy would not consider that observation as a scientific "fact.")[4] Investigators' texts do not necessarily exclude all wonder, but they increasingly segregate statements of wonder from those of fact.

Alan Gross, Joseph Harmon, and Michael Reidy argue that nineteenth-century scientific writers generally move away from the "tolerance for emotional expression" that characterizes seventeenth- and eighteenth-century scientific prose. Scientific writing from the seventeenth through the twentieth centuries increasingly "disdains emotional or overtly literary expression" (as they put it).[5] Daston and Peter Galison argue that by the nineteenth century not just wonder but also sensibility and judgment were (temporarily) discouraged; investigators were encouraged to strive for a mechanical objectivity. Such trends emerged patchily, of course; medical case histories, for example, continued to explore unusual biological phenomena well into the Victorian era, often just downshifting their language from what was "marvelous" or "curious" to what was "singular."[6] But by the end of the century, H. G. Wells argued that even popularizers should avoid the "vulgar wonders of the 'millions and millions and millions' type," publishing instead "problems for [readers] to exercise their minds upon."[7]

Although I have been, so far, emphasizing the embodied, affective flows that entrain the observer in preferred forms of microscopical wonder, Victorian microscopy was, of course, shaped by these movements toward plain speech and objectivity. The surge in nineteenth-century microscopy chimes with a remarkable series of developments in scientific practice and scientific language. It spans from a time when the word "scientist" itself was new and little used, to a period when universities boasted laboratories and the government funded science and science education directly.[8]

But Victorian microscopists exemplify the commitment to plain speech only unevenly. Even working microscopists did not agree on professional norms, nor on the proper work of the instrument or its societies. Many of them would have identified as naturalists, botanists, entomologists, or physicians (for instance) more than microscopists. Victorian microscopy developed out of a thriving tradition of natural history and flourished in overlapping communities, some scientific and others better described as social, educational, theological, or domestic. These communities did not all share the same approach to microscopical wonder; but they all variously use tropes of wonder to describe microscopical observation as an embodied practice, and to shape observers' affective experiences of that practice.

As Anne Secord has shown, influential early Victorian naturalists and publishers were concerned about how learners were affected by wonder, specifically by beauty. They wished not to eradicate students' pleasure in viewing appealing images but to direct it in specific ways—to "relocate the sites of pleasure" toward "rational pursuits and philosophical contemplation rather than the passions and sensuality."[9] The physician C. R. Goring similarly warned that microscopical beauty could seduce observers into a purely affective, undiscriminating response. In *The Natural History of Several New, Popular, and Diverting Living Objects for the Microscope* (1829), the first of a series of influential books he and the optician Andrew Pritchard published at the outset of Victorian microscopy, Goring both emphasizes microscopic marvels (as is evident in the title) and warns against misusing them. He argues, "Great disgrace has been brought on microscopic science by the manner in which it has been perverted to the support of preconceived opinions and hypothetical views, as well as to a spirit of wonder-making." He calls for microscopists to usher in "a new and golden age of observation," where workers would be rigorous in their practice.[10]

Surprisingly, given these admonitions, Goring poses wonder and rational knowledge as necessary partners. He suggests that the microscope helps the observer move past wonder to understanding, as in his discussion of the transformation of the mayfly (*Ephemera marginata*): "we are astonished at the apparently sudden change...but when, by the assistance of the microscope on these transparent objects, we are enabled to detect their gradual formation in disguise, a great portion of our wonder vanishes."[11] That is, despite Goring's strong words of disapproval regarding the "wonder-making" microscopists of the day and his efforts to demystify the objects he discusses, he does not eschew wonder altogether but sees it as an important steppingstone to knowledge. Even in examining the mayfly, he recalls his observation of the mosquito [*Culex*], where he noted that "the transformation of this animal from the larva to the pupa is one of the most singular and wonderful changes that can be conceived." He exclaims,

> The circuitous manner in which the Creator appears to [form] this species of gnat, and many other of His smaller productions, is truly wonderful....They may be deemed by Him the choicest and most exquisite of His productions. These mysterious, creative operations of nature, as detected and unravelled by microscopes, are surely GRAND AND CAPITAL SUBJECTS FOR OBSERVATION. I should pity the spirit of the man who scorned to be amused by inspecting these MARVELLOUS METAMORPHOSES.[12]

As Goring suggests, then, the act of wonder is valuable but must not stand alone; the observer must both appreciate and "inspect" the microscopical image, partnering with the instrument's agency as it "detects and unravels" these mysteries. Wonderful sights remain important to every nineteenth-century microscopist, but the observer should consider wonder as only one element of the path of knowledge.

Mid-century writers echo this concern. In *The Microscope*, an 1858 guidebook for novices, Louisa Lane Clarke mocks the flurry of wondering metaphor voiced by undiscriminating observers. She says,

> We not unfrequently hear the section of an Echinus spine pronounced "very pretty, exactly like a crochet pattern"—the Echinus itself being an unknown thing. Spicules of Holothuria or Gorgonia are brilliant little clubs and crosses; but what a Holothuria is they cannot imagine. The foot of a Dytiscus, with its cluster of suckers, is like the eye of a peacock's feather; cells of spiral fibre nothing more than coils of variegated wire; and the head of Rhingia, with its wonderful eyes, is looked at as a beautiful piece of network.[13]

Clarke here imagines a lay observer—perhaps female, given the attention to "crochet" and "network"—distracted by beauty and failing to learn. Despite this critique, her book is saturated in tropes of beauty, with as many references to lovely or delicate form as there are pages. Clarke emphasizes the beauty of objects and uses inventive visual metaphors to characterize them. But like Goring she criticizes the novice's tendency to focus on beauty and to form analogies for each exquisite or curious form without seeking knowledge about the thing itself. In the scene she sketches out, the wonderful microscopic world is reduced to a cabinet jumbled with unrelated objects: a crochet-pattern, clubs, a peacock's feather, coils of wire, and a piece of netting. Like Goring, Clarke advises students not just to acknowledge beauty but to use it as an entry into greater understanding. She urges her readers to consult "better books" for further reading and observation.[14]

However, contrary to Goring's suggestion that beauty invites better knowledge, Clarke worries that novices' ignorance limits them to a merely aesthetic engagement with the slides. She explains, "If we are wholly ignorant of the structure of plants, their uses, their variety, and the secret mechanism by which their life is renewed day by day, we are apt to look at these slides for mere amusement, for the lust of the eye, pleased as a child or as a savage with strange forms or brilliant colours." Ignorance of structure renders observers likely to see only beauty in the object. Clarke depicts this approach harshly, as an impoverishment or perversion of observation ("lust of the eye") and a sign of immaturity or even barbarism ("pleased as a child or a savage"). It is not simply the appreciation of "strange forms or brilliant colours" that primarily concerns her; indeed, Clarke's own text many times remarks on some "strange" or "brilliant" image. Rather, she wishes observers to educate themselves *before* or *while* observing. For example, she explains of plants that "few [objects] lose more in being hastily looked at, as merely pretty objects, without that knowledge of flower-life which alone can enable us rightly to appreciate them."[15] She suggests that, while the sight of beautiful objects can spur readers to take up microscopy, one must study structure and function in order to truly appreciate the beauty of those objects. Clarke's stricter interpretation of beauty's role may reflect her historical moment. Where Goring primarily needed to reintroduce readers to microscopy in order to convince them to buy his new serial, Clarke wrote at a time when interest in microscopy was surging and workers were becoming increasingly concerned about raising standards of practice.

It is tempting to declare that later microscopists eliminated tropes of beauty and wonder in order to discipline the observer's gaze. Especially among investigators, these tropes do become less frequent later in the century, and—as I have noted in earlier chapters—generally restrained in rhetoric and textual placement. As the microscope became increasingly a professional scientific tool, research texts became more restricted, even formulaic, in style and content. Some workers even voiced scorn for slides prepared as exhibition objects, designed for maximum visual effect. The longtime President of the Reading Microscopical Society, Frederick H. Lang (a retired Army captain), wrote in the *Monthly Microscopical Journal* (*MMJ*) in 1870 about these. He notes dismissively that "Selected diatoms of the more showy sort are occasionally to be obtained at the opticians, even arranged in patterns as pretty 'Oh law' objects for the microscope." An *"Oh law" object* is, then, any obviously pretty or striking object suitable for exhibition, at the sight of which visiting ladies would presumably exclaim, "Oh law!" The trope is not common in the microscopical record, but the idea, and Lang's mockery, would have been instantly recognizable to any working microscopist. Typically, however, he also solicits admiration of diatoms as "these exquisitely delicate and beautiful forms"—he was, in fact, a diatomist, who dedicated his studies to these quintessential *oh law objects*.[16] His example provides a useful reminder that investigators do not entirely abandon romantic tropes of beauty and wonder even late in the century, and even when they overtly declare their opposition to wonder in microscopy.

A. D. Michael, in his 1895 presidential address to the Royal Microscopical Society (RMS), tells an anecdote about Dr. James Scott Bowerbank that clarifies working microscopists' complex approach to beauty. Bowerbank was known for his "Monday evenings"—weekly open houses (1844–64) for aspiring microscopists and fellow workers, whom he called his "Band of Brothers." These "evenings" influentially helped build the Victorian microscopical community during its early decades, but Bowerbank did set some limits on these gatherings. He reportedly followed a simple determinant upon welcoming a new guest. Michael explains,

> Bowerbank kept a small collection of showy and striking slides in a special box which he called his "goodness gracious box,"...if the visitor asked some

> intelligent question about structure the host was delighted to go on showing and explaining; but if the first remarks were unappreciative Bowerbank would hand him over to some one else to exhibit "the beautiful things in that box."[17]

The trope of the *goodness gracious box* echoes the dismissive tone of Lang's *"Oh law" objects,* although unlike Lang's phrase it does seem to have been used by other writers. It similarly dismisses overtly beautiful objects as trinkets for ignorant or silly visitors (again suggesting ladies in this role) and implies a distrust of sheer beauty.

Despite these concerns, microscopists did not drive out wonder from their practice. Bowerbank does find beauty valuable, as a spur to learning about structure (as Goring suggests) and as a means of winnowing out pretenders, those who show little interest in learning. Michael, speaking three decades after the end of Bowerbank's "evenings," tells this anecdote in order to push this premise further, into an appreciation of beauty for itself. Michael continues,

> [T]he two classes into which [Bowerbank] divided slide-seeing humanity still exist, the lovers of structure and the lovers of the "goodness gracious box" are with us to-day; personally I do not see why the latter should not get all the pleasure they can from the face of Nature. When Carpenter wrote somewhat contemptuously of "mere beauty," he...missed one of the most charming aspects of Nature....It seems to me that he who can admire both the structure and the beauty to the eye is the truest student of Nature and derives most pleasure from the study.[18]

Michael uses William Benjamin Carpenter here as a bit of a straw man. I have not been able to identify the source of the deprecating remark about "mere beauty," but Carpenter not uncommonly makes remarks that distinguish beauty from scientific value while still marking beauty as valuable in itself. For example, regarding "the stems and roots of Plants," Carpenter comments, "sections of these...are objects of peculiar beauty, independently of the scientific information which they afford."[19] Carpenter does, however, offer a caveat; microscopists should recognize beauty and wonder as an enjoyable element of microscopy but not allow it to direct their study—because "the operations of Creative Wisdom" are "not less wonderfully displayed among the forms of Animal life which are accounted the simplest and least attractive, than in those which more conspicuously solicit the attention of the Student of Nature, by the beauty of their aspect or the elaborateness of their organization." He suggests that students not limit their observations to beautiful objects but rather choose to follow out some methodical work. "If those Microscopists who spend their time in desultory observation, and in looking at some favourite objects over and over again," he says, "would but concentrate their attention upon some particular species or group, and work-out its entire history with patience and determination," there would be "infinite advantage to Science."[20] For Carpenter, beauty should be an acknowledged but not a limiting factor in microscopical observation.

Michael's own position in telling the story of the "goodness gracious box" is to push back against Bowerbank's use of beauty as a simply useful quality—to defend the right of beauty-lovers to simply "get all the pleasure they can from the face of Nature." Michael endorses the love of beauty as equal in value to the love of structure; he lauds the observer who can combine these as "the truest student of Nature." Even as Michael serves as an outlier in defending beauty for its own sake, he is aligned with his colleagues in discouraging thoughtless wonder.

Beauty in some cases actually becomes central to microscopical practice. In 1883, Albert McCalla defends the art of beautiful staining, saying, "Some look on staining as only a refinement of dilettantism, a thing of mere looks, like coloured varnish rings and ornamental labels. But these staining fluids, as this use is now developed, differentiate the various tissues from one another, and are a most invaluable help to exact knowledge." He turns to literary analogies to emphasize this point about the importance of beauty as a skeptical tool. He remarks,

> As the presents of sword and spear and shield, offered, along with the jewelry and costly robes, to the daughters of Lycomedes, by the crafty Ulysses, in the old Homeric story, served to discover the young Achilles in spite of his womanly disguise, so do these chemical staining fluids serve to disclose to us by their selective power the different tissues and organs in substances otherwise alike transparent and invisible.[21]

This defense is a little confused—Achilles selects *away* from beauty, choosing arms over jewelry, which is what unmasks him as a man—but McCalla's rolling, periodic prose is effective. He here describes the compelling selective power of microscopical stains by comparing their action to a dramatic scene drawn from the classical myth "Achilles on Skyros," which had long been popular in visual art and opera as an occasion for meditations on hidden or true nature—very appropriate for his subject matter. In both cases, a form of beauty serves as a tool aiding disclosure.

McCalla makes this argument in his presidential address to the American Society of Microscopists, reprinted in the *Journal of the Royal Microscopical Society* (*JRMS*), a genre that can be unusually tolerant of wonder and figuration; but similar claims appear in research articles also. Arthur C. Cole, a developer of microscope apparatus and specialized procedures, in an 1895 monograph directs readers in how to prepare specimens. He argues that the presence of beauty and wonder provides helpful feedback about the quality of that preparation. In well-made slides, he says, "beyond mere beauty (which, in visible nature, is inexhaustible) there are revelations of structural form, quaint elegancies, mysterious changes of tissues...under radiances." Here the preparer's actions upon the object and the object's actions upon the preparer reverberate upon each other. Cole recommends that the microscopist prepare "the head, and surrounding parts of an insect" so it "discloses a combination of form and colour of surprising beauty...[with] the most exquisite tints...bathed in refulgence."[22] Such florid expressions of wonder can strike a modern reader as excessive or even risible, but the text holds no suggestion that would support such a reading. Rather, Cole points to this beauty as a technical marker indicating that the preparation has been properly executed. Tropes of beauty and wonder thus continue to appear throughout the century and across diverse genres. They are potent as channels of affective response, as an encouragement to further study, and as aids to skeptical practice. And, as I show in Part IV of this book, Victorian microscopists experimented with many different ways to integrate wonder and skepticism in their writing.

The Pleasures and Persistence of Wonder

> I still recall with pleasure the stimulating excursions that I made fifty years ago...with my teacher Ehrenberg, and a few other pupils....Equipped with fine nets and small glasses, we fished in the ponds of the Zoological Gardens and in the Spree [in Berlin], and caught thousands of invisible micro-organisms, which then richly rewarded our curiosity by the beautiful forms and mysterious movements they disclosed under the microscope.
>
> Ernst Haeckel, "The Controversy over the Soul," 1905

In the epigraph above, the eminent zoologist Ernst Haeckel recalls his early experience of wonder and awe at the beauty, mystery, and romance of microscopic worlds. Many Victorian microscopists valued such experiences and sought to induce them in readers; tropes of wonder shape texts for both working microscopists and general audiences throughout the nineteenth century. But why did Victorian microscopy leave a place—albeit a limited place—for mystery and wonder, in a historical period known for its rapid modernization, crisis of faith, literary realism, industrious professionalization, and materialism? One reason might be that, like sciences such as botany, microscopy grew out of a natural history practice strongly shaped by natural theology. Natural theology writers often turned to microscopy for examples of design and of God's infinite power and reach; the two practices (natural history and natural theology) mutually reinforced each other, and microscopy provided important evidence for both. As noted in Chapter 3, despite the temptation to acquire vast amounts of gear, research in microscopy demanded relatively little space and expense compared to, for example, astronomy. It was thus well suited to the parson-naturalists who would be likely to endorse a tradition of microscopical wonder. Theologies of nature appear in popular microscopy through the end of the century. Charles Kingsley and Philip Henry Gosse, for example, were known not only for their popular microscopical texts but also for their use of discourses of design. Many of the tropes of wonder examined here linger as echoes of the natural theology tradition, although some—like beauty—serve secular purposes equally well.

Thomas Henry Huxley's scientific naturalism was the strand of British science approaching closest to the methodological ideals of a self-restrained, secular, mechanical objectivity as described by Daston and Galison. Huxley and his colleagues argued for empirically derived, material, and natural (not supernatural) explanations for phenomena.[23] They fought to secularize Victorian intellectual work and build a professional culture of expertise, forming a dining club (the X-Club) dedicated in part to these goals. While the scientific naturalists gained influence in British science and education during the second half of the century, they still shared the wider ecology of the Victorian study of nature with other groups. At mid-century their views—congenial to the transmutation of species and a physiological explanation of mind, among other controversial theories—were decried as rank materialism and associated with atheism and class unrest. Some British naturalists vocally opposed them as a result.[24] However, as Peter Harrison notes, "By the end of the century, there was an almost universal, if tacit, understanding that the term 'science' excluded the aesthetic, ethical, and theological."[25] Scientific naturalism was an important force in the development of a twentieth-century research culture that expects a restricted rhetorical range in published research reports; research articles today in scientific journals are unlikely to use the descriptor "beautiful" or other tropes of wonder even in passing.

It would be reasonable to expect that the tenets of scientific naturalism might inflect Victorian microscopical culture. Many of the scientific naturalists were accomplished microscopists. Huxley engaged directly with communities dedicated specifically to microscopy. He became a member of the council of the Microscopical Society of London (MSL) in 1857 and published two articles in the Society's journal, and he served as President of the Quekett Microscopical Club (QMC) from 1877 to 1879.[26] However, Huxley does not often appear as an author or discussant in reports of Society meetings. Instead, he and his colleagues often presented their work with other audiences—in broader conversations, as with the Royal Society or British Association, or narrower ones, as with societies and journals centered on particular sciences. George Busk, a naval surgeon, zoologist, physiologist, paleontologist, and ethnologist, was an X-Clubber with Huxley. He was an authority on the *Polyzoa* (tiny marine invertebrates, now *Bryozoa*) and served as MSL President in 1848–9. British microscopists were familiar with the work of these men and other scientific naturalists, in part due to their shared participation in other societies and journals. For instance, physiological psychologists like Carpenter and George Henry Lewes adopted the language of the conservation of energy, one of the tenets of scientific naturalism.

I must emphasize that, despite their desire to anchor scientific work in empirical observation and experiment, scientific naturalists were not necessarily hostile to beauty and wonder, even in their own work. Many scholars have shown that, despite their secular and materialist aims, these investigators engaged with spiritual and aesthetic structures of thought. For example, Adrian Desmond terms Huxley's philosophy a "reverent agnosticism" or "reverent naturalism"; and Bernard Lightman notes that Huxley "perceived himself as having a deeply religious sensibility" in counterpoise with his skepticism.[27] Indeed, although Huxley advocated for the separation of theological and scientific spheres, he can be found using tropes of wonder in his microscopical writing. In 1851, delivering one of his two MSL papers, he describes the rotifer *Lacinularia* using familiar tropes of wonder such as *beauty*, *analogy*, and the *object as jewel* to express and presumably solicit admiration. For instance, he notes that "its edges [are] richly beset with large cilia, which present a very beautiful wheel-like movement."[28] Much later, in the Preface to a popular 1880 guide to zoology, he approvingly quotes an eighteenth-century entomologist: "'Common and lowly as most may think the crayfish,' well says Roesel von Rosenhof, 'it is yet so full of wonders that the greatest naturalist may be puzzled to give a clear account of it.'"[29] Evidently there was still a place for wonder in Huxley's descriptions of the natural world. Similarly, Charles Darwin's work, as George Levine has shown, is suffused with wonder.[30]

Other scientific naturalists maintained other links to wonder. Michael Reidy has studied the interest in mountaineering among the scientific naturalists as both a kind of laboratory and a quest for the sublime.[31] The physicist John Tyndall, an avid mountaineer, was a vocal advocate for this multivalent approach to intellectual work. In his "Address to the Students of University College, London" in 1869, he commented that "even the intellectual action of a complete man is, consciously or unconsciously, sustained by an under current of the emotions," and warned, "It is vain, I think, to attempt to separate moral and emotional nature from intellectual nature." He suggested that "round about the intellect sweeps the horizon of emotions" and urged his listeners to "keep this horizon open."[32] Tyndall here frames scientific work as a conjunction of moral, emotional, and intellectual experience in constant interrelation, supporting and adjusting one another. He contended that "the vocation of the true experimentalist may be defined as the continued exercise of spiritual insight, and its incessant correction and realization."[33] Indeed, Ruth Barton has argued that, although Tyndall gave up organized religion, he did not abandon spirituality, turning instead to a Romantic idealism. And Lightman has shown how, despite framing science and religion in oppositional terms in his famous Belfast Address of 1874, Tyndall—like Huxley and Spencer—borrowed Christian structures of thought in his scientific work.[34]

Overall, as Lightman has argued, Tyndall concluded that "ontological materialism was contradicted by the most recent research in physiological psychology."[35] While scientific naturalists advocated for a professional, secular, and empirical Victorian science, Tyndall and his colleagues leavened their materialism in practice.

A continued tolerance for wonder, if evident among the self-proclaimed scientific naturalists, is even more pronounced in the leaders of the elite British microscopical community, the MSL (later the RMS). Busk aside, most of the Victorian-era Presidents of the MSL/RMS resisted a strictly secular and material empiricism in diverse ways. Many of them represented medical or faith communities (both Anglican and Dissenting) that complicated Huxley's attempts to build a scientific tradition around secular biology as a rival center of authority.[36] The MSL/RMS thus developed a culture that, while dedicated to microscopical science, remained open to the judicious use of wonder.

A brief review of some of the most active Victorian-era Presidents of the Society shows how influential British microscopists variously preserved a role for beauty, mystery, and wonder in their lives and often in their practice as well. Richard Owen, when he served as the first President of the MSL (1840–1), was Hunterian Professor of comparative anatomy at the Royal College of Surgeons and turning toward German idealist morphology.[37] Owen constructed a theory of development based on a divine archetype under which humans enjoyed unique status. His battles with Huxley are well known. Thomas Bell, a dental surgeon (President of the MSL in 1844–5), later in life spent fifteen years producing an edition of the *Natural History and Antiquities of Selborne* (by the parson-naturalist Gilbert White). This text, first published in 1789, is notable for its deeply observed portrait of a locality; it is a classic of old-school, holistic natural history.

The work of later Presidents of the Society, after its Royal Charter in 1866, suggests an overt openness to diverse forms of wonder despite the improved standing and influence of scientific naturalism in the latter half of the century. During the 1860s and 1870s, when the X-Clubbers had achieved leadership positions in the Royal Society, the Royal Institution, the Linnean Society, the Ethnological Society, and elsewhere, the MSL/RMS largely chose otherwise. The surgeon Charles Brooke served twice as President (1863–4, 1873–4); Lionel Beale's obituary notice for him in the *MMJ* remarks that "through life he was opposed to materialist doctrines."[38] Joseph Bancroft Reade (1869–70) was an Anglican clergyman, representative of the parson-naturalist tradition that Huxley was challenging. William Kitchen Parker (1871–2), a surgeon and man of fervent faith, was known for his unrestrained, enthusiastic, and poetic lecturing style. Other Presidents during this period dedicated significant time and energy to faith-related work: Brooke published a popular treatise on natural theology in 1872. The geologist Henry James Slack (1878) spent years editing the work of the Unitarian William Johnson Fox (published 1865–8).

Lionel Beale (1879–80) was a physician and an outspoken vitalist, who argued in the *MMJ* against Henry C. Bastian's materialist theory of abiogenesis (spontaneous generation).[39] The schoolmaster Charles Thomas Hudson (1888–90) in 1886 published with Gosse a monograph on the *Rotifera* (wheel-animalcules), which he opened with the comment that "the study of these animated specks…irresistibly leads the mind to the contemplation of Him, whose almighty hand is as visible in an atom of this animated dust, as it is in the myriad sparkles of the starlit heavens."[40] And the botanist William Carruthers (1900–1), who had wanted to be a Presbyterian minister, edited the *Children's Messenger* (English Presbyterian periodical) for forty years.

The complex cross-cutting of microscopical wonder with empiricism is especially evident in the three Presidents who were perhaps most influential in the development of Victorian microscopy: William Benjamin Carpenter, Edwin Lankester, and William Dallinger. Carpenter (1854–5), a physician and physiological psychologist, was accused of materialism with his 1839 attempt to set up a system of natural law governing physiology, but his physiology was far from atheistic; it was deeply informed by his Unitarianism.[41] I have noted how he draws deeply on tropes of wonder in *The Microscope and Its Revelations* (1856), which became a standard text. But Carpenter also became friendly with Huxley, J. D. Hooker, John Lubbock, and others who would carry the standard for scientific naturalism. Then again, he refused membership in the X-Club in 1864, albeit "probably for health reasons."[42] Carpenter was dismayed by the extreme positions of Huxley, Tyndall, and William Clifford in the 1870s and he considered one of his greatest works to be developing a model of "theistic science" that could offer a "middle ground" between Victorian theologians and scientific naturalists, between intuitionists and empiricists.[43] In a further turn, he later became known for his exposés of spiritualist quackery.[44] Carpenter's influence on British microscopical culture can hardly be overstated, given the importance of *The Microscope and Its Revelations* as a core text for Victorian microscopical practice.

Lankester (1858–9) offers another fascinating case. A surgeon and public health reformer, he enjoyed a long working partnership with Busk. They were friends of Huxley, and together Lankester and Busk founded the *Quarterly Journal of Microscopical Science*, co-editing it from 1853 to 1868 (Lankester through 1871). However, while Busk was a member of the X-Club, devoted to a naturalistic science free of religious dogma, Lankester was a devout Congregationalist whose faith informed his work. His popular guidebook *Half-hours with the Microscope*, written for a general audience, often admires beautiful form. His research articles adhere to a more restrained vocabulary, but he does occasionally use tropes of wonder to power effective description. For example, his 1863 "Notes on Raphides" (plant crystals) uses tropes of *analogy* (needles, dumbbells) to describe these forms.[45] Lankester's work shows how a single microscopist might use tropes of wonder differentially to engage different audiences and aims.

Working with the physician John James Drysdale, William Henry Dallinger (1884–7) pioneered life-cycle studies of microscopic organisms—a line of research helping to disprove Henry Bastian's materialist claims of spontaneous generation (abiogenesis) and proving the evolution of monads under adverse conditions.[46] Darwin, Tyndall, and Huxley complimented Dallinger's painstaking, important experiments, eager to dissociate scientific naturalism and evolutionism from the controversies surrounding strict materialism at the time. Dallinger was also a well-known Wesleyan Methodist preacher and popular lecturer. He was adept in negotiating the rhetorical demands of diverse settings. In attending the 1884 meeting of the British Association for the Advancement of Science (BAAS) in Montreal, Dallinger avoided explicit mention of religion when addressing the BAAS, as was conventional, but lectured on the necessity of religion when speaking to general or religious audiences.[47] However, the *JRMS* report on Dallinger's BAAS address shows that while he stood with the scientific naturalists in defending evolution and dismissing abiogenesis, he also used tropes of wonder drawn from the discourse of design. For example, Dallinger remarks on "a world of minute living things so ultimately little that we cannot adequately conceive them, which are, nevertheless perfect in their adaptations and wonderful in their histories." He puts guardrails on scientific inquiry by directing his listeners to ask not how life originated but whether spontaneous generation currently occurs; he explains, "The mystery consists solely in how these elements can be so combined as to acquire the transcendent properties of life [48] In addressing the BAAS, Dallinger defines the question so as to exclude religion. But the lacunae in his address, and his oblique references to "mystery" and the "transcendent properties of life," leave space for wonder in scientific work—a space that (as Ciaran Toal shows) he explicitly fills with religious belief in other addresses.[49]

Finally, Frank Crisp was never President of the RMS, but he influenced the late-Victorian Society as much as any of its presidents. A banker by trade, he was an accomplished microscopist, fluent in optical theory, and a very active member of the RMS. He was an officer of the society for over twenty years, serving as (variously) treasurer, secretary, the editor of its journal, and vice-president (1893–7). He was the primary author on a comprehensive summary and explanation of Ernst Abbé's theories and the aperture question published in the *JRMS* in April 1881, and he translated Karl Wilhelm von Nägeli and Simon Schwendener's influential optical treatise *Das Mikroskop, Theorie und Anwendung* (*The Microscope in Theory and Practice*). But with this skeptical commitment to improving the science of microscopy, he maintained a sense of whimsy and willingness to play around with wonder, as is evident in the extensive, fantastical gardens he built at Friar Park, a fairytale setting where he offered visitors lively gnomes, romantic grottoes, caves stocked with ghostly illusions, even the chance to (apparently) walk on water. While those can readily be separated from his scientific work, Crisp is also said to have been also a spiritualist and a believer (with Sir Arthur Conan Doyle) in the Cottingley fairies.[50]

Of the figures I've surveyed here, many were clergy or medical men. These were among the most active members of the Society and their election as leaders of the MSL/RMS suggests that the Fellows generally respected their views. No doubt this pattern among the Society leadership owes much to the value of microscopy in the practice of religion and medicine, and to the dearth of other career options for science-minded men throughout much of the century.[51] These were superb microscopists but their vocations required consideration of the non-empirical. Clergymen, of course, cannot entirely eliminate a theological perspective. And although the study of anatomy could lead either toward or away from religious faith, and medical students were often criticized for their coarse materialism, physicians and surgeons were expected to balance medical science and medical art, and to recognize the role of intangibles—such as the patient's mindset—in the disease course. Of course, as microscopists these men might still strive to ensure their scientific work was grounded in fact. However, their vocations and interests, taken as a group, suggest an ongoing openness to wonder, an openness that is also visible in the print record of the MSL/RMS. This community long maintained an openness to aesthetic and affective response despite its commitment to empirical work. While other societies doubtless included such men among their active members, the MSL/RMS demonstrates a pronounced and suggestive pattern: even as they strove to improve microscopical science, the members of the Society chose leaders whose perspectives extended beyond a strictly materialist biology.

To be clear, I am not arguing that the MSL/RMS was isolated from broader scientific trends. Indeed, most of the nineteenth-century Presidents of the MSL/RMS were Fellows of the Royal Society as well, indicating recognition of their work among the wider community of science. Research reported to the Royal Society and other societies in Britain and worldwide percolated through microscopical meetings and publications; and members of microscopical societies engaged regularly in larger scientific conversations such as debates over species identity and spontaneous generation. Not just Huxley and Tyndall but also Dallinger and Carpenter served as President of the BAAS, one of the most prestigious posts in British science. And there were a few MSL/RMS Presidents from this era who were strong advocates for scientific naturalism. Busk is one, serving in the 1840s, and Lankester's son, E. Ray Lankester, served as President in 1909. He was an ardent rationalist and aggressive supporter of Huxley and Darwin; he would do important work in exposing spiritualist deceptions. Regardless, it is striking that many of the Victorian leaders of the RMS, while promoting rigorous microscopical enquiry, largely preserved in their work or life a willingness to acknowledge or explore wonder and mystery.

Tropes of wonder served Victorian microscopical communities in many different ways: enabling the identification of species, enhancing personal faith, attracting book-buyers and exhibition-goers, revealing wonders in ordinary life, enticing novices to the study of microscopy, signaling the importance of a particular theory, practice, or shared route of investigation—and, of course, they help Victorian microscopists entrain readers with themselves in favored affective responses to the microscopical image and the objects it represents. Daston and Galison have argued that working microscopists sought to supplement the scientific virtues of mechanical objectivity with compatible virtues: curiosity, patience, discipline, determination, care, attention, humility, honesty, self-restraint, self-correction, and cooperation.[52] These qualities supported a sense that one's life and work had meaning and purpose. Victorian microscopical writing suggests that to this list of contemporary virtues we should add the capacity for particular forms of wonder.

Wonder features in other Victorian sciences and technologies as well, of course; the print record provides many examples of exuberant language on topics ranging from botany to paleontology, from the telegraph to the Roentgen ray. But perhaps because so much microscopic observation takes place in virtual space, hedged about with obscurity and doubt, and reliant upon speculation, Victorian observers seem particularly willing to gesture toward values beyond what is empirically knowable and quantifiable in the material world.[53] The strangeness of many microscopical objects and the disorienting microscopical image can be difficult for observers to read. These unfamiliar images can arouse a confused affective response in observers. Beauty, mystery, vastness, sublimity—all elements of wonder—spark both humility and transcendence, and offer access to meaning whether theistic or non-theistic. Tropes of wonder enable writers to emphasize the significance of microscopical work, contextualizing it within familiar patterns of meaning such as natural theology, the sublime, and the Oriental tale. These tropes solicit particular responses, redirecting observers' unsettled feelings into favored ones—not anxiety, frustration, or disgust but awe and admiration, surprise and curiosity. In general, working microscopists seeking new converts and social support for microscopy seek to direct these affective flows away from threatening or frightening tropes and toward tropes emphasizing the beauty, delights, pleasures, and uses of microscopy. Threatening tropes such as the phantasmagoria appear most often in texts with primarily recreative aims, addressing audiences who are seen as viewers rather than observers, or in texts hoping to rouse fear in order to motivate some course of action: faith, sanitation, medical treatment.

The epigraph from Haeckel, above—"I still recall with pleasure"—suggests another reason that wonder persists. Texts across diverse genres rely on tropes describing the microscopical world not just as beautiful, mysterious, vast, or sublime, but also as charming, whimsical, fantastical, mythical, exotic, or phantasmagoric. Many of these tropes of wonder allow microscopists to identify and amplify particular *pleasures of microscopy.* Pleasure, both intellectual and physical, had been an important element of microscopy since Robert Hooke wrote in his Preface to *Micrographia*, "And I do not only propose this kind of *Experimental Philosophy* as a matter of high *rapture* and *delight* of the mind, but even as a *material* and *sensible Pleasure*."[54] John Hedley Brooke and Geoffrey Cantor emphasize that natural theology, too, appeals not just to rational argument but also to sensibility—to imagination and the heart.[55] Amy King also emphasizes the positive valence of the natural theological affect, arguing that in reverent natural history, "empirical observation was married with reverence and joy."[56] Darwin's letters similarly express not just the labors and frustrations of his microscopical practice but his enjoyment, perhaps shaped by his memory of receiving a microscope as an anonymous gift in 1831, when he was just out of college; decades later he remarked, "I can hardly call to mind any event in my life which surprised & gratified me more."[57] He calls one microscope a "splendid plaything" and crows that his work is "really delightful" with a new "charming lens."[58]

The communities of Victorian microscopy were capacious enough to foster many forms of wonder. Tropes of wonder record and inspire moments of emotional resonance, when the observer is struck into correspondence with the microscopic world. In Chapter 10, I quoted Hudson's description of a rotifer as a "tiny speck" that he sees "whirling" with "the exquisite happiness of living." That speck metonymically serves as a spark to and a microcosm of his own delight in microscopical vision. Working Victorian microscopists were generally committed empiricists, but, like Hudson, they often made room for wonder to persist—with some restrictions—and to inspire. Again and again, Victorians used tropes of wonder to direct and solicit affective engagement with the strange microscopical worlds around us and in us.

Writing Embodiment in Victorian Microscopy: Beautiful Mechanism. Meegan Kennedy, Oxford University Press.
 DOI: 10.1093/9780198940623.003.0011

PART III
DIFFICULTY AND ERROR

12
Difficulty and Error

Falsehood is so easy, truth so difficult.
George Eliot, *Adam Bede*, 1859

Introduction

I have shown how tropes of wonder infuse microscopical texts throughout the nineteenth century, rousing and channeling affective energies into preferred forms. However, microscopists' recognition of observation as an embodied practice meant that they considered not just what was marvelous but also what was difficult in their work, such as optical limits on human perception and the compound instrument, or the likelihood of error in their observations. As with beauty and wonder, writers on microscopy developed skeptical tropes that circulated across genres. These skeptical tropes refute unhelpful narratives about the microscope—for example, that it was a *mere toy* or that *seeing is believing*—by modeling a meticulous microscopical practice through phrases like *patient labor* and *true structure*; challenging naïve trust in *natural sight* and the *naked eye* even as they hold out hope of *true structure*; training novices in rigorous methods by *training the eye* and *scrutinizing vision*; and valorizing the instrument as a *weapon against disease*. Tropes like these encourage readers to become disciplined observers who can produce reliable and valuable observations.

While many of the tropes of wonder examined in this book are adapted from well-established literary conventions, skeptical tropes are very much of their moment in that they reflect nineteenth-century physiological and psychological research on the instability of human perception and techniques for mitigating scientific error, as noted in Part I.[1] The unique conditions of microscopy exacerbate these difficulties of perception, acting as an exemplar of unstable vision in the Victorian period. This role has been well canvassed by scholars. Kate Flint argues that microscopy promoted a narrative that "knowledge and control over the natural world could…be obtained," but she also points out that "nothing showed up the limitations of the eye so much as technological developments" such as the microscope.[2] And Isobel Armstrong depicts the microscope as "insist[ing] on the illegibility of the object," noting Edwin Lankester's description of microscope-assisted vision as a practice built on "defects, correctives, and supplements."[3] Martin Willis concludes that Victorian "visual technologies just as often undermined visual authority, or came to stand as metaphors for vision's fragility."[4] The microscope offers a superb site from which to examine such concerns. Working microscopists discussed empirical principles with sophistication and nuance, grappling with the limitations of Victorian optical instruments and the instability of human perception. They understood that vision, especially microscopical vision, does not always serve as a reliable authority for empirical claims, and they strove to find practical solutions for these difficulties in individual and community-based practices. The constrained visual encounter of observer, instrument, and object exacerbates the complex relations that structure Victorian vision. The skeptical tropes I examine here seek to stabilize this contingency and difficulty of microscopical vision.

The historian Jutta Schickore, examining microscopical error, argues that nineteenth-century microscopists were interested in pragmatic technical questions more than theoretical or philosophical ones. She examines second-order practices, such as test objects and micrometers, that allowed microscopists to compensate for particular weaknesses in observer or instrument. I show here how, in texts available in English at mid-century, microscopists use skeptical tropes to manage the fraught engagement between observer and instrument. One major set of challenges arose from observers themselves. Contemporary research warned that errors and misreadings could result from undisciplined work or fatigue, from differences in the physiological and psychological makeup of different observers, and from innate slippages of human perception. Another major set of challenges concerned the physical structure of microscopes. Despite significant advances in the technology, nineteenth-century working microscopists still worried about the possibility of optical variation and error. In fact, more powerful microscope technologies required ever more precision and care in use, bringing Victorians to the very edge of their abilities and the capacities of the instrument. Just as the microscope seemed to bring new worlds within Victorians' gaze, it also revealed the limitations and vulnerabilities of that gaze.

The status of microscopy was complicated by these difficulties. Many disputed points—especially debates over structure—could not be resolved simply by the procedures most closely associated with nineteenth-century empiricism, such as objective measurement, quantification, and experiment. In that sense, microscopical work challenges T. H. Huxley's famous dictum that "science is simply common sense at its best: that is, rigidly accurate in observation, and merciless to fallacy in logic."[5] Microscopists studying disputed forms using high powers, working at the far reaches of microscopical resolution, cannot rely on a single careful observation. They do use a "merciless logic" when they factor in the optical properties of the lens and object before interpreting the image. But they must also adopt multiple forms of intervention with the object and reconcile conflicting results. And they must pursue elusive glimpses of structures, momentary shifts in the image that cannot reliably be replicated. The pursuit of the invisible brought these investigators to the very borders of materiality itself.

The resulting practice became as much art as science. Particular observers were widely celebrated for their *trained eye* and *tact* (sensitive manipulation of object and apparatus), as they strained to multiply and sustain these peeks beyond Nature's veil—to determine true structure. Leeuwenhoek is often cited for his ability to achieve impressive observations through microscopes that were little more than a bead of glass in a holder. And in the nineteenth century, Christian Ehrenberg made remarkable contributions to the study of the infusoria (tiny freshwater organisms) despite a mistaken thesis about the complexity of infusorial bodies and his stubborn commitment to outmoded instruments until his death in 1876. His refusal to use a compound microscope shows how suspicion of that instrument lingered for decades after significant improvements to compound lenses in the 1830s, and also suggests that he, like many workers, felt a kind of affection for and loyalty to his old favorites. As the *Monthly Microscopical Journal* (*MMJ*) observed in its 1877 obituary of Ehrenberg, for good or ill, "one cannot but feel that the eye at one end of the tube is, after all, the most important part of the whole optical arrangement."[6] The quest for true structure proved an attractive game for ambitious observers even as reliance on individual skill complicated attempts to professionalize microscopical science.

Expert microscopists used skeptical tropes to strengthen weak points in microscopical observation and enlist readers in particular, favored work habits. In urging ambitious novices to mimic particular protocols, these tropes help standardize microscopical practice. Echoing contemporary principles of physiological psychology, the call to form certain habits attempts to direct the unruly, entangled embodiment of observer, instrument, and object. These tropes model and seek to evoke particular forms of practice. They draw attention to the vulnerabilities of microscopical work—specifically, the interrelation of imperfect observers with a changing technology—in order to secure the authority of that work. In each case, a mitigating strategy is suggested, usually some version of *testing* the observer or microscope, *training* the observer, or *sharing* the image. I focus here on how skeptical tropes support the strategy of *training*; microscopists also developed test objects and new modes of representation to support the strategies of testing and sharing. Skeptical tropes train readers by guiding them in preparing the object, manipulating the apparatus, observing the image, and interpreting that observation. Microscopists focus on each point where the observer interacts with the microscope or object, changing the image, which in turn modifies the observer's work as well.

It would be difficult to overstate the role of the skeptical strain in nineteenth-century microscopy. Victorian microscopists were insistently aware of the constraints on microscopic vision, as indicated by the number, length, and rhetorical force of the references in the print record. These concerns appear well before the 1870s, when Ernst Abbe and Hermann von Helmholtz published their work on the deceptive images of diffraction and the theoretical limits of the optical (light) microscope. Skeptical tropes find an obvious purpose in texts emphasizing research and instruction, addressed to workers and students. Texts like Woods' *Common Objects of the Microscope* show that even a cheap (1*s.*) introductory text might use them to model careful, disciplined practice for a broad audience. However, these tropes also sometimes appear in recreative texts. Entertainments like an oxyhydrogen show or soirée might feature difficulty and error largely as a performance of the operator's skill in addressing a technical mystique. The texts I examine here propose a mitigated or working skepticism: they generally offer both a pessimistic view that difficulty and error pervade and constrain microscopical work and an optimistic view that the careful observer can prevent or moderate many of these problems through disciplined practice. As the critic Rick Rylance says of George Henry Lewes's epistemology, for Lewes "knowledge is both creatively adequate and radically provisional."[7] This exactly describes the pragmatic approach that allowed microscopy to gain strength as a tool of research and leisure in this period.

While most British microscopical manuals from mid-century onward demonstrate lively concerns about difficulty and error, some texts present typologies of microscopical error. Despite typological distinctions between difficulties of observer and microscope, microscopists consistently frame observation as an entanglement of the actions and qualities of observer, instrument, and object. Although recent scholarship on nineteenth-century microscopy has emphasized its relation to debates on visuality, microscopists were acutely aware of the embodied nature of their work and theorized the body as part of the microscopical apparatus. For example, writers advise readers how to position their body at the instrument, since bodily fatigue conditions vision and attention. They circulate skeptical tropes that encourage observers to gain physical and mental discipline in particular domains: acquiring physical *tact* in manipulation, a *trained eye*, and the skill of cognitive *appreciation* that enables a discerning response to the image. This part of the book enlarges upon these efforts.

Philosophies of Experience

> All the uncertainty, and mistakes of humane actions, proceed either from the narrowness and wandring of our *Senses*, from the slipperiness or delusion of our *Memory*, [or] from the confinement or rashness of our *Understanding*.
>
> Robert Hooke, *Micrographia*, 1665

Victorians understood that our knowledge of the world develops from human sensory experience and its effect in the brain. They developed new practices of empirical investigation rooted in Francis Bacon's model of inductive reasoning. Philosophers and investigators like William Whewell, John Herschel, and John

Stuart Mill worked to revise Baconian induction to suit the changed conditions of nineteenth-century science. Chief among these were a renewed emphasis on the frailty of human perception, which undergirds all empirical work.

The skeptical empiricist David Hume argued that the ideas produced in the mind by sensations do not provide a reliable ground for making definite claims about the nature of external objects. For him, the failures of sense are so well known as to be "trite"—giving as example "those which are derived from the imperfection and fallaciousness of our organs, on numberless occasions; the crooked appearance of an oar in water; the various aspects of objects, according to their different distances; the double images which arise from the pressing one eye; with many other appearances of a like nature." Microscopic vision, of course, derives its power from refraction (the "crooked" oar in water). Hume takes these falsehoods as a reminder that "that the senses alone are not implicitly to be depended on; [and] that we must correct their evidence by reason."[8] That is, a naïve empiricism must be supplemented by rational means.

Nineteenth-century researchers in neurology and psychology, supported by the physiological researches of David Brewster, Jan Evangelista Purkyně, Johannes Müller, Hermann von Helmholtz, Charles Wheatstone, and others, closely examined these powers and failures of human perception. Much of this research analyzed the physical and mental components of vision, increasingly understood to be flawed and deceptive. Brewster studied human vision in *A Treatise on Optics* (1831) and microscopical optics in "Treatise on the Microscope" (1837). Victorian microscopists adopted an empiricism structured by Brewster and Hume, elaborating upon both the mechanics and the vulnerabilities of vision, and recognizing how these glitches and lapses of perception threaten human knowledge.

While Hume found sensory fallibility to be "trite," he engaged seriously with the failures of perception. He specifies, "All belief of matter of fact or real existence is derived merely from some object, present to the memory or senses, and a customary conjunction [a habitual association] between that and some other object."[9] As to the first element in this process, *sensory experience*, he warns, "nothing can ever be present to the mind but an image or perception, and that the senses are only the inlets, through which these images are conveyed." These images "are nothing but perceptions in the mind, and fleeting copies or representations of other existences, which remain uniform and independent."[10] As to the second, the "*inferences from experience*" grounding the inductive process, he argues that they are all "effects of custom, not of reasoning."[11] That is, we establish a relation of causality (like resemblance or contiguity) through a kind of habitual association rather than rational or volitional act, seeing causality simply when objects are regularly conjoined in ways that encourage our mind to link them as cause and effect. We must be cautious about not only the evidence of the senses but also the evidence of our own judgments regarding sensory experience—our use of inference and interpretation.

Hume's work was far from uncritically accepted by Victorian thinkers. His skepticism and religious agnosticism drew censure throughout the nineteenth century; his historiography and political economy were also challenged; and his approach to empiricism through an associationist psychology was critiqued by idealists and Scottish common-sense adherents alike. Among Victorians known to use the microscope, some writers—Thomas Dick writing as a "Christian Philosopher" and Alfred Russel Wallace as a spiritualist—deplored Hume's skeptical resistance to supernatural truths. William Whewell similarly criticized Hume's attempt to ground mathematical truths in fact.[12] Still, the utilitarianism of Bentham and Mill and the evolutionary biology of Darwin and Huxley owe much to Hume's empiricism; Huxley even wrote an admiring biography of Hume in 1879. And despite Hume's critique of materialism, his associationist epistemology—his view of perceptions as dynamic, unstable, interacting—also shaped the work of the physiological psychologists Spencer, Lewes, and Bain.[13] The controversies over Hume's work did not prevent it from becoming widely recognized; even the novelist Margaret Oliphant, author of the *Chronicles of Carlingford* and other works of domestic realism, defended him in 1869 in the pages of *Blackwood's*, where many of the debates over perception and cognition had been discussed over the previous three decades.[14]

Nineteenth-century researchers adapted Hume's challenges to experience and to induction. John Herschel, in his *Preliminary Discourse on the Study of Natural Philosophy* (1830), warns that sensory impressions may be misread by our own faulty judgment. He describes an array of them and argues,

> though we are never deceived in the *sensible impression* made by external objects on us, yet in forming our judgments of them we are greatly at the mercy of circumstances.... [W]e can only regard sensible impressions as signals conveyed from [sensible objects] by a wonderful, and, to us, inexplicable mechanism, to our minds, which receives and reviews them, and, by habit and association, connects them with corresponding qualities or affections in the objects; just as a person writing down and comparing the signals of a telegraph might interpret their meaning.[15]

While Hume argues that sensory experience is merely a series of impressions, Herschel claims that even those sensory impressions are not immediate but require registering—decoding—by the judgment, leaving open the possibility that we misread their signals.

These concepts were generally known among educated Victorians. For example, Lewes elaborates on "the *subjective* nature of vision" for a broad audience in his two-volume *Physiology of Common Life* (1859–60), originally serialized starting in 1858 as instructive essays in *Blackwood's Edinburgh Magazine*. He explains, "We can no longer suppose that we see the objects themselves. Our visual sensations are simply excited states of our sentient organism." He offers examples of spectral illusions, fixed (lingering) impressions, and color blindness, and he insists upon "the strange blending of Thought in the act of Perception," explaining, "We fancy that we see the solidity of an object, when in truth we infer it."[16] As Peter Garratt notes of these thinkers, "Experience... that most crucial concept in empiricist philosophy, carried with it by the middle of the nineteenth century an urgent new sense of multiplicity and instability."[17] This was especially true of microscopical experience.

Typologies of Microscopical Error

Victorian writers wrote about the familiar issue of *optical aberrations* in microscope lenses. But they increasingly wrestled with new concerns as well. For example, *difficulties of the observer* include *difficulties of practice* involving preparation and manipulation, *vulnerabilities of the eye* affecting human perception, and *errors of judgment* such as faulty interpretation. And *difficulties specific to the microscope*, such as diffraction, raised grave concerns about the image. Some writers defend the microscope, seeing human carelessness as the source of most inaccurate observations; they argue that the difficulty of microscopical vision warns the observer to adopt painstaking habits of work. In fact, many authors suggested particular training exercises, hoping to improve microscopical science by guiding the movements and habits of its workers. Such efforts recognized the entangled embodiment of instrument and user, and helped to regularize microscopical practice.

Among the most basic of these were optical aberrations that had been solved in telescopes for some time but had defied microscope-makers due in part to the small size of the lenses involved. During the eighteenth century, compound microscopes (with two or more lenses) could produce stronger magnification but were dogged by spherical and chromatic aberration in the images. In the 1830s, Joseph Jackson Lister, a wine merchant who worked in physics and microscopy, collaborated with the instrument-maker William Tulley to produce a reliable achromatic compound microscope using a triplet—a lens formed of three stacked glasses.[18] John Quekett, in his 1848 *Practical Treatise on the Microscope*, emphasizes the complexity of the problem and the magnitude of opticians' achievements, arguing that "perhaps there is no single problem that has ever yet engaged so much of the attention of the learned, in all countries, as that of achromatism" and that "of all the triumphs in science that have been achieved by... the [combined] labours of the mathematician and the workman, no one can outvie in delicacy of construction... a well made achromatic combination."[19] Victorian opticians and microscopists were able to manage (minimize or compensate for) spherical and chromatic aberration, although the systems to correct for them sometimes introduced their own problems. Furthermore, vision through the microscope is still restricted by the laws of optics. Carpenter explains that any instrument represents a tradeoff between working depth and sharpness of focus.[20] An increase in magnifying power reduces the depth of focus, the field of view, and the light available to the eye, presenting a narrower, shallower, and darker image.

British microscopists such as Brewster and Quekett had discussed these problems in the 1830s and 1840s, but they did not present these discussions as formal methodologies of practice. When Edwin Lankester translated the botanist Matthias Jakob Schleiden's *Principles of Scientific Botany* (1842) into English in 1849, he omitted much of the methodological frame but did include Schleiden's system of practice, where he asks observers to act in specific ways to compensate for the inherent limitations of microscopical vision. Schleiden, an accomplished microscopist well known in England for his work on cell theory, notes that microscopical vision lacks important characteristics that help us determine true structure using natural vision. In ordinary life, "our eye is moveable, and we may wander about with it among the objects," collecting various impressions of them. Indeed, we can "move ourselves or the objects" in order to, again, "gain quite different views of one and the same object."[21] Also, we usually enjoy binocular (dimensional) vision, and our other senses also supplement vision, thus providing a broader base of input about an object.

Lacking these powers, the microscopical observer, confined to a narrow view of a decontextualized object, is thus unusually vulnerable to error. Deprived of the ordinary aids to vision, the microscopist must strive to "increase the number of elements" (sensory impressions) of an object in order to rightly perceive its form.[22] This approach was not necessarily new; Tita Chico argues, in a reading of Robert Hooke's method in the seventeenth century, that "[t]he observed particular is ultimately a process and a compilation."[23] Nineteenth-century investigators, including microscopists, invested new energy in developing and formalizing scientific methods of observation. Although (as Schleiden's treatise shows) Continental workers had taken the lead in such matters, British investigators also followed the debates between Herschel, Whewell, and John Stuart Mill, who published his *System of Logic* in 1843. In the 1849 "Translator's Preface" to Schleiden's 1842 text, Edwin Lankester recommends two books—Herschel's and Whewell's—"as general introductions on the principles involved in scientific inquiry."[24] These treatises laid out the precepts of modern inductive science. Herschel and Whewell differed on some matters such as whether fundamental ideas emerge from experience or are a priori, or whether hypothesis derives from intuition or inference. But they both insisted that observation must be systematic, endorsed speculation (hypothesis) in scientific work, and rejected a naïve empiricism, underscoring the mind's role in the inductive process.

Schleiden, who was interested in these matters and also well regarded in England, formulated a number of rules of engagement for microscopical observation. He asserts, as a "fundamental rule," that the microscopist "must observe frequently and with the most profound attention," specifically urging observers to "obtain from [an] object...as many views as can possibly be obtained from it, in order to construct a clear image from the combination of the individual conceptions." These strategies include turning the object in forceps, mounting it in various fluids, sectioning it, and dissecting it. Schleiden expands upon this approach, which he terms *many-sided comprehension* (*Vielseitigen Auffassung*) by identifying four investigative methods enabling an observer to submit an object to "as many different circumstances as possible, in order thereby to increase the number of points of view." These methods, which he terms "reagents" (*Reagentien*) are "optical, mechanical, chemical, and physical auxiliaries" enabling investigation via different forms of illumination, pressure, chemicals, and heat or electricity.[25] Schleiden's point is that microscopical seeing requires both repetition and difference, the observer finally collating many diverse observations into a single conception of the object. Although Schleiden's phrase *many-sided comprehension* did not catch on, this approach helped microscopists reach more reliable conclusions, and it was echoed in many later guidebooks.[26]

Schleiden's framing of microscopical vision as an explicitly creative act aligns well with Whewell's description of the inductive process: the microscopist collects discrete sensory impressions which the mind collates into facts; then, using existing fundamental ideas like number, space, and symmetry, or modified versions of these ("conceptions") such as the typical form of a particular species, the observer "colligates" the facts into a new unity, in this case a mental or theoretical model of the structure of a microscopic object.[27] As Whewell explains, "Before the inductive truth is detected, the facts are there, but they are many and unconnected. The conception which the discoverer applies to them gives them connexion and unity."[28] Over time these new unities are confirmed, maturing into facts on which to build future work. Whewell presents this stepwise methodology for use in science generally, but Schleiden's emphasis on the building up of ideas from disparate kinds of observations shows how well Whewell's approach fits nineteenth-century microscopical method, a method that seems laborious but is necessary due to the challenges of microscopical vision.

Schleiden's process of many-sided comprehension also translates well to Herschel's model. Describing "the observation of facts and the collection of instances" in empirical science, Herschel directs the observer to "accumulate a sufficient quantity" of "recorded instances" (distinct impressions), specifying "that the more different these collected facts are in all other circumstances but that which forms the subject of enquiry, the better."[29] Herschel later argues that induction works by both connecting and creating objects. He explains that the mind enjoys a "peculiar constructive or plastic faculty" which, "out of a series of dots" (discrete observations), will construct "an outline" that bodies forth an idea. That is, after many observations, "the particulars are brought so close, that the attention glides easily, and is, as it were, conducted from one to the other." The mind will thus "fill up by its own act the intervals between" the examples in hand, as it "constitutes an object" out of collected instances or experiences of it.[30] This cognitive process, as Omar Nasim shows, corresponds well with Herschel's experience in sketching nebulae.[31] But this model also correlates well with the observational method proposed by Schleiden: collecting many different kinds of "views" of a microscopical object, then incorporating those disparate aspects into a reliable mental model of that object.

British microscopical guidebooks from the mid-1850s, notably Lionel Beale's *The Microscope and Its Application to Clinical Medicine* (1854), followed Schleiden in working to systematize difficulties of practice. William Benjamin Carpenter, in his classic manual *The Microscope and Its Revelations* (1856), devotes many pages to the varieties of microscopical error and how to avoid them, identifying a comprehensive typology of error. He also identifies a new category of "misinterpretations . . . peculiar to objects thus viewed" such as diffraction lines, restricted focus, the effects of refraction, and the like. These *difficulties specific to the microscope* had been noted before, but Carpenter poses them as a unique challenge.[32] Despite drawing such distinctions between various types of difficulty, Victorian microscopists repeatedly emphasize that the actions of observer, microscope, and object jointly shape the microscopical image and the conclusions drawn from it.

Texts addressed to novices or young audiences generally demonstrate less concern over microscopical errors, but the topic is not entirely absent. Such texts most often instruct readers on sundry *difficulties of practice*, while generally encouraging more microscopical study. The broad range of casual or occasional microscopists, the expected audience for popular texts of instruction and recreation, would not develop a systematic approach toward error as an experienced microscopist might, but they would know to use the microscope with caution.

Mitigating Skepticism

How trustworthy is microscopical work? Many factors—Humean empiricism, microscopical optics and technology, and human physiology and psychology—suggest that the microscopist may not be able to see "true" images of a "real" existing world. In today's world, Ian Hacking provocatively asks, "Do we *see* through a microscope?" and concludes that we do indeed see "with," if not actually "through," the microscope; that we can often confirm whether its images correlate to actual structure, although the images are not necessarily "real," and that "our conviction [in the microscopic image] arises partly from our success at systematically removing aberrations and artifacts."[33] Victorian microscopists reached a similar consensus that there was indeed a world of invisible objects and structures that could be accessed by the microscope. Hume lays out general principles for empirical work; microscopists developed specific techniques to discipline the practice of microscopical science. They agreed that while microscopical observers could not avoid many optical and practical errors, these problems could be managed: investigators could often ascertain true structure, and practical work with the instrument could generally be relied on. Even the debates over diatom structure questioned only the "striae" or "beads" imaged, not the diatoms themselves. Working microscopists largely tempered their skepticism with a pragmatic approach marked by judicious optimism about the progress of microscopical science, to be achieved not simply by technological improvement but also, importantly, by changes to microscopical practice. They worked to safeguard microscopical observation by testing individual instruments, training observers in a standardized process of observation, and comparing results, generally by sharing the image in diverse ways.

Nineteenth-century microscopists developed a modern scientific method informed by insistent awareness of the difficulties of vision and by Humean skepticism. But Hume offered a compromise between extreme skepticism, which offers a strong philosophical case, and lived experience, which denies it. He proposed "a more *mitigated* scepticism" which results when "*excessive* scepticism" has "its undistinguished doubts…in some measure, corrected by common sense and reflection."[34] Toward this goal, he proposed concrete steps:

> To begin with clear and self-evident principles, to advance by timorous and sure steps, to review frequently our conclusions, and examine accurately all their consequences; though by these means we shall make both a slow and a short progress in our systems; [they] are the only methods, by which we can ever hope to reach truth, and attain a proper stability and certainty in our determinations.[35]

Hume further sets some general guidelines, saying, "there is a degree of doubt, and caution, and modesty, which, in all kinds of scrutiny and decision, ought for ever to accompany a just reasoner." And he recommends that investigators should "consider the imperfection of those faculties which they employ, their narrow reach, and their inaccurate operations."[36] These strictures—caution, doubt, reason, restraint, and reflective review—helped Victorians enact inductive empiricism in practical terms.

Microscopists throughout the nineteenth century widely understood it to be their duty to be skeptical about what they saw through the microscope. For example, in 1875, the American John Michels (a journalist who would found *Science* five years later) wrote an article focusing on "The Microscope and Its Misinterpretations" for *Popular Science Monthly*. Michels' theme of "the difficulties which beset the path of the microscopist" spoke to contemporary anxieties, and his article was reprinted in the *MMJ*, *English Mechanic*, and elsewhere. He begins, "The old adage, 'seeing is believing,' has long been exploded." It was well known among workers that "the utmost caution is required by those using optical instruments for the elucidation of scientific problems or ordinary research." However, novices, who are "fascinated by the wonderful and beautiful appearances presented," frequently "congratulate themselves upon the ease with which they have handled the instrument," only later discovering that "the instrument requires the most delicate manipulation, and that much practice is necessary" to avoid "the most contradictory and perplexing results." Indeed, Michels admits, we see "the most contradictory statements from men whose powers of observation and skill in the use of the instrument are admitted."[37] While some optical errors were due to microscopical construction, many other errors—involving preparation, manipulation, vision, and interpretation—were produced by observers' interactions with the instrument. Although—as Michels notes—even expert observers could err, training was central to this effort, mitigating even challenges innate in microscopical vision, such as the decontextualized image or diffraction lines. Skeptical tropes helped them counter familiar, simplistic narratives like *seeing is believing* by revising the privilege accorded to *natural vision*, codifying the principles underlying new disciplinary techniques, and inducing observers to practice these techniques.

Victorian texts from a range of microscopical genres present microscopical work as inextricably entwined with difficulty and error, possibly endangering scientific progress. The observer, struggling to wrest knowledge out of doubt, is both a threat to microscopical science and its protector. As a result, many Victorian microscopists warn of human error and the difficulty and limitations of the apparatus. These caveats do not derail the work but rather shape it, spurring and organizing the observer's actions through a systematic, pragmatic practice. Skeptical tropes help establish a disciplinary structure through which the microscopist operates, meticulously checking and recording with each step and emerging with a provisional assertion of fact. Overall, this part of *Writing Embodiment in Victorian Macroscopy* demonstrates how skeptical tropes help working microscopists manage concerns about embodied observation and the linkage of observer, instrument, and object. These tropes promote particular optical, mechanical, and cognitive practices in microscopical work. I analyze these concerns in the context of nineteenth-century discourses around how perception, attention, and fatigue contribute to discovery, learning, and error. I show how these cautionary tropes anchor both research and instructional writing and how popularizers use them, often performatively, in recreative settings.

In Praise of Difficulty?

The challenges constitutive of microscopic vision—its unnatural quality, the constraints of the optical system, the need for a habitual skepticism—crucially shaped Victorian microscopical culture. But Schickore argues that "discordant results did not engender overall distrust in the microscope. Rather, they stimulated second-order discourses and practices," which in turn grounded a kind of "epistemological optimism" that she later qualifies as "mitigated optimism" and I call "mitigated skepticism" or "working skepticism."[38] Indeed, Victorians developed and used a number of techniques—test objects, stains, club meetings and instruction, written reports and guides, improvements in illustration and photomicrography—to render the microscopical image more trustworthy. Most of these techniques offer some way either to discipline the eye, hand, and mind of the observer or to share the image. These strategies to enable skeptical work greatly supported the growth of microscopy. They ensured that investigators could make reliable claims and also encouraged the formation of microscopical communities: observers gathered in group settings like meetings and journals to share the image, train novices, compare instruments, and review results.

The debates about the adequacy of microscopical images did have consequences, though. While Schickore contends that veridical perception was not generally doubted, expressions of concern do appear in the literature.[39] For example, the diatomist W. J. Hickie in 1876 penned a frustrated response to a recent article about diffraction phenomena. "If the results of assisted vision be so utterly untrustworthy," he asked in what was not quite a rhetorical question, "what becomes of the microscope as an aid in scientific research?"[40] As this chapter will show, microscopists working at the boundaries of visible structure knew that they could not rely on any necessary relation between object and image.

However, an optimistic view was also voiced, often by the same writers. In this view, the microscope was not at fault; errors are always attributable to human observers; and the difficulty of microscopy actually helps prevent error by putting observers on their mettle. Schickore argues that for early-nineteenth-century British investigators like Thomas Young, David Brewster, Michael Faraday, and Peter Mark Roget, "insights into the mechanisms of illusion strengthened rather than weakened [their] confidence in the process of vision."[41] She argues, for example, that Young's discussion of what he called "microscopical fallacies" put the onus on the observer to avoid or compensate for known sources of error; and that Brewster presented his experiments on visual illusion as a way to allow observers to prepare themselves against potential errors in their work.[42] She notes Faraday's remark that visual error "ought to be considered, rather, an *error of the judgment* rather than the sense" and Roget's "epistemological optimism" in arguing that human vision, though riddled with error, was still functionally well adjusted to the purposes for which a Creator had intended it.[43]

This need for skeptical precautions could be reframed as a benefit, as in a passage by the literary critic and science popularizer George Henry Lewes, in his *Sea-Side Studies* (published in 1858, also from a series of essays in *Blackwood's*, 1856–7). Lewes wrote lively, entertaining prose that served largely as community-building and recreation, informing an educated lay audience about current work in science with examples often drawn from his own microscopical work. He references the trope of the innocent microscope and argues that the need for caution in observation is actually helpful. He notes, "It is...said, that microscopic observation is apt to be very erroneous, and that we can see whatever we wish to see. Undoubtedly men often do see what they want to see." But, he counters, "this is not the fault of the instrument. So far from the Microscope being in itself deceptive, I maintain that it is *less* so than the unassisted eye; and for this reason: all vision is mainly *inferential*." Lewes here presents all vision as destabilized, "inferential" vision—this "holds of the eye as well as of the Microscope"—but he argues that inference is less deceptive in microscopy because "the microscopic observer is on his guard against fallacies of interpretation."[44] He evidently considered this an important point; he added this explanatory passage for the book publication, then elaborated upon the nature of vision the following year in *Physiology of Common Life*.

Stephen Jacyna has noted, in discussing the difficulty of microscopical observation, that "seeing had become a matter of contrivance" with the adoption of microscopy for medical and scientific work at mid-century.[45] His choice of the term *contrivance* points to the double-edged sword of microscopical difficulty. Microscopists were frequently reminded of the limits and flaws of both eye and microscope and of the care they must take to mitigate these difficulties in their work. Yet just as William Paley praised the contrivance of natural objects as proof of God's ingenuity and delicate craftsmanship, microscopists adopt the trope of *contrivance* to reference the mechanisms invented by their colleagues to overcome some problem with microscopical sight.[46] Beale, for example, notes that in the construction of the microscopical apparatus, "many beautiful contrivances have lately been devised by various observers."[47] Willis has persuasively pointed to the difficulty and spectacularity of Victorian microscopical vision as evidence that the mystification of vision produced by eighteenth-century and Romantic optical technologies actually persists through the Victorian period.[48] But Victorians also sought to demystify microscopy: developing earnest guidance and clever contrivances addressing the difficulties of microscopy; emphasizing the value of microscopical practice for research, recreation, and other purposes; and employing strategies to mitigate the skepticism that both undercut and authorized their work. Such practical advice cut both ways. It framed microscopy in terms of ordinary physical tasks like collecting, cleaning, cutting, mounting, illuminating, and the like, reframing "revelations" as products within reach of any student willing to slog through the training manual. And it reinscribed the mystery of microscopy by insisting only laborious, hieratic practice could yield these microscopical visions.

Overall, Victorians agreed that microscopical vision is more difficult and prone to error than ordinary vision. Some defenders of the microscope argued that the difficulty and vulnerability of microscopical vision serves a valuable function in putting the observer on guard. The primary problem in microscopical vision, in this view, is not the instrument but its careless users. In either case, Victorians emphasized that the eye, hand, and mind were inextricably part of the optical and physical system of microscopy; and that while the microscope itself presents significant challenges in use, the observer represents both a significant problem in microscopy and a solution to that problem.

Writing Embodiment in Victorian Microscopy: Beautiful Mechanism. Meegan Kennedy, Oxford University Press.
 DOI: 10.1093/9780198940623.003.0012

13
Trained Hands and Tact

The eye and the hand require alike to be educated, before even the most ordinary observation can be made with accuracy.

"Review Essay on the Microscope," *Natural History Review*, 1855

Many microscopists have become good observers by their own perseverance; but the years they have devoted to the task, the difficulties they have had to overcome, the mistakes they have made, are known only to themselves.

"Wonders of the Microscope," *Chambers's Miscellany of Useful and Entertaining Tracts*, 1847

Introduction

In this chapter, I examine the errors most traceable to the observer, specifically *difficulties of practice*, which were of great concern to Victorian microscopists. They recognized the human observer as an integral element of the optical and physical system of microscopical work. As the physician H. E. Fripp writes in 1877, "the knowledge and skill of the microscopist...constitutes...part of the performance of the microscope."[1] Some, indeed, pointed to the observer as the most significant source of microscopical error. Even those concerned primarily with optical problems inherent in the instrument exhorted observers to train themselves in a patient and painstaking work method. Tropes about patient labor, vibration, and fatigue, drawing on contemporary research on these subjects, direct operators on specific physical movements and body placements when working at the microscope—especially of the movements and placements of the hands, but in truth of the whole body.[2] Some of the terms for different portions of the instrument frame it, through analogy, as a counterpart to the human body, with its own body, eyepiece, nosepiece, arm, collar, diaphragm, and foot.[3] The popularizer Arabella Buckley, writing for young readers in *Through Magic Glasses*, even labels an optical diagram as the "skeleton of a microscope."[4] In acting upon the instrument, then, the human body of the observer enters into a kind of dance with that of the microscope. Overall, this chapter will show how Victorian microscopists used skeptical tropes to encourage observers—in this context microscopists also call them *operators*—to execute this dance as a careful, disciplined practice, giving specific practical directions and general rules about how to minimize error and secure microscopical knowledge.

Victorian microscopists wrote at length about how an operator's carelessness, ignorance, or misunderstanding in preparing the slide or manipulating the apparatus can allow contamination or some optical phenomenon to produce a misleading appearance in the image. John Quekett's attention to preparation and manipulation in *Practical Treatise on the Microscope* (1848) and translations of work by Continental researchers prompted concerns that British observers needed better training to prevent undisciplined work. Thus the late 1840s and 1850s brought a wave of advice on avoiding these *difficulties of practice*. These texts formalize physical scripts directing microscopists' actions; they explain how to execute specific preferred methods of microscopical work. Quekett prefaces his *Practical Treatise* by noting the dearth of "treatises of a practical nature" since the work of Brewster, Goring, and Pritchard. He surveys the history and kinds of microscope, "the methods of applying same to scientific inquiry," and "the various plans of preparing, mounting, and examining" a wide range of substances.[5] Six years later, Jabez Hogg's manual *The Microscope: Its History, Construction, and Applications* offered detailed instruction on "the management and manipulation of the microscope" as well as the "varied wonders and hidden realms of beauty" accessed with it.[6] Hogg included a seventy-page chapter on how to use the instrument; most pages offer cautions about particular challenges in preparing objects and manipulating the apparatus (by which I mean adjusting not only the instrument and slide but also the condenser, light source, and other accoutrements).

Hogg's chapter on management exemplifies the tendency of Victorian microscopists to identify a specific material component, procedure, or arrangement of the microscope system, often named after its inventor, as a solution to some problem of microscopical work. An example of this convention, which we might term *the personified contrivance* (because they are so often identified as an "ingenious contrivance") is the "achromatic gas-lamp" developed by "Mr. S. Highley, of Fleet-street." Highley's lamp corrects the "yellow glaring light" produced by "ordinary gas lamps," which is "injurious to the eye and the definition of the object."[7] Hogg's chapter summary lists thirty-three individual attributions, from "Mr. Brooke's double reflector. Lieberkuhn's condenser. Mr. Highley's achromatic lamp" to "Ruysch's and Breschet's method. Goadby's process. Dr. Schacht's re-agents and preservative fluids."[8] While this pattern likely just represents a convenient shorthand, identifying the gadget and providing attribution to its developer, the effect is to imprint the inventor upon the technology. Highley "owns" his lamp but the lamp also invokes Highley, pairing their two bodies in mutual ownership.

These innovations themselves required new skills. In 1861 the biographers of the botanist Edward Forbes remark,

> The trained hand and eye soon perceive how their common tool may be made more perfect, and the tool thus perfected trains them to a higher skill. This in its turn realizes a still more perfect tool; and so acting and re-acting on each other, the microscope and the microscopist have brought us to our current advancement.[9]

This passage celebrates the entanglement of observers' bodies with new technologies in a productive chain of difficulty and innovation. However, sometimes these very improvements introduced new errors. For example, concerning the reagents developed for use in specimen preparation as a preservative or stain, Hogg passes along the "useful warning" of "Dr. Parkes," that "some of these re-agents" can form "unusual crystalline forms…under the microscope," and that "without…a perfect recognition" of these forms, "errors in micro-chemical research cannot but occur."[10] Hogg's book presents at once a romantic narrative about heroic individual researchers overcoming obstacles, and a realist narrative about a community of observers working together to solve a series of problems of increasing complexity. Both these narratives start from an assumption that observers working with microscopical apparatus inevitably encounter, solve, and also generate difficulties of practice.

The mid-century surge of guidebooks and research texts swell a flood of published instruction guiding readers in the physical, mechanical actions of microscopical practice. These instructions insistently place the operator's body—hands and eyes especially—within the scene of operations, functioning as agent but also as part of a dynamic physical system. A book review in *Fraser's* invites "Madam" to "unveil the nervous system" of the marine bristle worm *Eunice sanguinea* "with the aid of your lens and your microscope, a delicate pair of pincers and a cataract needle"—but sardonically warns her to "take care of your taper fingers" as the "sharp and curved teeth" of this little worm "will soon, if you are not careful, incarnadine your lily hand."[11] Here the hand adjusts the scene of observation but also becomes part of it, even endangered by it. The microscopist's hand does not serve primarily as an incarnation of human identity or a symbol of God's thoughtful design, as it does in other literary and scientific traditions.[12] The hand of the microscopist is just an ordinary physical appendage; it is capable and trainable but not inerrant. Even the *trained hand* (or *educated hand*) enjoys only unstable and shared agency.

Indeed, many problems in microscopy derive from clumsy technique in object preparation or instrument manipulation, due to the difficulties presented to the operator by instrument and object—their implicit material resistance to observation. Microscopical guidebooks seem to discuss these difficulties endlessly. Problems arise first in preparing the specimen. For example, Schleiden discusses sectioning as one of the interventions observers can perform to obtain *many-sided comprehension* of an object. He offers techniques for very delicate sectioning, which, he says, must be done by hand to ensure more sensitive handling than with a Valentin (double-bladed) knife or microtome. However, operators still risk tearing or crushing soft organic objects with the blade, and manual sectioning "does not suffice for very accurate investigations." Some objects can be dissected instead of sectioned, he says. Here again, though, the simple (dissecting) microscope usually used for such work has a "great disadvantage," namely "that the object is reversed [in the image]" and "our hands are so far removed from the eye as to render their movements uncertain, so that a tearing or crushing of the object at random is scarcely avoidable."[13] Worse, the forceps, needle, and blade themselves are "the greatest hindrance," since "no point is sufficiently fine" for accurate work.[14] Microscopists warn readers that their own fingers and the imperfect tools they work with can both reveal and damage the specimen. Such cautions also appear in cheap instructional texts for general audiences or children. The Rev. J. G. Wood, in *Common Objects of the Microscope* (a 1s. "elementary handbook" published by Routledge) endorses this complaint about the awkwardness of dissection, remarking, "Under a low power the stage forceps are almost indispensable, but for the higher can hardly be employed at all, as the light and the focusing are both so difficult of management that the comparatively coarse forceps cannot be successfully used except by a very practised hand."[15] Here Wood is not just recommending particular dissecting tools but also demonstrating how microscopical vision relies upon the operator's dynamic interactions with object and apparatus. Only a patient, long-practiced operator can work productively with "coarse forceps" under high powers, where the light and focus become stubbornly "difficult of management."

Similar difficulties arise in slide-making, leading to further errors. Operators are called upon to either prevent or recognize and discount these errors, which are figured as interpolations of unwanted materiality into the image. Hasty preparation risks reorienting the image around the wrong kind of matter—a strew of debris drawing attention that rightly should be directed to some specific object. Similar methodological errors doomed Emanuel Edward Klein's 1874 "discoveries" of the germs for typhoid fever and sheep-pox. The physician and germ-theory skeptic Charles Creighton showed that the so-called bacteria were simply coagulations formed from his reagents, a reversal that damaged the claims of the microscope to be a reliable scientific tool.[16] In these cases the materiality of the slide preparation intervenes in the observation, capturing and misdirecting the observer's gaze.

While Victorian sources do not agree upon a particular phrase for this recurrent concern, the portrayal of dirt as *matter out of place* is a contemporary and relevant trope.[17] This description is especially appropriate as the so-called contaminants were themselves objects of study in other contexts. Quekett warns, "In wiping the slides and the covers ... be careful not to employ a substance likely to leave any nap or down behind, as coloured filaments, derived from table-covers, pocket-handkerchiefs, &c., have, more than once, been mistaken for highly organized structures."[18] William Benjamin Carpenter advises the novice to learn to prevent, clear away, or ignore the air-bubbles that look like important structures under the cover-slip and the "fibres of silk, wool, cotton, or linen [or] grains of starch" or dust that may contaminate the slide.[19] Schacht recommends that students "gain an acquaintance" with a number of "extraneous objects": air-bubbles, "fibres of paper, or of linen, woollen, or silk textures" left from the cleaning cloth or the hairs of a camel-hair brush used in preparing the specimen; "granular particles of dust"; the "common forms of infusoria," yeast, and mould; "epithelial cells" brought into the preparation when "the brush is drawn through the mouth"; and pieces of cork or "small fragments of the skin of the finger" cut off in section-making. (His phrase "gain an acquaintance" points to a kind of agency among these contaminating factors.) Two decades later, Robert Koch notes that the aspiring bacteriologist may be misled by "organisms which are derived from the fluids used for staining, washing, &c."[20] It is striking how often the untidy materiality impinging upon the image is, ironically, residue from the operator's efforts to clean and prepare the slide—sometimes residue from the operator's own mouth or finger, traces of the self and its agency, inadvertently presented for scrutiny. The body of the microscopist prepares, adjusts, and views the microscopical object, but it also sometimes becomes that object.

Popular texts addressed these issues also. Mary Ward, writing for a middle-class lay audience, claims that she designed *Microscope Teachings* (1864) as a "useful" book that would "attract those readers who, unversed in microscopic marvels, might possibly feel repelled by a complete and lengthened treatise."[21] Despite this address to an impatient reader, Ward emphasizes the need for the microscopist to take pains in practice. She recalls an embarrassing error in preparation that was inadvertently disclosed "in an illustrated periodical," when the author noted "singular-looking objects armed with hooks" distinguishing his object; Ward readily recognized them as bits of feather "probably flown from the end of his quill"—another example where the pursuit of order and method (in this case, notetaking) leaves traces that ironically thwart reliable observation.[22]

All too often in these scenarios, glass becomes less than transparent. While the magnified image relies on material properties of the lenses, the materiality of glass can obtrude into the image in unwelcome ways. Confounding elements include condensation or scratches on the lenses, or the "disturbing effect of the glass cover" of the slide.[23] Glass also provides a necessary material frame for microscopic objects, affixing them helpfully onto a solid structure that can be readily handled and stored, although their near-invisibility requires the worker to label slides carefully to prevent mistakes. Wood notes indeed that haphazard labeling of slides can "cause great confusion and loss of time, and often renders a valuable collection quite useless."[24] Some microscopists avoided these errors by purchasing ready-made slides from an optician. Novices hoped to see more than they could skillfully prepare; experienced operators might not be able to manage the delicate work of injections or sought specimens of objects not available to them, such as those from distant sites or dredged from the ocean floor. Moritz Pillischer, a Hungarian optician working in London, in 1873 offered a twenty-five-page "Classified List of Choice Microscopic Objects, comprising Rare, New and Interesting Specimens...prepared by the best English and Foreign Artists." The list offered "nearly 3000 objects" for sale, starting with three pages of *Diatomaceae*; Pillischer was clearly angling for the custom of the "diatomaniac." But he also draws "special attention...to the *transparent injections*, which are prepared expressly for me by a foreign (German) gentleman, which for their superiority cannot be equalled."[25] While simpler slides cost only 1*s*. or 1*s*. 6*d*., injected preparations cost 1*s*. 9*d*. or 2*s*. 6*d*. and transparent injections, "specially prepared...by a celebrated German Professor," up to 3*s*. 6*d*.[26] Pillischer sells the skills of his Continental associates as much as the objects themselves.

The scale of the objects examined in microscopy means that often the worker must prepare a tiny object without a clear view of it, or prepare (for example, section) a visible object so as to expose invisible structures. The historian of science Hans-Jörg Rheinberger notes that "As a rule things tailored to the lens cannot themselves be seen during the process of preparation, at least not in the detail that is the key to the success of the enterprise....This requires that the preparer focus on the *regularities* of the production process, which must so to speak function blindly."[27] Here again, training is needed to counter an effective loss of vision in the delicate work of preparing the object for observation.

Once slides are prepared, the operator must select, set up, and manipulate components from an extended apparatus including objective, eyepiece, diaphragm, focus, condenser, polarizer, micrometer, and the like. In the 1840s and 1850s, writers increasingly offered practical advice about how to do this work, eventually standardizing many common practices (these, too, often identified with the innovator). For instance, one should increase magnification by changing the objective, not by choosing a stronger magnifier for the eyepiece; Quekett points out that "any error, however slight...is further magnified by the eye-piece."[28] Similarly, an 1847 essay on the "wonders of the microscope" in *Chambers's Miscellany of Useful and Entertaining Tracts* (available for 1*s*. per volume), recommends low or medium powers for ordinary work. Addressing "the novice, who is fond of seeing 'fleas as big as elephants,'" the author explains that in such a grossly magnified image "the beauty of the markings and fine colour is lost, and only a minute fraction of the object can be viewed at one time."[29] In this gently mocking passage the author ventriloquizes the voice of the eager tyro, boasting of his "fleas as big as elephants" in language reminiscent of the oxyhydrogen show.

J. W. Griffith and Arthur Henfrey position their *Micrographic Dictionary* in 1856 as a response to improper observation, arguing that careless action imperils the future of microscopical science. They remark that "the microscope has given rise to most absurd errors," such that "the very utility of the microscope as a means of investigation has been called in question." Still, they conclude optimistically, "all the known errors committed in the application of this instrument have been corrected also by its use." Their *Micrographic Dictionary*, they contend, provides a valuable service by "entering minutely into what may be called microscopical manipulation and the special education of the eye."[30] Their commitment to "entering minutely into" matters of manipulation shows how the microscopical *aesthetic of the detail* extends beyond the object to the work itself, here prompting not observers' wonder but their skeptical practice.

Working microscopists must skillfully select and adjust not just their microscope and objective but also their illumination. Chris Otter has traced Victorian debates over how well gas and electric light comparatively permitted discernment of "color, distance, and detail."[31] These and other visual capacities, and technologies designed to adapt diverse forms of light in their service, were even more central to microscopical practice. Apparatus affecting the quality and passage of light included various kinds of lamps, mirrors, condensers, illuminators, diaphragms, reflectors, prisms, polarizers, tinted shades, colored filters, and more, tuned to control the intensity, heat, tone, color, flickering, direction, glare, and polarization of daylight or artificial light. Even small changes in illumination could create very different images under the lens. As the skilled American microscopist Col. J. J. Woodward notes in a letter to John Mayall in 1875, "I can get *a great variety of appearances by modifying the illumination*."[32] Treatises on microscopy devoted whole chapters to the pleasures and frustrations of various lighting systems—Carpenter's 1891 edition of *The Microscope and Its Revelations* dedicates some forty pages to such matters—as manufacturers increasingly offered a range of lamps and other apparatus for purchase.

But the intricacies of lighting were not easily mastered. Andrew Pritchard comments in 1832 that difficult objects "require the most perfect instrument and precise tact in the management of the illumination."[33] The term *tact* here acts as a trope, playing off the word's usual denotation, sensitive social action. Social *tact* was a separate skill, thought to be primarily displayed by women.[34] In contrast, microscopical *tact* describes an (implicitly male) operator's manual dexterity in handling the microscopical apparatus with precision and sensitivity. Microscopical *tact* does seem to reference the nuanced, feminized judgment of social *tact* but combines it with an informed physical action—a *tactile* skill in the sense of touch and the delicate control of the muscles. The quest for tact informs many of the instructional manuals of the 1850s and beyond. It was a critical skill for original investigators, since the image under high powers was excruciatingly responsive to even fine adjustments. Carpenter explains in *Revelations*, "A minute alteration of the focus often causes so entirely different a set of appearances to be presented, that, if this alteration be made abruptly," the relation between these images "can scarcely be even guessed-at."[35] Tact was equally important in delicate preparations such as sectioning, staining, and injecting tissues.

The adept microscopist's hands thus became a valuable component of the microscopical system; they even occasionally appear in illustrations of the microscope and its tools.[36] While the inclusion of the operator's hands was an established iconographic convention in representations of scientific experiments or medical procedures, this was becoming less common in the nineteenth century.[37] The appearance of the operator's hands in some microscopical illustrations echoes and amplifies writers' frequent discussions of the placement and actions of the hands as an integral element of microscopical practice.

Microscopists suggested diverse physical exercises for training the hands and acquiring tact. The surgeon Arthur Edward Durham, giving a presidential address to the Quekett Microscopical Club in 1868, argues that "[n]o one can become a successful microscopist without acquiring in a very high degree delicacy of touch, and lightness, accuracy, flexibility, and steadiness of manipulation." He tells a story about a surgeon—perhaps himself—who is discovered dissecting a fly as a means of honing that sought-for delicacy of touch.[38] Another medical man, John Hughes Bennett, advises using manual controls whenever possible, for more precise management of the instrument and to train the fingers in making tiny, delicate adjustments. He explains, "[T]he habitual employment of artificial methods of moving about the object, as by the screws of a moveable stage, will prevent your acquiring that dexterous use of your fingers and accuracy of manipulation which is at all times so useful." He scoffs, "Nothing, indeed, can be more amusing than to see a man twisting his screws, pushing his heavy awkward stage about, and laboriously wasting time to find a minute object which another can do in a moment, and without fatigue, by the simple use of his fingers."[39] Here the supposedly convenient artificial aids prevent the operator from developing tact and, in the long run, create more labor. Using one's fingers is, Bennett explains, a constant element of microscopical work. Arthur Cole, a preparer and seller of microscopical slides, argues that gaining "manipulative skill... sometimes proves to be the most arduous portion of the work" and that "[p]ractice and perseverance... are the only royal roads to manual dexterity."[40] The trope of the *royal road* refers to the broad highway constructed by Persian king Darius the Great in the fifth century BCE, and is generally used to refer to a trouble-free or easy mode of approaching a difficult task. It is usually used in the negative: there is no royal road to expertise. Cole uses the phrase ironically here, to make the point that patient, painstaking, diligent practice is paradoxically the easiest (or even the only) way to attain manipulative skill.

It took years of skill to master proper management of a complex microscope and its objects. The American J. E. Smith, indeed, believed that the manipulation of microscope instrumentation "is a study, a profession in itself, and an accomplishment that must be fully mastered at the outset, before one can investigate with precision, or have claim to the confidence of others." He warns that "the most expert manipulator will constantly have something to learn—there is no landing place where one can stop and rest."[41] Many of these technical difficulties were exacerbated at higher powers, so when microscopists worked at progressively stronger magnifications, their practice had to become more and more exacting.

In the instances above, what makes tact necessary is that the physical apparatus and object present a series of implicit difficulties to the operator. Even inanimate substances can seem to engage with the operator's actions insofar as they offer (or microscopists perceive them to offer) resistance to their actions. A late-century American article instructing students on apparatus remarks, for example, that "[a]t first it will be somewhat difficult to illuminate the field properly as the mirror will seem to have a stubborn disposition of its own, but it will finally respond to the microscopist's slightest touch."[42] This phrasing calls upon humor to ease students over those first frustrating moments, but the locution draws upon a longstanding sense that operators must learn how to evoke the best response from their equipment, and that the non-human elements of this partnership, like the human ones, help determine the success or failure of the endeavor.

Texts addressed to investigators also frequently note the physical or optical challenges that objects and apparatus so often present to the operator. In 1854, for example, the engineer F. H. Wenham, discussing the fine focal adjustment, warns that Smith and Beck's "very fine 4–10th" objective with a 90-degree working aperture "will not well show the markings on a severe test object such as the *P. angulatum*, until it is brought to the exact point of adjustment."[43] (This phrasing accommodates either a passive or active reading in that the objective could be unable to show, or refusing to show, the markings on the diatom. In either case its materiality forms a limiting factor in the observation.) Learning how to adjust the focus in this way requires coordination between operator, apparatus, and object. Human interventions with the object and apparatus will yield changes in the image; these changes then direct further human action, in a feedback loop. For instance, referencing an iconic test object like the *Podura plumbea* (springtail) scale, Wenham explains that "if the dots have a tendency to run into lines... when the object is within the focal point, the lenses must be further separated."[44] The object, instrument (fine adjustment dial), and operator must work together to bring the details of the object into view. Only the trained hand can elicit this graceful cooperation, and even then not reliably. Wenham, a frequent innovator in microscopical apparatus, designed a fine adjustment mechanism that made this careful dance easier, but other versions could defy even the trained hand. He complains, "I can generally adjust my own glasses in a few seconds, whereas I sometimes most uncomfortably spend several minutes in wrenching round the adjusting collars of our professional makers, to obtain the same result."[45] For Wenham, hand, glass, and fine focal adjustment can accommodate one another nicely, but too often they force a wrangle. Implicit in his statement is the notion that preparation and manipulation should be frictionless and transparent; but all too often these processes, and their recalcitrant materiality, impede observation instead.

One might say, using Bill Brown's terms, that microscopical instruments and objects become visible not as *objects* (in the general sense) but as *things* in the course of this resistance. Brown posits that "We begin to confront the thingness of objects when they stop working for us…when their flow within the circuits of production and distribution, consumption and exhibition has been arrested, however momentarily."[46] After all, the much-vaunted "difficulty" of microscopical work inheres in the many little obstacles that microscopists must overcome—and there are multitudes of these. Each of those difficulties, minor or major, can be understood as a resistance inflecting the dynamic network that observer, instrument, and object co-create. Beings, body parts, and substances first become microscopical objects when they are entered into the process that so often characterizes microscopical work—where they are collected, selected, prepared, manipulated, and finally observed. But integral to the notion of microscopical work is this notion of difficulty or resistance; so that microscopical objects become visible as such only through a kind of fog of difficulties. That is, since difficulty and resistance are constitutive of the category of the microscopical object, they serve as microscopical objects only through a process where they almost always function as things as well.

Beale and Patient Labor

Lionel Beale, a physician and professor of physiology and anatomy at King's College, London, helped to institutionalize this trend, offering an early, influential book focused entirely on technique. He originally presented the chapters in *How to Work with the Microscope* (1857) as lectures for medical students and frames this book explicitly as a textbook of microscopical training, published as an act of evangelism, "to assist in diffusing a love for microscopical inquiry."[47] The structure of Beale's guidebook reflects this practical and material orientation. The first half of the book describes types of instruments, tools, and other apparatus and instructs readers on the illumination and mounting of objects; after two chapters on observation, the last quarter of the book focuses on special preparations (injections and chemical reagents). Beale eschews optical explanation altogether, directing readers to a list of other authorities. He concludes with his own prescription for acquiring tact: eight "Tables for Practising the Use of the Microscope and Manipulation"—a formal series of exercises addressing common errors and, as a whole, a system of instruction in microscopical manipulation.[48] Beale's work was lauded as thorough and well grounded, and it became a standard text not only for students but for working microscopists also, reaching a fifth edition in 1880. The *Journal of the Quekett Microscopical Club* reports in 1869 on the coursework of "Mr. Suffolk's Class," which includes elements mentioned in Beale's program of study and cites him specifically.[49]

Beale grounded his work, as he thought every microscopist should, on patience and painstaking care. Novices were often caricatured as hasty, careless observers. Oliver Wendell Holmes, in his 1877 address to the Boston Microscopical Society, jokes,

> if a mistake in looking through a microscope is within the bounds of possibility, the young student will make it. If there is an air-bubble, or a crack in the glass, or a speck of dust, or anything of any sort which has no business in the field of vision, his eye seems instinctively to fasten upon it, as the needle leaps to the magnet, as lovers' hands fly to each other the moment the train enters a railway tunnel.[50]

Holmes's crack about the novice's "instinctive" eye that "leaps" or "flies" to the wrong object suggests that microscopists must learn to be deliberate; observation is necessarily patient and painstaking work.

Drawing on his own early frustrations, Beale argues that a system of training encourages more beginners to aim for original research while disciplining them in best practices. The challenges of microscopical work were daunting. Beale writes, "[M]any…are deterred from attempting any branch of original investigation solely by the great difficulty they experience in surmounting elementary detail and mere mechanical operations."[51] He urges novices to turn the meticulous eye of the naturalist onto microscopical method; he promises that, with meticulous attention to detail and tact, the merest tyro can hope to become an adept and contribute to microscopical science. Beale's books and articles advocate for what he calls the "patient prosecution" of mechanical exercises from simpler to more complex, inspiring James Glaisher, in his 1866 presidential address to the Microscopical Society of London (MSL), to say, "[T]his *patient labour* proceeding step by step, from the lowest to the highest…constitutes…the difference between a trained and an untrained observer." Indeed, Glaisher continues, "many beautiful details can be seen by the carefully-trained eye, and traced by the trained hand, of which not a trace is even suspected by observers wanting in this educated eye and equally important educated hand."[52] The trope of *patient labor* here combines optical and physical work to yield hidden wonders.

Beale designed *How to Work with the Microscope* to promote this ethos of methodical, patient labor. He frames this strategy as a departure from earlier authorities and sets himself to shifting audience expectations, urging a commitment to the detail and to tact. (These were principles required by his own particular areas of interest; his "bioplasm" version of germ theory was influential in the 1860s.)[53] Although his original listeners were medical students, they may well, he says, "find me tedious, and feel that I am trespassing too far upon your patience." They "may, perhaps stigmatize many little points…as merely mechanical" or "unimportant and hardly worthy of attention." However, Beale notes that his own "difficulties…at every step in original investigation" had taught him "not to undervalue the importance of what have been termed mere mechanical operations."[54] He argues, "I feel very strongly how intimately success in microscopical... inquiries, is connected with a readiness in surmounting comparatively small difficulties and with the possession of mechanical dexterity."[55] Indeed, he contends, skill in microscopical manipulation is an "absolute necessity... to all who work with the microscope."[56] As with Griffith and Henfrey, Beale's concern is not so much for the hapless novice or even a misdiagnosed patient, but for microscopical science itself. As he says about errors of preparation and judgment, "many mistakes have arisen in consequence of sufficient care not having been taken to prevent the introduction of various substances by accident."[57] Insufficient attention to detail and method dooms not just the observer but all the other workers who try to build on that erroneous observation, and it imperils the reputation of the instrument as an investigative tool.

Beale's later work even more strongly advises the reader to master detail and method through patient labor. In the 1868 (fourth) edition of *How to Work with the Microscope*, he first acknowledges that "manual dexterity" is "subordinate to many higher mental qualifications." Then, however, he doubles down on his argument: original observers "who are most fully conversant with elementary detail and most clever at demonstration, will be the most successful" in difficult problems. Indeed, he proclaims, "[n]o man ever did perform real work until he had himself mastered minute practical details."[58] In the 1880 (fifth) edition, he even more aggressively argues that these skills are needed to not just secure but police microscopical science. A disciplined practice assures our own ability to do "real work" and allows us to recognize and reject "arbitrary dogmas and false statements" by those—presumably the anti-vitalists he was battling at the time—who "deceive and humbug" others.[59] These later editions of the book powerfully convey Beale's conviction that the willingness to take pains with "apparently unimportant practical details" is central to microscopical practice.[60] If anything, he adds new emphasis in 1868 on how a patient address to detail and method "invariably... expedites true scientific progress." Yet, Beale laments, his approach was not shared by all: "Many even express contempt for mere elementary work and mechanical skill."[61] His tone is one of disappointment and frustration with the resistance to his program of improvement.

By 1880, however, Beale prophesies that, although the subject matter may be "reproachfully stigmatized as mechanical" by some, "those who understand real work will not find fault" with his approach.[62] Indeed, Durham had even in 1868 endorsed such concerns, telling the assembled Queketters, "No man can be a successful microscopist who has not a well-nigh inexhaustible fund of patience." He exclaims, "In the collection, the preparation, preservation, and examination of objects how much patience is required! It is not enough for the microscopist to 'learn to labour'—he must 'learn to wait.'"[63] Working microscopists knew that often only patient, well-trained hands could coax good images out of the instrument and object. They saw the progress of knowledge as deeply rooted in these embodied habits of microscopical practice.

Beale modeled his own tenets of patient labor, attention to detail, and manipulative tact in his own work. Charles Brooke, in his presidential address to the MSL in February 1865, praises an article Beale had presented that year "on the structure and formation of striped muscle." Brooke points to Beale's "admirable methods of preparation" and references the 1865 (third) edition of Beale's *How to Work with the Microscope*. Brooke notes that Beale had managed to resolve "many points of structure" previously unknown, but his observations on muscle contradicted the consensus of Continental workers. Brooke comments, "It has...been suggested that some of the appearances described by Dr. Beale may be the results of manipulation." That is, because manipulation can cause artefactual errors, Beale's own reputation as a master manipulator may have worked against him with some listeners. However, Brooke continues, "I can only express my own conviction, that in this case 'seeing is believing;' although I am fully aware that microscopic illusions are neither rare, nor difficult to produce."[64] As Brooke knows, the prevalence of "microscopic illusions" constantly threatens to undercut the common trope *seeing is believing*. Brooke's performance of mitigated skepticism here ("in this case," "I am fully aware") defends Beale's reliability as a manipulator, even as these comments overall acknowledge that the operator's hands can both prevent and cause deceptive appearances.

Beale's research complemented his commitment to detail and manipulation. In fact, when he gave the Croonian Lecture at the Royal Society in 1865, he spoke on how nerves communicate action to the muscles, in a talk based on his work with very high powers. He argued that the finest sections of nerve do not terminate but form "nerve-circuits" running through muscle; that "the fundamental arrangement of a nervous apparatus is a complete and uninterrupted circuit"; that "a current probably of electricity is constantly passing through all nerve-fibres"; and that the muscles "are influenced by the varying intensity of this nerve-current."[65] This image of a charged neural network in constant communication with the mechanical structure of muscle provides a useful model for Beale's (and Carpenter's) vision of microscopical work, in which operator and apparatus are enmeshed in a continuous discourse of sensitive action and response.

As workers labored to acquire tact, individual components seemed to take on agency and even personality, requiring careful management of their quirks. Such descriptions play on the established trope of the *instrument as companion*.[66] The diatomist William James Hickie, a skilled observer, notes in 1875 that the objective he was reviewing

> had only been in my hands some thirty-six hours,—too short a time to enable one to become acquainted with all its little whims and peculiarities. For lenses

> too have their little whims as well as human beings; and these have to be studied, and humoured, if one is to succeed in making a lens do its very best.[67]

Hickie uses humor and anthropomorphism here to describe his sense of the lenses' agency in enabling or hindering the work of observation. His affectionate phrases ("little whims," "have to be…humoured," "do its very best") also express an emotional bond with his instruments and objectives, reflecting the many hours spent figuring out how best to adjust them in various situations. They show how, especially over time, the committed microscopist perceives an affective life in apparatus as well.

Experienced workers did develop longstanding relationships with their favorite instruments and lenses. Famously, Christian Gottfried Ehrenberg used the same instrument from his student days in 1832 until his death in 1876.[68] While his familiarity with this instrument enabled his groundbreaking work in micropaleontology, his loyalty to it likely contributed to his errors in arguing that all microorganisms contain complex internal organs.

Hickie would have been well aware of the implications of his language in the passage above; he taught linguistics at Cambridge. His fond, indulgent tone in this *Monthly Microscopical Journal* review invites readers to reframe microscopy as a collaborative venture with the instrument not unlike child-rearing or animal-training, suggesting a trope of *microscope as pet* as well. He invokes a vision of community in which observers recognize the agency of the apparatus, if only whimsically, and share an affective and physical entanglement with their recalcitrant lenses.

Vibration and Fatigue

Not just the observer's hands but the whole human body and indeed the whole working environment becomes a functioning part of the microscopical research apparatus. Microscopists, whether novices working with cheap instruments or original investigators working with fine instruments at high powers, often had to contend with vibration.[69] Because of its great magnification, the microscopic image was exquisitely sensitive to vibration from loose or worn parts, or from poor design; an operator adjusting the stage, slide, or objectives or leaning too heavily onto the instrument could jiggle the field of view.[70] Screw adjustments and other innovations were developed to minimize unnecessary jolting to the apparatus, in what was termed the "coarse movement" and the "fine movement," but these could introduce new problems like "spring," "loss of time," and "twist."[71]

Microscopists using instruments without sturdy stands might often find the image trembling. J. G. Wood explains the problem with a memorable trope, as is typical in his lively, introductory text: in the microscope, "motion is magnified as well as substance, so that if the instrument trembles in the least degree, the object becomes almost invisible, seeming to flutter before the eyes of the observer like the whirring wings of a hovering fly."[72] Here the image becomes lovely and whimsical, though impossible to study. A more common trope was *dancing* objects or stand, which endangered vision and research. Carpenter in 1856 explains that a poorly designed stand communicates a vibration to the image, "giving a dancing motion to the objects."[73] Similarly, the *Pharmaceutical Journal* advises that "the object should never 'dance' under a moderate power," and the American journal *The Microscope* warns "even a slight tremor becomes somewhat appalling if the stand is a dancer." The author concludes by rating various stands according to whether or not each is "a dancer."[74]

Vibration could arise from even seemingly inconsequential events, such as a heavy vehicle rumbling by in the street outside.[75] In an 1841 lecture to Edinburgh medical students, Bennett asked the audience to refrain "from applause, or any thing which may induce vibration, or occasion dust," because "instruments delicately adjusted" to exhibit difficult structures "are liable to become deranged by the least vibration" and dust on the object glass may "considerably mislead" the eye or "even obscure the object altogether."[76] A decade later, Bennett explained to a similar audience that with "sensible vibration" the object may be "thrown altogether out of the field of view."

However, it was not a foregone conclusion that the largest, heaviest stand secured the best physical conditions for observation. Bennett, for example, rejected the "large London instruments," which were designed for steadiness but "at so much cost and increase of bulk as to render [them] almost useless."[77] Twenty-five years later, J. J. Plumer seems to argue otherwise in his little book on choosing a microscope; he prompts readers to buy the largest and most expensive London stand that the purse can bear. "*Begin by procuring the best Microscope that the best Optician can give you*," he urges. "*Be satisfied with nothing short of No. 1*."[78] He reasons that only the biggest and best instruments accommodate important technical components and provide a firm, unmoving foundation for microscopical work. He acknowledges the affective elements of this question as well, noting "the pleasant sense of security" secured by the "steadiness" of a fine stand; a good instrument is a pleasure to work with. But even Plumer admits that the large, complicated instruments bring physical drawbacks as well. For instance, Pillischer's "No. 1 stand," with its expansive traversing stage movement, results in "the inconvenience of a bulky Microscope, and...the loss of the circular rack"; Powell's No. 1 boasted a "circular rack movement" that could "make an entire revolution," but as a result suffered from "a limited range of stage motion, and a somewhat unwieldy stand."[79] All of them are heavy and bulky, necessitating the purchase of a smaller instrument for travel.

Some elite microscopists went to great lengths to test their stands or reduce vibration. In the 1883 edition of his *Revelations*, Carpenter pits two iconic technologies of the century against each other, advising, "The Author has found no test for steadiness so *crucial* as the vibration of a paddle-steamer going at full speed against a head-sea."[80] The great microscopist E. M. Nelson built a basement studio and worked into the small hours of the morning in order to avoid any "chance of tremor" caused by traffic passing his Hampstead residence. An obituary for him notes that "he used to declare that it was not possible to obtain satisfactory photomicrographic results in London until well past midnight." The many hours Nelson spent in this underground workspace produced his "wonderful photographs of difficult, and hitherto unknown, diatomic and other structures" but likely also (the obituarist says) "induced [his] severe rheumatic troubles" and "permanently impaired health."[81]

Even motion as slight as the observer's pulse or breath might derange the focus of a sensitive or poorly made instrument—or threaten the body's ability to see distinctly.[82] The elite microscopist attempting difficult objects must learn to regulate digestion, respiration, and other bodily energies as well as the emotion that might exaggerate them. The great American naturalist Louis Agassiz remarked upon the need for observers to modulate their physiological rhythms in order to achieve and sustain crisp vision. In an introductory textbook on methodology, he says,

> There are subjects of microscopic research so obscure that the student must observe a special diet before undertaking his investigation, in order that even the beating of his arteries may not disturb the steadiness of his gaze, and the condition of his nervous system be so calm that his whole figure will remain for hours in rigid obedience to his fixed and concentrated gaze.[83]

Agassiz's admonition about embodied vision echoes the tenets of physiological psychology, where body, mind, and spirit necessarily act on and through one another.

As the problem of vibration demonstrates, Victorian microscopists' embodied entanglement with the instrument extended beyond hands to the whole human body. Eye, head, neck, hands, wrists, arms—even heart, lungs, torso, stomach, and feet—are drafted into microscopical practice. Microscopists argued that the body is fatigued by this work and imperils it.[84] Books instructed readers not just on how to use their eyes and hands in observing through the microscope, but also on how to position their body vis-à-vis the microscope in order to minimize fatigue, discomfort, and other sources of error. For example, the eye, which (as I will show) forms part of a theoretical optical system with the microscope, also functions as part of the physical apparatus. An observer can cause error simply by approaching the eyeglass in the wrong manner. Hogg's 1854 *The Microscope* suggests, "In winter it will be proper to slightly warm the instrument before it is used, otherwise the perspiration from the eye will be perpetually condensing on the eye-glass, thus greatly impeding vision." At times, the optical and physical roles of the eye combine, as when Hogg recommends that readers "place your eye quite close to the eye-piece, otherwise the whole field of view will not be visible."[85] Again, this concern is registered in elementary texts as well. Maria Hack's science primer *Harry Beaufoy, Or, The Pupil of Nature* (1821) includes a humorous incident where ten-year-old Harry, looking down a microscope for the first time, is taken aback by the "legs of a great, black spider" coming across the scene. His father is puzzled, but "Mrs. Beaufoy, who guessed the cause of his perplexity...laid down her work" and instructed Harry "to touch the instrument with his hand, so that he might obtain a resting-place for his cheek." This would prevent his head from "waving about" and causing his "eyelashes [to] brush against the glass."[86] By positioning himself in the wrong way, Harry caused the instrument to magnify—and transform—his own body rather than that of the tadpole under the lens.

While it is obviously important to align the eye and eyepiece correctly, microscopists also address the positioning of the hands. Wood says that "it will be found very useful" when using a dissecting microscope "to support the hands by means of a book or piece of wood on each side of the glass cell, as the handling becomes very awkward and fatiguing without some such precaution."[87] Fig. 13.1 from Hogg, showing hands working without such a support, does suggest how fatiguing such a position might be. Worse, an ill-positioned hand can lead to error. In the ninth (1878) edition of the *Encyclopedia Britannica*, Carpenter recommends avoiding the Continental style of dissecting microscope because "the resting of the hands on the supports are apt to depress the stage in a degree that affects the focal adjustment." Similarly, he prefers a compound microscope with a tube that can incline. He explains, "The instrument should be so constructed [that]...the arms may rest conveniently upon the table," giving "such free play to the muscles of the hands that movements of the greatest nicety may be executed by them, and the fatigue of long-continued observation is greatly diminished."[88] Such comments ask the observer to place the hands in certain positions in order to reduce errors in adjustment and manipulation, likely improving mental alertness and attention as well.

Often such advice shifts from directing the reader's hands to addressing other parts of the body. In David Brewster's 1837 article on "Microscope" for the seventh edition of the *Encyclopedia Britannica*, he frequently tells readers where to place their body and hands in performing a particular technique in order to work better and avoid fatigue. Discussing the polarizing microscope, for example, he says, "The observer thus carries the polarizer and the object in his left hand, and in his right the magnifier and the analyser." In ordinary work, the observer should position "his doubled arms upon a stool...while his chin rests either upon his arms or upon his breast."[89] Brewster depicts microscopical observation as a full-body endeavor: "Unless particular arrangements are made by the observer for his own comfort," he says, "there is no bodily fatigue to be compared with

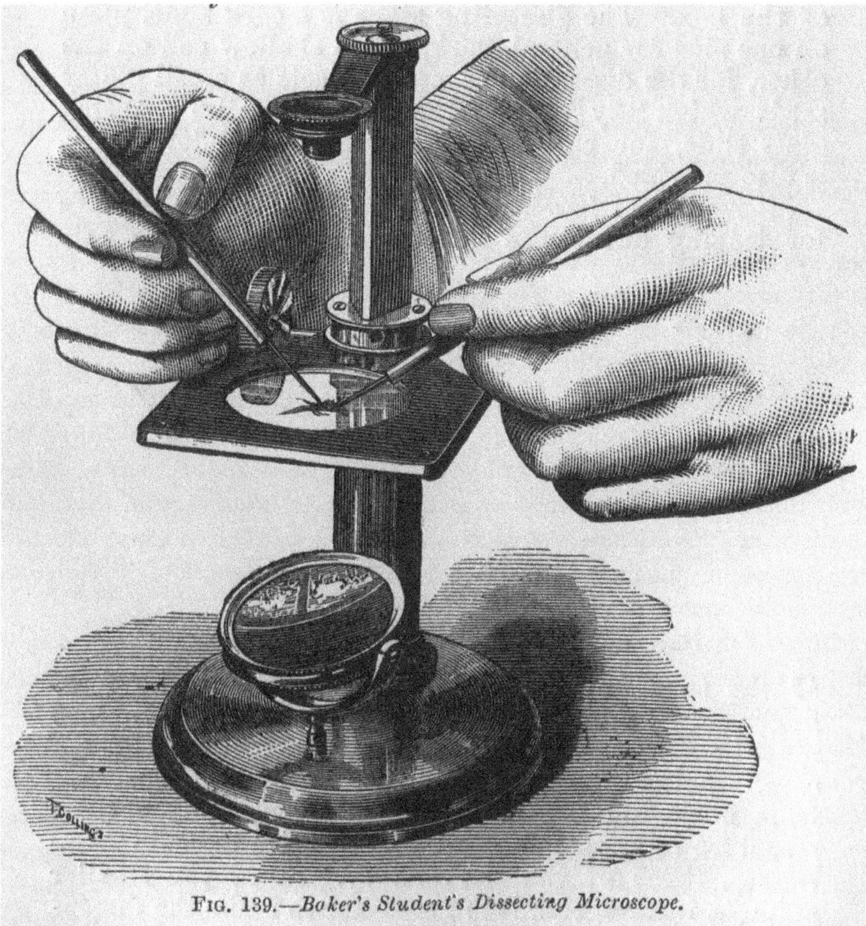

Fig. 13.1 Baker's Student's Dissecting Microscope.
Source: Jabez Hogg, *The Microscope*, twelfth edition. London: George Routledge and Sons, 1886, p. 200.

that of the use of the microscope. The eye, the mind, and the whole frame, are on the stretch."[90]

In 1850 Bennett similarly moves from fatigue of the arms to the whole body. He warns of "the necessity of continually turning the large [coarse adjustment] screws affixed to most microscopes," which "becomes fatiguing in the extreme." As for the fine adjustment, he recommends its placement "near the hand which rests on the table," as it was in Oberhaeuser's medical microscope designed with Bennett's guidance. Bennett complains, "in some London instruments, it is placed on the top of the pillar, so that you must raise your hand and arm every time" you use it—or "it is placed in front of the body, so that you must stretch out the arm and twist the wrist to get at it. No one could work long with so inconvenient a contrivance," requiring the operator to embrace the instrument for every adjustment. Furthermore, the "large London instruments," he complains, require the observer either to stand during observations or to tilt the microscope body toward the eye.[91] Indeed, the physician Heinrich Frey notes in 1872 that English microscopes generally enable an inclined body tube "to prevent the neck from becoming tired, or a flow of blood to the head."[92] However, this triggers a chain of additional difficulties: the stage is then tilted at an oblique angle and requires the observer to stretch forward to reach it, making it difficult to manipulate the object. More complex instruments offer screw movements to facilitate adjustments, but then—Bennett grumbles—"in this way complexity has been added to complexity, until a mass of brass work and screws is accumulated, to the advantage of the optician [the seller], but to the perplexity and fatigue of the observer." Tilting the microscope's body alleviates the need for inclining the observer's body forward, but "by no contrivance is it possible to avoid the aching arms" that such an instrument "invariably produces" in those who work with "such a cumbrous machine."[93] A good microscope design will anticipate and, in a sense, meet the operator's movements; it will enable an easy, fluid collaboration between operator and instrument.

Discussion of the embodied nature of microscopy continued and even intensified over the course of the century, following an increased cultural and scientific fascination with all manner of laboring bodies and their fatigue, as noted in Chapter 5. These concerns inflect not just factory work but microscopical practice as well, where they appear well in advance of the late-century surge in fatigue studies. The Danish pathologist Adolphe Hannover is among the first to consider the problem of the operator's physical fatigue and its effect on microscopical observation, in his 1847 treatise *Om Mikroskopets Bygning og dets Brug* (1847; translated into English by the Scottish anatomist John Goodsir as *On the Construction and Use of the Microscope* in 1853). Hannover warns that "if a person be engaged during several hours continuously, he will be liable to suffer from weariness of the system generally, and more particularly from pain in the muscles of the neck, and oppression of the chest." Hannover instructs observers on positioning the body; sitting and resting one's elbow on the table provides a surer touch on the screw and reduces fatigue. However, "the breast must not be supported against the table, lest the pulsation of the heart against it should interfere with the observation."[94] The observer must chart a course between too much support (which aids manipulation and prevents fatigue, but vibrates the apparatus) and too little (which protects the instrument but, taken to an extreme, disrupts the work with exhaustion and pain).

Microscopists demonstrated concerns about physical and mental fatigue well before it was culturally widespread, but textual references to fatigue do increase over the final decades of the century. The 1891 edition of Carpenter's *The Microscope and Its Revelations* (edited by W. H. Dallinger and E. M. Nelson after Carpenter's death in 1885) warns that a "round table...is most undesirable"; it lacks "support for the arms on either side of the instrument," which is "not only a serious, but an absolutely fatal defect" for a working microscopist. And the Belgian diatomist Henri van Heurck, in his manual *The Microscope*, another standard text (first edition 1865; published in English in 1893), provides a list of dicta for microscopists. He urges them, "Do not make observations directly after a meal" and recommends an inclined microscope to prevent "the passive congestion of the head...produced by the compression of the blood vessels of the neck."[95] Thus even the stomach and neck contribute their bit to microscopical vision. Van Heurck's practical advice aligns well here with the theories of physiological psychology about the effect of bodily rhythms and flows on perception and cognition.

The American writer Albert Schneider wrote a guide to *Microscopy and Microtechnique* (1899) that goes even further in laying out remarkably detailed guidance on posture and circulation. He expends two full pages on the positioning of the observer's body in relation to the compound microscope. Schneider suggests, "The chair should have a narrow back and no arm rests" to enable "free movement of the arms," while still providing support to the back, because, he explains, a "long continued forward inclination of the head and body produces fatigue." He specifies a recommended posture in detail, from the feet up: "the seat of the chair should...permit the feet to rest flat upon the floor," while "the body should incline forward...from the hips up and should be supported by resting the elbows upon the table on either side of the microscope." Schneider's advice is designed not only to assure the stability of the apparatus (including the body) and prevent fatigue-driven errors but also to safeguard the health of the observer. "Stooping forward or sagging of the body is pernicious," he warns, "as it retards the heart's action as well as free respiration" and produces dyspepsia, "stooping shoulders and curvature of the spine." Inclining the head forward too long will "produce congestion of the conjunctiva and interferes with free cranial circulation," as do "high and tight standing collars." He also warns against observing through the microscope too soon after a heavy meal or for too long. In fact, "the worker" should "rise occasionally" to execute "lung and muscle gymnastics" in order to "invigorate the body, retard fatigue and make vision more acute" and "the brain more active."[96] His advice, strongly echoing the arguments of the physiological psychologists, complexly rebuts the old trope of the *ascetic* (or *disembodied*) *philosopher*. He retains traditional images like the philosopher's refusal of food but for new reasons, insisting on the material embodiment of both work and worker.[97]

Schneider also proffers unusually detailed advice on the placement of human body parts in relation to the instrument. For example, to use the fine adjustment, "the thumb and first finger of one hand should grasp the milled head...and move it back and forth"; this "constant but careful manipulation" is like "the manipulation of the rudder of a ship." He warns against letting the eyelashes "rest upon the lens," because "moisture from the surface of the eye-ball may condense upon the cooler lens, producing...disturbance of vision."[98] And he concludes with a long set of instructions on the care of the eyes and protection of one's vision.[99] He unnecessarily specifies that, to remove or adjust objectives, the worker should "grasp the collar between thumb and first finger of the left hand" and warns that "at no time should the hold upon the objective be lost, otherwise the objective may fall upon the stage, the floor, or the table."[100] Schneider's minutely detailed directives on the placement of various body parts may seem comical, but his admonitions demonstrate a grasp of postural mechanics and the physical foundation of mental work; and they reflect contemporary research on physiology, fatigue, and attention.

The fatigue of observation is just one of many physical rigors of microscopical work. Guidebooks often offer advice on how to collect specimens, a task frequently requiring not just a ramble but also the management of awkward or slippery footing, contorted positions, and the like. Margaret Gatty, in her popular guide to *British Sea-Weeds* (1872), warns of "the risk of cold and destruction of clothes." Female readers should wear a specific outfit, especially "a pair of boy's shooting boots" treated with "neat's-foot oil, such as is used by fishermen" to waterproof them. She explains,

> It is both wasteful, uncomfortable, and dangerous to attempt sea-weed hunting in delicate boots. Wasteful, because a guinea pair will scarcely last a week. Uncomfortable, because to walk on some rocks in thin soles (the slate edges of those in Douglas Bay, for instance) is so painful, that it very soon becomes impossible. Dangerous, because you must be wetted through by...moist sand.

Here the environment of seaweed-hunting threatens the hunter, who also (Gatty notes) runs the risk of falling into a pool or being trapped in the liminal zone by a sneaky tide. But, appropriately shod and dressed, the reader should "*feel* all the luxury of not having to be afraid of your boots.... *Feel* all the comfort of walking steadily forward"! Gatty's vision is of a body—a woman's body—protected from harm and reveling in the sensation of healthy effort. Indeed, she opens the book with the promise that "[e]verything but what a man labours for becomes wearisome to him after a time" but "a cherished occupation, never." After a long day when the seeker has "tired himself in its pursuit," he wakes "from the blessing of rest to the still higher blessing of exertion."[101] For Gatty, the effort and fatigue of microscopical work will tire the body and mind, but for the true student they are translated into sheer physical delight.

Some microscopists, framing the observer and instrument as joined in a complex mechanical system, developed a kind of ergonomic thinking. They contended that the instrument's body should meet the human one at crucial points to reduce fatigue and error.[102] Bennett, drawing on his own body's experience, recommends that physicians use an inexpensive microscope he had designed with the Continental optician Oberhaeuser, because "when it is placed upon a table, its height is almost on a level with the eye," and "the stage is elevated," which "enables the two hands" to reach it easily even while "resting on their external edges" for steadiness. Bennett sometimes refers to the microscope "body" in a manner that can be ambiguous, given this mirroring of microscope body to operator's body. This is emphasized when he uses the definite article *the* (rather than *his* or *your* or *its*) in discussing both human and microscopical body parts. He complains that the fine adjustment in London instruments is often "placed in front of the body," but this means the operator must "stretch out the arm and twist the wrist to get at it." These instruments are so large that "a movement is given to the body" so that the operator's eye meets the eyepiece, but this means the hands must reach out away to the stage.[103] Bennett's phrasing becomes confusing, and the flow of practice is impeded, when a poorly designed microscope's body does not connect to the observer's body as it should.

Automaticity and the Body

The careful tuning of eye, hand, illumination, slide, and microscope, linked as a single technology, appears to the beginner as a daunting array of interconnected adjustments but with practice becomes an almost automatic process. Victorian physiological psychologists wrote extensively on automatic actions, building on the work of David Hartley, an eighteenth-century philosopher who sought to combine associationist psychology with physiology. Hartley wrote in 1749 not only on "automatic" actions such as the senses or heartbeat but also on physical actions that have been learned, such as walking, talking, and the like, that we can perform without conscious attention to them. He termed these "secondarily" automatic actions because their automaticity is acquired as a result of habit or training.[104] Carpenter and Lewes cite Hartley's insight directly (Bain and Spencer do so in other works), and it grounds their work on consciousness, unconsciousness, reflex action, attention, and automaticity as embodied functions. They cite secondarily automatic actions like handwriting, playing musical instruments, and, unsurprisingly, the delicate adjustments of microscopical work.

Each of these writers leverages the topic of secondarily automatic actions toward somewhat different ends. In *The Emotions and the Will*, Bain uses this idea to emphasize the essential embodiment and contingency of mind, one of his core ideas. He describes how "the habitual actions, including some of our most difficult displays of power and skill, tend to become unconscious." These secondarily automatic actions, he says, provide "another illustration of the accidental, temporary, and intermitted presence of the mental property, and the indispensable and perennial character of the corporeity in giving origin to moving power."[105] Spencer uses microscopy to show how rational actions become automatized through association and a process of mental evolution, since the most effective action is most likely to be repeated over time and become habitual and eventually automatic. He illustrates this process by referencing a common task that is automatic for "the practised microscopist." He explains,

> Everything which [the microscopist] places under the object glass, is seen reversed. All adjustments of the stage, and all motions of his dissecting instruments, have to be made in directions exactly opposite to those which the uninitiated eye would dictate. Yet by practice, this reversed manipulation becomes as easy as ordinary manipulation—it becomes as unnecessary for him to take thought how he shall move his hands, in the one case as in the other.[106]

Spencer's example here emphasizes the initial difficulty of the task and the operator's eventual ability to perform it without thinking.

Carpenter's *Principles of Mental Physiology* (1874) would help forge the modern theory of the adaptive unconscious. He discusses at length the "habitual 'training' of the Automatic mechanism to a certain sequence of movements" triggered by context, "whilst the Attention is wholly directed elsewhere."[107] Citing a common example, he notes that "a Musical performer will play a piece which has become familiar by repetition, whilst carrying on an animated conversation, or whilst continuously engrossed by some train of deeply interesting thought."[108]

However, as Vanessa Ryan shows, while contemporary definitions of the *automaton* might emphasize mechanical action, they did not necessarily presuppose a loss of free will.[109] For example, when Lewes discusses secondarily automatic actions in *The Physical Basis of Mind*, he uses the same metaphor to push back against concerns that such theories frame humans as automata, arguing that such phenomena do not render organic life machinic. While a "violinist may perform while completely 'unconscious,'" executing "delicate and changing movements" with "thoughts and attention…elsewhere," the human musician differs greatly from an "automaton…constructed to play on the violin." Lewes explains, "The wheels, levers, springs, and valves of a machine" cannot learn or adapt as living beings do; the "automaton…will only play the air to which it is *set*, it cannot vary the performance: cannot play a false note, or throw in a *crescendo* here, a *largo* there, according to a caprice of feeling."[110] The microscopist's "delicate and changing movements" can similarly become automatic but adjust to events as needed.

This automaticity in tuning the apparatus acts as a pivot point in H. G. Wells's story "A Slip Under the Microscope," first published in *The Yellow Book* in 1896. The protagonist Hill, during an exam when the students must identify a specimen based only on the bit of slide displayed to them, "sat down on the little stool before the microscope, turned the mirror to get the best light, and then, out of sheer habit, shifted the slip."[111] Hill's inadvertent adjustment, an unthinking movement preparatory to observation akin to brushing a lock of hair away from the face or resettling glasses on the bridge of the nose, both demonstrates his experience as a microscopist and leads to his dismissal from the program. Wells, having studied biology at South Kensington under Huxley, understood how a skilled operator often tunes the apparatus automatically, just as a driver steers a car around a bend or a writer drives a pencil across the page. The philosopher of science Don Ihde calls this *embodiment relations*, where the operator of a technology works fluidly through the instrument without having to concentrate on each action. The instrument becomes transparent and functions as an extension of the body.[112] Superior instruments make this flow of agency seamless and automatic; the chemist George Wilson argues that an expert microscopist relates to his instrument with "the same feelings as an enthusiastic musician regards his Cremona violin."[113] The phenomena of secondarily automatic actions and embodiment relations further demonstrate Victorians' experience of microscopical observation as a practice linking the bodies of observer, instrument, and object in a single dynamic system. Painstaking training and patient labor might, they hoped, render that system more automatic, and more stable.

Writing Embodiment in Victorian Microscopy: Beautiful Mechanism. Meegan Kennedy, Oxford University Press.
 DOI: 10.1093/9780198940623.003.0013

14
The Trained Eye

The eye sees no more than it brings with it the power of seeing.
Thomas Carlyle, used by Louisa Lane Clarke
as epigraph to *The Microscope*, 1858

The trained eye can see distinctly, and the educated hand measure accurately, that which the uneducated eye cannot see at all.
James Glaisher, "The President's Address for the Year 1866"

Introduction

In this chapter I examine the range of Victorian discourses around the eye and argue that skeptical tropes helped microscopists promote and regulate their practice. During the nineteenth century, visual tools and toys inspired by research into optics—like the kaleidoscope, stereoscope, zoetrope, and of course the microscope—explored human perception and produced phantasmal images that were marvelous but could be unreliable. As with other historical technologies like the camera obscura and magic lantern, the post-1830 compound microscope allowed Victorians to bring old kinds of visuality into use in new ways. Victorians enjoyed an explosion of new technologies, many of them optical or visual in some way. They built a culture that was visually modern in ways we would recognize today, with a vibrant array of screen technologies—such as the phantasmagoria, oxyhydrogen microscope, panorama, aquarium, and various forms of cinematography—as well as other arresting visual scenes enabled by new lighting technologies (oxyhydrogen, gas, electric arc), color lithography, mass advertising, brilliant aniline dyes, plate glass, railway journeys, the Roentgen ray, and more.[1] While the microscope, especially the projection microscope, offered thrills on par with any visual spectacle, it also presented perhaps the most intense and challenging visual and mental experience. Writers on microscopy warned that using a table microscope required a trained eye and trained judgment, even more than in ordinary life or in other sciences. On the other hand, they implicitly promised that microscopy would prove transformative for the student, both from the awe-inspiring sights viewed under the lens and from the intensive training regimen designed to remake body and mind.

Many microscopical tropes about the human eye adapt familiar tropes to microscopical contexts. For instance, microscopical texts commonly present the human eye as an *optical instrument*, a common trope in natural history and natural theology. In microscopy, however, that instrument (the human eye) joins with the microscope in an *integrated optical system*. (This notion echoes Herbert Spencer's claim in *Principles of Psychology*, discussed in Part I, that the observer using a tool like the microscope becomes a kind of hybrid entity.) The physical flaws and weaknesses of both eye and microscope thus constrain visual performance. Microscopists manage these difficulties with tropes like *many-sided comprehension* (multiple perspectives) and the *trained eye*. Workers meet the challenge of the microscope's *unnatural image* by reframing the prosthetic lens as necessary in tropes like the *naked eye* and *unarmed eye*, and by embracing the challenge in tropes like the *bewildered student* in need of a master guide and the quest for *true structure* rather than superficial appearances.

One Integrated Optical System

Whether using an expensive Ross binocular or a penny microscope, the observer was as integral to the optical structure of the instrument as the light or lenses. The eye functions as part of the optical system of a compound microscope (an idea that also sometimes appears in writing on the telescope). Especially earlier in the century, it was common to demonstrate the optical principles of the microscope via a diagram showing the eye as the terminus for the image. This convention can be found late in the century in introductory guidebooks also, as in the "skeleton of a microscope" diagram that Arabella Buckley presents in *Through Magic Glasses* (1890) (see Fig. 14.1). Jabez Hogg (1854) provides diagrams both with and without the eye; but even when the eye is not present in the diagram, it is referenced in the text as a part of the optical system. Hogg says, for example, "to take the pencil *l*, for instance, as it enters the eye at *r b*."[2] The trope of the *pencil of light entering the eye* attributing agency to the ray of light with a cringe-inducing physical metaphor, is also conventional, no doubt in an attempt to provide definition to what is actually a group of light rays. The physician H. E. Fripp explains that "delineating pencils of light…pass out of the microscope in a concentrated bundle of intensely bright rays" and that, when "thrown upon the front of the eye" these rays must "undergo a fresh series of refractions and reflections."[3] Indeed, in 1864 Franciscus Cornelius Donders, a Dutch ophthalmologist, published influential research on how the eye refracts light, a reminder that the organ functions as just one more element in an extended construct of reflected and refracted light.[4]

However, the eye illustrated in those diagrams is not actually the final destination for the microscopic image. Victorian microscopists describe the eye as both

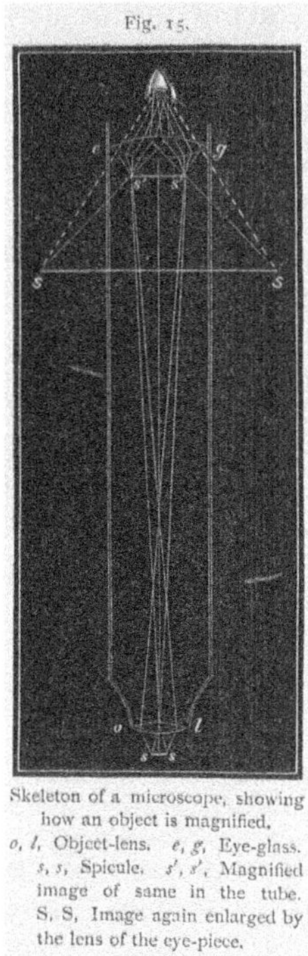

Fig. 14.1 "Skeleton of a microscope." Optical diagram of a compound microscope. Arabella Buckley, *Through Magic Glasses and Other Lectures*. London: Edward Stanford, 1890, p. 37. Image courtesy of The Huntington Library, San Marino, California (498357).

the *screen* on which the image finally rests—the pencil of light is "thrown upon the front of the eye"—and a *lens* through which it passes when the pencil of light "enters the eye" and undergoes "a fresh series of refractions and reflections." In the latter case, the cornea is acknowledged as a lens through which the image passes to the retina and eventually to the brain—a model that draws upon physiological research into the neural mechanics of vision. This trope portrays the eye more accurately, as the final lens in a series of lenses that refract and focus the pencil (ray) of light.[5] David Brewster flatly asserts that "the cornea…must be considered as a lens, and as part of the microscope."[6] In this view, the retina, then, serves as the screen on which the image is ultimately formed. The retina-as-screen is clearly specified in the old trope of the *eye as a camera obscura*, which like the *eye as optical instrument* or *eye as wonderful contrivance* is remade for the microscopical context. William Carpenter cites this trope succinctly in saying, "All vision depends…on the formation of a picture of the object upon the retina of the Eye, just as the Camera Obscura forms a picture upon the ground glass placed in the focus of its lens."[7] Similarly, in *Microscope Teachings* (1864), the popularizer Mary Ward explains the retina as "exactly answering to the slightly concave round table inside a landscape camera" (a walk-in camera obscura).[8] Whether the eye is understood as lens or screen, it is portrayed as an optical component in a larger system, where the microscope serves as an intermediary or vehicle. Overall, Victorians understood that in any microscopical observation, including with projection microscopy or photomicrography, the pencil of rays passed—and, in the compound microscope, repassed—through the body of the instrument to a receptive surface, which could be eye and brain, screen, or photographic plate.

If the retina serves as an element of the optical system, its composition must affect the quality of the image just as the lenses do. Fripp suggests this when saying that "the retina is not so sensitive as photograph paper." This comparison implicitly cites the trope of the *eye as camera obscura* since "photograph paper," carrying the image transferred from a photographic plate or negative in a camera, serves in place of the ground glass or table in a camera obscura. He concludes that "the capacity of the eye as an optical instrument has a direct relation to the performance of the microscope."[9] Fripp elsewhere calls the eye "an optical instrument through which the microscope image must pass to reach the perceiving organ." He describes "imperfections in the retinal image" as functioning like "aberrations" formed within the microscope.[10] Here the eye is both part of the microscope and itself like an optical instrument. Fripp's remarks echo the arguments of physiological psychologists, shown in Part I of this book, that the optic nerve and the brain are integral elements of visual perception as well. The eye functions automatically (albeit not perfectly) within this system, in another example of embodiment relations joining operator and instrument.[11]

With object, light, lenses, eye, and brain all integrated into this optical system, microscopical writers sometimes metonymically invoke some other element of the system as an "eye." Brewster, for example, sometimes describes *the observer as an eye* with phrases like "if an eye placed at E looks at a man *a b*."[12] Alternately, many authors refer to *the eye of the microscope*, a trope that—whether describing the objective or the instrument as a whole—claims visual power and agency for the technology.[13] Such tropes insist on the centrality of the eye in this chain of representation and reinforce the sense that microscope and observer are working together, paired off, body to body, eye to eye.

If the observer provides the endpoint of an optical system, the performance of the final "lens" of the system—the eye—must be taken into account. In unpublished work from the 1830s, Joseph Jackson Lister emphasized this innate limitation of any microscopical system.[14] Much later, in 1875, Fripp argued a similar point, saying that "resolution of minute objects is not limited by characters of physical light," as Ernst Abbe and Hermann von Helmholtz had showed, "but by physiological incapacity of the eye."[15] While many concerns about visual vulnerabilities could be managed with careful practice, Victorian microscopists debated whether human vision or optical and technological constraints presented a greater limit on microscopical work.

Even with a perfect microscope, individual differences in physiology mean that different observers would see different images of the same object. The physician George West Royston-Pigott, who was fascinated by microscope technology and its effects on the visual field, notes in 1876, "Acuteness of vision varies so considerably in different individuals as to render an average estimate somewhat difficult."[16] Such differences were exacerbated when microscopists worked with higher powers. In 1874 J. E. Ingpen presented a theory of *personal equation* in microscopy (modeled on a similar concept in astronomy).[17] This concept addressed individual differences in visual perception, sensibility, and interpretation, aided by the formalization and standardization of vision tests and "normal" vision during this period.[18] The personal equation promised to flatten out differences in observation to achieve a kind of aperspectival objectivity, so that observations could be collected and shared through collective empiricism.[19]

The personal equation identifies individual variations in vision, but all human eyes share some vulnerabilities. While the trope of the *eye as an optical instrument* often appeared in support of theologies of nature, it could equally undergird critiques of the eye. For much of the nineteenth century, popular writers on the senses demonstrated the continued influence of natural theology, especially the familiar trope of the *eye as a wonderful contrivance* and the exaggerated praise for its structure and abilities as perfectly designed for human needs. At the same time, however, the instabilities of this construct became more noticeable as investigators like Brewster, Charles Wheatstone, and Helmholtz studied the performance of the eye and extended in physiology a philosophical tradition, elaborated by Hume, Locke, and Berkeley, that questioned the reliability of the human senses and, through these, empirical science. This skeptical tradition appears in Fripp's comparison of the retina to photographic paper. He warns that this comparison fails due to the physical limitations of the retina. Here, in a reversal of the trope of *created vs. constructed objects*, the natural object (the eye) is inferior to the artificial one (the camera). The eye is, then, a wonderful but not necessarily perfect contrivance. Fripp's conclusion that the "resolution of minute objects" in the microscopical image "is not limited by characters of physical light but by physiological incapacity of the eye" sets the blame squarely on the coarseness of this contrivance—this element of the optical apparatus—the human eye.

The limits and errors of human vision fascinated many nineteenth-century investigators and inventors. Brewster's 1832 treatise *Letters on Natural Magic* in particular reported on many "ocular illusions" relevant to microscopy, including the blind spot on the retina, the indistinctness of oblique vision, the eye's "delirium" in feeble light, the loss of focus in darkness, the appearance of "phantasms" from physical pressure on the eye, the persistence of vision, and the phenomenon of ocular spectra (accidental colors).[20] In contrast to the natural-theology trope *created or constructed objects*, which praises the perfect contrivance of natural objects while disparaging the microscopic coarseness of human-made artifacts, concerns about the performance of the eye present the reverse scenario: the eye is inferior to the instrument in that it degrades microscopical performance.

Brewster's work contributed to a flurry of nineteenth-century research on the senses, whether investigators worked within optics, physiology, neurology, psychology, philosophy, or other related fields. Johann Müller's *Handbuch der Physiologie des Menschen für Vorlesungen* (1833–40) presents his doctrine of specific nerve energies and shows how much human vision relies on what Jonathan Crary characterizes as "a fundamentally arbitrary relation between stimulus and sensation" in that the sensation of light could be produced by various external or even internal stimuli.[21] Crary contends that Müller's model of perception was "potentially... nihilistic" in that it made vision arbitrary, severing the image from its apparent referent.[22] Schickore suggests a different narrative, pointing out that Faraday, at least, considered optical delusions to be potentially curious, amusing, and instructive.[23] And Chris Otter points out that while that nineteenth-century researchers understood the eye as "a variegated zone of mediation" between world and self, marked by incessant "perceptual flux," Victorians also explored many new ways to measure, manage, and protect visual imperfections through medicine and technology. Overall, Otter positions Victorian "visuality [as] always a synthetic act, involving bodily, discursive, and material factors," because human vision occurs within particular cultural and physical environments.[24]

Victorian microscopists and investigators themselves argued that, if not prevented or adjusted for, these vulnerabilities could endanger the authority of vision. The physician and researcher Hermann von Helmholtz directly confronts the conflict between the tropes of the eye as wonderful contrivance or defective tool. "Of all our members the eye has always been held the choicest gift of Nature," he muses in a popular lecture on vision, published in 1868. "No wonder then that, according to a wide-spread conviction, the eye is looked on as an optical instrument so perfect that none formed by human hands can ever be compared with it." Helmholtz sets himself at odds with this tradition of the eye as a "perfect" instrument and a proof of design by other than "human hands." He famously declares that the previous ten years of research had proven so many flaws in human vision that "if an optician wanted to sell me an instrument which had all these defects, I should think myself quite justified in blaming his carelessness in the strongest terms, and giving him back his instrument."[25] Similarly, in an 1873 popular lecture, John Tyndall also challenges "the common delusion that the works of Nature, the human eye included, are theoretically perfect." Tyndall continues, "A long list of indictments might indeed be brought against the eye—its opacity, its want of symmetry, its lack of achromatism, its absolute blindness, in part."[26] Such concerns would not invalidate visual evidence but served as a caution to investigators.

Victorian physiological psychologists, of course, argued that vision was materially inflected not only by nervous structure and function but also by the actions of muscle, mind, and the organism as a whole. Carpenter, for example, remarks in 1842 that "The sense of Vision depends," of course, "on the transference to our minds of the picture which is formed upon the retina" but also on "all the ideas respecting the real forms, distances, &c. of bodies"—that is, on our "perceptions, either instinctive or acquired," many of them derived from our physical encounters with the material world via the sense of touch.[27] By 1874, he was phrasing it more boldly: "Every *acquired* visual Perception" of an object "may be regarded as the *resultant of our whole previous experience*" of that object, combining with the retinal image in a reflex action. Thus "the action of *ideational* states upon the Sensorium can modify or even produce *sensations*."[28] Even simple absent-mindedness, he explains, can produce ludicrous results, citing anecdotes about the Scottish economist Robert Hamilton, who, engaged in thought, would mistake cows and posts for persons in the street.[29] If the optical system, strictly speaking, involves the eye and retina as "lenses," the system of vision generally speaking reaches even to the brain and body.

Mid-Victorians wrote in the wake of both natural theology and optical research on the limits of vision. Ordinary readers would have encountered tropes of the eye as both perfect and integrally flawed, and they would have known that vision is an embodied process requiring scrutiny and correction. Brewster's *Letters on Natural Magic* found a wide audience in eight editions and multiple reprints through 1883. And Tyndall's lectures at the Royal Institution in London and across the United States were well attended, drawing huge crowds in New York despite brutal winter weather.[30] The popularizer Dionysius Lardner drew largely on Brewster's work when he published an extended discussion of "The Eye" in his encyclopedic series *Museum of Science and Art* (1855). In this largely instructive text, Lardner presents a similar negotiation of perfection and error. Having termed the eye "the most beautiful example of design, structure, and contrivance," Lardner goes on to list numerous flaws built into the structure of the eye, such as the phenomena of double vision, the persistence of vision, and the blind spot on the retina, as well as individual peculiarities such as farsightedness, nearsightedness, and color blindness.[31]

Victorians worried that the eye's powers might render its illusions more dangerous, because more persuasive. In an earlier essay on "Popular Fallacies" in the *Museum of Science and Art* (1854), Lardner argues that our sense of the eye as "exact and unerring" leaves us uniquely susceptible to its "popular fallacies." He suggests a way to negotiate these competing narratives about the eye, explaining,

> Of all the organs, that which would seem to be most exact and unerring in its indications is the eye; and, although in a certain sense this is true, yet there are no impressions which more imperiously require the exercise of the judgment to

> adjust and rectify them than those of vision. By this sense we receive the perception, subject, however, to many qualifying conditions, of form, magnitude, brightness, and colour. There is not one of these qualities, however, which is not frequently mistaken or wrongly estimated.... The susceptibility of the organ of vision itself is liable to frequent and rapid change, and the same objects at different times produce upon it extremely different impressions.[32]

A romantic language of extremity and exaggeration ("most exact and unerring," "imperiously require," "extremely different") appears in this discussion as it does in natural theology texts that celebrate the perfection and contrivance of the eye—but here the human eye is remarked upon not for its superior qualities but for its flaws. Lardner notes the eye's many fallacious impressions; its susceptibility to frequent errors of every type; its unreliable and changeable communications. The very impression that the eye is the "most exact and unerring" sensory organ is shown to be itself an error, only true "in a certain sense." In fact, Lardner judges the eye to be instead the most deceptive organ, partly because its false sensory messages are so compelling; and he calls for "the exercise of the judgment" to correct its deceptions.

Similarly, the physician John Scoffern, in an article in *St. James's Magazine* in 1865, concludes that the human senses are "[f]aithful—yes, but not infallible; they want close watching. Do our very best, they will often deceive—vision especially."[33] The irony of Scoffern's advice to "watch" over "vision especially" points to the persistence of the trope in which "seeing" still figuratively denotes mental scrutiny even as the eye itself requires oversight.

Nineteenth-century investigators did not generally reject the romantic trope of *the eye as a wonderful contrivance* but rather revised it, offering instead a more tempered, realist version. For example, Helmholtz, despite his skepticism about the eye's abilities, goes on to praise the eye as intricate and beautiful, commenting that "the great performances of this little organ can never be denied," and that our appreciation for it should be the greater for having been more accurately informed.[34] Tyndall, too, equivocates, saying, "The degree of perfection of any organ is determined by what it has to do," and concluding, "The eye is not to be judged from the stand-point of theory. As a practical instrument...it must ever remain a marvel to the reflecting mind."[35] The literary structure of *counterbalancing*—the eye, with all its faults, is still a pretty impressive contrivance—provided a productive standpoint for discussions about the complexities of microscopical vision.

If, as Tyndall urged, the eye should be judged not in theory but in performance, microscopy provided a field of endeavor that pushed both eye and instrument to their physical limits. It was imperative for the operator of a microscope to understand and, to the extent possible, compensate for the unreliable embodiment of vision. Microscopy was affected by many of Brewster's well-documented "ocular illusions," and the image could be threatened in unexpected ways by the materiality of the visual organs. In 1837 Brewster provided an article on "Microscope" for the seventh edition of the *Encyclopedia Britannica*, reprinting it as a standalone book with fourteen plates. He explains that while "no general rules can be given" for adjusting the intensity of light, "the following rules may be laid down" about lighting and observation generally. Brewster frames these rules as a comprehensive and prescriptive methodology, using a potent literary form, the *list of general rules of action*. In fact, his rules offer a haphazard smattering of advice, but they proved influential regardless, perhaps due to the brevity and authority conveyed by the list format. Brewster's rules were cited by Quekett, Hogg, and many others; the literary form he chose (the list) featured in many Victorian manuals as well.[36]

Brewster's rules present a view of microscopy where the vulnerabilities of the human eye threaten the microscopical image. He emphasizes the need to "protect" the delicate eye and the tendency for vision to be "injured" by seemingly trivial conditions such as head position.[37] These rules do not mention the well-known deceptions or limits of human vision that Brewster had discussed in the *Letters*, but focus on physiological vulnerabilities of the eye affecting microscopical work specifically: the danger of damaging the eyesight with too-bright illumination, and the distortions of the image caused by a "viscid" corneal fluid that can form minute but vision-distorting lines and bulges. He later elaborates that viscous fluid on the human cornea is "pushed into a [horizontal] ridge" by the closing of the eyelids, and advises the observer to induce tears to wash his vision clear.[38] However, this more liquid lubricant (after tears) flows down with gravity; it forms multiple minute vertical columns through the vision. Corneal fluid also forms an irregular bulge over the cornea when the observer looks down, as through a table microscope. These deformities "obliterate and render indistinct all horizontal lines" in the image, though rendering vertical ones more distinct; overall, they "injure vision." Although Brewster praises the "beautiful contrivance, not mentioned by natural theologians," that the horizontal sill of the eyelid sweeps over and smooths the vertical "currents," the eyelids also push the fluid into that horizontal ridge, a state "incompatible with microscopical observation."[39] He suggests that the vision is least affected when "the observer lies on his back"—"a position, we admit, not favourable for research."[40]

Similarly, Helmholtz, assessing the threats to the integrity of microscopic observation, notes "opacities in the tissues of the eye" causing a "peculiar spotty appearance of the [visual] field" in the microscope.[41] John Phin, a Scottish lawyer and Shakespearean scholar who moved to the United States in his early forties, remarks in an 1877 introductory guide to the microscope, "The eye is a very delicate organ, and the slightest derangement of the stomach or nerves affects it to an extent that few persons realize." As an example he comments that "when in good personal health, we find no difficulty in resolving the *P. angulatum*" but "a very slight disturbance of the digestive organs, renders the lines perfectly invisible."[42] The physician Heinrich Frey, in his 1872 manual for physicians, even advises "avoiding" microscopical work "immediately after dinner."[43] Microscopists working at the limits of instrument and observer found that even trivial impairments of the eye or other organs could significantly affect performance.

Eyestrain

In suggesting that *microscopy injures the eye*, Brewster cited a longstanding trope, one that (uncharacteristically) seems to warn against taking up the practice. In this narrative, microscopical vision not only was potentially unreliable but could be downright dangerous. Injudicious or overlong use of the microscope was thought to strain the eyes, harming vision during observation and possibly forever. As Jane Nicholl wrote to William Henry Fox Talbot, her cousin, in 1816, "I hurt my eyes so terribly by straining them thro' the Microscope that I am afraid of doing them some injury."[44] The repetitions here ("hurt," "strain," "injury") describe the microscope as capable of causing both temporary and permanent damage.

The problem of eyestrain again shows microscopists navigating the material conditions of microscopy; it arises because an optical system yokes the instrument to a fallible body. Microscopists were not alone in their apprehensions about vision—astronomers had also noted the "hazardous and artefactual nature of seeing," as Isobel Armstrong puts it.[45] Such warnings were a consistent force shaping mid-Victorian instruction on microscopical practice. John Hughes Bennett, in his 1850 book of lectures to medical men, explains that even with "careful management of the mirror and diaphragm, the field of vision is often dazzling" and "presents a glare most detrimental to the eyes." In fact, he says, "I cannot employ Ross's fourth of an inch for fifteen minutes without feeling intense headach [sic]," and he warns of multiple acquaintances in whom damage to their eyesight was sufficient "to incapacitate them from continuing their researches."[46] Overuse was frequently cited as a cause. In 1856, Carpenter warns, "The habitual use of the Microscope for many hours together, especially by lamplight, and with high magnifying powers, has a great tendency to injure the sight."[47] And the surgeon Arthur E. Durham, in his 1868 presidential address to the Quekett Microscopical Club, explains "beyond . . . the *fatigue point*, exercise [visual work] tends to exhaustion; use becomes abuse," and the eye can suffer "lasting damage."[48]

Experienced microscopists, though, offered a variety of suggestions to counter the fatigue caused by long stretches of monocular vision. Brewster suggests in 1837 that an observer adopt the wooden "snow spectacles of the Greenlanders," each side with "a circular aperture the size of the pupil," to "protect his eye from all extraneous light." These spectacles should be custom-made to fit the face and lined with black velvet.[49] Brewster's concern seems largely to prevent the glare from affecting observations, not for ocular health overall, and this elaborate precaution does not seem to have been widely followed. However, observers did sometimes mount a black velvet eyeshade to the eyepiece to exclude peripheral light, especially for delicate work.[50] Carpenter later urges observers to "learn to use either eye indifferently" and "acquire the habit of *keeping open the other eye*" so as to "work much longer, without incurring unpleasant or injurious fatigue." The microscopist must also "rigorously... observe the simple rule of *not continuing to observe*" once the eye has become fatigued.[51] As so often happens in these discussions of vision, Carpenter's dictum is confused by the common use of *vision* as a metaphor for mental scrutiny: the microscopist must subject observation to oversight—must both observe and not observe, or not observe by observing. Victorian microscopists' caveats about human vision thus figuratively implicate mental "vision" as well.

The popularizer George Henry Lewes challenges these concerns in *Sea-Side Studies* (1858), arguing that "A very general belief of its 'injuring the eyes,' will be found even among microscopists.... By artificial light, indeed, I find it very trying; but by daylight... it does not produce more fatigue than any other steadfast employment of the eye."[52] However, the warnings continued through the end of the century. Carpenter's *Revelations* kept his warning in subsequent editions through 1883. In 1881, the eminent optical theorist Ernst Abbe was quoted in the *Monthly Microscopical Journal* warning, "It can scarcely be doubted that continuous one-sided vision by those who use the Microscope incessantly must gradually diminish the ability to use the eyes for ordinary vision."[53] Henri van Heurck, in *The Microscope* (1893), explains that "with proper precaution microscopical researches can never cause injury; on the other hand, they may cause real danger if recklessly conducted." He cites cases where researchers nearly lost their sight and recommends best practices in observing—for example, "as soon as your eyes feel at all fatigued suspend your observations at once," "avoid brilliant illumination," and "wash the eyes thoroughly every morning with warm water." (Countering Lewes, Van Heurck tells readers to "on no account use solar light for ordinary observation.")[54] Most Victorian references to eyestrain and fatigue endorse the narrative that certain physical conditions of work at the microscope (too long, too much light) endanger both ordinary and microscopical vision. This narrative, although it seems critical of the instrument, serves to regulate microscopical work (promoting regular work habits, not binges of observation) and helps frame the microscope as not a toy but a powerful tool that must be handled with care.

Unnatural Vision and the Trained Eye

Even if both eye and instrument were functioning well, microscopic vision was considered, by both enthusiasts and detractors, to be *unnatural vision*, frequently framed in opposition to *natural* or *ordinary* vision. For example, Fripp, arguing for a science of microscopical observation asks, "How are [its] revelations to be verified? How is microscopic observation to be raised above mistakes of eyesight exercised under unnatural conditions?" He remarks on "the peculiar actions of the microscope, with its peculiar excellencies and defects" and argues that "as in ordinary circumstances all seeing is not believing, so in the extraordinary conditions of microscopic vision the observer is even more responsible than his instrument for the true definition of what he sees."[55] Frey remarks, "The peculiar nature of the microscopic images causes some difficulties for every commencing observer.... In this kind of vision we are deprived of the aid which enables us, in ordinary vision, to judge rapidly of the shape of the object."[56] Pushing back against the simplistic notion that *seeing is believing*, Fripp instead represents microscopical vision as unnatural, peculiar, extraordinary. Like the narrative of eyestrain, above, the trope of unnatural vision acknowledges physical challenges inherent in microscopical work but also asserts special status for the instrument and its work.

Microscopists defend their *unnatural vision* by opposing it not so much to *natural vision* (where the comparison is unflattering) but instead to the *naked eye*. For example, Carpenter notes that in a "calcareous deposit" in Pampas, "particles of shell" were "distinctly recognizable by the microscope, though invisible to the naked eye."[57] The trope of the *naked eye* retains the distinction between "natural" and "unnatural" vision but reverses their status. The metaphor of the *naked eye* reframes the microscope as a normal supplement to vision—like clothing, it becomes metonymically part of the body—and redefines the "natural" eye as insufficient, perhaps even deficient. The eye's relative impotence and vulnerability becomes even more pronounced in the German phrase *unbewaffnete Auge* (*unarmed eye*).[58]

The trope that Victorians most frequently oppose to the *naked eye* is the *trained eye* of the naturalist and especially the microscopist. The close empirical observation of natural objects, as in natural history or medicine, was understood to be an art requiring careful training. Bennett remarks in 1850 that "the senses require to be educated before they can receive proper impressions," so that "what is obvious to the sight of an experienced practitioner" (or to the experienced doctor's ear or hand) "is overlooked by the student." Skill in observation, Bennett tells medical students, is "a kind of information which cannot be obtained from others; you must acquire it for yourselves" if you wish to master the medical art. The "clinical instruction" these students seek, then, is simply to acquire this skill—"the learning how to observe"—and anchor a "sound judgement" upon it.[59]

The botanist Phoebe Lankester (Edwin Lankester's wife) links this notion of the *trained eye* with the old trope *eyes or no eyes*, which compares the disciplined and meticulous scrutiny of the naturalist to the unobservant gaze of the layperson. In an 1861 *St. James's Magazine* article for young readers, Lankester explains that "some [people] see far more than others," because "the eye... needs training and teaching…perhaps more than any other of our faculties." She explicitly links *eyes or no eyes* to microscopical vision: having a microscope, she says, "makes all the difference of 'eyes and no eyes.' What eyes would be to a blind man, the microscope is to the unassisted vision"—or, as a child in her story says, "the microscope is like another pair of eyes."[60] In this analogy, a version of the *microscope as a new sense*, the microscope stands in not just for "eyes" but for trained eyes specifically.

For although the *trained eye* refers to any type of visual scrutiny, a trained eye was especially important in those sciences where vision was necessarily mediated by instruments, like astronomy or microscopy.[61] Carpenter says, for example,

> Even to observe well is not so easy a thing as some persons imagine. Some are too hasty, imagining that they can take-in everything at a glance, and hence often forming very erroneous or imperfect notions.…Others, again, are too slow and hesitating and fix their attention too much upon details, to be able to enter into the real significance of what may be presented to the vision.[62]

Skill here becomes as important a power as magnification technology, perhaps because of the unnatural conditions of that technology. Writers sometimes presented microscopical seeing as an extreme example of the difficulties of ordinary perception. The *trained eye* trope usually refers to activities requiring precise observation, but in fact our experiences in the course of ordinary life are all affected by the inexactness of human vision. What we see around us is often limited, erroneous, or conditioned by our expectations.[63] Physiological psychologists addressed these vulnerabilities of human vision at length (see Part I). Despite these known issues, seeing in ordinary life still feels like a natural or innate function.

In contrast, as Goring says, "all microscopic vision" is "a totally new way of seeing to the uninitiated," requiring special education.[64] Louisa Lane Clarke, writing for novices, suggests that a trained microscopical eye, once achieved, provides a valuable sensory education, analogous to that provided by other occupations. "As the *ear* is educated by the study of music," she argues, "so the eye is educated by a habit of observation.…The Microscope gives this training, if it is rightly used." Her analogies demonstrate the kinds of visual abilities she considers valuable: the artist's facility in discerning small inflections of form, the physician's power of reading the significance of visual elements ("every line in the patient's face or movement of the body is to [the physician] an index of the inner life"), and the naturalist's skill in noticing seemingly trivial details ("the folding of a leaf, the spot upon the wall…where the untrained eye sees nothing").[65] Clarke proposes the difficulties of microscopic vision as part of a range of human work demanding specialized training in how to see.

Carpenter, in fact, contends that training in microscopical observation is not just analogous to other kinds of visual training (as Clarke claims) but actually helps train observation and reasoning generally. In a lecture to the Society of Arts in 1854, he argues that "the powers of comparison, abstraction, generalization, in fact *all the reasoning faculties*" are exercised and improved in microscopical work, more than in "any other *single* object of pursuit."[66] The education of a microscopical "eye" entailed, then, the concomitant training of the brain as well.

Observers could avert some errors simply by learning to recognize and ignore erroneous structures. Brownian motion distractingly jiggles tiny particles suspended in liquid; certain oblique angles of illumination create a double image; and the object's own refractive properties can cause misleading light effects.[67] The botanist John Hutton Balfour warns of a "class of deceptions [that] originates in the eye itself." Some of these—the *muscae volitantes* (the illusion of little "flying flies" in the vision) and the *pearly spectrum* (the appearance of "twisted strings of minute pearly globules, hung across the field of view")—are named with tropes that help students identify and remember them.[68] While careful preparation can prevent some errors, as noted earlier, the microscopist must practice identifying and disregarding many artefacts and misleading images. Carpenter notes, "No rules can be given for the avoidance of such errors, since they can only be escaped by the discriminative power which education and habit confer."[69] As early as 1832, C. R. Goring recommends experimenting with "glass toys," "glass tubes" filled with objects and water, and "solid rods of glass." Beale argues that simply recognizing common artefacts will avert "the most ridiculous mistakes," such as the misidentification of air bubbles as structural bodies.[70] Common fallacies of structure include difficulty sorting out the routes of tiny vessels ("tubes"), apparent structures produced by dyes and reagents, and cell-like clusters of oil globules.[71] Thirty years later, Simon Henry Gage, a prolific Cornell professor, lists nineteen confusing conditions or objects and the instructions for replicating them for study "so that they may not puzzle [the student] nor lead him into error. The exercises also furnish an excellent drill in manipulation."[72] Programs like this emblematize Matthias Jakob Schleiden's "fundamental rule" that "to observe with success," the microscopist must pursue what he calls *many-sided comprehension*—that is, "must observe frequently and with the most profound attention." The microscopist may thus "gradually learn to see, for seeing is a very difficult art."[73] As always, readers are directed to embrace patient labor in order to train the body in particular habits of practice.

Other problems cannot be solved so easily. For instance, unlike ordinary vision, the microscope traverses scale more than distance. Writers commonly explained magnification, as the optician Andrew Ross does in the 1839 *Penny Cyclopaedia*, saying that it "enables the eye [the observing subject] to approach more nearly to its object," as one might do by "walking across the street to read a bill."[74] Despite this explanation, the microscope does not "bring the eye closer" to an object in quite the way that a telescope or camera zoom lens does. Those instruments optically reduce the distance between our eye and an object that we could see with the unassisted eye face-to-face if we could just get physically closer to it. Without a lens, we can never get physically "close enough" to access the microscopic structure of objects, and in some cases the objects themselves, even if they are actually resting on a hand that could be brought to the eye. Instead, the microscope provides access to many objects and structures that exist on other scales of being; it increases proximity virtually rather than physically. As Goring says, in microscopy "the object is in a manner placed in a most unnatural state of proximity to the eye."[75] His statement acknowledges this proximity to be metaphorical only ("in a manner"), and "unnatural." The notion that we *bring the eye closer* to the object—that magnifying it for study is like *walking across the street to read a bill*—draws on an everyday occurrence. It provides a familiar physical metaphor to reframe the unfamiliar, unnatural magnifications of microscopy. It rewrites disorienting magnifications in simple physical terms, as if the observer's eye is materially transported by the instrument to the very point of the lens, and reinforces the notion of the single optical system connecting observer, microscope, and object.

These shifts in microscopic scale are complicated by the abrupt transpositions between an image magnified at one power and an image magnified at a greater or lesser one. Even in elite instruments featuring coarse and fine adjustments and a rotating nosepiece with multiple objectives at different powers (a new feature noted by Carpenter in 1856), at each shift in power the image jumps rather than zooms, creating a temporary disorientation and dislocation with each bound. Each objective offers a different image of the object, yielding sometimes a startlingly different reality of form. It is then difficult to identify the desired structure or its relation to other elements in the object; Carpenter devotes four pages to a discussion of this problem.[76] Many guidebooks recommend beginning work with low powers in order to gain a sense of the microscopic landscape before plunging into deep powers that might leave the observer disoriented.[77] As the clergyman Dixon notes in *Household Words* in 1856, with high powers "you can only... look at a fragment, a claw, a tongue, a jaw, a proboscis, an eye, a petal, an anther." Better to ask a lower-powered objective to "act as the finder for the stronger" lens, or in the trope he offers, "*play the jackal to the lion*."[78]

Even at low powers viewing a whole object, the experience of viewing a microscopical image can be disorienting. Novice microscopists were confounded, for instance, by the reversal of the image. That is, as the clergyman Edmund Saul Dixon explains to readers of *Household Words*, "to make it go to the right, you must push the object-slide to the left, and to move it apparently upward you must direct your gentle touches downward." The unexpected relation between hand movement and image movement creates a kind of friction on one's practice that observers must learn to accommodate, despite bodily fatigue—for Dixon notes that "these little manipulations are not acquired without a fatiguing amount of practice."[79] An erector attachment can revert the image but does not work well at all powers.

More challenging, the image often is ungrounded: it hovers in space without any visible connection to its origin, with no ready correlation between an object's macroscopic and microscopic appearance. Flowers, water, seaweed, milk, blood, skin—many objects are unrecognizable under the microscope, with no perspective available that bridges back to their ordinary form. It is no wonder that popular exhibitions usually chose objects, like flies or infusoria, that look like familiar animals when displayed whole at low powers. Often, however, the microscopical image is divorced from the physical object (which is itself often too small to apprehend) and the student looking through a microscope sees nothing recognizable as form. As Lisa Cartwright says, "The representational microscopic image differs from the representational painting or photograph in that its accuracy cannot be confirmed against the human eye's view of its object."[80] The microscopic image seems untethered from the surrounding environment; it is abstracted and alien.[81] Thus, as Isobel Armstrong explains, in this "ungrounded perspectival world," although the image may present with unnatural clarity, "the object [apparently] exists in atopic space…freed from relational coordinates." Whereas the unassisted human eye can perceive a macroscopic object from afar, in its setting, and then examine it closer, as a whole, and finally home in on a particular detail, the microscopic eye sees "no part and whole, simply diminishing autonomous or segmented parts."[82] John William Griffith and Arthur Henfrey, in their *Micrographic Dictionary* (1856), note an egregious error caused by just such a contextless conflation of scale. One observer identified some unknown "rounded bodies" found in "human secretions," whereupon another observer pegged them as "spores of a fungus" that he had found in bread. He alerted "a Society that ought to have been learned in these matters," but embarrassingly, "no one detected" the fact that the mysterious rounded bodies were "twelve times the size" of the spores![83] Under the microscope, then, the observer sees not integrated nature but a succession of different views within a constantly disputed hierarchy of magnifications and illuminations.[84]

It is possible to see this disorientation as a benefit. Armstrong links the "intense allure" of nineteenth-century optical toys and tools, including the microscope, to this ability to provide playful "experiments with different subject positions, control, displacement, obliteration, power, centrality, powerlessness."[85] Others declared war on the problem, hearkening back to the German notion that the *unarmed eye* (without a microscope) cannot be trusted. Victorian medical men, already disciplined in a tradition of specialized observation, extended that tradition to the microscopic realm, with mixed results. An observer of Bennett's work with Edinburgh medical students recalled how he implemented the trope of *microscopists as an army* in practical terms: Bennett "drilled his class in the use of the microscope until every man knew his instrument as a trained soldier knows his rifle, and until in the handling of it he was as perfect as the veteran in the manual of arms."[86] A former assistant of his, William Rutherford, reiterated this trope of the *microscopist as soldier*. He modeled his own teaching at King's College, London on the Edinburgh model, explaining, "The plan I now adopt on all occasions is to treat my class like a regiment of soldiers, making every one do the same thing at the same time." He clarifies that it is not simply students but specifically medical microscopists who form this corps of observers: "The class resembles a little army of investigators—it is, in fact, a miniature of the histological world."[87] If these young microscopists are drilling like soldiers, the weapon they must learn to wield is, of course, the microscope.

This trope of the *microscope as medical weapon* builds upon the longstanding metaphors of *scientific workers as an army* and medicine as *a war with disease*, incorporating the microscope as a new offensive technology and celebrating the professional discipline and skill of these up-to-date medical men.[88] But Bennett's newly minted army did not always observe with rigor, as is evident in his concern that while "systematic instruction" was "raising up a host of experienced microscopists," yet "a few inexperienced investigators may still publish their erroneous observations."[89] Lionel Beale complains that "many of our modem researches are not carried out with the same patience, painstaking industry, and conscientious care" as in the past and thus "are likely to be but short lived."[90] Indeed, "this sort of false observation," he argues in a later edition, has led in some quarters to "complete scepticism of all observation," halting scientific progress. He warns that "years must be passed in patient investigation before a man can expect to be able to trust himself as an observer of facts." Beale even anticipates that the observer's eyes may start to fail from old age just when he finally masters the process of observation.[91] In this view, the microscope's unnatural vision could yield great scientific advances, but it will also push the observer to his physical and mental limits.

Those limits especially appear in discussions of the novice observer; microscopical manuals often refer to the *bewildered student* (in contrast, of course, to the authors' presumed mastery). The *bewildered student* trope promotes a useful narrative in which the awesome power of the microscope makes the author's guidance necessary. Wood remarks in *Common Objects of the Microscope*, "The quantity of objects which can be placed under a microscope is so vast, that without some guide the tyro flounders hopelessly in the sea of unknown mysteries, and often becomes so bewildered that he gives up the study in despair of ever gaining any true knowledge of it."[92] For example, he notes of "the petal of the yellow Snapdragon" that "the surface cells assume a most remarkable shape, running out into a variety of zigzag outlines that quite bewilder the eye when the object is first placed under the microscope."[93] Wood offers his own book, of course, as just such a guide. Similarly, Beale suggests giving each student of histology a briefer guide, which he details: a "card on which are printed the following points:—1. Shape. 2. Edge. 3. Colour. 4. Transparency. 5. Contents. 6. Size. 7. Effects of reagents." This "prevents the student from getting bewildered, and teaches him method and thoroughness" in the face of a confusing image.[94] These paper-based guides act as prostheses in extending an author's pedagogical presence into classroom or parlor.

Microscopists further considered what it meant that compound microscopy, while enabling a clearer and stronger magnification, removed direct visual access to the object. When we see something through the compound microscope, what we see is not a direct image of the object itself but a magnification of a virtual image of the object, formed inside the tube of the instrument. The English naturalist and optician Andrew Pritchard in 1827 balances compound microscopes' freedom from aberration against "the never-to-be-forgotten fact, that they only show us a *picture of an object instead of nature itself*." In contrast, "the simple unadulterated view of an object" in a simple microscope "must enable us to look farther into its real texture, than we can by any artificial arrangement whatever." Observing through a single versus a compound microscope is thus "like seeing an action performed instead of a scenic representation of it." He concludes, "the advantage of viewing an object by the interposition of but one magnifier"—that is, using a simple microscope—"instead of looking at a picture of it with an eye-glass (as is the case in all compounds), must surely be appreciated by every person endowed with ordinary reason."[95]

Of course, Pritchard makes these criticisms in the early years of compound microscopy. However, a half-century later, after the compound microscope had definitively overtaken the simple microscope for many tasks, Abbe also, though less dismissively, noted, "in microscopic vision we do not see the object itself, as in direct vision with the naked eye; we see the virtual image only."[96] That image has been produced by the magnifying power of the objective, reduced by the field-glass (which clarifies the image and provides a greater field of view), possibly inverted or flipped by an erector, and magnified again by the eye-glass. Fripp even says that "the microscope exhibits but a counterfeit (more or less accurate) of the object itself."[97] Pritchard's early ambivalence reflects a simple fact: each additional layer of mediation changes the image. However, the mediated nature of compound microscopy also offers opportunity: an observer unhappy with the image can choose from many possible interventions, whether by this means changing the microscope settings, the lighting or other apparatus, or the object itself.

It was difficult for some Victorians to accept this weird visuality, with its decontextualized, mediated view estranging familiar objects. The disorienting microscopic image roused resistance and skepticism in some observers. The Rev. William Tuckwell, who, like Wood, studied with Henry Acland at Oxford in the 1840s, notes that the older faculty would occasionally try out "the then unfamiliar microscope" but "could not force their minds into the new groove." Tuckwell recalls, "Dr Ogle, applying his eye to [the] microscope, screwed a quarter-inch right through the object." Another of the faculty, Dr. Kidd, upon viewing a particularly delicate structure, exclaimed "first that he did not believe in it, and secondly, that if it were true he did not think God meant us to know it."[98] This portrayal of the befuddled senior dons as out of touch with the modern microscope accords with Tuckwell's later persona as a Christian Socialist and "radical parson," but his anecdote shows how microscopic vision seems *unnatural*, even to educated men surely capable of grasping the optical principles of the instrument. Graeme Gooday points to this story to show that microscopy required both training in "operational skills" and a revised sense of "what counted as 'natural.'"[99] Overall, challenges to the microscope as a scientific instrument echoed the uneven adoption of other new technologies, like the stethoscope and ether.[100] Victorian microscopists' tropes of wonder and skepticism had to frame the challenges of microscopical vision in a way that could overcome the stubborn suspicion of established men of science as well as enlist the interest of a new generation.

Unnatural Objects

Just as the microscopic image is unnatural due to optical effects of the instrument, the object is often unnatural too. In many cases our encounter with the microscopic object occurs only after manipulating that object to make it or its structure visible to the microscope. At the very least, the object has been captured in a live-box or mounted on a slide, in an artificial setting far from its original home.[101] The object is not innocent: it has been dried, frozen, sectioned, stained, fixed, and/or mounted, all of which can change its appearance and its nature. For example, changing from a dry to wet mount could significantly change the image. It is evident from Fig. 14.2, taken from Hogg's *The Microscope* (1887), that mounting could require a considerable apparatus and intervention on the object. Also, each mounting fluid (water, oil, Canada balsam, or another substance) has its own refractive index, and the evaporation or mingling of fluids can also distort the image.[102] And the aniline dyes that proved so effective after William Henry Perkins' discovery of mauveine in 1856 can damage a specimen: they poison as they stain.

The microscopist's access to the object takes place as a kind of experiment on the object more than a simple observation of it. The observer must learn to handle the object appropriately, which means understanding not just what to do but what not to do. Alexis Julien, an American petrographer and slide-maker who was a Fellow of the Royal Microscopical Society, wrote an 1893 article in the *Journal of the New York Microscopical Society* reviewing slide preparation methods and warning of the changes to live objects in the process. In order to preserve "the lovely phenomena of life thrilling and pulsating within the field of the microscope," he says, the preparer must take "great care and skill" to prevent changes to "the normal forms and relationships of... structure." Julien lists a number of possible irritants: "any change of temperature, produced by the heat of the room or

Fig. 143d.—*Mounting Apparatus.*[1]

Fig. 14.2 Mounting apparatus.
Source: Jabez Hogg, *The Microscope*, twelfth edition. London: George Routledge and Sons, 1886, p. 210.

lamp, or the coldness of the metal of the stand; by lack or excess of moisture, oxygen, or light; by vibration or jar; by offensive vapors in the atmosphere, carbon dioxide gas from the observer's breath, or the volatile solvents…used in the laboratory." The threat of "carbon dioxide" here reveals one more way, beyond vibration, in which the observer's mere breathing can change the dynamic of observation. Many processes ("fixatives and stains, dissection and vivisection") are fatal to the object and, Julien notes, likely involve "morbid change" in the minute structures under the lens.[103] In a kind of prefiguration of the observer effect in physics, Julien and others caution that it seems impossible to observe an organism, living or dead, without affecting it in some way.

Lewes argues in *Physical Basis of Mind* (1877) that such interventions both enable and inevitably distort the observation of nervous tissue. Although investigators seek to "form a precise picture" of the elements and structure of "the living tissue," he says, microscopical investigation requires "that we should *alter* that arrangement, *removing* some elements of the tissue, and *changing* the state of others, without knowing what were their precise state and arrangement before the change." These necessary preparations can, for example, cause specimens to "lose their constituent water" and "become deformed." Often, he says, "we are unable to say what has disappeared." But investigation is impossible without intervention. He explains,

> Place a piece of nerve-tissue under the microscope, without having subjected to various mechanical and chemical operations, and you can see next to nothing of its structure. You must tear the parts asunder, and remove the fat and nerve-sap (plasmode) before you can see anything; you must coagulate the albumen, and otherwise chemically alter the substances before a thin section can be made; you must get rid of the tissues in which it is embedded, without knowing what are the connections thus destroyed.[104]

As Lewes demonstrates here, the much-vaunted access to "Nature" was paradoxically achieved only through a complex process of mediation—which, when it worked well, offered a transparency that was only illusory. As Gooday explains, "the use of microtechniques, natural lighting, stable working surfaces and robust mobile instrumentation were *conventions* employed by microscopists to effect a controlled 'naturalization' of their domestic working environment," thus "minimizing the apparent artificiality and situational idiosyncracies of the indoor location" in order to "claim a direct, i.e. unmediated, contact with 'Nature.'"[105] While these techniques were adopted to make the object visually more accessible, they also made it less "natural," for any intervention involves changing the specimen in some way. As Julien warns, a live object is likely changed by the act of observation; and "a mounted preparation . . . can only be a mummy, or a slice of a mummy, never a satisfactory substitute for the living protoplasm of the original tissue or organism."[106] Students must train the eye to appreciate these changes to the object—to notice and account for their visual effects on the image.

The historian of science Hans-Jörg Rheinberger argues that the manipulation of organic objects raises "the question of what was nature and what was artifact" within a single preparation.[107] He sees each conjunction of specimen and instrument as a necessary and discrete "intersection" of "epistemic things" manipulated within particular "experimental systems" constructed of both material and theory. These "intersections and interfaces" raise contentious questions, such as establishing the "small, provisional borders between what may still be considered biological nature and what may have to be conceived of as an artifact."[108] When stains, fixatives, heat, light, and more are required in order to see a particular organic structure, the dream of an unmediated glimpse behind the veil of Nature slips ever farther beyond the reach of the questing eye. Rheinberger contends, however, that this "dialectic between fact and artifact" actually enables the recursive process of scientific knowledge, in which objects of knowledge "can be left behind time and again" as they are remade and reimagined with each new theory or discovery; they are "made to be surpassed."[109]

Victorian microscopists also found effective ways to rebut the suspicion of the unnatural object. Beale admits that "methods of preparation" may change the specimen—that "structures [may even be] *formed* which have no real existence in [the object's] natural state." However, he rejects the idea that simpler techniques offer access to unmediated nature. He notes that the specimen is never seen in its natural setting and that probably "many of the more delicate tissues have never been seen by any one" in their natural, living condition.[110] Schleiden more expansively recommends a list of interventions despite concerns about interfering with the object. He adopts the term *reagents* (*Reagentien*), which usually refers only to chemical interventions (and which he uses elsewhere in its original sense), as a metaphor for optical, chemical, mechanical, and physical interventions with the object.[111] The trope of *reagents* aligns microscopy with chemical work in a laboratory; it argues against the simplistic view of intervention as contamination and promotes instead the view of these techniques as valuable tools aiding the dispassionate, meticulous, many-sided comprehension of the scientist.

For difficult objects requiring skilled interventions, microscopists adopted the idealistic trope of *true structure*, powerfully revising the natural/unnatural binary and defending their own histological skills by suggesting that hidden forms are truer than superficial ones. As Peter Garratt argues, the microscope sent "the paradoxical message that reality required alteration (enlargement, exaggeration) if it was to be known satisfactorily."[112] For instance, Griffith and Henfrey describe the varying appearance of a thin transverse section of fingernail when examined dry, immersed in water or oil of turpentine, or immersed in potash and heated. "Now which is to be considered as representing the structure of the nail?" they ask. "Undoubtedly" the treatment with potash, they reply; the increased intervention may not be "natural" but yields "true structure."[113] None of these appearances is false, but the more invasive approach reveals hidden form.

Microscopists must learn, however, to adjust for these interventions during observations. The botanist Schacht warns that since "we seldom see the objects under the microscope in their natural condition," the trained eye "must take into consideration the changes which we ourselves partly produce, either by the medium in which the object is placed, or by the use of the knife or other influences."[114] Similarly, Pritchard argues that we must assess the unnatural image in the context of microscopical optics; we must "perpetually consider the laws of the refraction and reflection of light, in all the conclusions we draw from the evidence even of the very best instruments, used with every possible precaution."[115] Thus, in dealing with unnatural vision and unnatural objects, the trained eye must work with an educated power of judgment (*appreciation*). Observers must acknowledge their own interventions and reason through them to an accurate reading of the image.

Writing Embodiment in Victorian Microscopy: Beautiful Mechanism. Meegan Kennedy, Oxford University Press.
 DOI: 10.1093/9780198940623.003.0014

15
Trained Judgment and Appreciation

> Certainly there is [no instrument] which requires more expert management in itself, or more caution in drawing conclusions from its employment, than the microscope.
>
> John Hughes Bennett, *An Introduction to Clinical Medicine*, 1850

> It cannot be too strongly or too constantly kept in view that the value of the results of microscopic enquiry will depend far more upon the sagacity, perseverance, and accuracy of the Observer, than upon the elaborateness of his instrument.
>
> William Benjamin Carpenter, *The Microscope and Its Revelations*, 1856

Introduction

Once the problem of optical aberration had been substantively addressed in the 1830s, the great English optical houses set about competing to produce more and more powerful objectives. These lenses offered observers the chance to scrutinize increasingly tiny objects but brought their own problems. High powers were exquisitely sensitive to adjustments in practice and produced images that were especially prone to phenomena of *unnatural vision*. These lenses featured strong magnifications but often very shallow planes of focus, hindering attempts to determine the internal structure of microscopical objects. After the 1870s, British microscopists also had to grapple with the implications of German physicist Ernst Abbe's work on light diffraction and how it distorts the image. Working microscopists had to combine *tact* in sensitively adjusting the instrument, a *trained eye* in reading the images thus produced, and the evaluative power of *appreciation* in constructing accurate mental models from those images. Observer, instrument, and object worked together to yield a legible microscopical image, in a process requiring both mental and mechanical focus that fatigued both mind and body.

Victorian writers offered page upon page of advice on the difficulties of microscopical practice, but they became equally concerned about errors of interpretation, which were committed when observers accepted the microscopical image at face value, forced the image to conform with preconceived beliefs, or allowed imagination to fill in any doubtful areas of the image. Margaret Gatty, a phycologist and children's science writer, argues, in the Introduction to her *British Sea-Weeds* (1872), that tact (a trained hand) is valuable but a trained eye and "delicate" judgment (*appreciation*) are more difficult to achieve. She says, "Were the mechanical part now described the worst difficulty in the examination of seaweed structure, all the world might be made learned by algological *durchschnitts*." (These were prepared seaweed sections on slides; Gatty later provides a list of "Rules for Preserving and Laying Out Sea-weeds.") "But," she continues, "the delicacy of eye and judgment requisite for understanding the meaning of what is seen, is a part of the matter not so easily taught or acquired." Gatty thus instructs her readers—many of them female hobbyists, given the attention to women's dress in her Introduction—on how to undertake certain procedures to discipline the judgment, suggesting, "[a] comparison of fresh-gathered and dried *durchschnitts* of the same *genera*; a habit, if possible, of making drawings of everything one sees; and a patient acquiescence with the necessity of being twenty times mistaken at first for every once one is right." These "will go a long way towards making a *scholar* in its secondary sense, out of a mere scholar or amateur."[1] Gatty's recommendation that readers acquire a habit of preserving and drawing specimens is a common one, acknowledging how the material familiarity with an object or image, achieved through hands-on preparation or sketching, helps us know it mentally as well. Perhaps more unusual, she recognizes that authoritative judgment derives from the willingness to fail and redress the failure time and again. Gatty wrote for broad audiences, especially children and women, but her guidelines on practice, above, demonstrate the concern for accuracy and caution that underwrote her correspondence with other naturalists and her work with the eminent phycologist William Henry Harvey.[2]

Gatty's example shows that even popularizers warn about errors of judgment, but such discussions appeared more commonly in texts for working microscopists. The German bacteriologist Robert Koch, for example, in an 1878 discourse on method distinguishes between certain situations—where "an error cannot easily occur…because the appearance" of grouped bacteria "is so characteristic that he who has once had this picture stamped on his mind will recognise it at any time"—and others, where bacteria "are scattered through the tissues"; they may then easily be mistaken for animal tissue structures and vice versa. Most of Koch's cautions here refer not so much to interpretive bias but to practical techniques with the apparatus enabling the observer to differentiate visually between the "structure picture" (showing distracting tissue structures) and the "colour picture" (showing up the stained bacteria).[3]

It is perhaps unsurprising that, over twenty years earlier, William Benjamin Carpenter repeatedly mentioned such problems in his standard text, *The Microscope and Its Revelations* (1856). Carpenter had dedicated considerable effort to the field of physiological psychology and was one of the first and most influential mid-century writers to discuss the problem of interpretation. He even denotes a separate category of "errors of interpretation" caused by lack of training or lapses in care or mental discipline.[4] He particularly cautions readers against preconceived ideas, when people "describe what they *believe* and *infer*, rather than what they actually *witness*." But Carpenter notes that this error of judgment is "a tendency common to *all* observers," not just microscopists.[5] To prevent interest and imagination from driving perception, observers should build up an ever-present, deliberate awareness of their own propensities, "to keep [their] observations free alike from the bias of preconceived ideas, and from the suggestive influence of superficial resemblances."[6] He proposes "*the suspension of the judgment, whenever there seems room for doubt*."[7] Carpenter warns that in microscopy we are particularly likely to misinterpret the patterns in the visual field "when an object is *least* clearly seen," as with inferior instruments, since then "there is *most* room for the exercise of the imagination," so he invokes "the rule of recording only what [one] can *clearly see*."[8] This skepticism informs Carpenter's approach to disputed questions of structure. For example, he declines to address the vexed question of the dots on the *Diatomaceae* in more detail since "no argument is likely to convince those, whose minds are prepossessed with a conception which influences their interpretation of what they see." Rather, contestants should undertake "the *education of their judgment*" by comparing coarser and more delicately marked specimens.[9] Indeed, he argues that "Microscopical study," with its constant call on the discriminative powers, helps students learn how "to keep [their] observations free alike from the bias of preconceived ideas, and from the suggestive influence of superficial resemblances."[10] Carpenter lays out the ethical implications of careless work, describing the observer's responsibility toward overlapping circles of duty. He says, "It is due to Science, that it should be burdened with as few false facts and false doctrines as possible. It is due to other truth-seekers, that they should not be misled, to the great waste of *their* time and pains, by *our* errors. And it is due to ourselves" not to "commit our reputation" to claims that might readily be falsified.[11] It is notable that Carpenter here joins together the effect of error on "Science" and on the microscopical community; microscopy requires careful interpretation to support scientific progress, to respect the time and effort of other microscopists, and of course to safeguard our own legacy.

Lionel Beale's *How to Work with the Microscope* (1857), another standard mid-century text, sees this problem of suggestibility as especially dangerous for microscopists. Beale warns readers, "You must not... be too hasty in deciding upon the nature of an object in the microscope; neither must you infer that what you have not been able to see, does not therefore exist." He continues, "Some fall into an error of the very opposite description, not less detrimental to forming habits of correct observation. Led away by their imagination, they think they see everything which has been delineated [in an illustration] or which they have heard described."[12] (In the fourth edition, he adds that the observer's "eye and mind will require much careful education before he can hope to be able to form a correct opinion.")[13] He warns, "No one engaged in the pursuit of any branch of natural science is more tempted to be led into too hasty generalization than the microscopical observer. It is his duty, therefore, to avoid drawing inferences until he has accumulated a vast number of facts to support the conclusions to which he has arrived."[14] Despite the confusion of tenses here, Beale offers a powerful warning about the risk that shoddy practice poses to individual work, to microscopical communities, and to microscopical science as a whole, and declares that microscopists must heed their "duty" to work carefully. He explains that through these failures of the observer, "errors have been propagated and strengthened to an extent almost incredible, and years of laborious investigation have been spent in overthrowing statements which had never resulted from actual observation in the first instance." Ultimately, "every false statement regarded as an observed fact, forms a terrible barrier to onward progress."[15] Beale here urges a trope of *duty to care*, showing how hasty inferences squander years of colleagues' and students' patient labor and threaten microscopical science generally.

In viewing visual experience not as a passive imprinting but as an actively constructed perception, Carpenter and Beale echo the caveats of thinkers like Hume and Herschel, and the work of physiological psychologists like Bain, Spencer, and Carpenter himself. Indeed, many Victorians called for caution when traversing the mental space between seeing natural objects and knowing them. Martin Willis argues that the Victorian "epistemology of vision held the actual and the imagined in fragile suspension."[16] As Peter Garratt says of this issue in microscopy, "The acts of seeing, imaging, and imagining began effectively to converge."[17] Carpenter particularly acknowledges the role of the mind, not necessarily the conscious mind, in nudging us toward the kind of image we desire or expect to see. Writing on spiritualism, mesmerism, ideo-motor action, and unconscious cerebration, he examines the power of unthinking actions and warns of the risk of unexamined thought.

Microscopical observation, where the image presented to the eye often does not track the true structure of the object, exacerbates Humean concerns about inferring from experience. The proficient microscopist must be well drilled not only in manipulative skill but also in the mitigated skepticism that helps educate the judgment into questioning both experience and interpretation. This need for a habitual but pragmatic skepticism appears everywhere in microscopical writing, addressed to general audiences or beginners, investigators and enthusiasts. The Rev. William Spiers, in *Rambles and Reveries of a Naturalist* (1890), warns readers to maintain a pragmatic skepticism about the evidence of the senses. He warns that

> intuitions or convictions, it may be said, cannot be trusted. Neither can the sight, nor the ear, nor the brain. Illusions and hallucinations come to every one. No day passes but we are reminded that things are not what they seem. The microscopist has to go through a long apprenticeship before he can interpret aright the revelations presented in his wondrous instrument, and there are appearances about whose meaning all but the tyro are uncertain. The senses need education, they need constant supervision, and their impressions have frequently to be corrected.[18]

Spiers himself had considerable microscopical experience; he was a Fellow of the Royal Microscopical Society (RMS) and editor of the *Journal of Microscopy and Natural Science*, and he had worked closely with Dallinger in the Wesley Scientific Society (a club for Methodist naturalists). The physician and surgeon Henry Edward Fripp, President of the Bristol Naturalists' Society (1876–9) and Bristol Medical-Chirurgical Society (1878–9), warns in 1877 that "the [experienced] observer, who knows how many causes of distortion of image may operate in the microscope, will often withold [sic] his judgment where the inexperienced will entertain neither doubt nor suspicion."[19] The validity of microscopical observation rests not just on physical and mental training but also on the cultivation of doubt—a doubt that the "tyro" crucially and dangerously lacks. The microscopist must reject the old trope seeing is believing—must always interrogate, correct, and interpret (not necessarily void) the "revelations presented in his wondrous instrument."

Early adopters of the new compound microscopes in the 1830s had had high hopes for the potential uses of the instrument, but its reputation suffered in the 1850s after a string of embarrassing misinterpretations. John Quekett's misidentification of specimens, noted in Chapter 1, was publicly criticized in the *Quarterly Review*. In his book, Beale chastises "a well known and otherwise highly valuable publication" for presenting a drawing of "what is evidently a portion of feather" labeled as "a representation of *lymphatic vessels*," and notes other examples where "vegetable hairs are described as *nerve fibres*," alongside "other errors equally unpardonable."[20] These and other publicized errors hampered efforts to promote the instrument as a scientific, medical, and forensic tool.

Some microscopists, however, persistently argued that the instrument itself was *innocent* of any error. The German botanist Matthias Jakob Schleiden as early as 1842 referenced a trope of the *unerring microscope* (and eye) to rebut a persistent suspicion of microscopy. Since "our eye is in itself an optical contrivance… founded, like that of the microscope, upon immutable mathematical laws," he argues, "errors consequently are only committed by the erring judgment in all observations, whether instituted with the naked eye or the microscope. The healthy sense and the optical instrument are always right.… The microscope is perfectly innocent of every thing of which it is accused." That is, even conflicting or obviously wrong images are not false; even if the image shows diffraction lines or the observer's eyelash, it accurately represents the optical patterns resulting from a specific configuration of magnification, lighting, focus, and angle of observation. Schleiden argues that false claims derive from "precipitation, superficiality,... scientific dishonesty" and "frivolousness"; that is, they are "not due to the falseness of the instrument, but to the untruth of men."[21] Carpenter similarly argues that it is not "just" for critics "to lay to the account of the instrument, what entirely proceeds from the fault of the observer, in recording, not what he sees in it, but what he fancies he can see."[22] In 1877, Fripp offered a similar view when he remarked, "The microscopist is himself more often on the wrong track than his instrument."[23] Although Fripp had recently translated Ernst Abbe's and Hermann von Helmholtz's theories on the limits of the microscope, he considered the instrument less fallible than its operator.

Discussions of microscopical error anchored Victorian microscopical culture in both skepticism and its mitigation: those practices designed to safeguard observation despite known challenges in the work. The 1850s mark an inflection point where microscopists, defending the practice of microscopy after these embarrassing errors, advocate for better training. Novice and elite observers alike were urged to develop an acute awareness of the sources of potential error in order to compensate for them if possible.

Microscopists use the trope of *appreciation* for this process of learned vision, drawing on a recently evolved usage of the term to denote sensory discernment.[24] Alexander Bain points to "the nicest sensibility to difference" as an indicator of "the intellectual aptitude of the individual." A nice sensibility may be demonstrated in diverse ways in microscopy, including some applicable to microscopy, such as "the delicate appreciation of degrees of muscular force, giving birth to dexterity of manual or other bodily execution" (*tact*) or "in the wide domain of the sense of vision," where microscopists sought a heightened sensitivity "to very minute shades of difference" (a *trained eye*).[25] While individuals may demonstrate innate aptitude for such discriminative power, *tact* and the *trained eye* can also be trained—as can *appreciation.*

In the microscopical context, *appreciation* crucially implies a blend of sensory and cognitive discrimination. Frank Crisp, a longtime promoter of Abbe's work, in 1878 defines *appreciation* as the "more or less laborious reasoning process before we can recognize what it is we really see" under the microscope, adding that "long experience and habit have enabled us (apparently) to dispense with any such reasoning in the case of ordinary vision." Crisp reminds his colleagues that "in the case of the microscope seeing is not believing" and criticizes the "tendency" of observers "to merely *view the object* rather than to investigate its true nature."[26] *Appreciation* refutes the static trope of *seeing is believing*, replacing it with a realist narrative that warns against gullibility and celebrates vision as a skill that must be paired with judgment, and that must be acquired through painstaking training, inevitable failures, and eventual achievement.

This trope rewrites difficulty as achievement. An observer practicing *appreciation* is able to discriminate between what is and is not salient in the image, discern what is most significant in the image, and construct a mental model of the object's structure by working out the relation between this image and other, remembered ones. For example, Carpenter advises students that "the best method of learning to appreciate the class of appearances in question" (air bubbles and the like) is to undertake a series of comparison of oil, water, and air bubbles in different substances.[27] Similarly, he says, "It is always desirable for the purposes of research, to employ the *lowest* power with which the details of structure can be clearly made-out; since, the lower the power, the less is the liability to error from false appearance, and the better can the mutual relations of the different parts of the object be appreciated."[28] Similarly, Beale encourages "all those who are likely to prosecute researches upon . . . any particular tissue or organ" to study it in lower lifeforms first, after which "our mind will be better able to appreciate" (recognize and study) the more complicated version of these structures in the human body.[29] Even once microscopical vision is learned, every observation remains a process and a challenge, since (as Carpenter warns) "even the most practised observer is apt to take no note of such phenomena as his mind is not prepared to appreciate."[30]

Attention and Automaticity

Central to appreciation is attention: the ability to set the mental focus on one thing to the exclusion of others, such as noticing one sensory input and ignoring others. In discussing "volitional Attention" in *Mental Physiology*, Carpenter turns to a simple but common microscopical example: when "the practised Microscopist, whilst applying one of his eyes to his instrument, and determinately giving his whole Attention to the visual picture he receives through it, can keep his other eye open, without being in the least disturbed by the picture of the objects on the table which must be formed upon its retina, but which he *does not see*, unless their brightness should *make* him perceive them." Lewes references this phenomenon also.[31]

The physiological psychologists argued that this ability can be trained and promised that it improves mental as well as sensory function. Carpenter argues in *Human Physiology* that "[t]he Visual power is susceptible of extraordinary improvement, through the habitual direction of our attention" to it. Indeed, "the habit of [visual] attention to any particular class of objects, sharpens the discriminating power for that class alone," although "it is usually rather the mental than the corporeal vision which undergoes improvement." For example, "the Microscopist, who is constantly on the outlook for the various forms of organic structure with which his mind is familiar, discerns these without difficulty or hesitation, where an ordinary observer sees nothing but a confused jumble of tissue."[32] In *Mental Physiology* he concludes, "The power of immediate and acute Perception is one eminently capable of being increased by habitual Attention."[33] In *The Emotions and the Will*, Bain emphasizes mental rather than visual focus, but to the same end; readers should train themselves in a "habitual control of the Attention… against the diversions caused by outward objects."[34] He explains, "The concentration of the thoughts upon one subject as against wanderings, digressions, and chance solicitations…is a part of the apprenticeship of every highly intellectual profession," for "[n]o man can think consecutively, or execute any systematic work of mental study, without a habit of concentrated attention."[35]

This precept reinforces the notion that scientific work requires training both body and mind. Carpenter, too—fascinated by how mental and motor activity blend in learned (secondarily automatic) actions—encourages readers to form "*methodical habits* of mental labour."[36] Such habits allow investigators to maximize the power of automatic thought. For unconscious, automatic actions are not limited to muscle movements; even thinking and creating may be unconscious and automatic. Carpenter's understanding of the relation between mental activity, willed actions, automatic actions, and underlying bodily physiology must certainly have been informed by his decades of experience integrating fine muscle movements with mental attention to the microscopical image. Microscopists' interventions with the instrument and object exemplify how operators entwine automatized actions and mental manipulations.

Carpenter introduced the concept of *unconscious cerebration* in 1853 to describe automatized thought processes.[37] He muses in *Human Physiology* that "the Cerebrum may act…*without any consciousness* on our own parts." For instance, he says, "an expert calculator will cast his eye rapidly from the bottom to the top of a column of figures, and will name the total, without any conscious appreciation of the value of each individual figure." In such a case, the Cerebrum has "automatically evolved the results without our consciousness," a phenomenon that he names "*Unconscious Cerebration*."[38] To provide a fuller example, he tells a long story about Francis Wenham, "an amateur Optician of great ability" who was noted for his innovations in engineering, aeronautics, and microscopy. Carpenter sets up the tale by reviewing the technical shortcomings of the Nachet binocular microscope, which "was perfect in theory" but suffered from loss of light and increase of error due to the additional optical surfaces in its construction. He says, "Now it occurred to Mr. Wenham, that it might be possible so to divide the cone of rays, that one-half of it should go straight on without any interruption, while the other half alone should be deflected by a single prism," which should obviate these flaws. In Carpenter's telling, Wenham

> thought of this a great deal, without being able to hit upon the form of prism which would do what was required; and as he was going into business as an

> Engineer, he put his microscopic studies entirely aside for more than a fortnight, attending only to his other affairs. One evening, after his day's work was done, and "while he was reading a stupid novel," thinking nothing whatever of his microscope, the form of the prism that should answer the purpose flashed into his mind.

Carpenter concludes, "[I]t has been on this plan that all the "binoculars" hitherto in ordinary use in this country have been since constructed."[39] It is ironic that a "stupid novel" provided the setting enabling Wenham to leverage unconscious cerebration and reach this valuable epiphany.

Hand, Eye, and Brain

Just as Victorians discussed the *trained hand* and *eye* as grounded in physiological structure, they were very much aware that *appreciation* emerges from a physical brain. Fripp calls the eye "an optical instrument through which the microscope image must pass to reach the perceiving organ."[40] This trope of the *eye and brain as optical instruments* figuratively diagrams this dynamic of influence. Fripp traces their relation within a single physical system that processes the pencil of light and its image.

Carpenter's *Principles of Mental Physiology* (1874) goes further, placing will itself as part of this physical system. Carpenter argues that there is a necessary relation between the will and "the Automatic activity of the Mind" that depends "upon conditions which bring it within the *nexus* of Physical Causation." He notes that even "Spiritualists" (anti-materialists) acknowledge that the body links up with the brain; that the body sends nervous impulses (such as the microscopical image) via "Organs of Sense" into the brain and carries out "Muscular effort" (such as adjusting the apparatus) directed by the brain.[41] He argues that, more than this, a "vast body of Physiological evidence . . . proves the direct and immediate relation between Mental and Corporeal agency" and cites philosophical authorities who argue "that our succession of Thoughts and Feelings is in great degree determined by antecedent conditions which are intimately related to those of our Physiological Mechanism."[42] In short, Carpenter asserts "the relation of the Will to the Automatic activity of the Mind, and of both to the Physiological Mechanism."[43] Carpenter proposes this relation to be an essentially dynamic one, since we know the world (matter) via "Force" and since "*Mind*, like *Force*, is essentially *active*; all its states are states of change." He compares this dynamic flow to the chain of events triggered in vision: "the formation of a luminous image upon the Retina of the Eye" produces a "Nervous expansion" like the impression of an image "upon a sensitive Photographic surface." This "excite[s] the activity of the Optic Nerve," which is communicated to "the Optic Ganglion" in a manner like "the transmission of a Telegraphic message."[44] Carpenter here updates the old natural-theology trope of the *body as machine* with reference to new technologies (a camera, a telegraph, and later, a "Galvanic battery"). These tropes emphasize automaticity in the sensory flow between material object and mental event even as they leave room for the human will to perform other functions.

Carpenter warns of "*False Perceptions*" caused when "the action of *ideational* states upon the Sensorium...modify or even produce *sensations*." Indeed, he adds, "the action of pre-existing states of Mind is still more frequently shown in modifying the *interpretation* which we put upon our sense-impressions."[45] That is, the inaccuracy of our sensory information leaves our interpretation of it vulnerable to influence from our pre-existing ideas. This is not a microscopical problem alone. Carpenter argues, "Every picture *must* be essentially defective" in some way; it can only "*suggest* to the mind an idea," which the mind "completes for itself by [a] constructive process, so as to form an aggregate which may or may not bear a resemblance to the original."[46] But the inherent challenges of microscopical vision render our interpretations especially susceptible to pre-existing influences.

The eye mediates between (and connects) the mental expectations of the observer and the image in the microscope. Problems arise when the flow of this system is misdirected. Significantly, Carpenter understands this connection as facilitating bidirectional movement; either end can influence the other. Indeed, he insists that "The Correlation between [mental] Mind-force and [material] Nerve-force is shown to be complete *both ways*, each being able to excite the others."[47] Albert McCalla, in an 1883 address to the American Society of Microscopists that was reprinted in the *Journal of the Royal Microscopical Society*, refers to this bidirectionality in describing the interacting optical projections extending from object through microscopical apparatus through eye to the brain—and potentially back. Discussing the "verification of microscopical observation," McCalla warns that "even the well trained eye is in danger of projecting the mental preconceptions of the observer into the focal plane of the objective."[48] Here the undisciplined observer threatens to reverse the flow of imagery, inadvertently contaminating microscopical observation with their own hopes and expectations. An unruly eye or mind can cause a reverse-projection of error (preconception) back into the apparatus, through the sensory channel of the eye, down to the objective itself.

Surface, Depth, and Structure

The educated, disciplined power of *appreciation* is particularly necessary for reading images of microscopical structure, where observers cannot simply read form and texture, surface and depth, but must construct knowledge through actions both mental and physical. Most microscopists used monocular instruments that present a flattened image of a three-dimensional object. In physical terms, of course, many objects of investigation were tiny enough that our eyes perceive them as essentially unidimensional, and larger objects were often sliced in preparation or squashed down by a top slide, making them flatter, practically if not technically two-dimensional. But the monocular microscope (and some binoculars) further flattened the image, resulting in the loss of depth perception. J. W. Griffith and Arthur Henfrey explain in their 1856 *Micrographic Dictionary* that for the microscopist "the following questions are constantly presenting themselves: Do certain spots, lines or other markings visible upon the surface, represent elevations or depressions? Are they cavities in the outer portion or layer of the object? Are they foramina or holes?"[49] A single monocular image was, troublingly, capable of multiple reasonable interpretations.

Stereoscopic binocular microscopes offered one solution to this problem, and an obvious connection to the fashionable toy. Inspired by Charles Wheatstone's invention of the stereoscope in 1838, stereoscopic binocular microscopes became available for purchase in the 1860s.[50] Microscopists could ask an optician to modify a monocular microscope with a binocular tube and prism or they could buy a purpose-built instrument. J. J. Plumer's little 1865 guide to choosing a microscope praises "Mr. Wenham's binocular arrangement" for providing "not a mere claptrap exhibition of the objects"; rather it "gives real relief to the eyes by calling both into exercise."[51] Plumer apparently uses the term "relief" simply to describe the operator's freedom from physical strain and fatigue, but of course Wenham's device also presented microscopical scenes *in relief*, reintroducing a sense of three-dimensionality by a contrivance similar to that used in the stereoscope.

Binocular microscopes allowed operators to add back in a degree of depth perception, but they presented new problems as well: they could reverse elevations and depressions (pseudoscopy) or unevenly divide the light between eyepieces, and the highest-powered binoculars (compatible with wide-angle objectives) generally lacked stereoscopic effect, simply offering more comfortable, two-eyed observation of a flat image. Not everyone saw binoculars as an improvement upon the monocular model. Discussions of this technology even recapitulated the dismissive trope of the *microscope as toy*. The publisher John Phin, instructing beginners in 1877, defends binocular microscopes; but he admits that "of the value of the binocular, there is a wide difference of opinion, some regarding it as a mere toy, and altogether beneath notice as an instrument of scientific research, while others consider it a most important addition to our means of investigation."[52] A year later, the author of the pamphlet *How to Choose a Microscope. By a Demonstrator* goes further, warning novices,

> As regards the scientific value of the stereoscopic binocular, I have grave doubts whether by an exaggerated appearance of relief, optical illusions may not, in

> some cases, creep in, and lead to errors of interpretation; but for all who require a microscope merely for rational amusement, in fact as a drawing-room toy, and extra cost is not a consideration, I recommend the binocular arrangement, as the aspect of suitable objects is marvellously beautiful and striking when seen in stereoscopic relief. But to those students of nature who seek only for the truth, I would advise them to stick to the MONOCULAR arrangement.[53]

This pamphlet scornfully rejects the binocular as "merely...a drawing-room toy" that could only amuse dilettantes seeking what is "marvellously beautiful and striking" rather than the study of nature. However, some original investigators did employ binoculars despite their limitations.

Even with a binocular, microscopic objects often presented insufficient contrast to allow observers to discern structure. One of the most important trends in nineteenth-century microscopy was the working out of methods and technologies to address this problem. Living microscopical objects are often transparent or nearly so, and very difficult to make out against the microscopic field. Even non-living specimens can pose this problem. For example, the German botanist Hermann Schacht warns, "[I]n the microscope, especially with high powers, we see *surfaces*, not *bodies*," and often, "in looking upon [transparent] objects, we are unable to make them out to be bodies until we have changed their position, and ascertained their dimensions in three different directions; this, in many cases, from the nature of the object itself, is a matter of great difficulty."[54] James Elkins, an art historian, notes that "the *manufacture of contrast* is a problem specific to microscopy."[55] Victorian microscopists experimented with stains and mounts to heighten contrast. They stained non-living tissues to bring out distinct elements of structure, but the dyes used in this new science of histology were often harsh, like the new aniline dyes developed from coal tar. Thus, staining of tissues was not well suited to observation of living objects. For live cells, microscopists tried different mounting media, since a fluid with a much different refractive index from the specimen would permit better contrast and thus observation.

The opposite problem in viewing microscopic surfaces, opacity, hampered other kinds of examination, as with the study of crystals and rocks. However, the Scot William Nicol around 1828 developed the first polarizer (the Nicol prism) and the technique of using polarized light on a ground-thin mineral specimen to show the inner structure of the mineral. This method can also be used to determine the makeup of fossilized woods. He published his method in 1831 but few followed up on it. In 1853, Henry Sorby brought Nicol's technique to the attention of petrologists.

The binocular and polarizer were just two of many supplemental tools (*contrivances*) developed to address microscopical difficulty. The Microscopical Society of London in 1858 declared a universal screw-thread standard, allowing observers to mix and match lenses and instruments from different makers, as part of an exciting array of optional or replaceable adjuncts that connect to or accompany the instrument: objectives, eyepieces, lamps, condensers, mirrors, filters, diaphragms, erectors, micrometers, and the like. As enthusiasts enjoyed an increasingly complex range of inter-working apparatus, the entity of "the microscope" became a complicated, modular system in which various elements could be added to or removed from the central "body" of the instrument or the scene of observation generally. These elements were not entirely fungible, since each contributed a distinct quality to the observation. Microscopes were sold with a few or many components but elite workers often also mixed and matched elements from a few favorite makers, each chosen to address some particular need. Their often fervent attachment to particular objectives, displayed especially in writing about difficult tasks, suggests an understanding of the operator–instrument relationship as a reciprocal one.

This understanding of the microscope as a modular instrument is evident in the letters written by the American physician Francis West Lewis to his friend James Lorimer Graham about building a model microscope from components purchased from various makers and even different countries. Even his definition of the instrument—"a certain combination of brass work with *bits* of *glass*"—lends itself to this piecemeal quality. He recommends Zentmayer of Philadelphia for the stand but chooses five objectives from the London firm Powell & Lealand, to complement 4/10 and 1/5 objectives made by the American Tolles, with the eyepieces similarly split ("the 3 lowest by P&L of London and the 2 deepest by 'Tolles'").[56] The microscope here ceases to be a singular, stable object like a simple magnifier but rather admits multiple and shifting potential arrangements of components. Microscopical texts discuss the strategic advantages of one specific component over another; objectives ostensibly of the same magnification could actually perform very differently. Research reports increasingly specify exactly what components were used and in which arrangement, and elite workers were distinguished by their expertise in choosing and adjusting these microscopical components to gain particular ends in particular instances.

Each configuration of elements provides a different view of the object—even successive observations under exactly the same conditions could sometimes produce differing results—and the observer could then try to consolidate these disparate views into a mental whole. Associationist psychologists implied that we perform such consolidations almost automatically in ordinary life. Our ideas of objects "coalesce" from diverse aspects of those objects, associated together over many encounters into an aggregate identity. Bain, for example, cites our "rapid association" of disparate sensory experiences and their "quick adhesion" into a singular concept we know as "pebble" or "rose."[57] However, the disparate, disorienting nature of many microscopical images required effort to collate, while their miscellaneity threatens to radically unground visual representation itself. Any individual image may claim indexical relation to the object but, taken together, this diversity of images unsettles traditional notions of a stable visual relation to the real.

Microscopists' work with the discontinuous and conflicting images of a single object precede shifts in the nineteenth-century art world toward visual modernism. Jonathan Crary, discussing Manet's painting *In the Conservatory* (1879), describes how "a binding together of vision, an obsessive holding together of perception to maintain the viability of a functional real world" meets "a dynamic . . . of equivalence and substitution, of flux and dispersal that threatens to unmoor the apparently stable positions and terms" of a painting's subject.[58] Crary finds these "binding" effects within Manet's painting, but microscopists learned how to construct them well before the 1870s, as a kind of mental scaffolding external to the image.

As early as 1849, Schleiden recognized how central to microscopical work was this multiplicity of representation, where objects look entirely different under different conditions. However, he argues, this cause of confusion could be leveraged to improve the rigor of microscopical work. As noted in Chapter 12, Schleiden urges readers to deliberately *multiply their views* of an object by varying their work within four distinct methods (optical, mechanical, chemical, and physical), in a process he calls *many-sided comprehension*. These tropes encapsulate Schleiden's argument that microscopical observers should try to replicate the way we usually, in everyday life, can examine objects around us through multiple perspectives and using different senses. Thus, to vary the optical method, one would examine an object under different powers of magnification, using different kinds of light including colored light and polarized light. The observer should also examine the effects of mechanical pressure, various chemical reagents, and heat or electricity.[59] Observers would then inductively combine the information gleaned from multiple views, to form a more well-rounded understanding of the object.

Beale recommends a similar series of exercises, but directs them specifically to the novice, who needs to learn how to recognize the effects of different forms of illumination, different powers, and different mounting techniques. He asks the reader to examine objects by six different light arrangements (reflected, transmitted, or polarized light from different sources) and in different media (air, water, Canada balsam, or another highly refractive medium).[60] In the 1880 edition of *How to Work with the Microscope*, Beale adds a plate illustrating how various objects should appear under diverse conditions.[61] Interestingly, the skilled American pathologist Col. J. J. Woodward suggests in 1871 that Beale should extend this approach to his investigations of disease origin as well. Woodward comments, in a letter to the photomicrographer Richard Leach Maddox,

> Dr. Beale is a very ingenious investigator and a plausible writer, but I think he falls into a grave error on ignoring as he does the manifold methods of the German school, which give (in my hands certainly) very satisfactory results,

> and are most useful when mutually compared (e g silver staining, gold staining, perosmic [osmic] acid &c &c.)
>
> He seems to pin his whole faith on a single method (glycerine and carmine) and that unfortunately yields only temporary preparations. Then on the one sided series of facts thus accumulated he builds [S]panish castles in the air.[62]

Woodward suggests here that Beale's speculations are anchored in too narrow a set of facts, too restricted a set of observations—a problem that a more diversified approach would amend. (To be fair, Beale urged working microscopists to prune their observations, reporting only the most germane, so he might well have executed a many-sided approach in practice.) British microscopists do not copy Schleiden's term "many-sided comprehension," but observers at all levels do report using the process, which allows them to build up knowledge about an object by studying how its image appears under different conditions, and then judging each appearance against the others.

However, while Schleiden's *many-sided comprehension* promises that diverse approaches to investigation will lock together into a solid understanding of structure, the accumulation of incongruent images could propose simultaneous conflicting realities that are difficult to reconcile. Microscopists thus learned to distrust the field of vision presented through the instrument, and instead to infer structure both visually and logically, by interpreting how physical changes to the object or apparatus affect the image. A difficult microscopical observation required observers to combine skilled preparation and adjustment of the object and instrument (tact) with logical reasoning based on familiarity with microscopical objects, general knowledge of optics, specific knowledge about the likely material structure of a particular object, and the known affordances of the instrument and apparatus (trained eye and appreciation).

For instance, Griffith and Henfrey advise that when viewing tiny globules of both water and oil, "some will become more luminous as the object-glass is depressed…these are globules of water surrounded by oil" while the globules of oil "will become darker under the same circumstances," because "[t]he globules of oil, being more highly refractive than the water, act like little convex lenses" and vice versa.[63] In another example, they elaborate their reasoning for the reader:

> Foramina cannot be mistaken for elevations on the surface, because they do not become more luminous as the object-glass is raised, after their margin has been brought most distinctly into focus; in fact the reverse occurs: hence they so far agree with depressions; but they differ from these in their luminous appearance with high powers, and their not being rendered more distinct by oblique light, but the reverse.[64]

In 1873, J. W. Stephenson similarly proposes "determining the structure of minute organisms by means of the refractive index of the media in which they are examined." He reminds readers that convex structures mounted in air appear bright when the objective is raised and dark when it is lowered, whereas for convex structures mounted in bisulfide of carbon, the reverse occurs. Comparing the image of *Coscinodiscus oculus iridis* in both media, Stephenson found that the circular markings on the inner and outer layers of the frustule, examined in both media, suggest instead "that it must be either a perforation, or...a plate of silex which gives no lens-like action in any medium. That these spots are, however, openings, there can, I think, be little doubt."[65] Stephenson here models the work of the good microscopist, whose mental and physical work is entwined with the instrument; as he delicately adjusts the focus on this tiny diatom, he studies changes in the image and logically works through their implications.

The process of determining an object's structure through many-sided comprehension was further complicated because lenses of different powers—even lenses of the same power by different makers—could show very different images of a specimen. Carpenter explains in *Revelations* that when viewing the *Morpho Menelaus*, the pattern is not visible at all using low powers; is misleadingly apparent "with a good objective of medium focus"; and is only properly discernible by using a "high power" combined with a special condenser to purify the illumination.[66] And Louisa Lane Clarke, in her elementary guidebook *Objects for the Microscope* (1863), instructs "the young student" to look for particular elements at different magnifications: how to work with a section of monkey skin, for example, so as to display either the nerve fibers or the sweat glands.[67] When microscopists use lenses with an extended depth of field, the visual image is complicated with markings showing different planes of structure at the same time. In a discussion involving the "butterfly termed *Morpho Menelaus*," Carpenter explains that the reverse side of the scale, showing through from beneath, can fool the microscopist into making false conclusions about its structure.[68] J. B. Dancer, the photomicrographer, even warns in the *Monthly Microscopical Journal* in 1869 that "Microscopists are sometimes forgetful that thin and diaphanous objects have two sides."[69] However, many lenses present the opposite problem; their plane of focus is very shallow. The observer then moves through adjacent planes of focus, seeing at any time only a particular horizontal slice of the object on the slide. As Schacht explains, "The microscope only affords a view of one surface of an object" and "in a well-illuminated and perfect microscope the eye cannot perceive anything which lies above or below the focal plane."[70] The wide-angled objective lenses often used to study minute structure especially suffered from this limitation, as they had a very restricted depth of field.

Confusion and even controversy were fostered by the role of focal plane in determining the microscopical image. George West Royston-Pigott, a physician who was active in debates on microscopical optics, notes in 1875, "Microscopists are not always agreed in viewing an unknown object which is the correct focal plane. And... different observers with the same instrument and adjustment [may] view a different focal picture." He points to "the existence of two sets of false images, viz. the one above and the other below the focal reality" as "the most peculiar difficulty" faced by the observer, who must take care not to over- or under-focus the lens.[71]

Microscopists hoping to achieve nuanced and accurate readings of structure were thus urged to train the hands, the eye, and the judgment. "To take a very simple case," Carpenter says, "the transparent body of a certain animal being traversed by vessels lying in different planes, one set of these vessels is brought into view by one adjustment, another set by 'focussing' to a different plane; and the connection of the two sets of vessels, which may be the point of most importance in the whole anatomy of the animal, may be entirely overlooked, for want of a fine adjustment." Indeed, he warns, "Errors of interpretation arising from the imperfection of the Focal adjustment are not at all uncommon amongst young Microscopists." To complicate matters, he says, the relative depth of each reading can readily be misinterpreted; for "where different structural features present themselves on different planes, it will be difficult or even impossible to determine which of them is the nearer and which the more remote... unless, by the use of the 'fine movement,' it be ascertained, when they are successively brought into focus, whether the object-glass has been moved *towards* or *away-from* the object." He adds the caveat, "Even this, however, will not always succeed in certain of the most difficult cases, in which the difference of level is so slight as to be almost inappreciable;—as, for instance, in the case of the markings on the siliceous *loricae* [valves] of the Diatomaceae."[72] Carpenter's example is well chosen, given the miniscule scale of many diatoms and the confusing diffraction lines produced by their delicately patterned frustules. Even as lenses objectively improved over the decades, the question of diatom structure sparked debate, rising to bitter controversies in the 1870s.

As Carpenter's comments suggest, as early as the 1850s it was recognized that an observer can only derive a virtual model of deep structure by executing a series of successive visual "slices," achieved by combining careful manipulation with a focused mental process requiring sustained attention. The researcher builds up a mental image of a three-dimensional object by carefully focusing the lens at different levels, "slicing" the physical object visually just as one might section tissue specimens, and reviewing and mentally manipulating the images in retrospect. The pamphlet on *How to Choose a Microscope*, discussing how to investigate the structure of a tissue, describes this common and necessary process among working microscopists: "by focussing upwards and downward through the several layers or depths of a tissue, &c., and translating the various aspects observed, through the eye to the brain, a safer interpretation may be arrived at than by the questionable reading of a binocular."[73] "Essex" similarly explains in the *English Mechanic and World of Science* in 1877 that using low powers before high ones will help an observer "read a connected and logical discourse" from these discrete observations, instead of disconnected "patches."[74] The trope of *microscopical observation as reading*, implied by the terms "reading" and "discourse" here, points to the role of the observer in constructing the image through a learned process, not just registering it passively.

This process was arguably even more complex than reading. As Schleiden explains, through the microscope "we only see a mathematical surface," and "the power of accommodation of our eye is of no use in seeing what may be placed above or below this surface." Thus in focusing up and down through the object, "we are compelled to annihilate the one object of sight, and to substitute another in its place," afterward "combining the separate impressions into one corporeal whole."[75] Schleiden depicts the observer as "compelled to annihilate" each slice of the image with each adjustment of the fine focus, only afterward building a mental model of the "corporeal whole" out of the memory of these disconnected, lost images. This challenge is difficult enough if the microscopist can use the fine focus adjustment to scroll through consecutive planes of the object, but it becomes even harder if it is necessary to switch to another objective and refocus to decide questions of structure. The observer must determine which elements are spurious, ascertain the correct plane for each true element of visible structure, remember each image and its textures as it disappears from the field of view, and site each image in relation to others, mentally constructing a relation that is absent from the disjunct images. To adapt the terms suggested by Austrian art historian Alois Riegl at the turn of the century, microscopists work to build a governing optical (and imaginary) whole out of a series of "close" and distinct haptic images.[76]

From the mid-twentieth century onward, this process would be facilitated by confocal scanning optical microscopes. Medical tomography (such as a CT scan) similarly takes images of adjacent planes of the human body (which we now call optical sectioning), reassembling the "slices" via computer (image stacking) to form a three-dimensional picture (volume rendering).[77] However, nineteenth-century observers executed these manipulations manually and mentally. Victorian microscopists were materialists—in the sense that expert observation requires a constant awareness of the physical and optical properties of the instrument, specimen, mounting agent, source of illumination, and one's own body adjoining and interacting with the apparatus physically—but this process shows how they also worked very much in virtual space. These discussions accelerated in the 1870s and 1880s, when improved microtomes enabled more uniform sectioning; but as I have shown, microscopists were concerned about the need to synthesize dissociated, even conflicting images and tactile inputs as early as the 1850s, well before researchers in neural science and mental pathology such as Pierre Janet argued (in the 1880s and 1890s) that human subjects must construct a model of the world from sensory perceptions that are shifting, fragmented, and unreliable.[78]

Victorian microscopists were also among the earliest to worry that bodily fatigue damages the human ability to work, as I showed in Chapter 5. This concern obviously affects the observer's ability to perform physical manipulations, but it also degrades the work of the eye and judgment. As noted earlier, Brewster remarked, "there is no bodily fatigue to be compared with that of the use of the microscope. The eye, the mind, and the whole frame, are on the stretch."[79] Accordingly, Albert Schneider recommends that the microscopical observer take periodic exercise breaks to "invigorate the body, retard fatigue and make vision more acute" and "the brain more active."[80]

Discussing this issue of fatigue affecting judgment, Carpenter warns that the mental process of optical sectioning, image stacking, and volume rendering described above—the act of building "a connected and logical discourse" from discrete mental snapshots—is physically taxing. His warning resonates with his experience as a physiological psychologist, concerned that mental effort is as fatiguing as physical effort. As he explains in the 1868 (fourth) edition of *The Microscope and Its Revelations*, this "constructive effort"—the task of "realizing the solid form of the object" based only on a "*flat picture*" and "aided only by the use of the Focal Adjustment" to sample the planes immediately adjacent—produces a "sense of fatigue" that "is rather *mental* than *visual*."[81] A microscopist working with a wide-angled lens bears the cognitive demand of reconstructing the object from incomplete visual cues, by mentally assembling remembered, focused images and shadows from adjacent planes to build a new, virtual object out of glimpses of the original. Martin Willis observes that "[t]o look through the microscope was to engage…in a visual phantasmagoria" that "entangled" both "visual phenomena" (observed via instruments) and "imaginary experience."[82] The process of optical sectioning and volume rendering—the mental manipulation of remembered planes of vision to form a new virtual object—entangles real and imagined bodies. These discussions remind microscopists that their work is both embodied and virtual. Observers create microscopical images through nearly automatic physical adjustments to the fine focus, even as they mentally manipulate those images, treating them like physical objects that could be held in the hand, stacked, and re-formed into a "corporeal whole" that blends material and imagined objects.

Light

These microscopical images are borne on light. Illumination was central to microscopical work but, as noted in Chapters 13 and 14, lighting was a challenging task requiring tact and appreciation to manage well. The innovative microscopical techniques that enabled Koch to discern bacteria, for example, relied on technical skill not only in culturing and staining specimens but also in the careful management of light via diaphragms and condensers.[83] Light simply works differently in microscopical vision. While we see most solid objects using ordinary vision via reflected light, we generally perceive a microscopic image via light that is transmitted, refracted, absorbed, and diffracted. As Beale's exercises suggest, microscopists had to learn how to select and adjust different forms of light and identify their effects in the image. They especially had to learn to detect and ignore the spurious diffraction lines that troubled high-definition tasks. Even very experienced microscopists could be deceived by these effects, caused when light waves interact with very fine structures. Carpenter at first almost dismisses these phenomena, saying, "No rules can be given for the avoidance of such errors, since they can only be escaped by the discriminative power which education and habit confer. The practised Microscopist, indeed, almost instinctively makes the requisite allowance for diffraction."[84] This statement is over-optimistic, though; many of the most heated microscopical disputes of this era involved the effect of diffraction on microscopical structures that were so tiny that only the most experienced microscopists ventured to proclaim upon them. Indeed, Carpenter does note that deceptions due to diffraction are "easily mistaken for genuine resolution"; and in later editions he deletes his claim that "the practised Microscopist... almost instinctively" avoids such errors.[85]

Ironically, diffraction lines often deceived expert observers who were studying the most delicately patterned "test objects"—the very objects used to test and standardize microscopic vision. Just as observers built up a virtual image by mentally manipulating slices representing discrete planes of structure, those encountering diffraction had to construct an image by processing visible images through a deep knowledge of optics and the material qualities of the physical object. Both of these practices were supported by a series of delicate adjustments to the instrument, to enable different settings of the focus.

Diffraction lines were troped as *ghosts* that haunt the microscopic image. The American J. E. Smith, discussing the "spectral, illusory appearance" of these lines, explains that the working microscopist must combine "the finest object-glasses [with] technical skill in the adjustment," something that is "only acquired by long practice," and devote "a good deal of study to this ghostly parasite [the diffraction line], which too often infests, and to a greater or less extent defeats the unsuspecting observer."[86] Smith's comments, like many others, discuss diffraction lines not only as a technical problem that deceives the untrained or incautious observer, but also, in romantic, even Gothic language, as a "spectral... ghostly parasite." Such phrasing frames diffraction as a challenge innate to optical microscopy, requiring a trained eye and educated judgment to manage, but also as a form of phantasmagoria and a reminder of the mystery that persists in microscopical practice.

In the 1870s and 1880s diffraction was recognized as more than a simple annoyance or troublesome illusion. Optical researchers showed how diffraction actually affects all images, significantly distorts images of delicately lined objects, and even sets a hard limit on the power of the optical microscope. This new understanding complicated and unsettled the process of many-sided comprehension, which had compensated for the diversity and confusion of microscopical images of an object. As Abbe argued in 1873, in a canonical essay translated by Fripp into English in 1875, the coarser elements of a microscopic object, such as its outline, are easily visible because light shines straight by or through the object, or is slightly refracted (bent) in the process, or is absorbed by the object and creates shadow. This "dioptric" image, Abbe argues, if properly corrected for aberrations, presents a true delineation of coarse form.[87] In 1881, likely prompted by the work of E. M. Nelson, Abbe clarified that gross structure also produces diffraction rays but that they remain within a narrow cone and thus do not significantly deform the image.[88] Thus it is not difficult for an ordinary objective (lens) to transmit all or most of the rays involved in an object with only coarse structure, and the image of the object correlates well with the true structure of the object.

But with very minute objects or those with very fine structure, the diffraction of the light rays around the edges and through the slits of the object splays the light, creating spurious patterns of light and darkness. The objective cannot collect all these oblique rays, so the pattern shown by the rays does not directly correlate to the structure of the object. Wide-angled objectives can gather more of these marginal rays to collate more complete information about the object, especially the fine detail, producing a more accurate image. "The greater or smaller the admitted . . . portion of these [diffraction] rays," says Abbe, "the greater or less similarity between the image and the object."[89] But he warns that if the objective excludes any diffraction rays, then the image represents not the object (because some of its rays are lost) but instead a different structure, one that would produce only the diffraction rays and interference effects currently being shown. Diffraction is thus a "ghost" not just figuratively but literally; it is an illusion representing nonexistent matter.

At almost the same time, Hermann von Helmholtz, a German physician and optical researcher, independently reached similar conclusions using different methods of analysis. Helmholtz describes the limits of the microscope, saying that observations of fine structure at "extreme amplifications" are "more imperfect than through any other optical instrument, such as telescope or camera." Indeed, he warns, "diffraction necessarily and inevitably increases with the increase of magnifying power, and at length presents an impassable limit to the further extension of microscopic vision."[90] Abbe and Helmholtz showed that (as Ian Hacking has phrased it), "we do not see, in any ordinary sense of the word, through a microscope."[91] Victorian microscopists were called upon to grapple with the practical implications of this theoretical work. They learned to use their knowledge of the specimen, mounting medium, and lens system to consider how the diffraction of light interacts with these physical substances, then intuit from the pattern of observed rays back to the original material structure imaged.

The importance of the diffraction theory cannot be overstated, but many British microscopists were slow to accept it.[92] John Ware Stephenson, Treasurer of the RMS, revisited it in an 1877 paper presented to the Society.[93] He advocated for Abbe's ideas by reframing them in simpler language and providing visual illustrations of the theory from Abbe's 1876 lectures. But Abbe's theory probably reached a greater audience when it was discussed in a posthumous edition of Carpenter's widely read *Revelations of the Microscope*. Nelson is thought to have interpreted Abbe's work for W. H. Dallinger's 1891 revision of the *Revelations*, although he was not officially credited for his work.[94] Nelson provides an influential and readable summary of Abbe's work in a passage that Hacking famously quotes:

> It is demonstrated that microscopic vision is *sui generis*. There is, and can be, *no* comparison between microscopic and macroscopic vision. The images of minute objects are not delineated microscopically by means of the ordinary laws of refraction; they are not *dioptrical* results, but depend entirely on the laws of *diffraction*.[95]

Nelson ably interpreted Abbe's work for reluctant Victorians. In any compound microscope we see not the object itself but an image of its image, but the 1891 *Revelations* explains that at least for coarse structure "a purely dioptric image answers point for point to the object on the stage, and therefore enables a safe inference to be drawn as to the true nature of that object." However, "the diffraction or interference images of minute structure stand in no direct relation to the nature of the object."[96] For example, upon using a diaphragm to block selected areas of the light beam, thus "varying the combinations of the spectra," the microscopist can produce "different figures of varying size and positions...all of which cannot of course represent the true structure" of an object.[97] When we look at a fine grid under a microscope, changing the settings of the microscope changes the image significantly even though the object remains the same. Any particular image is only a pattern of spectra, which can result from varying causes—that different structures can look alike when their diffraction patterns are similar, and that similar ones can look different when their diffraction patterns differ.[98] In short, the visible image has no necessary relation to true structure but records the effect of diffraction patterns—patterns which are governed in part by the form of the object (diffraction being especially misleading in finely lined or gridded tiny objects requiring high-powered lenses) and in part by the arrangement of objectives, stops, and lighting. The patterns viewed through the microscope, says Nelson in the 1891 *Revelations*, "cannot be interpreted as *images of material forms*, but only as *signs* of material differences of composition" that require interpretation.[99] To maximize the reliability of the microscopic image, then, observers must arrange the microscopical setup to include as many rays as possible.

Abbe's theories led commentators to rework some conventional tropes of microscopy. For instance, Victorians discussing diffraction theory reinscribe the tropes of microscopical vision as *unnatural vision* requiring a *trained eye*. McCalla introduces his discussion of microscopical error by reminding listeners of the various theories about diatom structure: "ridges, or furrows, rows of knobs, of areolations, or of minute perforations...hexagons, or circles, or as a wicker basket pattern." He continues, ominously, "[N]ow we have learned from Abbe's researches how little reliance is to be placed on any of these appearances," and argues, "The eye in ordinary vision needs, more than the other senses, to be trained to see aright, and when so trained surpasses all the rest in the fullness of its revelations. So still more does the microscopic vision require careful training that we may be able to safely judge of the reality from the appearances that it affords us." To build a habit of skeptical observation, the observer must learn to "interpret the appearances that present themselves and discriminate the causes that produce them, and so trace back the effects of the lenses themselves, of the diaphragm, of the obliquity of the light, and the effects due to the real structure of the object under examination"—if you know it.[100]

One of the first microscopical guidebooks of the Victorian period, Gideon Mantell's *Thoughts on Animalcules* (1846), reflects at length on this difficulty. He insists, "It is...absolutely necessary that the observer should previously have a general notion of the objects to be inspected." Mantell quotes Basil Hall's advice that "[t]he secret often lies in knowing exactly what to look for," in "learning how to adjust, not merely the focus of the eye, but what may be termed the focus of the judgment, so as to be able to pitch the understanding into such a key that the information may be understood when it comes." Hall's examples derive from astronomy and microscopy: a double star, and a bottle of zoophytes that appeared to novices "perfectly uninhabited," even "absolutely transparent." Once instructed, though, Hall says, "our understandings began to co-operate with our eyesight in peopling the fluid." He wryly remarks, "The wonder then was, how we could possibly have omitted seeing objects now so palpable."[101] This anecdote dramatizes the need for a trained judgment—probably one reason Mantell quotes it. However, Hall's precept of "know what you're looking for" takes no cognizance of either Carpenter's warnings about preconceived ideas or the practical difficulties in examining unknown substances and organisms.

Similarly, diffraction theory inspires Fripp to revise two tropes used by defenders of the instrument: microscopical images as *revelations*, which had been used in an analogy to the inerrant "revelations" of the Bible, and the microscope as an *unerring* instrument whose errors are really due to careless observers. Fripp explains, "[S]ince the microscope exhibits but a counterfeit (more or less accurate) of the object itself, its 'revelations' depend upon our mental grasp of what we see, or seem to see." His use of quotation marks around *revelations* ironizes this familiar romantic trope, while *counterfeit* warns of deception and fraud in the image. Indeed, he explains, "actual objective certainty is here only attainable by a reasoning observation substituted for direct sight of the object." Microscopical practice is thus "more liable to error both of sight and reason" than ordinary vision.[102] Even though later in this essay Fripp endorses the notion of the unerring microscope, here he flips these tropes, putting the microscope—which displays only "counterfeit" "revelations"—in the wrong. He contrasts its illusions against the agency and reliability of the empirical observer, whose "mental grasp" and "reasoning observation" can build on but not naïvely trust the image.

It is evident from such comments that diffraction theory bolsters opposition to the trope that *seeing is believing*. McCalla warns, for example, "The things seen [through the microscope] must not too readily be taken to be the invisible realities."[103] He neatly reverses the expectations built up from ordinary vision and ordinary realities here: in microscopy, what you see is often fiction and should not be believed; "reality" is "invisible." Observers gain a purchase on "real structure" only when they "interpret," "discriminate," and "trace back" the images visible through the lens. Victorian microscopists thus also revise contemporary notions of the passive observer; Fripp suggests that the microscopist has "a problem to solve rather than a picture to read off."[104]

In discussing beauty and wonder earlier, I argued that Victorian investigators strive to practice mechanical objectivity even as they make written references to wonder. The directive that seeing is *not* believing—that the microscopist must resist passive vision—challenges mechanical objectivity in another way, hewing closer to a later approach. The process that Victorian microscopists refer to as *appreciation* partially resembles the twentieth-century approach that Lorraine Daston and Peter Galison, in their work on scientific atlases, call *trained judgment*. In trained judgment, an expert's "trained eye" produces an "*interpreted image*" that may "synthesize, highlight, and grasp relationships" in the object.[105] Both concepts depart from mechanical objectivity in seeing a trained eye and interpretation of the image as necessary to perceive structure.

However, the *appreciation* of Victorian microscopists significantly differs from *trained judgment* as well. Daston and Galison emphasize how, in the 1930s, observers are encouraged to exercise an "expert, trained intuition," to "smooth, refine, or classify images" or pursue the "recognition of family resemblance" via "leaps of the imagination." Furthermore, their discussion sites judgment within a "trained eye" as the organ actually executing this intuition, a "seeing eye" that identifies patterns "at a glance."[106] In contrast, Victorian microscopists continue to distrust leaps of intuition or imagination; and while their trained eye may have learned to invert the image or discount air bubbles, its sensory message must still be processed by a skeptical mind. The microscopists I study see intervention as necessary but they do not endorse intuition. In fact, as this chapter shows, they warn that observers at all levels must question and test their sensory experience, their expectations, and their interpretations.

Rick Rylance has argued that *agnosticism* was in the 1870s "more of an epistemological issue than a religious one," entailing a sense "that error and limitation are structural features of human knowing."[107] It is, then, in its essential agnosticism—this sense that knowledge is possible but always fallible—that *appreciation* diverges from both mechanical objectivity and trained judgment. In Victorian microscopists' *appreciation*, the observer must acquire the habit of skepticism as well as dexterity in mental manipulation, constructing virtual models using a series of filtered and corrected images. Far from being a flash of intuition, this intervention is another form of disciplined, patient labor; it is difficult, fatiguing work that must be checked at every step.

Even before Abbe's work on diffraction, the ideal of mechanical objectivity was complicated for working microscopists, whose efforts to attain a disinterested, nearly automatic perception must be counterbalanced against the need to prepare, manipulate, and analyze the image.[108] Fripp indeed argues that observers should "recognise the immense advantage" experienced microscopists enjoy in "being able to correct our impressions and guide them by a reasoning experience."[109] In this view, only an educated judgment may yield knowledge of real structure, after exercising a hard-won discernment and skillful manipulation of the image—and with, always, the humbling possibility of error in mind. Astronomers also wrestled with questions of true structure—of an image potentially distorted by flawed apparatus, optical limits or illusions, observational setting, personal skill, careless reporting, and faulty interpretation.[110] In microscopy, tropes of embodied practice like *tact* and *appreciation* provided frequent reminders that revelation is by definition hidden from view; it is never certain and must be earned. The microscopist must learn how to regulate the physical and mental actions entangling observer, instrument, and object.

Writing Embodiment in Victorian Microscopy: Beautiful Mechanism. Meegan Kennedy, Oxford University Press.
 DOI: 10.1093/9780198940623.003.0015

16
Minute Description

Our nimble souls
Can spin an insubstantial universe
Suiting our mood, and call it possible,
Sooner than see one grain with eye exact,
And give strict record of it.

George Eliot, *The Spanish Gypsy*, 1868

Introduction

Victorian working microscopists developed textual strategies enabling them to model for readers how to manipulate the microscope and its apparatus, and how to observe the object once it was resolved by the lens. They honed two valuable and influential approaches to reporting one's observations: *minute description*, a prose representation of the image and object, and *articulated practice*, a narrative form that chronicles the working process. While these terms are my own, they identify textual patterns that Victorians used to build a specialized scientific prose style, and they became central to nineteenth-century microscopical culture. These textual forms model the caution, precision, and doubt inherent to skeptical microscopical practice. Jutta Schickore notes that British microscopists did not attend much to the explication of method until the 1820s, when they became more concerned about what caused divergent results in different microscopists' observations of the same object.[1] Their skepticism, paired with second-order attention to quality and method in scientific tools and practice, produced an efflorescence of written instruction and clarification. In such passages, writers attempt to manage the difficulty, error, and privacy inherent to microscopy by recording their practice in print, modeling it for other workers.

Both minute description and articulated practice, often found linked in a single observation report, require precision of language. The writer must balance the need to write fully with the imperative to do so clearly. More important, they model a meticulous, skeptical habit for the reader to emulate, although they serve different immediate purposes: one records a view and the other records the process used to get there. Minute description and articulated practice allow the writer to establish microscopical authority and solicit assent from the reader.

Formalized procedures like this appear in other endeavors as well, especially where observation is difficult or must be painstaking. For example, Omar Nasim describes Victorian astronomers' observation as a process executed through notes and, especially, sketches. He argues that these practices "were integral to the gradual discerning and systematic stabilizing of something barely visible."[2] His analysis shows how the process of observation could fix a series of glimpses into something much more defined. Since astronomical objects generally move across the field of the telescope and are often quite difficult to discern as well, this process prolongs access to the image for study and argues for its true form. In microscopy, where similar difficulties challenge the observer, the literary forms of minute description and articulated practice serve these needs also. But they also explicitly address other observers, attempting to form and direct the reader's embodied work with the image, by modeling good practice and by overtly directing the microscopist's eyes, brain, hands, and body in their optical, mechanical, cognitive, and affective interchanges with the apparatus and object. At the same time, they promote (again, through both modeling and overt direction) quintessential moral virtues of Victorian microscopy such as *patient labor* and the *duty to care* about future readers and observers. These function as, to use James Secord's term, paper instruments.[3]

Minute description and articulated practice became integral to the Victorian practice of microscopy, versions of them appearing in texts ranging from the most elite working communities to the popular guidebook. As these passages present information about an object or environment, they also offer instruction in that they stage microscopical practice for readers. They chronicle how working microscopists execute an observation, showing how skepticism guides but does not paralyze the work. By speaking practice in this way, microscopical texts model new methodologies for their contemporary readers and preserve them for later ones. Such representations were especially valuable as reference objects and to instruct readers in microscopical practice, where the observer's ability to produce particular images of the object might signal having achieved a set milestone, as in the work with difficult test objects.

My discussion of these forms centers on texts published in the middle decades of the century, a period of remarkable growth in cheap print, sparked by the repeal of the last remaining taxes on knowledge and technological innovations like stereotyping and the steam-driven rotary press. A surge of inexpensive and mid-priced books on microscopical topics began in the 1850s. These mid-century guidebooks called upon working microscopists to regularize their practices and their reporting. Those books addressed to novices often explicitly encourage readers to take up the work of original investigation, reflecting contemporary arguments for a broader, more inclusive path to participation in scientific work.[4] They use minute description and articulated practice to elaborate on favored methodologies at all levels of expertise. These attempts to standardize microscopical practice bolstered the status of experienced workers as scientists and helped discipline the proposed influx of untrained workers.

Nineteenth-century observers using the table microscope often required some means of sharing the image, whether for instruction, research, or even just discussion with a colleague. However, truly collective, simultaneous witnessing of the same image was difficult to attain without shifts in perspective (in microscopes with multiple eyepieces) or intervening mediating technologies (in projection instruments, or in illustration, photography, engraving, and printing). The sharing of microscopic images ranged dramatically over the course of the nineteenth century, from viewing the image itself through a double-head instrument or solar microscope to viewing a delineated (pictorial) representation of that image in wood engraving or photomicrography. Victorian microscopists also shared the image through written reports drafted in lab notebooks, then revised and circulated in lectures, journals, and books.

Writing was one of the simplest and most accessible means of sharing the microscopical image. A drawing—another widely available option—could delineate a specimen, but not all microscopists could sketch well. Drawings were difficult for private individuals to reproduce, and even for publishers, illustration remained costly.[5] In contrast, words could also replicate a specimen and could be more easily and reliably copied and circulated. By mid-century, microscopists were turning their attention toward regularizing the notes they took during observations, forming a consensus that descriptive notes on specimens should be complete, clear, accurate, and precise. Minute description aims to reproduce the microscopical image in textual form. It is "minute" in its extreme particularity of detail, its meticulousness and precision, and of course in its importance in writing about microscopical objects. However, microscopists did not agree on one best method for achieving this goal. They developed different models of scientific description—which I term *taxonomic character*, *inductive description*, and *defined description*—to address different concerns.

The model of *taxonomic character*, commonly found in atlases, briefly describes natural forms using scientific terms and a distinctive syntax. Victorian microscopists, like naturalists in all fields, defined taxa following a recently established scientific convention. For example, George Johnston, in his 1838 *History of British Zoophytes*, defines the "character" of the genus *Bowerbankia* as "*Polypidom confervoid, horny, creeping; the cells sessile, unilateral, irregular, the inflected portion with a spinous or filamentous rim:—Polypes ascidian.*"[6] By eschewing verbs and leaving the subject only implicit, this pared-down barrage of Latinate adjectives and stacked phrases produces an impersonal, percussive rhythm that promises precision, efficiency, clarity, certainty. Of course, the taxa themselves were far from the definite objects that this syntax suggests. Not only did individual species shift between classifications (often from one genus to another, but sometimes even between kingdoms), but for much of the century naturalists also disagreed about what elements should make up the character in botany and other fields, resulting in a "discord of competing classifications."[7] However, the character of a taxon provides a *general definition* of an organism; what I am interested in here is how Victorian microscopists reported their observation of a *particular* object.

The two writers I examine more closely in this chapter address different communities (in field of interest and level of expertise), yet they both strive to regularize microscopical note-taking. Hermann Schacht, writing "a guide for research in the field of scientific botany," hopes to train readers who are already fellow investigators.[8] Lionel Beale gave a series of lectures to medical students, seeking to enlist them as potential researchers. Expanding upon these lectures, his *How to Work with the Microscope* became a classic introduction to microscopical work useful to the general (non-scientific) reader as well as the practitioner; a review praised it as "suited to the wants of the student by its great simplicity, while, from its ability and research, the more experienced practitioner will gladly turn to its pages as a welcome guide."[9] These men writing for different audiences and with different aims not surprisingly reach different conclusions about the best way to reach their shared goal. Schacht is most concerned with how investigators may secure a comprehensive record of work to supplement their own faulty memories. He emphasizes a detail-heavy method that I call *inductive description*, which ensures that a full record of the image is available to meet unanticipated changes in the investigative plan. Beale, on the other hand, considers how published research can enable or hinder future workers; where Schacht insists on fullness, Beale counsels judicious editing. His *defined description* seeks to read the object closely but not too closely, offering a clear but selective record to ensure that the microscopical record remains widely accessible. Indeed, Alison Winter, writing on the diversity of scientific practice, observes, "there may have been as many ways of defining proper science as there were constituencies for science."[10] Although these books address distinct audiences, each sets up guidelines on how to write about microscopical work, and each—as also occurs with articulated practice—shapes this advice to meet the needs of its imagined readers.

Despite their differences, these texts agree on certain elements of microscopical practice. Although Schacht and Beale differ on how to achieve it, they both promote a realist, functional, and pragmatic description of the object, powered by a clear, accurate, and largely disinterested prose that pairs quantification and measurement with a qualitative report of form and structure. These detailed modes of description typically slow down the forward motion of the text, although they can still display narrative qualities by reporting on changes in the specimen or by shifting readers' attention through various elements of an image. Both of these texts thus register the entwined embodiment of the observer with the apparatus and object.

Inductive Description: Schacht

The German botanist Hermann Schacht, who published *Das Mikroskop und seine Anwendung inabesondere für Pflanzen-anatomie und Physiologie* in 1851, was among the early wave of writers developing basic protocols for microscopical practice. In 1853 the British lawyer and mycologist Frederick Currey edited, expanded, and translated Schacht's book into English (*The Microscope in Its Special Application to Vegetable Anatomy*) to launch Samuel Highley's series of inexpensive textbooks, "Highley's Library of Science and Art." The book was priced at 5*s*., including woodcuts. Schacht advocated for investigative method, saying, "if the method is accurate, the result will be valuable; if... the method be erroneous, the result will prove nothing."[11] Schacht's instructions on note-taking endorse a detail-rich practice strongly oriented toward the inductive method of investigation. The method he describes, a flexible but full principle that could be called *inductive description*, likely reflects his experience working from 1847 to 1850 with Matthias Jakob Schleiden, one of the originators of cell theory, at the University of Jena; it is especially well suited to Schacht's own research interests in plant anatomy and embryology. In fact, most of Schacht's book provides detailed instruction not just on general principles of investigation but also on how to observe particular plants and parts of plants.

Schacht proposes a method of investigation that moves from a general overview to specific elements, then builds back up to an inductive conclusion about the whole. He recommends two preliminary steps to establish context before plunging into work on a specific object. To begin, the microscopist should "obtain a general acquaintance with the object to be investigated." He should also "make himself familiar, as far as possible, with all the recent observations upon the matter in question." Only after that should he employ "the method of induction," working through the aspects of the specimen—again in a particular order, beginning with general work to provide context and set guidelines for interpretation. Schacht defends these preferences. Although induction moves "from the part to the whole," he says, microscopical work requires "a superficial general knowledge of the object" first in order to prevent being misled by the image. Thus "investigation must begin with generalities, must pass from generalities to details, and lead through the details" back to generalities, "to an accurate acquaintance with the whole." Working microscopists were often counseled to adopt a similar approach, using lower powers first to attain a view of the whole object and its basic structures before focusing in with higher powers to study individual parts, smaller structures, and cells. Schacht specifies beginning with scrutiny of "the entire perfect plant... by the naked eye, or... a magnifying glass," which he calls "a superficial knowledge, in opposition to the more accurate knowledge which is obtained by a complete examination, within and without, of the individual parts with different magnifying powers."[12]

Microscopists, whether investigators or novices, should further begin work with simpler plants and progress to more difficult ones.[13] While the inductive investigator will combine details from various aspects of observation to produce final conclusions, different elements of the investigation will require distinct approaches. Schacht explains that investigators will need to be flexible in their approach, adjusting tools and "mode of proceeding" in order to examine different aspects of the specimen: "outward form," "different portions of plants," or "the development of the cells."[14] The investigator will further need to adopt distinct "points of view" for different kinds of work: "morphological, which relates to [the] outward form" of objects, and "anatomical, which relates to their internal structure."[15] Schacht here again formalizes common principles of good microscopical practice. For example, in noting that the investigator must vary both the apparatus and the investigative approach in order to build up a full understanding of the object, he echoes the advice of his mentor Schleiden. Throughout all these changing modes of observation, the microscopist must never "los[e] sight of" the "principal question"—although the question itself may change as work progresses and "collateral questions" are answered.[16]

Schacht recommends particular methods of capturing or sharing the image to support this inductive work. He emphasizes that the microscopist should hand-draw "all the objects which…may appear to him interesting or important," commenting that "too much cannot be done in this way." The investigator should also "add short notes of everything which cannot be accurately represented by the drawing." While Schacht does here subordinate note-taking to "making drawings and preserving objects," writing was an important part of the methodical investigation that he proposes. The observer should record certain facts about each observation, such as the date and the magnifying power used, perhaps the eyepiece as well. Other details should be captured as well; specifically, "[i]t is a good plan to make notes at the time of everything that appears to be important." Here he endorses a selective principle in note-taking, capturing first what "appears to be important"; but then he continues, "and even of matters which may not at the time seem to be of much value."[17] This exhaustive approach reflects the uncertainty and changeability of the experimental process, for "during an investigation it is impossible to tell in many cases what influence small matters may have over the result." Schacht here identifies what the sociologist of science Karin Knorr-Cetina describes as "the frequent splitting of activities into different strands [of research], and the possible displacements of one strand by another." This "angularity of epistemic practice, its continual lateral divergence from itself" in the branching off of investigative foci requires an investigator to adopt a mode of work that can readily adapt to new questions.[18] Schacht distrusts human memory as a backup during this process, saying, "It is unsafe…to rely upon the memory; by so doing many things will be forgotten, and many things insufficiently or inaccurately described." To preserve a full record of one's work, then, "[s]hort notes should always be made at the latest the same evening, of the things which have been observed during the day."[19] Schacht's genre of microscopical note-taking thus advocates unfiltered notes taken at the scene plus notes elaborating on doubtful points of the sketch; he characterizes the latter as "short" but at the same time presses microscopists not to leave anything out. This practice was perhaps derived from his botanical work, where plants growing in the field appear very different from specimens retained for further study. Although Schacht's work involved close examination of those specimens, field notes were still valuable clues to context and natural form.[20] The note-taking method that he recommends reflects a skeptical, realist, and flexible view of the experimental process and of human capabilities in drawing and memory.

Schacht's book is filled with exhaustive instructions on what details to examine and notate regarding each particular part of various plants. He emphasizes quantification and measurement as well as qualitative description of form and structure. For example, "in the Conifers the length of the medullary rays must... be observed," and "in examining the pollen [of any flower] particular attention must be paid to the number of its coatings."[21] He occasionally indulges in aesthetic remarks, such as "The fluid contents of the cells, which in the petal are frequently of such beautiful colours, must be particularly noticed."[22] These instructions act as a kind of template directing the observer in what to note down, their passive construction modeling a disinterested tone. Despite his interest in complete and accurate notes, though, Schacht does not offer much detail about his preferred style of writing. He focuses instead upon the question of when to take notes—at the time, or that evening at the very latest—and a general principle of completeness rather than selection.

Although he elsewhere urges readers to acquire a *trained eye* that can identify important structures within a confusing microscopical image, Schacht's inductive description embodies a principle of comprehensiveness that Lorraine Daston and Peter Galison have discussed under the rubric of *mechanical objectivity*.[23] Since microscopical descriptions should be comprehensive, accurate, and precise, they explain, Victorians considered that the virtuous researcher would record the visual field fully, without filter or frame, and embrace the disinterested, self-denying labor required to do so. Schacht, driven by his concerns about human memory and awareness of how experimental aims can shift in practice, here emphasizes the first part of that definition—the benefits of an unfiltered note-taking practice. He urges naturalists to simply report everything that is before them, rather than screening out any information based on what they currently consider relevant or important.[24]

Defined Description: Beale

The physician Lionel Smith Beale also wrote about how illustrators can achieve a precise and accurate delineation of the microscopic image and recommended drawing as a means of gaining skill in close observation. However, Beale and

Schacht reached different conclusions on how the goal of precise and accurate depiction should be achieved in writing, and on how writing relates to memory. In 1856–7 Beale, a professor of physiology and general and morbid anatomy at King's College, London, gave a course of lectures on microscopy for medical students. In 1857 Beale was elected a Fellow of the Royal Society and published his lectures under the title *How to Work with the Microscope* (1857) with John Churchill, a medical and scientific publisher. As with Schacht's book, *How to Work with the Microscope* was priced at 5*s*., but unlike Schacht's book it was initially published without illustrations, which were added in later editions. By publishing, Beale could reach a broader audience; he targeted specifically the "many who…are deterred from attempting any branch of original investigation solely by the great difficulty they experience in surmounting elementary detail and mere mechanical operations."[25] As the new editor of the *Archives of Medicine*, Beale apparently hoped not just to increase the number of microscopists but also to improve the quality of microscopical practice and writing.[26] He had a reputation for personal combativeness, which complicated his professional friendships.[27] But his textbook was well received and reached a fifth edition in 1880, becoming an influential guide for working microscopists generally. While Beale's style of description—honed, practical, goal-directed—draws on a long tradition in scientific and medical writing, he offers explicit instruction on it in a book that became a standard of Victorian microscopy. Specifically, Beale argues that the true microscopist is an altruistic man who evangelizes for the science; who pairs written description with delineation (illustration); who writes with caution, patient labor, purpose, and due proportion; and most important, who writes clearly in order that others may follow.

Beale's great priority—expanding the ranks of working microscopists to support scientific progress—emerges in his advice here, as does the importance of accurate but restrained microscopical writing in achieving this goal. Beale sees increased access to microscopy as the path to scientific progress. His institutional environment at King's—like University College, London—was notable for its embrace of the modern curriculum (including scientific subjects) and for its tolerance of nonconformists. It is not then, perhaps, surprising that in his first chapter, Beale encourages readers to set aside any diffidence stemming from their station in society or financial state. He assures them, "Microscopical inquiry may be undertaken by persons in almost any position," who can use the "numerous cheap and excellent microscopes" then available on the market.[28] Determined to recruit more working microscopists, Beale defines useful notes not for their ability to capture everything about the microscopical image (as Schacht does) but rather for their accessibility to new users. Reprising the trope of *the duty to care*, he considers primarily how to ensure that notes are useful to others, in contrast to Schacht's concern that observers can review notes to supplement and correct their own fallible drawings and memory. Beale explains, "He who feels a real love for his subject will try all he can to enlist others in the same cause; he will try to remove all difficulties of investigation, and endeavour to express what he has learned himself in language which shall be intelligible to all."[29] He argues here that the true microscopist can be recognized by his evangelism: his efforts to recruit more workers of all classes to the cause. That is, a streamlined, efficient reportage does not oversimplify or dilute microscopical science, as one might fear, but rather strengthens it, by raising up an army of competent observers.

Beale asks observers to report only what is necessary, so as to provide uncluttered records for future colleagues. He recommends an efficient style of microscopical note-taking that is empirical, selective, and succinct, rather than Schacht's just-in-case preference for full, unfiltered notes. Schacht seems most concerned that the notes should capture all the details that might be lost to memory, in case these are needed later as the research question shifts in practice. But Beale's greatest concern is that notes should be brief and straightforward; he comments, "The notes should be short, and should consist of a simple statement of points which have been observed."[30] We might term his influential method *defined description*: it completely and meticulously surveys the specimen (defines it using quantification, measurement, and qualitative description of form and structure), yet it expresses only what is certain (definite) and characterized by a curtailed (defined) focus.

So while Beale argues that microscopical description should be definite—certain and well defined—he warns against description that is *too* well defined. In an early departure from mechanical objectivity, he does not endorse the unfiltered capture of every detail; rather, microscopists should write concisely and efficiently. He explains, "*Exactness of Description* should always be aimed at," but it should be "obtained with the use of a small number of words." He warns, "That appearance of precision which is often attempted by employing long useless descriptions cannot be too much condemned." Unlike Schacht, Beale endorses a selective principle of description: microscopists should not elaborate on all elements of the image indiscriminately but should reserve detailed description for what they know for certain and know to be useful. Beale argues, "An extreme minuteness in description is by no means a proof of accuracy of observation." Rather, such ponderous reports "encumber" science "with unnecessary words" and—just as important—"intimidate" the "earnest students" who might carry on the work.[31]

Beale further specifies that notes should be marked as much by their doubt as their certainty. He warns,

> Inferences should be carefully avoided, and nothing should be stated without the observer being thoroughly satisfied of its accuracy. If he is not quite certain of any observation he should express his doubts, or place a note of interrogation after the statement. The use of indefinite terms should be avoided as much as

> possible, and whenever any particular word is used a definite meaning should be attached to it. Much confusion has arisen from the use of terms which have not been well defined.[32]

In his anxiety to avoid, and to have his students avoid, ambiguity, Beale gives this rule in several different forms, telling students to reject both "inferences" and "indefinite terms." But unwarranted certainty is also to be avoided; they should signal any doubts using words or a "note of interrogation." His repeated reminders to express doubt where indicated show how skepticism was becoming an increasingly important element of microscopical practice.[33]

But while Beale endorses a screening process in writing down exactly what one sees, he rejects the unnecessary obtrusion of self into one's observations. He condemns the microscopist who "think[s] more of himself than his subject," but ironically he locates that error in the over-scrupulous pursuit of comprehensive note-taking. He rejects the pursuit of precision for its own sake and disparages the quest to scrupulously, even mechanically, produce every detail, presenting that as self-important posturing. This substantively revises the figure of the self-abdicating observer. Beale understands that the observer is unavoidably involved in the production of knowledge. Indeed, just as he argues that microscopists must gain the skill of *appreciation*, he argues that scientific workers must learn how to define—that is, judiciously curb—over-meticulous description so as not to flood the text with distractions. He critiques those who "describe[e] every object in the field of the microscope without the smallest knowledge of any one of them," elaborating their work in a vain attempt to parade their meticulous observations, without the judgment and discipline that should accompany microscopical skill. Such observers, he says, have "brought microscopic observation into great disrepute." For example, "Some have thought to gain the credit of being accurate observers by carefully measuring every object they see in every diameter, and putting down in numbers the results of this useless ceremony." Beale opposes this practice not only because it pointlessly clutters up the description and hampers others' work but also because he considers this kind of performative note-taking to be motivated by personal ambition rather than the disinterested dedication to microscopy that marks the man of science. "Such reports," he scoffs,

> show that the author is thinking more of himself than his subject. He attempts to acquire a character of extreme minuteness of observation, instead of striving to advance the real interests of the science which he professes to serve—and instead of endeavouring to excite in the mind of the reader a desire for more extended knowledge and a wish to take part in a similar investigation, he perpetually gives undue prominence to himself.[34]

Beale here challenges the practice of mechanical objectivity, arguing that while overly comprehensive note-taking ostentatiously claims to remove self from observation, in fact it is obstructively self-promoting. The bad microscopist almost theatrically pushes forward his "extreme minuteness of observation," giving both it and himself "undue prominence." A good microscopist, on the other hand, steps back, magnifying the work and the needs of the reader instead. In a discipline centered on extremes of minuteness and great differences in scale, the good microscopist is the one who understands the role of scale and proportion in all things, observing and writing included.

Beale also condemns the pretentious claims and vague writing that obstruct the progress of microscopy; he turns to the language of romance in judging such writing to be theatrical and deceptive. He mocks the affectations in microscopical reports, commenting that "[a] certain mysterious air pervading the description of an observation,—an evident desire to coin new words,—and exaggerated statements of the importance of the facts observed, are quite misplaced where all should be clear, simple, and intelligible to every one."[35] Beale's scorn for the trumpery pretensions of these *poseurs* rejects familiar qualities of romance (mystery, desire, novelty, exaggeration, grandiosity, ornamental overelaboration) and damns them as useless and exclusive. This rejection of the stigmatized genre of romance implicitly bolsters the status of microscopical science, just as he welcomes novices from a range of backgrounds in a gesture that could threaten that status. He says,

> Nothing, I believe, has been productive of so much pain and sorrow to earnest men...or retarded the real progress of scientific enquiry, more than that affectation of precision, and minute verbose and pompous style of description, which has been fashionable among some microscopists....All this is mere pretence, and not real, earnest, useful work,—distasteful to every scientific man and discouraging to every student.[36]

Beale here reiterates his association with "every scientific man" and his concern for "every student," thus reasserting his popularizing interest in expanding the ranks of microscopists as well as the code of conduct assuring their respectability and worth. He depicts their work as manly work; the apparently universalizing tone of "scientific man" is implicitly gendered by its echo of "earnest men." In contrast, he paints bad note-taking as essentially frivolous. The "affectation of precision" and "minute verbose...description" feminize bad writing, which hypes a "mere pretence" rather than producing "real, earnest, useful work." Twice endorsing what is "earnest," twice referencing the well-trained microscopist as a "man" and his work as "real," Beale here endorses a masculine ideal of disinterested realism over the feminized spectacle of self-promoting romance, even as he himself draws upon a romantic and emphatic language of excess to defend this ideal.

Beale does promote one important aspect of mechanical objectivity: the laborious practice of the observer. He argues that microscopic description (perhaps especially his concise yet comprehensive note-taking technique) requires significant training and practice to master, just as drawing does. Beale offers more detail about general note-taking than Schacht did. He dedicates a section of *How to Work with the Microscope* to the topic of "*Taking Notes of Microscopical Observations*," instructing his audience in specific habits that should produce clear, concise, and accurate notes. Although some observers do show "greater facility" at the task than others, Beale says, it requires practice. He explains,

> It is of great importance to acquire the habit of describing in words, the appearance of objects under the microscope....Accuracy in describing microscopical specimens can only be acquired by practice, and I think it a most excellent rule to take notes of the appearances of every object submitted to examination.[37]

In urging students to build a "habit" of note-taking, Beale emphasizes the need for regular, repeated observations.

This practice, Beale says, not only builds observational skill but also helps the microscopist remember what is observed. "The time is well spent," he explains, because "everything so described is retained in the memory." That is, while note-taking for Schacht supplements memory, for Beale it builds memory. In 1865 Beale expands upon this notion, advising observers

> *not* to make too minute notes during the progress of the investigation, but to retain, as far as may be, the facts observed, in the memory; and when the whole matter is made out, but not before, to begin writing and recording the observations. Otherwise, imperfectly observed facts are liable to be set down as actual facts, and argued upon as truths.[38]

He recommends particular times and methods for this note-taking rather than encouraging notes during observation, as Schacht does. Whereas Schacht directs students to make notes "at the time" since "[i]t is unsafe...to rely upon the memory," Beale prescribes not only a regular habit of note-taking but a retrospective one.[39] To be clear, Beale's actual practice might largely align with Schacht's; Schacht recommends description during observation but allows note-taking after the fact, no later than that night; Beale's advice to delay writing until the "matter is made out" could mean a wait of half an hour or a week, depending on the observation. But it is clear that these writers' views on note-taking reflect somewhat different views on memory. While Schacht considers on-the-spot sketches and written descriptions to be a safer archive than memory, Beale suggests that sketching, writing, knowing, and remembering all reinforce one another. His recommendation to make note-taking a regular but delayed habit formalizes this practice so that students may hone their observation, sharpen their memory, and produce accurate, focused reports of their work.

Overall, Beale trusted illustration more than prose description. Indeed, in an advertisement for *Archives of Medicine* he stresses that he intends to subordinate writing to illustration rather than the reverse. He vows,

> I shall introduce copious illustrations, feeling confident that drawings are really of much more use than long descriptions. I am anxious that the text should be as short as is compatible with a clear statement of the facts; and it is my desire, as far as possible, to substitute accurate representations of the objects for a minute description of their characters.[40]

He evidently believed that to convey information as efficiently as possible, the observer should favor illustration over writing, with written description "as far as possible" restricted. He must have been pleased that his *Archives of Medicine* was praised by the *Medico-Chirurgical Review* for providing "a large amount of valuable... information... compressed into a small space."[41]

Beale's own descriptions exemplify his preference for exact and appropriately skeptical but concise writing, giving each element proportionate weight, with frequent reference to illustrations. His 1857 *How to Work with the Microscope* focuses on manipulation, but his 1854 book *The Microscope and Its Application to Clinical Medicine* (another volume in "Highley's Library of Science and Art") models the kind of practice he encourages. This book does not feature many extended descriptions of specimens; Beale's goal is "not to describe minutely the appearances observed upon placing a specimen under the microscope, but to consider the practical methods by which the investigation may be most successfully carried out." Thus, he will primarily "show *how* the structures are to be submitted to examination, rather than... describe *what* are the appearances observed."[42] He does include some representative descriptions, however, to aid in diagnosis. For example, discussing "cancer of the lip, tongue, &c.," he reports,

> interspaces will be observed (fig. 130), from the walls of which the cells appear to grow. The cells often seem to be arranged in laminae; they do not vary so much in size and form as the cells of true cancer; the nuclei do not differ much in size; they rarely contain many nucleoli, and usually adhere to each other by their margins; frequently three or four, or more, will be found united together. In fact, these cells very nearly resemble in their general characters the ordinary epithelial cells of the surface, upon which the growth is developed.[43]

Similarly, on the *Achorion Schoeuleinii* (a fungus of the hair follicle), Beale says that it

> usually appears as elongated vesicles, of a more or less oval form (fig. 135a), many of them being rather irregular and varying much in size, but often joined

> end to end so as to form branches.... The *favus* consists of a little cavity filled with spores of the fungus, granules, and epithelial cells (fig. 136). One or two hairs usually pass through the centre of the favus.[44]

Both of these descriptions provide succinct description with the features Beale values. They exhibit ambiguity in terms like "often," "much," "rarely," "usually," "frequently," "very nearly" "many," and "rather." While this might tell against the aim of precise description, these passages describe general phenomena, not specific ones, and show Beale's adherence to his own rule: don't write doubt as certainty. While meticulously indicating doubt or variance, Beale's descriptions still provide a wealth of information in compact, accessible, workmanlike prose. We learn about the shape and structure of objects, their formal arrangement, their connections to one another and to other objects, their usual number, contents, variability, and the like. He often suggests a useful comparison and occasionally takes note of a "beautiful" form; for instance, of "the uriniferous tube" of the newt, which is fully "lined with ciliated epithelium," he exclaims, "A more beautiful object under a half-inch object-glass, can scarcely be conceived."[45] As for size, Beale had specified earlier in the book that, "unless stated to the contrary," descriptions refer to observation "with a quarter-of-an-inch object-glass (about 200 diameters)."[46] A reader can thus determine "the dimensions of all the drawings" (the size of delineated objects) with the use of a key to scale that he provides, which superimposes the lines of a micrometer (1/1000th of an inch and 1/5000th of an inch) over an assortment of microscopic objects.[47] This system reflects Beale's interest in concise but accurate writing; by standardizing the scale of magnification and illustration, he can provide the necessary measurements without cumbering each description with the information.

Beale's research papers are also detailed yet concise, with frequent links to illustrations. In 1863, in the midst of the vitalist debates, he reported on his enquiry into "the so-called intercellular substance of cartilage." Here he refutes the notion that the intercellular substance has "formative power" or that it is essentially distinct from the cell.[48] Facing interlocutors of the stature of Virchow and Huxley, Beale's descriptions of microscopical observations must fully support his argument, but he maintains a crisp, efficient style. Describing "fresh cartilage of the frog," he reports, "In the preparation passed round (Prep. 4), magnified 700 diameters" appear "numerous oval or spherical masses of a granular appearance." He draws attention to "many of a half-moon shape... and others varying very much in shape, and with so ragged and irregular an outline that no one would be disposed to call them cells, or would maintain that they *possessed a cell-wall* (Figs. 2, 3, 4, 5, 6, 7, 8)." He continues, "At the edge of the specimen, where it is exceedingly thin, some of these angular masses, composed of granular matter, with the central part deeply coloured [stained] with carmine (nucleus), can be seen to shade *uninterruptedly into the matrix*." He allows himself to repeat this crucial point: "The granular, ragged edges gradually pass into, and are continuous with, the clear, transparent matrix."[49] Beale adopts a periodic prose style here to accommodate his need for precision and nuance, but it is not pompous or turgid; every word tells. His own work aligns well with his advice to microscopists: strive for correct and complete note-taking but without hasty inferences or over-meticulous description, which inflate the author and impede the novice.

Beale's work significantly influenced Victorian microscopical practice. As I showed in Chapter 4, *How to Work with the Microscope* was considered a standard text well into the latter decades of the century. I don't wish to overstate his differences from Schacht. In fact, he recommended Schacht's book as a general resource in both *The Microscope and Its Application to Clinical Medicine* and *How to Work with the Microscope*, and the two writers share many goals.[50] Both advocate for clear and accurate description written in (mostly) disinterested prose. Both endorse quantification and measurement where appropriate, paired with qualitative description of form and structure; and both are cited by their peers. Both adopt a realist, functional, and pragmatic mode of description. But there are important differences. For Schacht, primarily concerned about investigative practice, the vagaries of experiment and of human memory require a just-in-case fullness of description. Beale, worried about how published errors or prolixity impede later workers, privileges access and accuracy over fullness of description, recommending greater selectivity in what is reported. Beale prioritizes readability rather than flexibility or reserve: notes should be definite but not overelaborate, precise but not ornamentally fastidious. He proposes a clear and direct textual aesthetic that eschews fussiness as much as fantasy.

Minute Developments

Passages of minute description represent an object in close detail, but they are not static. The roving *trained eye* of the narrator moves between different perspectives as the *trained hand* makes adjustments to focus, magnification, illumination, or the slide itself. The narrative focus shifts from one part or feature of the object to another even when the image remains the same, thus modeling what Lorraine Daston calls "the selective and focused exercise of attention."[51] These movements, physical and cognitive, reflect the operator's actions, which are implicitly—sometimes explicitly—noted in the text.

Minute description thus presents a virtual storyspace to readers, where the operator's actions are featured alongside those of the instrument and object, in a fluid, ever-developing scene. In Beale's article on the structure of cartilage, for example, he says,

> If the young cartilage be broken up, portions of the formed material are seen to be continuous with the germinal matter (fig. 2); and in specimens coloured with carmine we see, first, the smooth and very finely granular and transparent

> "matrix" or "formed material" colourless; next, a layer faintly coloured; and lastly, it is observed that the intensity of the colour increases as the matter situated more or less centrally is approached.[52]

In phrases like "be broken up," "coloured with carmine," and "is approached," the text registers the actions of the microscopist-as-operator (breaking up the specimen, staining it with carmine, adjusting the slide and apparatus to view the center of the specimen). The text also records the experiences of the microscopist-as-observer ("are seen," "we see," "it is observed"). Despite reliance on the passive voice here, the passage reports changes to the image when the unseen hand of the operator adjusts the apparatus and object. The operator's trained hand provides narrative momentum in this text even if it is not explicitly named. Such passages pair well with *articulated practice*, which explicitly narrates process and method for readers to follow. Together, *minute description* and *articulated practice* offer dynamic storyspaces in which readers can take up particular, favored forms of microscopical practice.

Writing Embodiment in Victorian Microscopy: Beautiful Mechanism. Meegan Kennedy, Oxford University Press.
 DOI: 10.1093/9780198940623.003.0016

17
Articulating Practice

> Experience has constantly shewn that nothing short of *written and specific* directions are sufficient to enable individuals of ordinary calibre to make these instruments *do their best.*
>
> C. R. Goring, "Exordium," *Microscopic Illustrations of a Few New, Popular, and Diverting Living objects*, 1830

Introduction

As shown in the previous chapter, microscopists publish much advice on how to describe the elusive, sometimes deceptive, and often ephemeral microscopic image, thus building up genres of *minute description*. However, they do not as frequently address how to record the manipulation required to obtain that image, perhaps because so much (though not all) of "practice" consists of ordinary movements in macroscopic space. Yet a genre developed around this task nonetheless, as microscopists considered how best to depict their actions: their work preparing the observation, their manipulations during the observation, their intellectual progress through the observation. They model these actions in an influential Victorian genre that I call *articulated practice*. Articulated practice is a meticulous form of explanation that stages (sequences and virtually performs) scientific practice, instructing readers on how to perform various tasks and warning them of difficulties they may encounter. It uses a (largely) dispassionate voice to describe objects and actions precisely. More important, the prose is constructed of stacked and ordered instructions; the writer "articulates" a scene of practice using a step-by-step movement through the elements of an observation. The writer rhetorically distinguishes each step, enacting it as a discrete performance, but also strings them together in a narrative emphasizing causal relation. Such passages depict the microscopist as an actor within the microscopical setting. While *appreciation* teaches observers to interact with the image virtually, as though manipulating a physical object in the hands, *articulated practice* stages the observer's body within the scene of observation, performatively executing the actions necessary to evoke a desired image. Minute description and articulated practice provide virtual storyspaces where writers represent their own bodies interacting with the microscope and object, capturing microscopical practice and image for others to experience virtually and later reproduce. Benjamin Morgan argues that Victorian scientists' "aspirations toward impersonality and disinterest were often barely tenable responses to a crisis in which the bounded stability of mind, soul, and self was rapidly eroding."[1] Those aspirations in microscopy quite often were paradoxically voiced in forms that also invoke and direct an entangled embodiment of observer, apparatus, and object.

This chapter tracks the circulation of microscopical culture through the published guidance, especially in overtly instructional texts, on how to perform specific tasks. I argue that microscopists attempt to shape a developing practice through such guidebooks and manuals, and through modeling their work. In microscopy, notebooks and published books and articles, taken together, present a valuable record of practice. The published materials that I focus on incorporate a process of editing, collation, and revision in order to produce a finished text suitable to share. Frederic Holmes has argued that we need to recognize how "writing, thinking, and performing experiments" are "mutually interacting," and that writing is "an integral aspect of the shaping of the investigation." He explains, "Far from being a distorted [and public] version of something the scientist has already done, the scientific paper is an essential phase of the investigation itself. Writing the paper is a central act within the creative process."[2] While writing is part of the process of investigative work, it also models and codifies scientific procedures more generally. Through long passages of minute description and articulated practice, Victorian microscopists both communicated their research to their peers and reported, modeled, and eventually formalized their ideals of microscopical practice. To study these forms of writing is to study not just the published record of nineteenth-century scientific investigation but also how workers negotiated microscopical skepticism and inscribed their practice: how they warned readers about doubtful or unreliable actions and interpretations, how they managed these doubts in practice, and how they gained readerly assent to their work.

In guidebooks for lay audiences, articulated practice explains and demonstrates microscopical work as a clear, manageable series of tasks; it enlists the reader as a practitioner and it guides and disciplines the work of that reader. In treatises and journal articles addressed to original investigators, articulated practice supports a research argument being made. In joining its particularity to a research argument, articulated practice asks all readers to model the author's work mentally and assent to its results. It further calls upon the most interested readers (for example, fellow diatomists or physicians, those engaged in debates on vitalism or other topics, those instructing students in the field) to replicate this work materially as well. In every case it presents its textual rendering of microscopical work as mediating between two embodied practices, one as origin of the process on the page, the other as an attempted realization of it. Finally, it serves as a priority claim on particular methods by recording them in print for posterity, albeit often subordinated to an experimental or theoretical frame. The functions of articulated practice overlap to a great extent with those of the scientific periodical generally.[3] By allowing writers to preserve, encapsulate, and circulate their observational and experimental work, articulated practice contributed to the development and success of nineteenth-century scientific periodicals.

Obviously, articulated practice appears commonly in research texts and manuals of instruction. But it features in many popularizing articles in general-circulation or family magazines as well. These provide directions for work at home, as in Edmund Saul Dixon's 1856–7 "how-to" articles on microscopy in Dickens's *Household Words*. Texts addressed to energetic self-improvers, such as the *English Mechanic and World of Science* or even William Clarke's 1885 *Boy's Own Book: A Complete Encyclopaedia of Sports and Pastimes*, are especially likely to feature extensive passages of articulated practice. Of course, many recreational texts do not include much articulated practice or even minute description; they simply promote the wonders of microscopy, offering information and entertainment more than practical instruction. For example, in articles like "The Microscope and Its Marvels," which was first published in *Chambers's Papers for the People* in 1850, the writer distills some centuries' worth of microscopical work into just a few pages, resulting in a stimulating draught composed of the most curious and startling anecdotes in microscopy but little instruction in using the instrument or making observations.[4] Or, as in the report on "A Microscopic Entertainment" in an 1854 issue of the Religious Tract Society's penny weekly *Leisure Hour*, the writer—like the lecturer in a soirée or a lantern microscope show—virtually displays and remarks upon a series of interesting objects viewed with a table microscope and a lantern microscope. However, even in such settings the operator at times describes the work, commenting on specific actions; this performative instruction allows non-microscopists to gain some sense of what the practice involves, and it also likely sets up the favored response of curiosity and wonder.[5]

Setting the Scene

The microscopical reader (whether novice or expert) is often imagined as sitting at the microscope, working at it, with book nearby for instruction or reference. This was not a new concept, but the flurry of nineteenth-century scientific publication popularized it.[6] It is obliquely referenced in guidebook series such as Routledge's "common objects" books, in the literary form of the soirée, and in images of microscopists seated with microscope and book before them on the table (see Fig. 17.1). As the popularizer J. G. Wood says in his 1861 *Common Objects of the Microscope*, "[T]he best plan will be for the reader to procure the plants, insects, &c. from which the objects [in the book] are taken, and follow the book with the microscope at hand."[7] Philip Henry Gosse models this behavior throughout *Evenings at the Microscope* (1859). Looking for the nose of "the great Eatable Crab," he turns to a textual guide referenced earlier, saying, "Our friend Mr. [Spence] Bate must assist

Fig. 17.1 Seated man with microscope.

Source: J. Edwards Smith, *How to See with the Microscope: Being Useful Hints connected with the Selection and Use of the Instrument*. Chicago: Duncan, 1880, frontispiece. Image courtesy of the National Library of Medicine.

us again. He will tell us to look at the outer or lower antennae. We will look accordingly, magnifier in hand, while he makes it clear to us that these are a pair of noses."[8] As this narrator demonstrates, the microscopist often shifts back and forth between guiding text and active observation.

Despite Gosse's whimsical tone, replicating the work reported in a book was a form of deep reading, a more active version of the close or comparative reading appropriate to serious books.[9] A writer on "The Story of Microscopical Discovery," published in 1860 in the *Universal Review*, argues that "it is impossible to form an idea of the greatness of Swammerdam unless you have dissected and observed *after him*"—not just conducting a "mere reading" of his work but "yourself going over some of the ground" he covers in his *Biblia Naturae*.[10] E. Ray Lankester's review of Huxley's *The Crayfish* (1880) similarly frames that book as designed for "serious zoological study, for those who will turn over its pages, crayfish in hand, and carefully verify its statements as to details of structure with scalpel and microscope."[11] Indeed, some genres of texts are designed for use alongside the microscope: handbooks, textbooks, manuals, which can be addressed to audiences from novice to working microscopists. These appear in periodical as well as book form. While the Quekett Microscopical Club ran a series of introductory workshops for members, mid-century introductory manuals sought to provide a similar program in print form. Beale's 1865 (third) edition of *How to Work with the Microscope* appends a notice to the Table of Contents urging "the beginner to 'read' the book by practising the tables...or if he has not patience for this,...at least go through the processes in SS 135, page 73."[12] Other books, such as Margaret Gatty's *British Sea-Weeds* (1872), offer a list of rules in the introduction to guide the reader's practice while working through later chapters. The microscopical text is not necessarily, then, to be read straight through, nor dipped into as a reference work, although readers can do so. But like a cookbook or travel guidebook, it is usually designed to direct particular actions and observations; it does so via minute description and articulated practice.

Some guides even prescribe specific arrangements of furniture, apparatus, and light to facilitate good observations. This scene-setting encourages readers to replicate the author's physical environment as well as actions. The 1891 edition of William Benjamin Carpenter's *The Microscope and Its Revelations*, which was edited and expanded by W. H. Dallinger and E. M. Nelson after Carpenter's death in 1885, describes how a laboratory table should be arranged, with prescriptive diagrams of the ideal microscopical setup. The microscopist should arrange a "suitable table" for microscopy—a rectangle 4.5 × 3 feet in size, covered in morocco leather—in a room "sacred to his purpose." The book even diagrams the scene in a view from above that shows the recommended placement of an array of objects including, among other things, two different microscopes, two different lamps, a cabinet of slides, and other apparatus (Fig. 17.2). The areas nearest the observer should be left clear of any "valuable appliance" that could be inadvertently swept to the floor, and otherwise reserved for "the work immediately in hand." Closest to the observer, however, the diagram places a book to the left and a writing pad to the right.[13] The siting of the book and notebook close to hand not only point to the common practice of note-taking during observation—providing the earliest versions of the textual material I discuss as minute description and articulated practice—but also remind us that, as Jonathan Topham puts it, "Reading practices, too...are embodied."[14] Another diagram shows a separate dissecting table with its apparatus as well. This advice clearly addresses a particular kind of microscopist: a "worker" with a panoply of microscopes and apparatus, among which the book and writing pad are integral components.

This care in specifying the best placement for each of the delineated objects further reminds us of the embodied nature of microscopical work. The reader who kits out a table precisely as the book suggests will be setting his or her body into the author's own pattern of movements—leaning forward to observe, turning to the left to read, turning to the right to make notes, reaching forward to change the illumination—along lines organized by the placement of the table and

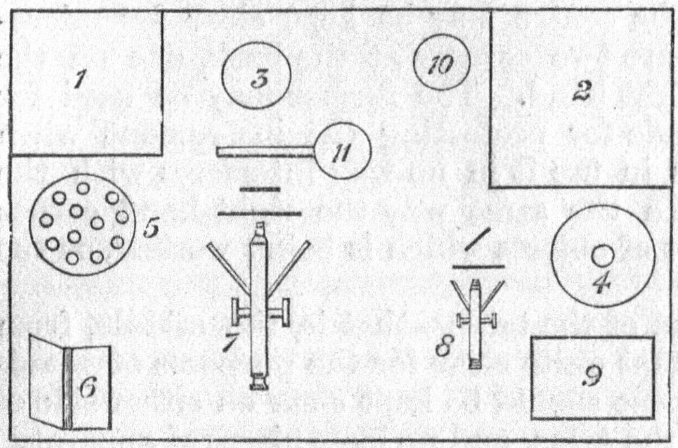

FIG. 287.—Microscopist's table.
(Scale, ½ inch to 1 foot.)

1. Case for microscope; 2. Cabinet for objects; 3. Microscope lamp; 4. Lamp with shade; 5. Stand of apparatus; 6. Book; 7. Large microscope; 8. Second microscope; 9. Writing pad; 10. Bull's-eye stand; 11. Light-modifier.

Fig. 17.2 Microscopist's table, with apparatus laid out.
Source: William Benjamin Carpenter, *The Microscope and Its Revelations*, seventh edition, edited by William Henry Dallinger. London: J. Churchill, 1891, p. 342.

the objects upon it. The table diagram shows how even routine microscopical practices were often carefully developed and shared, and it indicates the range of the late-Victorian interest in formalizing microscopical practice down to the smallest details, even if microscopists are likely to have tweaked such instructions to suit particular needs and eventually developed their own version of their most commonly performed procedures.

Beyond Virtual Witnessing

The genre of articulated practice has roots in the seventeenth- and eighteenth-century form of the experimental report. Influential readings of the work of Robert Boyle and the early Royal Society have emphasized how the inclusion of particular, realist detail in the experimental report lends the report not just verisimilitude but also the kind of argumentative weight otherwise accorded to either ancient authorities (texts) or immediate personal experience. Charles Bazerman finds that the development of the experimental report in the

Philosophical Transactions is marked by increasingly formalized "methods" sections with "a precise account of methodology." He argues that as readers are increasingly distanced from the site of experiment, "the story of the experiment" comes to "serve [them] as a surrogate for the actual experiment," which requires "verisimilitude" in the print record.[15] Peter Dear argues that the experimental report shifts attention from generalized theories about the world to a historical record chronicling the observer's direct personal experience of "an actual, discrete event" that had occurred "at a particular time and place."[16] Finally, Steven Shapin and Simon Schaffer, writing on Robert Boyle, term the experimental report a "literary technology" that presents "virtual witnessing" to help validate an investigator's work.[17] Shapin and Schaffer explain that Boyle adopted a realist prose in order to "convey circumstantial details and to give the impression of verisimilitude." The text then "constitutes a visual source," analogous to the experiment itself, because it triggers "the production in a *reader's* mind of such an image of an experimental scene as obviates the necessity for either direct witness or replication."[18] The realist quality of the prose in these reports—marshaling precise details about observation and experiment—anchors their verisimilitude and enables them to stand in for the work itself.

Equally important, the experimental report was built on an underlying structure of narrative: it records a series of related actions and events. In virtual witnessing, readers' mental experience of investigative practice is strengthened by the identification and absorption provided by narrative. Mary S. Morgan and M. Norton Wise have argued for the *narrative knowing* that occurs through the writing and reading of narrative.[19] This narrative knowing, where the reader becomes absorbed in the text, accounts for the power of virtual witnessing. When the prose solicits the reader's mental placement within the series of events, it implicitly places the reader within the narrative as a virtual observer or even actor.

Articulated practice, a form of scientific narrative, revises and extends this virtual witnessing. Over the course of the nineteenth century, Victorians adapted many of the elements of early natural philosophy, creating new approaches to the natural world. They rewrote plain speech into scientific realism, Baconian induction into a judicious blend of empirical observation and hypothesis, decorum and civility into a new professionalism and a new set of disciplines. To bolster the ranks of science they gave up gentlemanly exclusivity to accept ambitious bourgeois practitioners, with the caveat that original investigators must be well trained in a disinterested ideal of scientific discourse and practice. And out of virtual witnessing, a practice that Shapin and Schaffer seat in the civil and political landscape of seventeenth-century England, the Victorians developed an influential and long-lasting form of articulated practice that would help British microscopists compete with their Continental peers.

Previous work on the experimental report has focused on its function as a verbal replication of practice that the reader experiences only mentally, while reading.

Dear states that "the procedure" reported "could always be repeated; the event could never be."[20] And Shapin and Schaffer emphasize that virtual witnessing replaces "direct witness or replication."[21] But while Victorians' articulated practice encourages readers to mentally reproduce observations and experiments, it calls upon the most dedicated to physically reproduce them as well, by clearly setting out the path and sometimes expressly directing readers to follow it in person.[22] Dear argues that "an account of an experiment…is a piece of social currency."[23] This is perhaps especially true of articulated practice, which offers to direct the reader in not just a mental but a physical replication of its actions.

Articulated practice centers on materially specific, staged, sequenced, and directed action oriented around the manipulation of the microscopical apparatus and the natural object. It is what Frederick Holmes calls a "dense narrative account."[24] That is, articulated practice meticulously records a wealth of particulars, such as the setup of apparatus for an observation including the specific type, power, and manufacturer of the objective used; the type and arrangement of illumination (reflected or transmitted, direct or oblique); the preparation of an object to view; the manipulation of the apparatus including condenser, diaphragm, polarizer, coarse and fine focus, and the like; and the approach to take when observing, with attention to specifics of material objects, physical setting, affordances of equipment and observer, potential obstacles or snares, and the anticipated and actual result. During actual practice, the microscopist may omit, overlap, or cluster some of these actions, or rely on work completed earlier; published texts more often take apart the flow of microscopic practice and reassemble it in distinct sequenced steps or actions, "articulating" it for the reader. Articulated practice typically integrates these actions with commentary: the reasons why particular elements are necessary or preferred, the difficult elements or sites of potential error, and the anticipated and actual outcome of the process. It thus provides a fully realized scene and narrative, enabling the reader to participate in the investigator's work.

Readers' actual efforts to reproduce a task (virtually or actually) may vary significantly from the anticipated ideal, but by recording the details of an investigation, articulated practice accelerates the mid-century formalization of microscopical work. Hans-Jörg Rheinberger argues that these "ever more circumstantial descriptions of preparation techniques" are "supposed to ensure exact repetition" of delicate procedures on objects that are often difficult to see.[25] For example, Stephen Jacyna argues, such a description by Martin Barry sought to "establish protocols" for readers trying to replicate his work on the spiral fibers he claimed to have discovered in living matter.[26] Articulated practice demonstrates that microscopical observation, though inevitably private, is yet a disciplined and accountable space. Its performance of detail and system helps to authorize individual work in the face of nineteenth-century investigators' skepticism about microscopical vision. And it appears, with beauty and wonder, as a constitutive characteristic of nineteenth-century microscopical writing.

Articulated practice undergirds the instructional guidebooks seeking to attract and train novice microscopists from mid-century on.[27] It also powers the experimental reports that became increasingly formalized in nineteenth-century scientific journals, although other elements—such as literature review, argument, and conclusions—become important to this genre as well. The examples I discuss here appear within a longer period of development of the scientific article in English, from the beginnings of the Royal Society to the present. Alan Gross, Joseph Harmon, and Michael Reidy have examined and tracked this development using longitudinal studies, sampling articles across the sciences at different time periods and analyzing the changes in them using rhetorical, linguistic, and historical analysis.[28] These scholars find a shift in scientific articles from a person-oriented to an object-oriented discourse, showing a decline of personal references and an increase in a passive, nominalist, overtly disinterested syntax; with a more formal presentation of structure and scholarly apparatus (section headings, citations, and the like); and with increased quantification and description of method—although any individual periodical develops an independent style responsive to cost, anticipated audience, and the stylistic preferences of publishers and editors.[29] Bazerman explains, "Formal definitions, expected features, institutional force, impact, and understandings of [a] genre vary through time, place, and situation.... Each new text... reinforces or remolds some aspect" of it.[30] Thus the articulated practice of the nineteenth century, which was often integrated throughout a research report, becomes the segregated methods section of the twentieth.

Narrating Embodied Practice

A more important difference may be the narrative presentation in articulated practice. This is focalized through a researcher-protagonist and can render an experimental report more coherent and memorable. Articulated practice records the actions of a narrator-investigator who works to achieve an image and directs the actions of a reader-investigator who tries to replicate it.[31] Gross, Harmon, and Reidy argue that eighteenth-century scientific writing in Britain displays "narrative and epistolary conventions" marked by "the explicitly personal and social" and a "tolerance for emotional expression"; in contrast, they state, by the end of the nineteenth century, scientific articles often adopt a depersonalized style that avoids personal narrative and action.[32] Articulated practice flourishes somewhat earlier, in the middle decades of the nineteenth century (1840–80) and characterizes a style of scientific prose that is more systematic than eighteenth-century reports but still strongly narrative and focalized through the actions (sometimes also the emotions) of an investigator-protagonist who is also narrator. This narrative approach to argument diminishes greatly by the mid-twentieth century.[33]

Indeed, articulated practice develops very much from the moment when Victorians sought to both formalize and popularize scientific work. The physician Lionel Beale, lecturing to medical students in 1856–7, emphasized practical matters of manipulation rather than philosophical musings on the microscopic world. In *How to Work with the Microscope*, a guidebook collecting and expanding these lectures, he discusses how to record microscopical work, giving explicit directions on how to work the microscope and describe what you see. As shown earlier, Beale repeatedly exhorts microscopists not to pile detail upon detail when describing an object. But when he discusses step-by-step directions on method, he reverses his prohibition on overly precise writing. Just as C. R. Goring recommends "*written and specific* directions" in my epigraph, Beale remarks, "Directions for working cannot well be too explicit and precise, and the more simply they are given, the more useful they are likely to be."[34] While he continues to value simplicity (as with descriptions of objects), he sees "directions for working" as so important that they should be not just simple but also "explicit and precise." In fact, he refuses to set any limit on this explicitness and precision, although he recognizes some practical constraints; in the fourth (1868) edition he notes that "these processes are capable of being modified in many ways" and thus he has not provided "minute recommendations" on every possible technique. He concludes, "A few experiments will teach the observer more than the most minute directions."[35] Directions on method differ from descriptions of specimens, of course. While objects can be delineated in drawings or photographs—Beale considered illustration more valuable than textual description—before the advent of scientific films the microscopist could not depict workflow in any form other than words.

Beale records the injection of a frog in one example from his book. This passage displays typically dispassionate tone, sequencing of steps, precise description of objects and actions, including the placement of his hands, and warnings about potential difficulties or errors. He begins, "I will commence with the frog. An incision is made through the skin, and the sternum divided in the middle line with a pair of strong scissors; the two sides may easily be separated, and the heart is exposed." Articulated practice often appears in this depersonalized third-person passive, which begins to appear more frequently in mid-century scientific and medical texts. Beale occasionally also speaks in first person, as in "I draw up a little of the injecting fluid"; third-person plural, as in "We will now examine these injections"; and second person, as in "you will not fail to observe that all the capillary vessels are fully distended with fluid." This may reflect these chapters' origin as lectures, but these forms of address can be found in other authors' articulated practice also. Beale punctuates each step with an indication of its placement in a time sequence: "*before* I insert the pipe," "the thread is *now* tied," a syringe "is *then* connected with the pipe," "a little of the injection being *first* forced into the…pipe," "the pipe and syringe being *still* held with the left hand."

He describes the embodied relations of operator and object, precisely designating the placement of objects and of his own hands: "The needle with the thread is next carried round the vessel and the thread seized with forceps, the needle unthreaded and withdrawn," the pipe "firmly held by the finger and thumb of the left hand." Beale's articulated practice not only enunciates each step of the injection process but also warns the reader about inflection points where judgment or care is required. For example, "the piston is slowly and gently forced down with a slightly screwing movement with the right [hand], care being taken not to distend the vessel so as to endanger rupture of its coats."[36] The reader using the book as directed would not just mentally envisage Beale's procedures but physically act out his practice, moving their hands into the positions marked out by Beale's virtual hands, inhabiting his imagined body and following his movements as closely as possible.

As noted earlier, Beale further standardizes the microscopical studentship by printing "tables" at the back of his *How to Work with the Microscope* (1857), each with an itemized list of steps in "the most useful branches of manipulation" so that a reader might "practice for himself" and master them; he specifies that "each table should occupy the student about two hours."[37] Each step is cross-referenced with the page holding the relevant guidelines for practice and observation. For example, in Table VII, "On Injecting with Opaque and Transparent Materials—Prussian Blue Fluid for Injection," item number 49 reads, "*Frog.*—Insert an injecting pipe into the aorta of the frog in the manner described in page 81, and slowly inject the fluid."[38] In the second (1861) edition of this guidebook, Beale promises, "any one who will patiently submit to the routine and trouble, will be amply rewarded by the practical knowledge he will thus gain of the use of the instrument."[39] A reviewer for the *Edinburgh Medical and Surgical Journal* shows how microscopists valued Beale's focus on manipulation in this influential book. The review identifies Beale as "among our best original observers"; he is "a practical man" whose books "are tersely and clearly expressed, and bear the imprint of being a record of a man's own hard won experience." The review singles out for praise the articulated practice ("full directions are given") and the "tabular resumé" in this volume.[40] Beale's tables of instruction most likely draw from his experience running various courses of study in "microscopical investigation" at the "pathological laboratory" he had set up in 1852 in Lincoln's Inn, Holborn, not too far from the site of his professorship in physiology and anatomy at King's College on the Strand. For fees ranging from one and a half guineas for a two-month elementary course in microscopy to twelve guineas for a three-month course on advanced investigation, students could gain training via lectures, demonstrations, and supervised work with the laboratory microscopes.[41] Interestingly, W. T. Suffolk published a book *On Microscopical Manipulation* in 1870 that not only frequently cites *How to Work with the Microscope*, saying "The beginner is strongly recommended to go carefully through the tables at the end" of that volume, but also provides another appendix of lessons to follow on from Beale's.[42] These not only provide fine examples of articulated practice but also show the influence of Beale's work in formalizing microscopical practice.

Passages of articulated practice share these basic qualities whether found in a technical manual like Beale's or a report on a new investigative technique in a specialty journal. For instance, the American physician Arthur Mead Edwards, writing for the *Monthly Microscopical Journal* in 1874, instructs his readers on how best to find, collect, prepare, and mount diatoms. We find here the (mostly) flat tone, sequencing, precise description of objects and actions, and flagging of difficult elements familiar from Beale. Edwards says,

> If the microscopist wishes to mount a few slides of recent diatoms just to show what diatoms are, nothing is easier. It is only necessary to boil a small mass of them in strong nitric acid in a test-tube over a spirit lamp, and, when the acid has ceased to emit red or yellowish fumes, wash them thoroughly with clean water, allowing them to settle completely. Then a little of the clean sediment, consisting almost entirely of the shells of the diatoms, is taken up by means of a "dip-tube," and placed upon the central portion of a glass slide. Here it is dried, and the slide warmed over a lamp; then a drop of Canada balsam is permitted to fall upon the diatoms. As soon as all bubbles have cleared off from the balsam, a warm cover of thin glass is carefully laid upon it and permitted to settle into place. When cool, it is ready for examination by means of the microscope, any balsam which has exuded around the cover being washed off with alcohol.[43]

It is necessary for the author to decide what elements of method are unfamiliar to the reader and must be articulated, and which are well known, lie outside the scope of the article, or can be explained by citing another authority, such as Beale on stains. Speaking to an audience of working microscopists in the excerpt above, Edwards is able to elide some details (the exact strength of nitric acid, how much clean sediment) that an introductory guidebook would have to spell out, and he indicates a delicate stage of work with the single word "carefully." But even this eleven-step process is, Edwards warns, not sufficiently exact to "satisfy the student of the Diatomaceae," for whom "more elaborate methods are necessary." He follows up the simple method detailed here by providing an eight-page explication of method for the specialist, although even here he declines to describe the method for making cells on slides, saying, "how these are made cannot be gone into here, as the description would occupy too much space and time."[44]

While articles like Edwards' are explicitly centered on method, process, or apparatus, research reports written by investigators for investigators often shift in and out of articulated practice, including only what is needed at the moment and presuming some base knowledge on the part of the reader. A glance at the 1881

Journal of the Royal Microscopical Society (*JRMS*) provides a wealth of examples. For instance, a brief article on "Sensory Nerve-endings in Skin of Insects" remarks,

> By treating the skin with osmic acid the nervous fibres and cells are demonstrated. The method is to treat the demo-muscular coat with 1 per cent. osmic acid, and then for ten minutes with 25 per cent. formic acid; it is then left twenty-four hours in 1/50 per cent. chloride of gold solution in the dark; the tissue is finally exposed to the light in a 25 per cent. solution of formic acid. When the cuticle is removed, a great number of nerve-fibres without Henle's sheath appear between the hypodermis and connective layer.[45]

Or even more briefly, in an article on the "Anatomy of the Liver-Fluke" by Dr. F. Sommer, "The tubes can best be made out by filling the system which they compose with colouring matter, and this can introduced by the excretory pore, or, still better, injection" and later, "it is well to place the animal in carmine, then treat it with alcohol, and finally clarify it."[46] While the example from Sommer does not demonstrate as much exactness in describing his material, he provides the same narrative model, with a staged series of physical steps accompanied by the occasional explanatory or cautionary aside. Sometimes the sequence occurs as a simple record of the actions taken while executing a past experiment. A brief notice on "Chemical Composition of the Ink of Cephalopods" states,

> After drying the ink, treating it first with alcohol, then with ether, then with glacial acetic acid, subsequently washing and leaving for some time under the action of alkaline carbonate of potash, and again washing and digesting with hydrochloric acid, the extractive, albuminous, and mineral matters and the mucin were removed, and a homogeneous black powder with a greenish metallic lustre remained, insoluble in water, alcohol, ether, the alkalies, and the acids.[47]

The characteristic time-sequence terms appear prominently here (after, first, then, subsequently, again) as well as the focus on specified material and action that anchors articulated practice. These are typical examples of a rhetorical approach that saturates the journal, with an average of about every other page including some overt reference to "method" or "process," and many additional examples of articulated practice presented without naming it in this way.[48]

Of course, we would expect original investigators to include such detailed directions, since this helps to establish new procedures and authorize their conclusions. But popularizers also turn to articulated practice, primarily as a means of instruction for their novice readers. The Rev. J. G. Wood, for example, explicitly directs his elementary guidebook to readers who will practice these skills. Wood minutely instructs readers on how to examine the circulation of a frog. Addressing readers directly using the imperative mood, he says,

> The mode of arranging the foot so as to exhibit the object without hurting the frog is simple enough. Take an oblong slip of wood—my own was made in five minutes out of the top of a cigar-box—bore a hole about an inch in diameter near one end, and cut a number of little slits all round the edge of the wooden slip. Then get a small linen bag, put the frog into it, and dip him into water to keep him comfortable. When he is wanted, pull one of his hind feet out of the bag, draw the neck tight enough to prevent him from pulling his foot back again, but not sufficiently tight to stop the circulation. Have a tape fastened to the end of the bag, and tie it down to the wooden slide. Then fasten a thread to each of his toes, bring the foot well over the centre of the hole, stretch the toes well apart, and keep them in their places by hitching the threads into the notches on the edge of the wooden strip. Push a glass slide carefully between the foot and the wood, so as to let the membrane rest upon the glass, and be careful to keep it well wetted. If the frog kicks, as he will most likely do, pass a thin tape over the middle of the leg, and tie it down gently to the slide. Bring the glass into focus, and the foot will present the appearance so well depicted in the engraving.[49]

This is just one—but a typical one—of many such passages of instructions in the book. Wood also explains in detail how to make a dissecting microscope and dissecting needles; how to prepare a specimen for dissection under water; how to manipulate a compound microscope; how to construct various kinds of mounts for various types of objects; how to use a camera lucida or a stage micrometer; how to prepare sections of hair, bone, or other objects; how to "fish" for *Rotifera* in a pond. In all of these, Wood uses the same style: simple, direct, and materially specific step-by-step instructions, with explanations of the reason for particular actions or choices (dip into water to keep the frog comfortable; pass a thin tape over his leg to prevent him from kicking). Wood allows the occasional excursus to address sourcing, preference of materials, or potential difficulties, but the passage as a whole is strongly focused, sequenced, and directive. Overall, the passage articulates each of Wood's actions as he models his process through each stage (setup of apparatus, preparation of specimen, manipulation of apparatus, visual approach to the object, outcome) in relevant and necessary detail.

Articulated practice does not make gripping reading; that's not its primary aim. Authors addressing broad audiences or novices usually leaven it with more interesting material. Wood balances his articulated practice with a humorous anecdote. He says,

> In many cases [insect antennae] are all the better for being soaked for some time in liquor potassae, then dried between two slips of glass, then soaked in turpentine, and lastly put up in the balsam. Otherwise, their characteristics will be totally invisible under the microscope, and the observer will be as bewildered

> as a gentleman of whom I heard, who lately purchased a good microscope, and returned it next day as useless. The maker who had guaranteed it naturally thought that it had been injured by rough treatment, but finding that it performed well in his own hands, he inquired as to the details, and especially as to the object which it would not show. The answer was, that it would not exhibit the crystals of sugar. "How large a crystal did you try?" asked the optician. "A lump out of the sugar-basin," was the answer.[50]

Both the earnest student of nature and the clueless gentleman may find themselves "bewildered" by the view through the lens, but Wood's clunky transition to this little joke draws attention to it, allowing the reader to feel superior to this frustrated gentleman and his obvious error. More important, he offers this story as a mock analogy to another potential error, in preparing insect antennae; and as a rhetorical balance lightening his discussion of that multi-step preparation, which requires considerable patience and dedication to microscopical inquiry.

The patter of popular science lecturers in this period mimics and theatrically performs this genre of instructive and professional narrative while actually following a more spectacular rhetorical imperative. Such lecturers would often elide dull or deceptive steps in their work with the specimen and apparatus. The popular American lecturer "Professor Morton" (Henry Morton, inventor of the "College Lantern") explains the manipulations of the lantern, saying,

> We will now make a change in the lantern, and I will show you the lens producing on the screen images of solid objects. The light will be turned backwards, and the object will thus reflect the rays which it receives. By the action of the lens, those reflected rays will be thrown upon the screen...with a very strong effect of light and shade.[51]

But in presenting the lively animalcules that were the centerpiece of his show, Morton details only the most interesting steps of his process, emphasizing instead verbal appeals to the emotional engagement of his audience—appeals which are registered by the narrator reporting the lecture. Morton remarks,

> They will be stirred up presently—as soon as the water is a little heated.... These fellows are very sluggish. I will try one other specimen—[a fresh plate of eels, leaping with galvanic starts and jerks, whereat—Oo! Ah! Ee! There he goes! See, see! *There* he goes! Tee-he!].... The little tank or vessel in which I have these animals is made by taking a plate of glass and grinding a slight hollow in it. Here is another. [Agh—agh—it's awful! (They *are* horrid-looking things.) Why they're fighting. See the fight up there? See? (They *do* appear to struggle) applause! Positively; applause of a bug-fight.][52]

Here the text indicates Morton's instructional remarks directly, and in a confusion of square brackets and parentheses, Morton's asides and the narrator's own observations. Morton alternates discussion of his process, necessary for the educational remit of his talk ("The little tank or vessel... is made by taking a plate of glass..."), with interjections of surprise, amusement, and horror. The audience, expecting amusement with their instruction, readily mirrors his affective response to the animalcules' actions despite the ridiculous notion of "applause of a bug-fight." Morton's report of process is disconnected and partial, and it is interrupted by these spectacular elements, but, as a shorthand reference to articulated practice, it still stands as a kind of security for his assertions and actions. He repeatedly refers to his own preparatory process before the show without detailing that practice, as when he says, "Next Thursday I anticipate having a great variety of these creatures. It takes some time in their preparation, and I have been too occupied to repair my stock."[53] An even more performative sleight-of-hand occurred with the oxyhydrogen microscope operators who crammed extra animalcules into what they deceptively described as a "drop of Thames water."[54]

Such popular performers gesture to the emerging genre of articulated practice while not actually enacting it. But in written work, the proliferation of articulated practice—of step-by-step guidance as in Wood's *Common Objects* or the *JRMS*—suggests a complicated role for microscopy. Articulated practice, especially in popular texts, may function for most readers more as an advertisement of the fascination and complexity of microscopical work than as actual instructions for a procedure they hoped to perform. However, it still offers a clear path through that procedure, should they choose to follow it, by showing how this complex process is composed of a series of ordinary actions. In other words, popularizing texts do often mystify microscopy, as Willis argues, but passages of articulated practice in these texts both mystify and *de*-mystify the instrument, seeking to encourage its study.

Other nineteenth-century contexts, recording activities as different as cooking, medical practice, or physical exercise, may feature similarly directive passages of explication. However, microscopical texts are especially likely to feature articulated practice, due to the acknowledged difficulties of microscopical vision, the contested nature of methods to mitigate those difficulties, and the address to an audience often imagined as active fellow microscopists, whether attentive novice or disciplined practitioner. Of course, this guidance does not necessarily govern all observers' handwritten notes or even published papers; some microscopists do not provide even the basics of articulated practice, especially early in the century. Schickore comments that pre-1820s reports rarely attend explicitly to the type of equipment used.[55] Although in the 1840s Hermann Schacht recommended that each observation should be tagged with the date and objective used, many microscopists did not include this crucial information even in published records.[56] However, norms around scientific reports materially shifted over the course of the nineteenth century. For example, Alex Csiszar discusses how the development of refereeing systems in elite British scientific societies in the 1830s replaced the "gentlemanly trust" undergirding the system of virtual witnessing.[57] Greater editorial oversight helped to nudge scientific writers toward a standard of reportage that was, just as Beale wished, explicit and precise.

In the Victorian period, then, increasing pages are devoted to exhaustive articulation of microscopical practice, directed to an imagined reader who wishes not just to virtually witness but also to understand the process and possibly even replicate the work. Articulated practice poses the reader as not just audience but also participant. In its detailed, materially grounded progress through the steps of a procedure or argument, it asks readers to inhabit the actions and perspective of the microscopical observer-narrator, mentally accompanying the speaker in his embodied movements and attention. During the nineteenth century, articulated practice comes to serve as an instructional demonstration of and invitation to microscopical work; a means of providing virtual witnesses to that work; a warrant of the rigor and correctness of the author's actions; a demonstration of working skepticism; a staged model inviting the physical replication of original investigation; a proof of the author's priority in original work; and, for historians, a valuable record of nineteenth-century microscopical practice.[58]

Management of the Object

The operator's actions detailed in these passages of articulated practice often provoke responses from the beings or even the substances involved. This does not require that we assume will, thought or emotion, or even consciousness, in the microscopical object. For instance, Carpenter in 1862 describes how to prompt the "interesting spectacle" of crystallization of salts, commenting that "[w]hen these are viewed with polarized light, some of them exhibit a beautiful variety of colours of their own, whilst others require the interposition of [a] selenite plate for the development of colour." In 1862, he adds, "some of the double salts give forms of peculiar beauty."[59] These actions are simple changes in the optical system in which the object is enrolled, but Carpenter characterizes them as originating in the object, which *exhibits* or *gives* particular forms back to the observer. As shown earlier in this book, Carpenter extensively discusses non-voluntary action (such as unconscious cerebration) in humans and non-humans. Here in discussing minerals, he uses language implying, if not *voluntary action*, at least an *optical response* in the object; even a non-response is enough answer to prompt additional actions from the operator, such as adding a selenite plate. Microscopists similarly instruct readers in how to prompt changes in physical or chemical states by interacting with objects in various ways. For example, in a discussion of the microscopic plant then called *Protococcus pluvialis* (a form of freshwater algae), Carpenter notes that its "thread-like extensions" can be "rendered more distinct by iodine, and can be made to retract by means of reagents."[60] Such phrasing, which is common, suggests a dynamic relation of interpolated responses between operator and object.

Even lay readers were encouraged to interact with microscopic objects in order to study their actions. The botanist Phoebe Lankester, in an 1861 article for young readers published in *St. James's Magazine*, describes a scene where a young boy and his father are exploring the use of the microscope. They have discovered a rotifer in the water-drop under the lens.

> "Now let us feed him," said papa. A tiny morsel of bright crimson paint was put into the water, which this active little fellow instantly swallowed down his mouth, which is between his wheels. Very soon we saw a bright red spot appear in his body, so by this means we found out the position of his stomach.[61]

Lankester describes the rotifer as an "active little fellow" and uses terms like "mouth" and "stomach" that George Henry Lewes would critique; but she is careful to avoid any pronounced anthropomorphism here even though she is writing for children. Similarly, in describing the diatom *Navicula*, for example, she says that it "glides backward and forward in the water, appearing to feed, and to turn aside from any stray bits of stick or dirt."[62] Her careful phrasing ("appearing to feed…and to turn aside") does not unduly assume will, thought, or emotion where none may exist. However, Lankester does record the diatom's apparent responses to its environment and instructs readers in how to interact with microscopic creatures, building a chain of relations joining macroscopic and microscopic space.

This management of living specimens was a skill in itself, closely related to other techniques such as slide preparation and illumination. Goring complains at length about the liveliness of the infusoria he hoped to sketch. In explaining why his examples omitted adult forms of these creatures, he comments that "no *perfect insects*…can be so easily preserved and managed; so that the larvae, &c. appear to me peculiarly and exclusively devoted to the consideration of microscopists."[63] Those presenting microscopical exhibitions might even study how to rouse, or appear to rouse, the most compelling behaviors from the tiny beings on display. As early as 1832, Andrew Pritchard recommends that, to present a "very amusing exhibition of animalcules" using a projection microscope, "[t]he main feature is the apparent control of the exhibitor over the actions of these minute beings, and their obeying his commands." Thus "[o]n the command of the exhibitor the animalcules will commence marching across the field of view." He explains that "to those unacquainted with the plan it will appear in obedience to the [exhibitor's] order" but the movement is "merely [due to] their desire to spread themselves."[64] Similarly, he recommends that the larvae of the *Dytiscus* (a diving-beetle) be "confined in separate vessels for a few days" before exhibition, in order to sharpen their hunger and provoke "the most fierce and obstinate combats." Pritchard encourages an anthropomorphic interpretation of the battle, comparing "the little animals" to "two well-trained pugilists" who display extraordinary "courage" in their willingness to attack combatants of twice their size. He suggests that this exhibition will excite "lovers of such sports" just as at a boxing match.[65]

But by asserting a commonality of feeling between human and non-human, such anthropomorphism can arouse uncomfortable moral questions, encouraging observers to take account of the suffering that operators may inflict on the objects of their study. The vivisection debates during the period from 1870 to 1890 have drawn much scholarly attention, but anti-vivisection efforts troubled British attitudes toward science and medicine throughout the nineteenth century.[66] Microscopical work was integral to the vivisection controversies during the latter decades of the nineteenth century in England, in that (alongside the scalpel and experimental instruments like the myograph) the microscope was an important tool in physiological laboratories. As Nancy Anderson has shown, vivisected animals appear in the microscope after having been segmented and transmuted into sections on slides for close study and archival; "in fact," she argues, "the breaking down of the body, the analysis of its component parts, is the point" of many manuals in physiology and medical microscopy.[67]

However, vivisected animals in the nineteenth century were generally larger ones such as dogs, monkeys, or frogs. This does not mean that microscopical practice on these animals occurred only after active experimentation on them had been concluded. The frog in particular regularly appears in popularizing works on microscopy; the demonstration of blood circulation in its foot was a standard soirée exhibit, as suggested in the example by Wood, above. Writers comment on methods to increase the frog's comfort or decrease its alarm, actions that have material effects on the physiological functions being studied; Lankester, for example, notes that "the alarm of the frog on finding himself in such a new position stopped the action of the heart, and consequently the circulation," threatening to put an end to the demonstration.[68] John Quekett writes in 1848 on this problem, describing how "in order to secure the animal, and keep its web open, various contrivances have been had recourse to." He offers three pages of articulated practice describing different approaches to this exercise, which entangles multiple bodies. The body of the frog is enclosed in a bag (itself a kind of body, its "mouth" and "bottom" articulated by tape and "a drawing string"). The bag is fastened to a brass plate anchored to the stage of the microscope. The leg, foot, and toes of the frog are spread out and secured to this plate by a series of strings, thread, tapes, pegs, and slits, "to prevent the foot from being pulled into the bag," care being taken at all times "not to stop the circulation" within the webbing. The physical and perhaps emotional responses of the frog are hinted at in such phrases (as they are in Wood's rendition as well). These responses become a focus of concern when Quekett remarks that some operators secure the toes with pins into a cork base, but this is "generally looked upon by the fair sex as a much more cruel act than where the threads are employed."[69] Even given the half-mocking tone of this reference to over-sensitive ladies, microscopical work is implicated here in debates over cruel practice, not just in its use for analysis of biological specimens post-vivisection but in some forms of active observation as well. Indeed, such practices would be outlawed by the 1876 Cruelty to Animals Act.[70]

It may seem that microscopic creatures are too small or simple to be considered subjects of vivisection. Lewes, like many nineteenth-century physiologists, actually argued that the "animals lower than fish" were not capable of feeling pain; others simply considered this a negligible factor.[71] However, debates on animal suffering do occasionally appear even in Victorian discussions of these tiny beings. Even the infusoria in dirty water can rouse observers' sympathy when "imprisoned" within a drop of water on a glass slide. The little boy in Lankester's story, looking about for something to view through his new microscope, suggests "some capitally dirty, green water out of the flower-vase." The narrator comments that the "buttercups and grasses" there are "all quite dead." The boy's father puts a "tiny drop of the dirty water" on a slide and, observing the infusoria within, exclaims, "Dead, indeed! The life is all in the water now! But we must set some of these fellows free, and put them into our cages, where we can better see them."[72] This passage admits to considerable confusion: the object of interest shifts from grasses to water to something in the water; the grasses are dead, but the water, or what's in it, may be dead or alive; the infusoria are imprisoned in the vase or on the slide, but "set... free" in the "animalcule-cage." This notion of a *drop of water, slide, or live-box as a cage or prison*, which appears in other (mostly popular) texts as well, certainly suggests an anthropomorphic view of microscopic beings and, implicitly, a responsibility to limit their suffering under observation.[73] But the slippages in Lankester's tale also obliquely acknowledge some contemporary scientific debates that microscopists contributed to—debates that she would know well, as the wife of Edwin Lankester and a botanist in her own right—over whether particular objects were animate or inanimate, animal or plant, and over the proper treatment of one's objects of study.

What is the observer's responsibility to the living things being studied, if they may be, or appear to be, susceptible of feeling? This question emerges, perhaps surprisingly, from a sensational account in the *Morning Post* of an 1835 oxyhydrogen microscope show. The article frames this, among the earliest of its kind in Britain, not as a spectacle but as a scientific demonstration, listing the professors in attendance and highlighting anatomical points of interest made visible by the new technology. However, the writer emphasizes not just the operator's physical interventions with one particular object (the larva of a water-skeleton) and the activities visible within the body of that object ("the circulation of the blood, and the peristaltic motion of the intestines"), but also the viewers' affective response to this interaction, which concludes tragically. The writer explains,

> The powerful heat was, however, too much for the subject. Magnified to a length of nine or ten feet, its throes and struggles, internal as well as external, were witnessed with painful distinctness, and many [viewers] were heard to utter involuntary comparisons between the dying pupa, and those of a human sufferer in his last agony.

The burning-up of the larva in this instance was not planned by the operator, though it could have been foreseen: oxyhydrogen "limelight" was notoriously hot. The appeal of the oxyhydrogen show—its ability to show microscopic creatures in lively animation—was here put to horrific effect. Later the writer even compares this death scene to one in *Richard III*, in an explicitly anthropomorphic analogy to the catharsis provided by dramatic performance. Perhaps aware that such comparisons may seem ludicrous, the writer defensively comments, "The characteristics of pain in any grade of animated nature appeal too strongly to our sympathies to be misunderstood."[74] Many in the audience might originally have argued that such a simple creature could not feel pain. In the end, the creature's death "throes and struggles" forced this writer, and reportedly the distinguished viewers as well, to grapple with the apparent suffering inflicted on this little creature in the interest of scientific demonstration.

Articulated practice chronicles and emphasizes the activity, the agency, and ultimately the moral responsibility of the microscopical operator in manipulating the instrument, illumination, and slide (and thus also the image). Any novice soon learned not only that the eye functions as an integral part of the optical system but that the observer's hands and body are necessary elements of the apparatus. Microscopical observation requires the operator to engage in a series of meticulously enacted embodied practices, steadying and adjusting the apparatus and interacting with the object. It cannot then be described as a purely "mechanical objectivity," unframed, unmediated, with a passive observer and object, although Victorian texts do demonstrate the strength and persistence of this trope throughout the century. Nor does microscopical observation accord precisely with a Humean passivity downplaying the will and agency of the observer, and stressing the extent to which the mind is associational and reactive. Instead, Victorian microscopical texts argue for the necessity, the pleasure, and sometimes the discomfort of manipulative and interpretative work at the microscope. These texts attend, and ask readers to attend, to the precise series of actions through which operator, instrument, and object co-create microscopical knowledge and meaning.

Writing Embodiment in Victorian Microscopy: Beautiful Mechanism. Meegan Kennedy, Oxford University Press.
 DOI: 10.1093/9780198940623.003.0017

PART IV
WONDER AND SKEPTICISM

18
Wonder and Skepticism

The Microscope not only assists studies, and develops objects of profound interest, but it also opens up innumerable sources of entertainment and amusement.... It... fills our minds with additional reverence and admiration for the beneficent and Almighty Creator.... [T]he Microscope is of paramount importance to the seeker after truth, the students, and the man of scientific acquirements; and, it may be added, of infinite charm and interest to all.

Jabez Hogg, *The Microscope*, 1854

The brilliant discoveries which stud... the career[s] of some men... are the legitimate offspring of habits of close observation, carefully and *systematically* pursued; and we believe that, to the earnest student, there is no department of natural science whose unknown treasures do not exceed the wildest imaginings of her most enthusiastic votaries.

"Review Essay on the Microscope," *Natural History Review*, 1855

Introduction

In George Eliot's 1868 narrative poem *The Spanish Gypsy*, quoted as an epigraph in Chapter 16, a Spanish minstrel reflects on how challenging it is for humans to "see one grain with eye exact / And give strict record of it." However, he continues, "Yet by chance / Our fancies may be truth and make us seers. / 'Tis a rare teeming world."[1] This endorsement of both skepticism and wonder, by a character equally realist and whimsical, recognizes humans' ability to access diverse forms of truth. As Eliot's use of a microscopical metaphor ("see one grain") suggests, this linking of skeptical and wonderful approaches to knowledge was common in microscopy. Victorian observers regularly engaged in both approaches to their microscopical practice: a skeptical engagement with instrument and object (marked by attention to error, method, and discipline) and a wonder-filled one (a primarily affective response). The pairing of skepticism and wonder appears also in the epigraphs above, from Jabez Hogg's popularizing guidebook and a natural history journal. In both cases, skeptical, methodical microscopical work—marked by "profound" scientific "studies" and "close observation, carefully and *systematically* pursued"—can also produce "entertainment and amusement," "charm and interest," even to "the wildest imaginings of…enthusiastic votaries."

This unexpected partnership continues throughout the Victorian period. As late as 1895, Albert D. Michael, a solicitor, botanist, zoologist, and President of the Royal Microscopical Society (RMS), advises the assembled Society that "he who can admire both the structure and the beauty to the eye is the truest student of Nature and derives most pleasure from the study."[2] Michael himself exemplifies this approach; in the first pages of his 1884 book on the *British Oribatidae* he alternates between correcting the errors of earlier investigators and admiring the beautiful, marvelous *Oribatidae*—mites. As he explains, some of the young are "amongst the most beautiful creatures" of an order that is "equally [as] interesting, beautiful, and wonderful" as any other biological order. Michael's commitment to finding both correct structure and beauty in these "glorious" little arthropods illustrates this conjunction of skepticism and wonder about the instrument and the world it depicts.[3]

This book has shown how Victorian microscopists, engaged in a newly scientific field of inquiry with uncertain status, push back against longstanding criticisms of microscopy as trivial, unreliable, or disgusting. Microscopical writers counter these critiques with memorable, persuasive tropes that, by representing an observer's embodied entanglement with instrument and object, shape readers' own experiences with microscopy. Some of these tropes convey a sense of intricate beauty, spiritual meaning, and sublime immensity; others direct readers in a cautious and skillful microscopical practice. Passages of microscopical wonder typically employ romantic language to express and elicit affective experiences of awe, admiration, even enthusiasm, while skeptical passages describe an intellectual response to the microscopical image, modeling cautious, realist detachment and soliciting methodical work from readers.

I am arguing that these ostensibly conflicting elements in fact enabled microscopists to generate and manage a productive tension in the practice of microscopy. In some cases wonder is prior or dominant; in others, skepticism. But Victorians thread microscopical wonder and skepticism together throughout the century and across a range of genres. As noted in Chapter 1, much of the scholarship on nineteenth-century microscopy emphasizes either the wonder or the skepticism expressed by contemporary observers. Because this scholarship usually arises from a focus on particular types of microscopical text, the effect is to imply that Victorian microscopists had developed a more rigid generic code than actually existed—suggesting, for example, that natural history articles for broad audiences effused over microscopic marvels without ever considering difficulties in the work, or that researchers' reports offered a consistently disinterested skepticism without reference to beauty and wonder. However, the print record suggests otherwise. In fact, Victorian microscopy is characterized overall by its negotiation of both wonder and skepticism, and individual texts are located within distinct communities by that negotiation.

Since the earliest writing on microscopy, writers have turned to tropes of wonder to describe their observation of the minute world, especially given the tradition of microscopy as an adjunct to the arguments of natural theology from analogy, beauty, and design. Although Victorian investigators worked within a scientific culture increasingly defined by its skepticism, wonder persisted in many instances of microscopical practice. While some individual writers demonstrate a rigorous exclusion of wonder from their scientific work, many do not. After all, even strong efforts to maintain a stance of detached observation do not necessarily prevent the arousal of experiences of wonder, awe, beauty, even the sublime. References to wonder do decline over the course of the century in research texts, and late-Victorian investigators generally avoid overt reference to religious faith and other such values in their scientific work. However, as Part II of this book has shown, for much of the century it was not uncommon for microscopists to evoke wonder and awe in research texts, albeit usually in more restricted ways than in texts largely dedicated to instruction, community-building, or recreation. That is, working microscopists in the Victorian period both experienced and expressed wonder as part of their scientific practice. One aim of this book has been to identify where and how such expressions persist; how working microscopists, trained to discipline their mind and body, integrate affective responses within a skeptical and methodical practice.

In this chapter, I study how texts across a range of genres and skill levels interweave wonder and skepticism; how they bear witness to microscopists' diversely embodied experiences of observation. The examples I examine here derive from texts addressed to a variety of audiences and published from the boom years of the 1850s through the skeptical 1870s, a period marked by a surge of microscopical publications, a drive to recruit more observers, and efforts to regularize practice.

Although wonder and skepticism do not present a logical conflict, their differences do propose a tension when juxtaposed within a single text. Microscopical writers manage this tension in multiple ways, even leveraging the contrast at times for effect. One common technique is to separate them into distinct locations in a heterogeneous text. Wonder is often showcased in paratextual elements such as introductions or conclusions to otherwise skeptical genres, moments when the writer steps back to speak about the project as a whole. Conversely, texts for broad or novice audiences may incorporate a discussion of method, inherently skeptical, early in what is otherwise a catalogue of wonders. Writers also often juxtapose wonder and skepticism in neighboring sentences or paragraphs, counterbalancing them as was common in texts assessing the value of new technologies. Sometimes writers entwine wonder and skepticism in a single thought. And finally, as I will show in later chapters, the writer may embed skepticism and wonder within a framing structure that accommodates their diverse aims, often using the literary form of a soirée or a narrative of exploration.

Wonder and Skepticism across Genres

> We would commend the whole insect tribe to the rigid scrutiny of the microscopist, abounding as it does in all that is wonderful and beautiful.
>
> Jabez Hogg, *The Microscope*, 1854

The marveling naturalist and the dispassionate researcher emerged as distinct roles over the course of the nineteenth century, defined by different practices, aims, and values. But Victorian microscopists did not themselves see immersive wonder and detached skepticism as incompatible, as is evident in the optician Andrew Ross's remarks in 1839 that the instrument "holds a very high rank among philosophical implements, while the transcendent beauties of form, colour, and organization which it reveals to us in the minute works of nature, render it subservient to the most delightful and instructive pursuits."[4] Writers adjusted their work, of course, to specific publication contexts; thus wonder might ornament a research report by appearing mostly in introductions or conclusions, or touched on only lightly in a passing remark; while in a text for novices it might thread prominently throughout the text. Nevertheless, it is not uncommon in this period to find wonder and skepticism appearing in tandem or even fundamentally entwined, as in Hogg's epigraph just above. Here, a "rigid scrutiny" both enables and is informed by the wonder and beauty that the microscope affords the observer in viewing an insect leg, proboscis, or wing. This conjunction of wonder and rigor echoes the old trope *instructive and amusing*, which so many eighteenth- and early-nineteenth-century textbooks for young people promise to be. But Victorian microscopical writers paired wonder and skepticism in texts that range across the century and address diverse audiences. A brief catalogue of such examples will illustrate the diversity of settings where writers combine wonder and skepticism to express microscopical experience.

One such writer is an ophthalmic surgeon and photographer with the unpromising name of Jabez Hogg who in 1854 published a hefty introductory manual on microscopy. Hogg was well integrated into elite microscopical communities; he would serve as the honorary Secretary of the RMS from 1867 to 1872 and the inaugural President of the Medical Microscopical Society in 1874. Hogg conjoins wonder and skepticism in the very definition of his book, explaining that his aim was to provide a "concise, yet sufficiently comprehensive, *popular* account of the Microscope, both as regards the management and manipulation of the instrument, and the varied wonders and hidden realms of beauty that are disclosed and developed by its aid."[5] His emphasis on both careful management and hidden beauty exemplifies this important pattern in Victorian microscopy.

As I have noted, the 1850s represent an inflection point in the uptake of microscopy. Mid-century microscopical texts often echo the participatory, popularizing, and instructive impulse of the early Victorian period.[6] Such texts hope to draw a broad audience of working- and middle-class readers to the joys of microscopy and eventually to expand the ranks of working microscopists. However, elite microscopists were also eager to formalize their research methodology and promote the pursuit of method and objectivity. Hogg's book, *The Microscope: Its History, Construction, and Applications*, was one of many mid-century texts to spark and support Victorians' growing interest in microscopy in a range of settings, from a sociable soirée to the solitary, focused investigation of natural structures. Hogg's blend of wonder and skepticism thus speaks both to the moment and the market, as in the epigraph that opens this chapter—which speaks of wonder, beauty, and awe as well as the microscope's value to students of science. Hogg's prose casually leavens scientific nomenclature, terminology, and definition with earnest appreciation, evident when he explains, "*Alcyonella Stagnorum* is a dark green, shapeless, porous, jelly-like mass, filled with tubes, which have a thin lid or cover. Rows of ova appear in the tubes to a most extraordinary extent."[7] Hogg's book, one of the most frequently recommended mid-century texts, abounds in such passages, which infuse the cautious, exact phrasing of method with the exuberant language of zeal.

While Hogg blends wonder and skepticism for aspiring microscopists who had dedicated 6*s*. to the purchase of his book (more for a cloth binding), Edmund Saul Dixon does so for a broader, less committed audience, in his 1856 article on "Microscopics" in *Household Words*, a large-circulation family periodical edited by Charles Dickens and sold for 2*d*. per weekly issue. Dixon was an Anglican priest and poultry-fancier who contributed many scientific articles to this periodical. In "Microscopics" he knits the marvelous together with limitation and difficulty, warning (for example) that "the weakness of our eyes and the imperfection of our instruments... are the sole reasons, as far as we know at present, why we do not behold infinity within the marvellous drop" of water.[8] He explains, "the more powerful and refined the instrument, the more difficult is its management." Indeed, "first-class microscopes... demand the most masterly skill" from both optician and microscopist. They "are affected by infinitesimally small derangements" and require delicate adjustment, although "these little manipulations are not acquired without a fatiguing amount of practice."[9] The romantic elements of delicacy and extremity appear here in Dixon's text as harbingers of bodily fatigue and error. In another memorable example, he cautions readers about selecting a reliable microscope, saying, "if the instrument be muddle-headed and cloudy, the stronger it is the worse it is." In a lens without "distinctness of definition," he says, "the object appears merely as a huge, misty phantom, like Ossian's cloudy heroes."[10] The romantic image here—Ossian was purportedly a legendary Gaelic warrior and poet—teaches readers a skeptical lesson: that poor lenses will lead them into fog rather than fact. Dixon's blending of wonder and skepticism suggests at once the endless wonders of the microscopic world and the physical limits of the observer and instrument working to access those wonders.

It might seem that, in contrast to these texts for general adult audiences, Victorian children's texts would seek to attract young readers by trumpeting only the delights of microscopy, glossing over its difficulties. While such texts do emphasize the marvels of microscopy, they also feature notes of caution and method. The botanist and travel writer Louisa Lane Clarke demonstrates this conjunction of wonder and skepticism in her 3*s.* book *The Microscope: Being a Popular Description of the Most Instructive and Beautiful Objects for Exhibition* (1858). She recommends the "many beautiful species" of the diatom genus "Meloseira" (*Melosira*) but also cautions that it is "very likely to be mistaken under a low power for a mass of confervae. In fact, it has been misunderstood even by eminent naturalists."[11] Similarly, in a passage describing the parts of flowers, Clarke explains, "To see the actual pollen-tubes in their passage down the style is a more difficult matter" than seeing the pollen-grains in a flower; "nevertheless, with care and a good glass it may be managed." Once having explained how to find and identify the pollen-tubes, however, she shifts from a discourse of difficulty into a celebratory theology of nature, saying, "reverently let us ever search into the mysteries of creation, and find new and deep delight in these revelations of His secret order, wisdom, and care."[12] As with Hogg and Dixon, in Clarke's text the skeptical practices of caution and method support wonder, and wonder inspires caution and method.

The texts quoted above are written by observers with varying degrees of microscopical experience and addressed to strata of the general reading public. Original investigators also juxtaposed wonder and skepticism when addressing their peers. Disciplinary norms, being formalized during the Victorian period, did increasingly restrict romantic language in research texts. Expressions of wonder become more fleeting or localized to settings inviting a heightened rhetoric and awareness of mission, such as introductions, conclusions, and ceremonial addresses. However, texts for working microscopists still exhibit an interplay of wonder and skepticism, as in a paper presented by Joseph Delves to the Microscopical Society of London, on "the Application of Photography to the Representation of Microscopic Objects." Delves takes care to specify the materials and method for this early and influential experiment in photomicrography. He details the size and position of the apparatus as well as the time of exposure (which "varies from five to fifteen seconds"). He concludes, however, by shifting briefly from skeptical discourse to the discourse of wonder: "by calling attention to some very beautiful specimens, which were exhibited, and afterwards presented to the Society."[13] Together, skepticism and wonder lend authority and meaning to Delves' work and to that of the Society as a whole.

Strategies: Separation, Juxtaposition, Combination

The act of seeing can spark intense emotion as well as the cognitive work of reasoning; sometimes the same object powerfully evokes both of these responses. Texts describing this phenomenon can address it in different ways, by separating, juxtaposing, or combining wonder and skepticism. For example, Harriet Martineau describes her complex relation to the magic lantern as a child. She explains,

> A magic-lantern was exhibited to us on Christmas-day, and once or twice in the year besides. I used to see it cleaned by daylight, and to handle all its parts,—understanding its whole structure; yet, such was my terror of the white circle on the wall, and of the moving slides, that, to speak the plain truth, the first apparition always brought on bowel-complaint; and, at the age of thirteen, when I was pretending to take care of little children during the exhibition, I could never look at it without having the back of a chair to grasp, or hurting myself, to carry off the intolerable sensation.[14]

Here Martineau tells how she engaged with the visual technology of the magic lantern in two distinct ways. During the day she puzzled over its optical arrangement, discovering the workings of its mechanism. At night, however, she experienced a visceral "matter of pure sensation, without any intellectual justification whatever," where the lantern's visual trickeries overcame her physically. Her experiences of the magic lantern differed in nature and in timing, but neither obviates the other.

Victorian writing on microscopy can also exhibit this apposition of different responses to visual technology. Many texts segregate expressions of wonder and skepticism into distinct sections of a text. In texts addressed to elite microscopists, such as articles in scientific journals, wonder is most likely to appear in the introduction or conclusion; the same applies in annual presidential addresses. These paratextual moments invite reflection on the overall meaning of their work. Sometimes, however, this segregation is dictated simply by the shifting focus of a long argument. For instance, Dr. Robert D. Lyons, lecturing to Dublin medical students on microscopic anatomy in 1851, endorses both wonder and skepticism but in widely distanced passages. Early on, he proclaims the marvels of microscopy:

> [I]t may be said that what the telescope is to the astronomer…annihilating the almost immeasurable intervals of space, the microscope is to him who would explore those countless, wonderful, and exquisite productions of nature, whose…minuteness of size…almost reaches the limits of comprehension; and, in the language of Ehrenberg, shows an unfathomableness of organic creations.[15]

Lyons here offers conventional expressions of microscopical wonder, drawing on the tropes of invisibility, the sublime, and the telescope and microscope, and using exaggerated language to marvel at the immensity of microscopical space. But he later warns his audience, in more quotidian language, "All parts of our science are imperfect.... [T]he life of the physician is one long studentship."[16] Although it might seem that his warning about "imperfect" science would negate his claims of wonderful microscopical explorations, in fact he presents these as compatible elements of persuasion, urging—in different ways—his audience toward a difficult but rewarding practice.

Wonder and skepticism can also appear juxtaposed in consecutive passages. In Arabella Buckley's 6*s*. children's book *Through Magic Glasses* (1890), a magician-professor lectures an assembly of students by noting both the marvels and the difficulties of the human eye, the microscope, and other "magic glasses." He ends his first lecture by addressing both these elements together, first expressing wonder, mystery, and magic: "Now, boys, I have let you into the secrets of my five magic glasses—the magnifying-glass, the microscope, the telescope, the photographic camera, and the spectroscope. With these and the help of chemistry you can learn to work all my spells." For example, he says, "You can peep into the mysteries of the life of the tiniest being which moves unseen under your feet." The professor continues, however, with a warning, emphasizing the need to proceed cautiously. He says, "[T]he value of the spells you can work with my magic glasses depends entirely upon whether you work patiently, accurately, and honestly. If you make careless, inaccurate experiments, and draw hasty conclusions, you will only do bad work, which it may take others years to undo." He discusses this risk at some length, explaining that "if you question your instruments honestly and carefully, they will answer truly and faithfully." Indeed, even then it is possible to err, but the magician promises, "You may make many mistakes, but one experiment will correct the other."[17] Such juxtapositions use the trope of *counterbalancing*, where the writer models an evenhanded, evaluative stance for the reader in assessing the properties of a new technology.

Lyons separates his skepticism from his wonder, and Buckley juxtaposes hers in neighboring passages, but authors often join them in the same page or paragraph, or intercalate them in alternating statements. For example, Hogg oscillates between the language of wonder and that of skeptical investigation in describing *Flustra avicularis* (a moss-like marine animal). He begins by noting its lovely appearance, writing, "This is another of the little beauties of the deep found usually on old shells"; then plunges into a passage using the stacked phrases, measurement, and Latinate adjectives of taxonomic characters.[18] The little creature is

> an inch in height, spreading itself fan-like, and of an ashy colour, deeply divided in a dichotomous manner into narrow, thin, plane segments; truncate at the end, formed of four or five series of oblong cells, capped with a hollow, globose, pearly operculum seated between the spines, of which there is one on either side of the circular aperture.

Hogg then returns to wonder and whimsy for a moment—"[t]he opercula are so numerous that they give to the upper-surface the appearance of being thickly strewn with orient pearls"—before concluding with minute description: "the under-surface is even and longitudinally striated, the number of striae corresponding to the number of rows in which the cells are disposed." His prose frequently alternates in this way between diverse approaches to the object.

This conjoining of wonder and skepticism can appear even in the research papers of elite investigators. The Rev. J. B. Reade, President of the RMS, alternates between admiration and cautious description when telling of an "exquisite" diatom in a paper presented to the Society in June 1869. Although such papers would most likely turn to wonder and heightened language in their opening or closing statements, Reade does not shy away from the language of wonder even in the conversational style that characterizes the middle of his presentation:

> It is needless to observe that deeper powers are required when attacking a valve like *N. rhomboides.*[19] Here the 1/12th is used, and I am sure that the exquisite beauty of this valve will be a treat to critical eyes. The acknowledged difficulty of resolving it arises from the extreme closeness of the very minute hemispheres in the longitudinal rows.... Powell's immersion lens would bear with admirable effect upon this exquisite object, and bring out the wondrous structure which, without the aid of the microscope, must have remained among the invisible things of Him who created all things.

Here Reade voices his admiration for the "exquisite beauty" and "wondrous structure" of the diatom, in phrases framing his expressions of doubt, caution, and method. Given the "acknowledged difficulty" of his task, Reade carefully details and measures the patterns of beading. He blends aesthetic appreciation with meticulous minute description, saying,

> Round the valve, and forming an elegant border, are three rows of beads or hemispheres gradually decreasing in size—then, on the semi-diameter of the valve through the centre, 14 rows of much smaller beads, numbering at least 80 in 1/1000th of an inch—and then the two "median lines."... [T]he boss or *umbilicus* pushes out the adjacent beads of the median rows into an oval form.

Reade balances his wonder at the "elegant" patterning of *N. rhomboides* with these restrained periods of minute description. His caution here may reflect the awkwardness of making this particular argument; he had argued for a very different structure of this deceptive diatom at the previous meeting. He explains, "In then describing the markings...I used the language of the craft, and spoke of 'dots as black as jet'; but this mere silhouette representation of *rhomboides*, an unnatural distortion of light and shade, I never wish to see again."[20] Reade suggests that he responded to a kind of peer pressure ("the language of the craft") in both seeing and enthusiastically describing the "dots as black as jet," but his new argument rejects the faulty model while adopting similarly dramatic diction. We now know that the "dots" of diatoms in fact represent not "beads or hemispheres" but tiny fissures or pores, distorted in the light microscope by diffraction. Given Reade's willingness to own up to and correct previous errors, it is likely that if he had been able to discover the true nature of these pores, he would doubtless have revised his views once again, using similarly romantic language.

As Reade's example suggests, microscopists sometimes entwine wonder and skepticism within the same thought. The Rev. J. G. Wood does so for a broad audience of novices, in his 1s. guidebook *Common Objects of the Microscope*. He renders a note of caution by proposing a fantastical analogy, reframing the beauty and delicacy so often admired in the microscopic world. He says,

> The heavy stand [of the Society of Arts Microscope] is made of iron, and affords a firm and solid support to the instrument, a matter of no trifling consequence when the reader reflects that motion is magnified as well as substance, so that if the instrument trembles in the least degree, the object becomes almost invisible, seeming to flutter before the eyes of the observer like the whirring wings of a hovering fly.[21]

Wood expresses a conventional warning about the vulnerability of microscopical vision by juxtaposing the solidity of microscopical forethought and labor with the instability of microscopic vision and the delicacy of imagination. He first grounds microscopical practice in the heavy materiality of the iron stand, then shifts to the image itself, describing it, in a whimsical use of metaphor, as fragile and evanescent like the insect wings so often examined under the lens.

A trace of wonder can sometimes be glimpsed within even the most skeptical texts. Henry Edward Fripp translated Ernst Abbe's pioneering work on the physical limits of the light microscope from German into English in 1875.[22] Doubt governs most of the essay. Abbe argues that, at the limits he identifies, "the actual powers of the microscope (in the strict sense of correct and useful power) are… exhausted." He explains, "There exists no microscope in which there has been seen, or will be seen, any structure which really exists in the object, and is inherent in its nature, that a normal eye cannot recognize with a sharply defining immersion lens magnifying 800 times."[23] Any appearance of structure beyond this level of magnification cannot be relied on. He warns, "However constant, strongly marked, and so to speak materially visible, such indications of structure may appear, they cannot be interpreted as morphological, but only as physical characters" (that is, not necessarily representations of real structure but images created by the interaction of light with structure and its material, according to the laws of physics). Thus, he continues, "Nothing more can be safely inferred from *the microscopic revelation* than the presence, in the object, of such structural peculiarities as [can produce] the diffraction phenomena on which [these] images of minute details depend."[24] The skepticism informing this argument could not be more evident. But tellingly, in a translation that generally hews closely to Abbe's language, the English translator Fripp here uses the term *revelation* where Abbe had used "*Befund*" (findings, result).[25] Fripp would have understood the distinction; he had lived in Germany for over a decade before returning to England in 1856.[26] The noun "revelation," even more strongly than the verb "reveals," aligns the passage not only with the popularizing trope of revelation but also with the narrative of natural theology and its revealed truth of Creation. That is, in the midst of Abbe's undermining of microscopical certainty, as he rejects the microscopic image of fine detail and warns of diffraction phenomena distorting our knowledge, Fripp nudges his terminology toward what is eternal, marvelous, and sublime. It is an ephemeral mention of wonder, but a telling one.

Such an interlacing of wonder and skepticism can add nuance to an ostensibly skeptical message. The physician C. R. Goring displays the persistence of wonder in comments from his and Pritchard's influential early book *The Microscopic Cabinet of Select Animated Objects*, marketed to would-be naturalists in 1832 at a price of 18*s*. Goring advises readers, "Long practice must after all determine our opinions, and scepticism should ever form a leading feature in them; we should *suspect rather than believe*."[27] It is tempting to read this as an unmixed endorsement of skepticism. After all, he later warns that in deep eye-piecing (using high powers in eyepieces, a practice that risks simply magnifying errors from the objective), "all certain vision fades away, and the imagination is left to its uncontrolled operation." Even with high-powered objectives, "if our eyes are strong, and habituated to their use, we may place great reliance on their testimony; but we must never allow them to persuade us to believe marvels which are manifestly impossible, or contrary to the known laws of nature and right reason."[28] Despite these strongly skeptical statements, though, Goring was already on record in 1829 advocating for an observational process that both interrogates and appreciates the wonders under the lens. Indeed, in each of the examples quoted above, he qualifies his skepticism by suggesting that some forms of imagination or wonder may indeed be permissible; that skepticism is a "leading feature" but not the only one; and that it is specifically the "impossible" marvels or their "uncontrolled" acceptance that should be rejected out of hand. Observers needs must learn how to manage their wonder, just as they must gain tact and good judgment in managing the illumination that provides the image, but wonders should not be rejected out of hand. Rather, he implies, they should be examined using the "known laws of nature and right reason." This balanced stance aligns well with Pritchard's statements in his Preface to the book, which opens by defending "the study of the works of the Creator," especially those instances of "beauty" and "delicacy" that inspire microscopical efforts to "penetrate deeper into their structure and organization."[29] Indeed, despite Goring's critique of unexamined wonder, his remarks actually suggest that wonder is ever present in microscopic observation, and should thus be part of the study; that microscopical writers not just *can* but *must* recognize both their wonder and their skepticism.

Writing Embodiment in Victorian Microscopy: Beautiful Mechanism. Meegan Kennedy, Oxford University Press.
 DOI: 10.1093/9780198940623.003.0018

19
Soirée Form

> [J]ust take your seat in front of this tank, and with a lens before your eye, watch the colony [of barnacles] which is seated on that piece of stone.... From one and another, every instant, a delicate hand is thrust forth, and presently withdrawn. Fix your attention on some one conveniently placed for observation... [where] an apparatus of delicate curled filaments is thrust quickly out.... Now, by putting this specimen into a glass trough, and placing it under a low power of the microscope, we shall see what an exquisite piece of mechanism it is.
>
> Philip Gosse, *Evenings at the Microscope*, 1859

Introduction

The informal gathering known as a *soirée*, also referred to as a *salon*, *conversazione*, or simply "evening," was a commonplace of Victorian polite science. Whether sponsored by a microscopical club or offered in a friend's parlor, soirées attracted a variety of attendees, who could blend sociality with a guided exploration of miscellaneous exhibits.[1] Of course, there had long been literary and artistic soirées as well as soirées exhibiting curiosities of natural history, but microscopical observations became increasingly common elements of such gatherings after technical improvements to the compound microscope in the 1830s made reliable instruments newly available not just for researchers and doctors but also for students, hobbyists, and families. As late as 1882, a report from the Liverpool Microscopical Society commented that "the popularity of the microscope as a means of both instruction and recreation is on the increase," such that the microscope was "now considered an almost essential part of such entertainments."[2] Many microscopical objects were ideally suited for soirées, given the convention that these instruments "revealed" hidden "wonders" to the trained eye and the fact that such objects could be readily turned to support scientific and theological narratives. In turn, soirées helped microscopical clubs and societies promote the instrument as a difficult but rewarding companion; at these events they could attract new members and solicit greater public support for such work. The soirée became such a common form of gathering that natural history authors adopted it as a trope for a particular literary structure as well. This chapter uses Philip Henry Gosse's *Evenings at the Microscope* (1859) as a case study to examine the microscopical soirée as a literary form built on an aesthetic of miscellaneity (especially of intricate or lively objects), an ethic of sociality, and an appeal to affect. This convention also readily translated into other visual media such as the photograph, lantern show, and cinema.

Soirées became a common means of performing Victorian ideals of self-improvement and progress. They could range from large events welcoming hundreds of attendees, as with the yearly gatherings hosted by microscopical clubs, to more intimate affairs held in a middle-class parlor for family and friends. Certain intricate objects such as diatoms or insect wings, or living objects like infusoria, became recognizable as typical soirée exhibits, associated together for their spectacular display value more than any logical or scientific relation. Naturalists adopted the soirée as a literary genre as well. The popularity of the soirée in Victorian culture, especially in Victorian microscopical culture, developed a long-lived and influential set of aesthetic, literary, and social forms combining science, technology, literature, and art.

The soirée, a sociable evening dedicated to literature or natural history, was remediated as a versatile and powerful structure for popularizing texts, book frontispieces, magic lantern slides, even microcinematic short films. Microscopists writing for broad audiences often adopt the form of the soirée to recruit readers and microscopists by discussing a variety of spectacular microscopical objects in a friendly, conversational voice. Writers using this form emphasize visual forms using the convention of being *instructive and amusing*. They sometimes adopt skeptical modes of minute description or articulated practice to guide readers' practice, but they also offer more conversational, wonder-filled passages—what I call *enthusiastic description*—which invite readers to mirror the narrator's affective response to the image. They frame the whole in a lively and entertaining fiction of microscopical sociality.

Recognizing soirée form as a literary and aesthetic convention helps explains the hybridity of style in texts like *Evenings at the Microscope* (1859) by the naturalist Philip Henry Gosse. Gosse's popularizing books draw liberally from his other publishing interests, theological treatises and texts for working naturalists. He was well positioned to interpret microscopy for a general readership. He was an experienced and respected naturalist, known for painstaking descriptions of species more than philosophical speculation, and famously described as an "honest hodsman of science" by T. H. Huxley.[3] From 1856 he was a Fellow of the Royal Society, although he marred his reputation with the 1857 publication of *Omphalos*, with its tortuous attempts to reconcile natural facts with biblical narrative. When he published *Evenings* he was also working on his massive *Actinologia Britannica: A History of the British Sea-Anemones and Corals* (1858–60). *Evenings* cost a bit more than other books for novices (8*s*. compared to 5*s*. for the technical manuals published by Hermann Schacht and Lionel Beale)—perhaps due to its length (some five hundred pages, compared to about one hundred pages) and its wealth of illustrations—but it was still within reach of middle-class readers. *Evenings* proved to be one of Gosse's most successful books, being reprinted multiple times, even into the twentieth century. It was also quoted extensively, for example in Eliza Lynn Linton's "Under the Microscope" in *Household Words* in 1859.[4] In *Evenings*, Gosse uses the overall structure of a fictional soirée to invite readers into the practice of microscopy. He frames them as guests taking part in a series of imagined explorations of microscopical objects. In actual soirées, guests move among a diverse array of objects and interests. Accordingly, soirée form—perhaps the quintessential example of a text that should be *instructive and amusing*—prominently features a shifting focus, which authorizes the miscellaneity of subjects and prose forms in *Evenings at the Microscope*.

Soirée form is a genre of popular natural history, a variety of the *narrative of natural history*. This type of narrative blends fiction and fact, using a personal, emotional voice to describe concrete particulars of wonderful natural objects. It often uses comparisons to familiar objects and either eschews theoretical explanation or references to theologies of nature. The soirée genre does this within the specific setting of a fictional frame (a collection of "rambles," "half-hours," or, as in Gosse, "evenings"), suggesting a relaxed, informal community of observers. Just as the instruction in an actual soirée is less structured than in a schoolroom, the soirée genre enjoys an apparently meandering, happenstance form, similar to the ramble or the periodical essay. The soirée setting allows the author-narrator to survey a varied, even random assortment of objects and to more easily negotiate the challenges in writing what is instructive and amusing, with the concomitant shifts in both content and tone.[5] Microscopical texts written in soirée form present an eclectic assortment of objects in a friendly, conversational tone. They often directly address readers and overtly solicit their affective engagement, describing the actions of an imaginary gathering of friends. And they offer to demystify microscopical work and its objects of study even as they double down on the mystifying elements of fancy, wonder, and awe. In addition, Gosse's version of soirée form often describes microscopical objects by comparing them, sometimes at length, to familiar or fantastical analogues, a common enough trope in microscopy (see Part II) but one that underscores the domestic, recreative setting of his imagined soirée.

Authority, Experience, and Prose Style

While the soirée form shares many qualities with the ramble or personal essay, it also overlaps with another popular genre using the narrative of natural history: children's science books. Published from the late eighteenth century, these often attempted to make their blend of scientific and moral instruction more palatable by casting it into a fictional frame, usually dialogic, so that a helpful adult figure could lead fictional children through a series of explorations of the natural world and its wonders. Aileen Fyfe notes that children's books from 1780 to 1820 were characterized by "moral lessons, the use of the conversational style, and recourse to examples from everyday life"—although she also notes that by the middle of the nineteenth century, "evangelical writers... began to write moral fiction" instead of didactic conversation.[6] Barbara Gates chronicles the rise and fall of the "Conversations on..." genre developed by Jane Marcet, which was adopted by many popularizers, especially women writers, during the first half of the nineteenth century.[7] Gosse's approach echoes the style of those earlier, turn-of-the-century children's books by offering "moral and religious lessons drawn from nature, drawing on easily understood examples from everyday life" and couched in "a conversational format."[8] However, he does not acknowledge any resemblance to those books, and his approach does differ in significant ways. His "conversation" is largely one-sided; Gosse only occasionally allows the imagined reader to speak directly. That said, the reader functions as an imagined interlocutor throughout, even occasionally pushing back against the narrator's presentation. The host-narrator frequently directs the observation so as to answer anticipated questions and dispel imagined misunderstandings. Further, Gosse's imagined "evenings at the microscope" are not just conversations but *conversaziones*—a playful genre inclusive of adults, which he evokes in passages more lively than the stilted catechisms that often structured children's books.[9] Despite these differences, *Evenings at the Microscope* was often given as a school prize, a marker of its suitability for youthful readers.[10] The fictional soirée thus helps to attract a range of readers, but its echoes of children's books threaten to undermine the rigor and authority of the text.

Accordingly, one concern of the soirée writer is that, unlike the presumably authoritative author of a research text, the host of an evening's entertainment could be dismissed as a dabbler or mere popularizer of others' work. Geoffrey Belknap argues that much of the work of nineteenth-century science occurred in "a mix of visual, verbal, and textual observation, parsed and reconfigured through a set of community-based activities."[11] However, the soirée may not represent quite the type of community gathering that Belknap references. The Royal Microscopical Society and other elite societies distinguished between their meetings, which among other things offered research-based lectures to members only, and their annual soirée, which displayed spectacular exhibition objects to a crowd of guests, including ladies and non-microscopists. "Rambles" (excursions) or "Science Gossips" such as those sponsored by the Quekett Microscopical Club, or the renowned "Monday evenings" given by Dr. James Scott Bowerbank for the group of microscopical observers that he termed the "Band of Brothers," might represent sociable spaces in between these extremes—informal but productive for fellow workers.[12]

Soirées are tagged as social spaces, but they position the host as an authoritative, instructive role model. In *Evenings at the Microscope*, Gosse authorizes his imagined soirée by referencing his experience as a naturalist. He asserts, for instance, that the book features "original observation" since "almost all" of "the observations ... have been ... verified afresh, or corrected." This willingness to duplicate others' scientific observations, he claims, renders his work distinct from that of "popular compilers," whom he condemns for their recirculation of "the opinions of others, often without the slightest knowledge" of the matter themselves.[13] Like Schacht and Beale, Gosse encourages his readers themselves to contribute to microscopical research. He optimistically promises that "the student who shall have verified for himself the observations here detailed, will be no longer a tyro in microscopic science, and will be well prepared to extend his independent researches."[14] Indeed, the Rev. William Spiers reminisced in 1890, "Many an amateur microscopist dates, as I do myself, his passion for the study of Infusoria from the perusal of Gosse's delightful *Evenings at the Microscope*."[15] Gosse's imaginary gathering yielded real-world effects.

A textual soirée embraces a more miscellaneous aesthetic than the systematic approach increasingly adopted by introductory guidebooks. However, like other such texts, Gosse's *Evenings* provided a pedagogical framework for novices who wished to work toward greater facility with the instrument outside the structure of formal classes. Readers could practice their observations over and over at home, improving their skills. At any time, the reader could open the book and call forth a guide and companion to accompany him on his microscopical journeys. In accordance with this devotion to original work, Gosse's popular natural history writing includes many rhetorical elements that typically anchor texts for investigators, some even drawn from his own published research. These passages underscore that his aim is to recruit microscopists to similar work.

In particular, Gosse presents many passages of *minute description*—a clear, precise, purposeful, and dispassionate depiction of a microscopical object using scientific terminology that I discussed in Chapter 16. For instance, when Gosse, speaking through the narrator, observes the *Laomedea geniculate* (a tiny marine animal like a jellyfish) under the microscope, he employs "first a low power and afterwards ... a higher," reporting,

> We now see that the creeping thread is a tube of horny substance, flattened on its under side, and that the erect stems and their branches are similar tubes, whose cavities are in free communication with that of the creeping root. The wall is thin, and perfectly transparent and colourless; the whiteness of the whole being dependent on a soft medullary core of living jelly.... Where the branches issue from the angles of the stem, the medulla, and consequently the horny sheath, is dilated into a knob; immediately above this there is a joint-like constriction in the tube, and the branch itself is insected by four or five such constrictions.... Its extremity then expands into an elegant cup.[16]

Such passages do not include the quantitative identification of magnifying power nor measurement of the specimen often found in minute description, but Gosse does use a flat, mostly disinterested tone and conscientiously details the form, material, and structure of the specimen. He occasionally pauses to note the elegance or beauty of an object, but he just as often uses specialist terminology such as "medullary core" or the Latinate "insected." These passages of minute description accord with Gosse's reputation as, in the words of his son Edmund, "a technical observer of unusually high rank" making the most of his "very absence of imagination."[17]

Gosse is often discussed as either a plodding literalist or a fabulist popularizer, but both are true. Creativity and ingenuity power many of his descriptions in *Evenings*; he interleaves passages of minute description like the one above with a much more conversational, whimsical, and admiring style of description. Such passages overtly invite readers to join in the author's experiences, drawing on the affective and narrative appeal of genres such as the memoir, the anecdote, the travel narrative, and the spiritual autobiography to sweep readers up in a current of feeling. I call such language *enthusiastic description* to recognize Gosse's microscopical work as an expression of his fervent nonconformism (the term "enthusiasm" was often used to describe such a faith) and his overt evangelism for the instrument (his zealous converts might be described as *enthusiasts*). He pursues both of these efforts with a fervent optimism, powered, as Jonathan Smith argues, in the happy expectation of the Second Coming of Christ. Smith comments that Gosse's beloved rock pools were, for him, "as close as we can come to a vision of the utter profusion and perfect happiness in nature prior to the fall" as well as "when Christ returns."[18] Gosse infuses his work under the microscope with the same ethos.

Like Schacht or Beale, then, Gosse engages in a cult of the detail; but where their minute description requires detached precision, Gosse revels in whimsy, wonder, and delight. His popular writing often uses romantic language to compare microscopical objects to ordinary or fantastical ones and it models his own curiosity and pleasure in microscopical work. Such appeals to readers' affective engagement were common in popularizing texts about natural history; they most frequently appear in recreative texts or in instructive texts for novices. Gosse's enthusiastic description—in its whimsy, affective force, and outreach to the reader—powers his use of soirée form in *Evenings at the Microscope*.

Gosse's prose is marked, above all, by expressions of both the faith and the pleasure he experienced in observing microscopical creatures. As noted in Chapter 6, he imagines that any observer of the *Rotifera* will "hail their appearance in his drops of water with pleasure, and linger over them with unwearied delight." Similarly, expressing a sentiment perhaps not widely shared, he exclaims, "The parts of the mouth in different insects afford an almost endless store of delightful observations."[19] These instances of enthusiastic description both express delight and solicit it from the reader.

Gosse's enthusiastic description foregrounds not just his own pleasure in observation and the spiritual associations of microscopical study, as might be expected from his personal investments in the subject, but also a range of emotional responses to microscopical observation. Jonathan Topham notes that "in many religious magazines the affective aspects of science were at least as important as the rational aspects," and the same could be said for popular books like *Evenings at the Microscope*.[20] Among the affective elements of microscopy, Gosse emphasizes wonder and awe, but he does not restrict himself to these stalwarts of natural theology, venturing also into humor, sensationalism, even horror. As he says in discussing some of his favorite zoophytes, "Some objects which I have to exhibit to you are altogether unique as to their appearance; and, if you are not as imperturbable as a Stoic philosopher, or a Mohawk Indian, will certainly excite your risibility and your wonder."[21] Stereotypes aside, Gosse's enthusiastic description seems designed as a textual form of social interaction, stimulating a variety of affective engagements with the work in order to attract readers to the community of microscopists. *Evenings* is an evangelical text in its missionary fervor for both God and the microscope.

Gosse overtly positions his role as instructive, but he does not recommend any specific method for the microscopical worker, nor does he model any single textual approach, toggling between disciplined, minute description and creative, fanciful discourse. While his guidebook is rich in detail, the reader encounters a meandering, conversational approach, to the extent that "a considerable amount of information" on relevant topics—on "microscopic manipulation," on "the selecting, securing, and preparing of the objects for examination," and on "the power to observer and to discriminate"—is, as he says, "incidentally scattered throughout the work" rather than presented in an introductory chapter or appendix, as was more common.[22] For example, he defines "achromatic" in the midst of a discussion on the scales of the Green Wrasse, relegating the definition to a footnote.[23]

The fictional frame of the soirée accommodates this eclecticism in style and content. Gosse says, "In order to relieve as much as possible the dryness of technical description, a colloquial and familiar style has been given to the work." As he explains, the book "has been thrown into the form of a series of imaginary conversaziones, or microscopical soirées, in which the author is supposed to act as the provider of scientific entertainment and instruction to a circle of friends." So while Beale urges his medical students to clarify and distill their notes through retrospection and self-editing, Gosse instead adopts the narrative persona of a host welcoming novices to his soirée. He thus avoids the pedantry or the self-aggrandizing elaboration that Beale criticized, but his discursive style is more chatty than concise, more discursive than direct. His narrator blends the air of a teacher with the spirit of a garrulous neighbor whose endless fund of reminiscences and stories overlap and blend with one another. Despite many passages of minute description, Gosse frequently returns to this more sociable voice.

Even with the support of his record of zoological work, Gosse must draw a delicate balance in how he presents that traditional pairing, instruction and amusement, to his fictional guests. Indeed, Gosse acknowledges this risk in his defensive claim that "the precision essential to science has never been consciously sacrificed" and his casually authoritative remark, "A master may be easy and familiar without being loose or vague."[24] A review of *Evenings at the Microscope*—most likely written by Huxley or one of his associates—explicitly references this balance in assessing the book. It praises Gosse for conscientiously "studying the objects themselves" and comments that such study evokes his best writing. In contrast, the reviewer pans his efforts to be amusing or philosophical.[25] The review rebukes Gosse's attempts to lighten the tone "by calling the frog he is observing 'froggy,' and by throwing gasps of conventional colloquialism into paragraphs crowded with technical terms." But "[s]et before him an animal," the review notes, "and his real enthusiasm is excited, his style becomes straightforward and clear." The "only way" to approach technical material is thus "to awaken the curiosity, and make the enumeration intelligible"; all "writers of popular books" should "devote their whole attention to the clearness of their descriptions."[26] The reviewer praises Gosse's original observation and clear description while heaping scorn on his more whimsical passages as unhelpful to the student. Despite this critique of soirée style, Gosse's chatty and fanciful passages were more likely to be quoted in other publications, suggesting that they did in fact attract readers.

Despite his intermittent use of minute description, then, Gosse does not address his book to original investigators, as Schacht does. Indeed, the author of an article in the *Universal Review* notes that Gosse's work contributes to observational natural history rather than experimental science; that is, he "is an indefatigable naturalist, if not a very philosophical one" and thus *Evenings* "wants a philosophical spirit…which would give unity" and purpose "to [its] multifarious details."[27] While *Evenings* may not present advances in microscopical knowledge, and its eclectic style is not meant to model scientific reporting, Gosse writes descriptions that can move readers to take up the practice of microscopy, and he instructs them in what to observe once they do so. Although *Evenings* does include some moments of "dry" quantification and classification (important to natural history observation as well as to original scientific investigation), Gosse largely limits these to his passages of minute description and frames his more restrained, scientific passages with pages of earnest encouragement, reminiscence, and whimsy.[28] Overall, he solicits general readers by staging science as a lively social evening, the reader guided by the essayistic meanderings of a friendly host. Any diffident or reluctant guests can hardly refuse to follow up their host's pressing invitation to observe.

Physical Presence and Affective Response

All books hail their readers in some sense, addressing them in virtual space; but texts written in soirée form make that relation explicit, striving to realize a physical community on the page. Soirée form invites readers into the practice of microscopy by, paradoxically, placing them already as actors in the scene, and it facilitates a tone of friendly, collective instruction. Gosse calls upon imagined readers, as if they were standing alongside him in his parlor, viewing objects in all their physical materiality instead of virtually, through a text. While minute description and articulated practice narrate the entwined embodiment of observer, instrument, and object in virtual storyspace, the soirée form overtly introduces readers into that space as well. Gosse introduces the first observation by saying "Let us then begin our evening's entertainment by an examination of a human hair.... Here, then, is a hair from my own head." He walks the reader through each step in the first person, as though explaining his actions to an audience: "I cut off about half an inch...I now apply a power of 600 diameters....You see, crossing the bright circular field of view, a semi-pellucid cylindrical object; that is the hair."[29] Gosse speaks to readers directly, using the second person to direct their attention ("you see, crossing the bright circular field..."). Gosse uses a similar conceit in the epigraph for this chapter, where he tells the reader how to sit and where to direct the eye and attention.

Gosse is far from the only popularizer to use this conceit. T. Charters White, similarly, in his 1887 guidebook *My Microscope and Some Objects from My Cabinet*, explains,

> It has been a matter of surprise to me when I have been enjoying the pleasure of showing "something under the microscope" to a sympathetic public at a *conversazione*...to find how very many of them have never looked down a microscope worthy of the name....In the following pages I propose, therefore, to take the reader with me...and to show him one of my microscopes, and a few objects from my cabinet.[30]

White opens his first chapter with just such a friendly invitation, proposing that "We may pleasantly commence our investigations into the minute with a walk in search of 'an object.'"[31] Chapter 2 of his book invokes the reader's materiality even more explicitly. It begins, "And now I will ask the courteous reader to place his eye with mine over the tube of science, and look at a specimen or two of a different class of nature's marvels."[32] White chooses to refer to his readers in the third person but vividly describes their physical propinquity ("place his eye with mine"). Just as these authors imagine the placement of the reader's eye, the reader will imagine the narrator's eye aligned with his and peering down the eyepiece.

Some books do not explicitly present as a soirée or series of soirées, but the narrative is implicitly structured by the notion that narrator and readers are working together in the same space. Edwin Lankester was an important figure in the community of working microscopists, an editor of the *Quarterly Journal of Microscopical Science*, who organized his popular guidebook *Half-hours with the Microscope* (1859) as a series of "half-hours" at work with the instrument. Lankester does not adopt the fiction of the soirée as elaborately as these other writers. His chapters generally plunge into discussion of "A Half-hour with the Microscope in the Garden," "...in the Country," "...at the Sea-side," etc. with no more than a general scene-setting: "On a visit to the sea-side, the Microscope is an essential instrument to all who would wish to study the wonders of the ocean."[33] And he does not, as Gosse and White do, narrate his process in the first person. But Lankester does refer to the physical setting within which he imagines himself and readers using the instrument. He calls up the image of his audience congregated around the instrument and says, "when the rain is falling, and the sea-side forsaken, and the country miserable-looking, we can still enjoy the long winter evenings with our Microscope in the house."[34] This fictional gathering evokes the soirée even if not named as one.

Agnes Catlow, a popularizing naturalist, suggests a fanciful type of gathering, a fantastical *ramble* right into the body of the microscope and out the other end. She explains, "My readers must fancy themselves spirits...and so pass with me through a wonderful brazen tunnel, with crystal doors at the entrance." The little group passes in imagination through the "tunnel" at the end of which are more crystal doors, "two or three in number" after which "we are in [a] new world."[35] This pleasant fantasy enables the notion that she and her readers share the microscopical visions she describes. As with Gosse, White, and Lankester, Catlow presents her readers' eyes as aligned with hers, alternating between "I," "we," and "you" in her discourse. In describing the *Volvox* (a form of algae) she says, "You see before you a hollow globe" and directs readers' gaze: "Now look at the green dots with which it is spotted."[36] She also invokes her readers' surprise and pleasure at such marvels, remarking, "And now I see your astonishment: your minds are bewildered with the variety of new beings and forms you behold."[37] Catlow's gathering shows how much the literary structure of the ramble shares with soirée form.

As suggested by some of these examples, the soirée form encourages readers to share the narrator's emotional experiences at the microscope. In Spiers's 1890 *Rambles and Reveries of a Naturalist*, one of his chapter-length "rambles" takes the form of a soirée, "an evening at the microscope." He writes, "Let us now take a dip into the aquarium and see what will come up....In a moment or two all is in readiness for the inspection of my visitors. And what a spectacle it is which they gaze upon! It will not be long before all sorts of ejaculations and expressions of amazement and delight will be uttered."[38] Like Gosse and White, Spiers imagines the reader's eye placed alongside his own. "Now look at [the Volvox] through this half-inch objective," he says, and presses the reader for a word of appreciation: "Did you ever see anything more lovely?" He describes a succession of typical objects—Desmid, Volvox, Daphnea, Hydra (all tiny marine species)—with tenacious fancy and admiration. The Desmid's "covering" is "sparkling with emerald points, and filled with diamond-like granules" to form "a vision of beauty surpassing that which Aladdin's lamp revealed." He glances at problems, such as the difficulty in preserving Desmids' color after mounting, and at doubtful points: "What gives [the Desmid] this power of independent movement?" But after a paragraph or so he always turns away: "But these are details into which I have promised not to enter now." Similarly, he writes of the "ever-beautiful Volvox" that it is "familiar enough, but enshrin[es] deep mysteries that even the restless eagerness of nineteenth-century science has not solved." He demurs, "Whether there is actual revolution of the young [Volvox] within the parent globe...is a point on which authorities differ, and...about which I am not prepared to pronounce dogmatically."[39] In such passages he both draws attention to microscopical doubts and difficulties, and neatly sidesteps them. Clearly the narrator's duty for this "evening" is to solicit enjoyment of this beautiful mystery rather than attempt to solve it.

Miscellaneity as a Teaching Method

Gosse leverages the conceit of the soirée—its eclecticism and its fiction that readers are standing by his elbow ready to take up the work—to teach particular concepts and skills in specific contexts where this issue might arise. For instance, the scenario of the soirée allows him to shift magnifications and perspectives, as he imagines readers doing in the epigraph for this chapter, where he directs them to observe barnacles first with the naked eye, then a hand lens, and finally a microscope. He repeats this directed shift from ordinary vision through versions of microscopic vision over and over again in *Evenings*. In recording the dissection of "the medicinal Leech," Gosse begins with first-person-plural narration ("we") to describe a shared view, in ordinary vision, of the leech's mouth: "[W]e see implanted in the walls on the dorsal region of the cavity three white eminences of a cartilaginous texture, which rise to a sharp crescent-shaped edge; they form a triangular, or rather a tri-radiate figure." Invoking the fiction of the soirée that governs the book, he implies corporate authorship of the description ("we see"), generously ignoring the implausibility that novice readers could compose such a statement, even as he gently nudges their imagined diction toward more precise scientific terminology ("a triangular, or rather a tri-radiate figure"). He dissects out one of these "white points," places it under the microscope, and again invokes his imagined collaborators: "Now I apply a power of 150 diameters, and we will look at it in succession." Gosse then dives into microscopical space to report what he and his virtual guests see under the lens, shifting into second-person narration ("you") to accommodate the singular view through the eyepiece:

> You have under your eye a somewhat pellucid mass, of an irregular oval figure, and of fibrous texture, one side of which is thinned away apparently to a keen edge of a somewhat semi-circular outline. But along this edge, and as it were imbedded into it for about one-third of their length, are set between seventy and eighty crystalline points, of highly refractive substance, resembling glass.... The manner in which they are inserted closely resembles, in this aspect, the insertion of the teeth in the jaw of a dolphin or crocodile.

Gosse's minute description here uses precise terms that may not be familiar to general readers but which succinctly indicate the properties of the substance ("pellucid," "fibrous"). He qualifies these terms so as not to indicate certainty where it does not exist ("somewhat," "irregular," "apparently").

This halting, cautious passage is overtly instructive; it has more in common with Beale's more formal textbook style than with the exuberant admiration and sudden leaps from one fanciful analogy to another that we generally find in Gosse's enthusiastic description. And although Gosse notes a resemblance to "a dolphin or crocodile," it is not the teeth themselves oddly being compared to exotic animals, as might happen in one of his more sensational passages, but rather their placement which is similar to that in other species—a conventional and useful exercise of comparative anatomy. And although he does denote the little teeth as "crystalline," he supports this description with a judicious explanation that they are "of highly refractive substance."

As this example suggests, Gosse models microscopical work as a series of shifts in perspective enabling self-correction. He describes the teeth of the leech as positioned along the edge of the mouth, then reverses his earlier statements, for "this appearance is deceptive." He records his (and readers') manipulation of the object—"By affixing the little jaw to the revolving needle, we bring the edge to face our eye"—to reveal the truth of the leech's jaw. After all, "[i]t is not an edge at all; but a narrow parallel-sided margin of considerable breadth. And the teeth are not conical points, as they seemed when we viewed them sideways, but flat triangular plates, with a deep notch in their lower edge." The vision of the jaw that he built and shared with readers must be revised: the "edge" of the jaw is no edge, and the teeth are not the "points" we thought them either. Tellingly, this meticulous performance of microscopical skepticism does not impair Gosse's wonder, nor ours. After a final sentence about the positioning of the teeth, he instructs readers (using the imperative mood), "Observe now how beautifully this apparatus subserves the purpose for which it is intended."[40] He guides readers in their wonder here as deftly as he earlier instructed them on skeptical practices of observation.

Such head-snapping pivots are a particular feature of soirée style. Gosse performs the move from admiration back to skeptical instruction just as rapidly as the switch from macroscopic to microscopic vision or the deft rotation of the jaw, adding a lesson about the virtues of juxtaposing disparate perspectives. In describing "the Sea-urchin (*Echinus miliaris*)" he says, "Every part is a wonder.... Take the spines first," then warns, "The whole surface appears to be fluted, like an Ionic column, but this is an illusion." Again, "When we look at [the spine] laterally, the appearance is such that we cannot but firmly believe, that it is grooved throughout with straight and deep longitudinal furrows." A change of perspective, as always, reveals new truths: "But if we break off the same spine transversely, and so exhibit it that the broken end shall be presented to the eye, we perceive that there are no grooves" but that, in fact, the "summits of the ridges" are actually valleys, and the "grooves" are actually convex. These deceptive appearances, says Gosse, "read an important lesson to the inexperienced microscopist, not to decide too hastily on the character of a surface or a structure, from one aspect merely. So many are the chances of illusion, that the student should always seek to view his subject in different aspects, and under varying conditions of light, position, &c."[41] Gosse here restates advice that had been offered by Schacht and, before him, the German botanist Matthias Jakob Schleiden, who advocated for what he called "many-sided comprehension" (*Vielseitigen Auffassung*) in microscopical observations.[42] In thus advocating for juxtaposing "different aspects" or perspectives in microscopical observation, Gosse also implicitly provides a useful metaphor for his own sudden shifts between minute description and enthusiastic description.

Analogies Familiar and Fantastical

While Beale welcomes novices into his text by tailoring his prose to ensure clarity, Gosse solicits readers' interest and caters to their microscopical ignorance with a very different strategy: evoking common experiences through familiar comparisons that highlight elements of natural beauty. Gosse's enthusiastic description is marked by both a "familiar" (friendly, conversational) tone and an emphasis on "familiar" (well-known) objects and experiences. These accord with the supposedly "familiar/family-based" (domestic) setting of his imagined soirée. However, these analogies also often veer into fantastical territories.

Introducing the shell of the cuttlefish, for instance, Gosse uses the second person to activate the readers' memories of seaside holidays, like a teacher encouraging the response of a diffident pupil: "You have seen many times, when walking along the yellow sands kissed by the rippling waves, the shell... of the Cuttlefish. You know that it consists of a shallow boat-shaped shell." He enjoins—more accurately, compels—the reader to participate with him in observing, to imaginatively inhabit the scene as a curious beachgoer or perhaps recreational naturalist. That reader's involvement extends to physical action, since "you know that…the hollow" of this shell "is filled with a white substance, which can be scraped away even with the finger-nail." And he links this holiday memory to a common household item, since this substance "is sometimes used as *pounce*, to rub on paper from which writing has been erased."[43] Gosse introduces another object with a similar, imaginary reminiscence: "When we were at the sea-side last summer we bought, you may remember, of a poor widow whom we met on the beach, a little basket of dried sea-weeds." The woman's figure rises from the page, a wraith from some sentimental novel, only to subside like the tide upon which her livelihood depends, leaving us holding her salvaged wares. Using the imperative mood, Gosse invokes the reader's physical participation in the scene, inviting us to "[l]et us take this flattened brown leaf.… [P]ass your finger over it, and you feel that its surface is evenly roughened." He identifies it in a way that makes it curious, both familiar and strange to us, re-viewing and renaming it over and over again. This seaweed is "Broad Hornwrack"—it is "Leafy Sea-mat"—it is "*Flustra foliacea*"—it is not a seaweed at all but an "animate being"![44] As host of this imaginary soirée, Gosse deftly uses this stream of imagined objects and contexts to direct the ebb and flow of his guests' interest and attention. His casual reminder of the shared experience of seaside memories, his insistence on natural beauty, his references to everyday household objects—all work together here to establish the physical setting of this soirée within an ordinary middle-class English household, and to evoke the sensory pleasures aroused by microscopical work.

Gosse's analogies often include objects associated with household tasks, many of which are gendered. He opens one chapter, for example, by declaring, "You have doubtless often admired the noble Dragonfly, with its four ample and widespread wings of gauze, hawking in a green lane, or over a pool in the noon of summer."[45] Here, the dragonfly itself is familiar, but Gosse also compares its wings to gauze, a material familiar to all for its use in fashion (dresses), household lighting (the wire gauze of lamps), and medicine (bandages). Similarly, he describes the "calcareous and siliceous pins (*spicula*) of a Sponge" with the remark, "If you have ever shaken up a box of dressing-pins, and have then endeavoured to take one out, you know how by their mere interlacement they adhere together in a mass, so that by taking hold of one you may lift a bristling group of scores."[46] He makes comparisons to a range of familiar tools and substances from high life and low, including shovel and pitchfork, silk and velvet, cup and wineglass, basket and vase. When he navigates the microscopic world by reference to these domestic objects associated with sewing, nursing, cooking, and flower arranging—or even non-gendered objects that would be familiar to readers of both sexes (watch and knife, pen and pencil)—Gosse implicitly includes women in his soirée.

Gosse frequently lingers upon elements of natural beauty but, again, describes them with reference to everyday, even domestic objects. He says, "If you are familiar with the pin-cushions which children often make with a narrow ribbon round the edges of these very Scallop-shells, you can scarcely fail to be struck with the resemblance borne by the living animal to its homely but useful substitute; and the beautiful eyes themselves might be readily mistaken for two rows of diamond-headed pins...the ribbon itself representing the satiny and painted mantle."[47] He elides any reference to how "the living animal" died or was killed for its pretty shell, deftly shifting focus to "its homely but useful substitute" the pincushion, in a rhetorical sleight-of-hand that keeps readers at their ease. At times, he describes microscopical beauty by referring to crystal or jewels, which are, if not homely, still well-recognized objects. In describing the eyes of "the Squin (*Pecten opercularis*)," a type of scallop, he directs readers' mental gaze to the beauty of this little mollusk: "as we peer among these slender threads, our attention is riveted by some tiny points...which glitter like brilliant gems....Even the unassisted sight can detect the gleam and glitter of these little specks; but it is only when we bring the lens to bear upon them that we see all their beauty. Then they look like diamonds or emeralds."[48] Here, the lens of the microscope, itself so often troped as crystal, performs the transformative magic that reveals the true beauty of the microscopic world, turning the "little specks" on a humble mollusk into glittering "diamonds or emeralds."

Stringing together such analogies, Gosse bridges the marvelous but unsettling microscopic world to the ordinary human one, implicitly referencing the trope that Amy King identifies as the *divine in the commonplace*.[49] Each comparison generally focuses on a significant detail of the observed image and imagines it as some familiar thing, either a common household object or a very beautiful or valuable (but recognizable) object like a jewel (drawing upon the trope of *microscopical richness*). This was a frequent strategy among popularizers; Robert Chambers invokes "familiar knowledge" at the opening of *Vestiges of the Natural History of Creation*, and George Routledge centered his popular science series launched in the 1850s around "Common Objects."[50] Jabez Hogg, in *The Microscope*, his massive 1854 guidebook for novices, similarly links to a range of common objects when he explains that "[t]he forms of the Infusoria are endless; some changing their shape at pleasure, some resembling globes, eels, trumpets, serpents, boats, stars, pitchers, wheels, flasks, cups, funnels, fans, and fruits."[51] Gosse endorses comparison as a form of investigation, saying that "in all branches of natural history, but perhaps pre-eminently in microscopic natural history,—owing to its greater liability to error from deceptive appearances,—we gain much information on any given structure by comparing it with parallel or analogous structures in other forms."[52] But in *Evenings* these descriptions are not just comparing one hair with another or one polyp with another but comparing microscopic forms to everyday objects. When his descriptions thus contextualize strange creatures using familiar, homely comparisons, readers are entertained and reassured of their own competence for microscopical observation.

Although these analogies draw upon familiar objects, Gosse uses them to dizzying effect, shunting the reader deeper and deeper into strange waters at each step. Isobel Armstrong finds a "leakage of violence and libido" in Gosse's prose, an "obsessive" straining after the limitless detail of God's Creation, provoking John Ruskin's critique of microscopic vision as (she says) an "ungrounded atopic image…lead[ing] to disproportion and grotesque virtuality."[53] To be fair, most of the descriptions in *Evenings* read as cheery, if fantastical, observations subject to abrupt shifts of perspective, but he does occasionally dip into the stuff of nightmare, heedless of where his enthusiasm takes him.

Gosse's comparisons, although helpful and entertaining, can produce a kind of startling dissonance or even humor when he suddenly returns us to the original context. The Broad Hornwrack presents alternating rows of cells "like the meshes of a net," each of these "shaped like a child's cradle" holding "minute white globules, which look like tiny pearls." At each point he holds the comparison up to our gaze, explicitly calling upon our imagination as he asks readers to "suppose" there is a "coverlid [coverlet]" on the cradle, to "suppose" there is a baby in the cradle. Just as each "pearl" features a tiny, "moveable door" underneath, each "cradle" holds "a baby, with its little knees bent up to its chin, in that zigzag fashion that children little and big, often like to lie in." Incredibly, "here is a child moving! He slowly pushes open the semicircular slit in the coverlid" of the cradle and "is raised half out of bed, when lo!" Gosse's fanciful scene goes horribly awry. The imagined baby's "head falls open, and becomes a bell of tentacles! The baby is the tenant-polype!"[54] After this barrage of exclamations, the sweet domestic scene of sleeping infants in cradles is abruptly recoded as Gothic horror and then again as instructive natural history, in a shift that Gosse clearly intends to be whimsical but which reveals the sensational or even grotesque elements of his work as well. Even in those moments that might well provoke fear, disgust, or shock, however, Gosse's own wonder and pleasure infuse the scene.

Gosse occasionally offers a comparison not to familiar objects but to extraordinary ones, as in his extended description of a colony of *Proboscidactyla* (hydroid polyps) ensconced on a marine worm, "that rather rare and very interesting *Sabella*, the *Amphitrite vesiculosa* of Montagu." The colony's tiny, human-shaped "guardian forms" around the edge of the *Sabella*'s tube "are incessantly bowing and tossing about their arms [tentacles] in the most energetic manner." Gosse carefully describes these forms and their position, but he lingers on the eccentricity of their movements, swaying forward and arching back, "toss[ing]" their arms "wildly upward" and down, "as if to mimic the actions of the most tumultuous human passion." These extraordinary gesticulations impel him into an extended fantasy about the little forms, which he has exhibited to his "scientific friends" with "much entertainment." He explains, "When I see them surrounding the mansion of the *Sabella*, gazing, as it were, after him as he retreats into his castle, flinging their wild arms over its entrance, and keeping watch with untiring vigilance until he reappears, it seems to require no very vivid fancy to imagine them so many guardian demons." Such moments engage not just analogy but also narrative; while Gosse here calls up a romantic vision very far from the bourgeois domestic scene that he evokes elsewhere, he links it to a myth that would be familiar to many. "The Lares of the old Roman mythology occurring to memory," he says, "I described the form under the scientific appellation of *Lar Sabellarum*."[55] Indeed, he offers readers their choice of another familiar story instead, saying, "You may, however, if it pleases you better, call them 'witches dancing round the charmed pot.'"[56]

Romance and Realism

Although Gosse seems to relish the whimsy of these fantastical travelogues, frequently calling upon romantic language and fanciful comparisons to build a picture for the reader, he insists that his narrative is, overall, fact rather than fiction. After the emergence of the "tenant-polype" in the Broad Hornwrack, he imagines the reader expostulating, "this is a very amusing romance." He responds, "Nay, it is no romance at all"—indeed, the comparisons "could scarcely be more apt."[57] Again, later in the book, he remarks, "It sounds almost like a fable to be told that the great Spotted Slug, which we sometimes find crawling in damp cellars, carries a tongue armed with 26,800 teeth!" He describes the marvels of the slug's tongue, "armed with thousands of crystal teeth," but concludes, "yet there is no doubt of the fact."[58] And in an extended description of the *Polynöe* (a marine worm), he compares its "bundles of spines of exquisite structure" to "sheaves of wheat," "the armoury of some belligerent sea-fairy," and "the arms of some ingenious but barbarous people from the South Sea Islands" as viewed in a "missionary museum." He describes the "weapons" as lances, scythe-blades, staffs, reaping-hooks, clubs, shark's teeth, spears, and/or a saw. After this dizzying string of analogies, he pauses to reflect, "Though a stranger might think I had drawn copiously on my fancy for this description, I am sure, with your eye upon what is on the stage of the microscope at this moment, you will acknowledge that the resemblances are not at all forced or unnatural."[59] Whether he imagines the reader as a dissident interlocutor or a fellow enthusiast following along with him (in person or at home with the book as guide), Gosse frequently leverages the marvels and exclamations of romance to make a flat statement of fact.

Gosse's writing thus shuttles between romance and realism, interpreting and authenticating the unknown and invisible world by insistently linking it to the familiar, visible one. Although "romancer" was hardly a term of approbation in Victorian times, the tropes of science as a romance and a microscopical fairyland were common among writers for children or general audiences, and Gosse was soon to embrace the term in titling his 1861 book *The Romance of*

Natural History, which features chapters ranging from "The Marvellous" to "The Doubtful."[60]

Not only does Gosse ping between skeptical and wonder-filled description, his plunges into romantic, even sensational flights of fancy often serve as a rhetorical bridge, ushering readers into a more analytic space. For example, when he imagines the marriage plots daily enacted in miniature within the tide-pool, his description of those plots tacks between the romantic and the realist:

> What little worlds are these rugged basins!... What arts, and wiles, and stratagems are being practised there! What struggles for mastery, for food, for life! What pursuits and flights! What pleasant gambols! What conjugal and parental affections! What varied enjoyments! What births! What deaths! are every hour going on in these unruffled wells, beneath the brown shadow of the umbrageous oarweed.[61]

Gosse foregrounds a romantic diction, in his use of apostrophe and exclamation, and he first imagines a scene rife with sensational acts—deception, treachery, struggle, flight and pursuit—in what Lynn Merrill calls a "wilderness."[62] He juxtaposes these, however, with more cheerful occupations—"pleasant gambols" and "conjugal and parental affections," suggesting (as King notes) the plots of domestic realist novels.[63] Finally he pulls back from this magnified view of the tide-pool and frames all these with a broader view, noting the (deceptively) "unruffled" surface of the water, which the naturalist sees darkened by "the brown shadow" of a plant that he—awkwardly falling back into pedagogical voice—identifies as "the umbrageous oarweed." Such frequent shifts in scale and style trace out an eccentric, but (given the success of this book) apparently effective, oscillation of narrative styles in his attempt to describe both the fact and the fancy he encounters in his observations.

Scholars often remark upon Gosse's pleasure in observing nature and the deep faith which animates his work, but they disagree about the appropriate way to classify his eclectic and compelling style.[64] Merrill, examining Gosse's popular writing, identifies his written style as romantic in approach, noting a blend of "highly Romantic diction," "the picturesque," and "a slightly archaic diction" (seen above in the phrase "when lo!"). She remarks that his religious worldview saturates his prose with images of light and transcendence.[65] King, reading Gosse's popular seaside books, identifies several ways in which he expresses his faith through literary realist techniques, particularly his abundantly detailed descriptions and attentive dilation upon those details, his interest in minuteness, and his appreciation for humble and common objects.[66] Terming this approach *reverent empiricism*, King finds that in Gosse, the "proliferation" of detail serves both a scientific and theological function, as well as enabling "the narrative effect of dilation," which she sees as "a kind of reverence" facilitated by the microscope.[67]

It is evident that Gosse's descriptive narratives defy easy categorization. Certainly there are moments of staid realism, but the frame of *Evenings* is unabashedly romantic. Gosse introduces the microscope as a touchstone of romance, invoking hidden worlds and the Oriental tale. The instrument is "the key which unlocks a world of wonder and beauty before invisible," working wonders like "some mighty genie of Oriental fable."[68] Gosse's narrative skill, expertise as a naturalist, and undergirding evangelism allow him to shift easily between fanciful scenes and minute description. Throughout all, his enthusiasm carries him, and readers, on toward a delightful reward.

In noting Gosse's affinity for detail and his dilation upon the details of minute objects, King captures an important aspect of this work. However, it can be argued that Gosse also departs from traditional literary realism in his rhetorical magnification of each detail. While minute description indeed undergirds literary realism by providing a Barthesian "reality effect," in *Evenings* the flood of detail can be overwhelming, even distracting. Isobel Armstrong has said that, for Gosse, "God is detail without end."[69] In this flood of infinitely regressing, fractal detail, Gosse holds up each element of the object as a little treasure, ornamenting his narrative with a series of brilliant sparks of fancy, bedazzling the prose in a way we do not ordinarily associate with realism. An obsessive particularity can read as romantic or even decadent, as George Lukács and Naomi Schor have argued.[70] So while Beale's defined description exemplifies the restrained use of only essential detail that has long been valorized as (in Schor's terms) "strong" and "concrete," masculine and real, Gosse's glittering cascade of details (a term often castigated in the plural) invites diagnosis as decadent and effeminate.[71] But Gosse's evangelical faith motivates and excuses this evident delight in a luxuriant proliferation of detail. For him the detail is never superficial, never superfluous, never noted for its own sake, but always treasured as a pointer to God's overarching plan.[72]

Overall, Gosse layers macroscopic and microscopic, real and imaginary, skeptical and wonderful, familiar and strange, present and past, realist and romantic scenes atop one another. In the latter, the page shimmers and dissolves like "paper from which writing has been erased," its march of printer's type replaced by the palimpsest of memory and fiction. These superimpose the imaginary parlor with its welcoming host and guide, the remembered beach with its shells and seaweed, the sentimental novel, the microscopical structure of the seaweed, and the rapid succession of imagined figures—net, cradle, pearl, coverlet, baby.

When Gosse voices the discursive narrator presenting the homely or sensational comparisons that keep flickering into view, he offers readers a literary version not just of the microscopical soirée but also of the *magic lantern show*. This genre would have been familiar to his readers. It was anchored in technologically mediated visuality, with a lecturer and operator screening a succession of transparencies for a presumably rapt audience. Gosse's changing images sometimes even replicate the transformation effects produced by dissolving slides, as when a seaweed is revealed to be an animal; a leaf becomes a net, then a web of cradles; nestled babies become tentacled polyps. Indeed, in discussing the "Corkscrew Coralline" ("Bird's-head Coralline... *Bugula avicularia*") he says, "You gaze; but you know not what you see.... I must act the showman, and tell you what to see."[73] Gosse's series of directive musings ("you will observe," "you will see," "you discern," "you will not fail to remark") provide a mesmeric accompaniment to the shifting series of scenes he presents to us. While Gosse's expertise underwrote his extensive knowledge of these microscopical organisms, his curiosity, creativity, and evident delight powered the "show" that attracted generations of readers to *Evenings with the Microscope*.

As Gosse's reference to the "showman" suggests, the soirée was but one of a crowd of Victorian visual entertainments. Its aims—scientific instruction, intricate or lively spectacles, playful illusion, miscellaneity, sociality, and the appeal to affect, especially pleasure—were shared in common with many of them. For instance, while the microscopist Mary Ward draws on the miscellaneity of the soirée in her books, she also positions them as another form of Victorian visual amusement, the panorama. Her first book had originated in microscopical soirées she had enjoyed; she recalls, "some years ago, when the beautiful microscope represented in our Frontispiece was a somewhat recent possession of mine, I took much pleasure in exhibiting its wonders to my friends, at the same time explaining the objects seen." *A World of Wonders Revealed by the Microscope* (1859) offers "an illustrated account" of such gatherings. She explains, this "little book, with its coloured plates, aided by minute descriptions, was intended as a substitute for the actual exhibition" of microscopical slides, and notes, "my object was... to present these wonders successively to view in the manner of a panorama" rather than to provide extended practical instruction in microscopy.[74] This metaphor of the book-panorama promises, she suggests, a survey of assorted sights, accompanied by an informative lecturer.

In her next book, *Microscope Teachings* (1864), Ward expanded her soirée-inspired panorama to include more purposive instruction. "My object," she says,

> will be to unite the provinces of the Guide Book and the Panorama; attending to the former, in the hope of making my remarks useful to those who are already in possession of a microscope, while I continue to preserve the latter—namely, the Panoramic method—selecting a few from the multitude of lovely scenes presented by the microscope, in order to attract those readers who, unversed in microscopic marvels, might possibly feel repelled by a complete and lengthened treatise.[75]

The microscope here takes on agency in that, as in a panorama, the "lovely scenes" are selected and "presented" ostensibly by the media form itself without reference to the hidden operator.

As Ward's comments suggest, soirée form was remediated as an implicit form structuring other visual genres—such as frontispieces, photographic displays, magic lantern slides, oxyhydrogen shows, and microcinematic short films—for a broad public. For instance, William Henry Olley, a photographer from Great Yarmouth, published a series of his photomicrographs in *The Wonders of the Microscope, Photographically Revealed by Olley's Patent Micro-photographic Reflecting Process.* This book, likely the first British book illustrated with photomicrographs, was issued in parts for subscribers from 1857 to 1861, then published in volume format in 1861. Olley's photographs present an assortment of objects that would have been familiar to soirée attendees: portions of a fly, spider, butterfly, and other insects; the stems of various plants; diatoms; samples of human blood and muscle. Similarly, a popular set of magic lantern slides ("A Peep into Nature through the Microscope," produced by York and Son of London around 1880) presents a comparable range of common objects with spectacular effect. The technological limitations of the oxyhydrogen microscope meant that oxyhydrogen shows developed their own miscellaneous but iconic set-pieces (fleas, cheese mites, wriggling infusoria), the lively antics of which were themselves remediated in early cinema.[76] The soirée form can be identified in these other media by their miscellaneity, sociality, and appeal to affect. These visual media, unlike table microscopes, allowed observers to view microscopical images in a truly collective manner. They provided the occasion for Victorians to gather in person over an assortment of conventional soirée objects.

Gosse often wrote within the tradition of detailed and quantified microscopical description, but he also modeled a different form of description, one that celebrated the detail by foregrounding his narrator's (and, implicitly, his own) wonder-filled response to it. This juxtaposition of approaches formed an influential convention of microscopical culture, and it likely contributed to the success of *Evenings at the Microscope*. Intermingled with the minute description and articulated practice that build out a skeptical microscopy, his passages of enthusiastic description propose wonder as a principle and a goal. Gosse beckons to new microscopists not only with detailed direction of their practice but also with a buffet of fancy and delight. Over and over again, his book invokes readers' embodied and especially affective participation in microscopy. Whether virtually, at Gosse's soirée, or physically, in their own parlors, he shows them fully engaged in the work: their eye at the eyepiece, hand on the focus, heart leaping with wonder and pleasure.

Writing Embodiment in Victorian Microscopy: Beautiful Mechanism. Meegan Kennedy, Oxford University Press.
 DOI: 10.1093/9780198940623.003.0019

20
Narrative Investigation

Such an "object" on the stage of our microscope tells its own tale.
T. Charters White, *My Microscope and some Objects from my Cabinet*, 1887

Introduction

Earlier chapters showed how Victorian microscopists combined wonder and skepticism *textually*, using literary structures like counterbalancing or the soirée, as Gosse does. The print record shows that these elements could also be *temporally* and indeed *causally* related: some Victorian writers suggested that skeptical practice could enable the observer to access a wonderful sight, or wonder inspire continued labors. Philosophers and scholars have often mused on the role of wonder in motivating skeptical investigation; Sophia Vasalou, for example, cites a strand of thinking that sees wonder as "a temporary irritant to be cured by explanation."[1] Philip Fisher similarly identifies "the drive within wonder toward curiosity, questioning, and the search for explanation" but argues that wonder can reappear later in the investigative narrative, in "a chain of experience built on ever repeated, small-scale repetitions of the experience of wonder."[2] Victorian microscopists also engaged with this question. They used narrative to connect skepticism and wonder, siting them as interrelated elements of a larger story about the entwined work of observer, instrument, and object. While conventions of scientific reporting generally turned away from longform personal narrative late in the century, passages of narrative continue to appear in discrete moments, providing (for example) a trajectory of action and causal explanation, as in articulated practice. These narrative threads also allowed microscopists to describe their experiences during observation. The hybrid literary form that I call *narrative investigation* enables writers to incorporate a range of tropes and practices: skeptical ones such as articulated practice (a step-by-step, meticulous record of the observer's actions) and minute description (a textual record of close observation that deliberately surveys the specimen and its structure), and wonder-filled ones such as theologies of nature and revelation. It also allows observers to depict themselves not only as physical actors embodied within the storyspace formed by the text, manipulating the material elements of apparatus and object; but also as virtual ones, narrating their visual, material, and intellectual interventions as if from actually within the microscopical image.

Albert McCalla, in a presidential address to the American Society of Microscopists in 1883 that was reprinted in the *Journal of the Royal Microscopical Society*, suggests a causal interrelation of skepticism and wonder when speaking on the topic of "The Verification of Microscopical Observation."[3] His concern was to promote a judicious skepticism about microscopical work, but he frames that skepticism by emphasizing its place in a larger narrative about what causes (inspires) skeptical work and what skeptical work itself causes. He suggests, in fact, that skepticism is the road to wonder, and wonder an important motive and reward for methodical work. McCalla takes full advantage of the conventional oscillations of style within a presidential address, a genre conventionally considered to be a rhetorical space set apart from the ordinary papers reported in meetings and journals of a society. Although such a speaker addresses scientific workers, the speech is expected to be itself more a "state of the field" review than a report of scientific work. This form allows—even encourages—McCalla to engage in heightened rhetoric when he invokes the microscopist as both researcher and visionary. For his address is not, as it seems from the title, simply about the need for skeptical "verification of microscopical observation." He argues that wonder adds meaning and purpose to the painstaking work of microscopy; more important, that skepticism brings us closer to, perhaps even into the presence of, the sublime. He says,

> At the present time the Abbe diffraction plate [a test plate] offers itself as a most fruitful field of study, and when we can learn to discriminate without hesitation the various appearances of its squares and rhomboids we can attack anew the mysteries of histology, resolve the diatom frustules in a truer and more perfect sense, investigate the bioplasm theory to a final and satisfactory conclusion, and perhaps discriminate optically between the septic and the pathogenic bacteria, learn the true structure of muscle and the real meaning of its striations, and in a thousand other ways approach a little nearer to an understanding of the mystery of life and the wonderful, beautiful symmetry of the structure of the universe of God.[4]

McCalla here reviews a number of contemporary research questions, each of which demands cautious, methodical microscopical work (signaled by reference to Ernst Abbe's test plate and the use of terms like "investigate" and "discriminate"). He suggests that this skeptical practice, paradoxically, brings the microscopist "a little nearer" to the "mystery of life" and the inherent beauty of Creation.[5] He tantalizes the reader with revelation ("final and satisfactory," "learn the true structure…and the real meaning") but also recognizes the limitations of microscopical work by framing some of its goals in relative, not absolute terms ("truer," "more perfect," "a little nearer"). Overall, he argues that while the microscopic image requires verification by many methods—indeed, it may not even depict real structure—even imperfect observation, if appropriately cautious, leads us toward the sublime. By thus linking skepticism and wonder in this temporal and causal relation, McCalla implicitly invokes an ancient literary device—narrative.

Many texts other than microscopical ones must integrate diverse approaches to focus and mode; narrative fiction, for example, excels in this kind of "gear-shifting." Although there are many approaches to narrative, in its simplest form it is a record of real or imaginary events told by a narrator, set within a temporal frame, and connected in a way that implies causality; a narrative protagonist (not necessarily the narrator) often faces some conflict or series of obstacles to resolve.

Change, momentary or extended, rachets all narratives forward, but writers of narrative retain great power in selecting and organizing the events to be reported. Indeed, the selection of events as relevant and appropriate or not suggests the skill and reliability of any narrator, and it helps determine the genre of any individual narrative. This is why research papers don't often begin, "On 1 May 1872 I mounted the following diatoms..." or "In October 1861 I purchased Nachet's 'petite modèle' microscope with the following objectives..." or "For my eighth birthday in 1850 my father took me to the new Reptile Room at the Zoological Gardens in Regent's Park, where I was fascinated by..."—although a scientific memoir may include such events to illustrate the genesis of a career, or a realist novel may include them to build out a fully developed, specific character and setting.

The writer of narrative also enjoys considerable leeway in structuring the order in which to reveal selected events and their relation. Mary Morgan has argued that time may not always be the most important ground of narrative order.[6] Nineteenth-century narrative investigation, for example, follows not only temporally ordered events in the material world but also mental events representing the investigator's progressive steps in logical reasoning and even, perhaps, emotional events experienced along the way. Time-bound events are often described by a narrative in an order different from their original chronological order; it may not only delay the revelation of certain events or relations but also elide many related but non-essential events. While narratives are of necessity linear, since they are constructed of lines of text that emerge into the reader's consciousness one at a time, narrators often play with a tension between the order of events as they happened and the order of events as they are reported, complicating the movement of readerly attention through time and space via flashbacks, narratorial asides, and elliptical projections forward to a goal or speculation.[7] Narratives thus readily enable an overlapping of diverse perspectives of time and space within the linearity of text. The complex relations between narratives and narrators, and between narratives and plots, have prompted a wealth of narratological scholarship.

Within the non-fictional frame of science writing, a narrative of investigative practice similarly permits the writer to manage different modes of inquiry, including pushing forward in enquiry or pausing for prolonged study, reflection, or appreciation. Narrative allows the investigator to demonstrate how a commitment to skeptical, rigorous method may lead to and underwrite beauty, wonder, and pleasure—which, in turn, help motivate further research. Scholars in literary and cultural studies, narrative studies, and even history of science have contributed to an increasing body of work on how scientific writers use story to structure inquiry.[8] Perhaps most relevant, the Narrative Science project headquartered at the London School of Economics (narrative-science.org) was founded by historians and philosophers of science to pull together relevant research across many disciplines; with the aid of European Research Council funding it has published two collections of essays on narrative as a form of scientific argumentation.

Narrative form undergirds and powers the virtual witnessing discussed by Steven Shapin and Simon Schaffer in *Leviathan and the Air-Pump* (1985). It allows scientific authors to address the branching of possible explanations as an investigator studies an object or phenomenon. John Beatty argues that narrative form allows us to model contingency by registering these "unrealized possibilities."[9] Narrative is also useful in scientific writing because it "works to create coherence between a variety of different elements that otherwise do not appear to hang together."[10] The capaciousness of narrative allows microscopical discovery narratives to enfold and link many different approaches, such as articulated practice (itself a narrative form), minute description (narrative when it describes movement or change), enthusiastic description (also strongly narrative in its use of analogies and anecdotes), theoretical or conceptual contextualization, logical argumentation, and more-or-less static moments of wonder, within a frame story about an investigator working on a problem. Brian Hurwitz, speaking of the medical case history, characterizes its narrative as composed of "controlled disclosures" structured to support a particular diagnosis. An experimental report similarly selects and organizes its elements to support a satisfying tale of mystery and revelation.[11]

It was not unusual for Victorian investigators to turn to the familiar structure of narrative to frame their research. In a study of *The Strand*, James Mussell has argued that some late-Victorian popular science writers present the scientific process as a kind of detective story, just as H. G. Wells had suggested when he praised Sherlockian fiction as a performance of step-by-step ratiocination and a good model for popular science. The difference, as Mussell puts it, is that "the scientific author must…act as both murderer and detective—must deliberately stage a manoeuvre that renders the familiar strange before resolving it through scientific knowledge."[12] This analogy allows both Wells and Mussell to spotlight the importance of the reasoning process, which Holmes calls "deduction," in both detection and science. Inductive reasoning is considered central to original scientific investigation, and the Holmes stories even figure the detective as a scientific researcher committed to rationality above all else.

There are some important differences between detective stories and narrative science. In the Holmes stories—which are, after all, a form of romance—the end (the solution) is foreknown and always achieved. This premise of solvability overstates the role of reason in Victorian scientific work, misleadingly portraying that work as an untangling of problems that can always be resolved, with clear answers that are just temporarily obscured. Furthermore, while both scientific work and detective work are often associated with puzzle-solving, scientific articles were not strongly driven by hypothesis until well into the Victorian period. Still, narrative permits the writer to organize diverse elements in a temporal relation that argues for causality.

Plot-Excitement and Narrative

Speakers often use narrative to evoke a sense of suspense. In *The Emotions and the Will*, Alexander Bain calls "Plot-interest" or "Plot-excitement," a form of "Pursuit" where spectators avidly follow the experiences of some protagonist, sharing the aims and emotions that they observe.[13] He notes that"[t]he narrative of a chase, a battle, an adventure, places the hearer under the dominion of the emotion before us"—indeed, the spectator "catches the fire of the actual witness." Bain finds this phenomenon in a wealth of domains—not just sport but also "a lawsuit, a contested election, a debate"—but the term *plot-interest* associates the phenomenon most strongly with narrative fiction.[14]

Bain emphasizes that plot-interest can seize us whether based on our own experience or imagined ones. He sketches a delightful scene from the schoolroom, where, "[c]uriosity once awake, the earnest student is all eye, all ear, until a mystery is unravelled, a truth demonstrated, a discovery achieved." The sense of "pursuit" also holds seasoned investigators in its thrall. Bain explains, "Original research introduces an element still more capable of arresting the forward regard." A research interest is "by no means a gentle excitement, but capable of making up a powerful life interest, and yielding moments of almost unnerving suspense."[15] As Lewes recalls of his first studies at the microscope, "The hours I spent thus, fled like minutes, and left behind them traces as of years."[16]

The investigator's absorption in microscopical work, where time and space seem to temporarily collapse or expand, in many ways echoes that of the novel-reader. This is perhaps especially true of microscopical work, which (like fiction) can immerse the observer in imagined worlds. Scholars have found this intense, intermittent form of attention in a range of Victorian literary genres. Leah Price remarks that "two paces of reading—a leisured appreciation of beauties and an impatient, or efficient, rush through the plot" characterize a variety of genres, beginning (she argues) with the anthology and developing through "genres like the gothic novel...the life-and-letters biography...and even the tourist guidebook."[17] Nicholas Dames similarly shows how Victorian novel-reading evokes a "wavelike, in-and-out, temporally restricted capacity" of attention that was also articulated by the physiological psychologists.[18]

In microscopy, this attentional rhythm is complicated by the changes in scale that structure observation. The wonderful scenes of the microscopic world, while they may invite the mind to drift, also require intent, focused study. And skeptical practice involves not only periods when the observer dives into close attention to the object, as in minute description, but also moments of frustration or error, or pauses to adjust the apparatus, which break the spell of concentration. Narratives of microscopy allow writers to navigate this uneven texture of attention and experience within a familiar, commonsensical frame.

Narrative Investigation

What I call *narrative investigation* elaborates the movement from uncertainty to provisional knowledge in a discovery narrative that lays out, among other things, the causal interrelation of skepticism and wonder in an investigator's process of action. Unlike Doyle's detective narrative, narrative investigation overtly acknowledges the importance of uncertainty and serendipity and accommodates the role of emotional and aesthetic experience in scientific work. It provides an overarching narrative frame that can shift back and forth between diverse approaches, often by spatializing the image—describing the flat microscopical image as a space within which the observer moves—a strategy that reframes the conventional trope of *argument as journey* to make it applicable to scientific investigation as well.[19] Here the observer variously negotiates movements of cautious action and moments of admiration, whimsy, or even the sublime, in just the kind of intermittent recurrence of wonder that Fisher theorizes. The record of these diverse movements reports the observer's uneven progress toward an unknown end.

Bain dedicates a section of *The Emotions and the Will* (1859) to the notion of pleasure in difficulty. He says, "When some very congenial exertion on our part produces an effect also very gratifying, the confluence of the two pleasures must needs beget an intense delight," adding, "It is this element of difficulty that imparts a meaning to skill in operations."[20] He frames intellectual work as a "pursuit" with inherent pleasures; as he says, "There is...something very interesting and engrossing in the working for an end."[21] Bain frequently locates such delight in scientific investigation. For "a feeling of elation may be generated by the discovery of new facts or principles, by the possession of extensive knowledge, or by the opportunity of promulgating ascertained truth. The pleasure thus derived is shown by the desire to get at new truths."[22] He lingers on the narrative suspense (the "desire to get at…truths") inherent in intellectual work, framing this work as a compelling journey where the investigator "pursues" some particular end. He reminisces, "The mere fact of gradual approach satisfies the mind and stimulates the exertion. As we come nearer, the interest is heightened, the attention becomes more arrested, and the voluntary activity more sustained. At last full fruition crowns the labour; and the pleasure is then made up of the actual enjoyment, and the elation of successful endeavour and difficulty overcome."[23] The language of "approach" here not only elaborates the trope of *investigation as a journey* but also allows Bain to play up the role of suspense and what he elsewhere calls "plot-excitement" in intellectual work.[24]

At times, knowledge readily unfurls under the lens as the *object tells its own tale*. Usually this trope describes an object for which the structure is easily legible to the microscopical eye or can be readily inferred. T. Charters White uses this trope in my epigraph to refer to a delicately engraved test plate produced by F. A. Nobert; the "tale" this slide tells, he says, is "of the skill and patience necessary for its production."[25] Similarly, the microscopical illustrator Tuffen West, in an article on how to mount various objects for best results, argues that "fish-scales" should be taken from just above and below the fish's lateral line and mounted accordingly. "Placed on the slide in this way," he says, "they tell their own tale" about the formation of the scales in relation to the anatomy.[26]

However, some pansies dissected by the American writer Sophie Bledsoe Herrick "told" a mystery seemingly without solution. Herrick had spent "many a sweet bright morning…in trying to penetrate the mysteries of the fertilization of our common garden flowers," *Viola tricolor* (wild pansy) among them. "Pansy after pansy was devoted to dissection," she recalls in 1884, "but all told the same [misleading] story," seemingly rejecting any possibility of "cross-fertilization by insect agency." Finally, Herrick relates, she found the answer in a print publication instead, "Sach's wonderful 'Handbuch,'" which explained how the "lower lip of the stigma" allows insects to cross-fertilize the blooms.[27] Herrick's chronicle demonstrates how a discovery narrative often acts as a frame story consolidating multiple attempts at knowledge, by showing the investigator working through one tale (explanation of structure) after another. Such narratives form a reminder that the microscopist must be willing to replace one tale with another, in a constant process of self-correction.

In puzzling cases like Herrick's, the final version of the tale is often told not by the object itself but by other actors in the story. J. Edward Smith, an American histologist, uses this trope in 1880 to commend some fine lenses for their quality. Discussing the well-known *Podura* test (the challenge of resolving the true structure on the wingscales of the *Podura plumbea* moth), he comments, "Hardly two of the 'real honest working glasses,' owing to the hap-hazard character of their corrections, will give the same appearances" of the *Podura* scale, "while the objectives of Spencer and Tolles will have but one and the same constant story to tell." The consistency of the image across diverse observations shows that these high-powered American lenses are well constructed—an important point during this period, when many British microscopists still resisted using the wide-angled American objectives. "With these [lenses], properly manipulated," Smith says, "it is impossible to get aught over the podura but the exclamation points" that were the best contemporary resolution of *Podura* wingscale structure.[28]

Shifting Scale and Suspense: Lewes

Of course, the investigator-writer usually plays an active role within the story. George Henry Lewes's study of color change in the squid makes particularly effective use of narrative framing, as his narrator shifts between macroscopic and microscopic scenes, and physical and intellectual movements. This appears in his popular book *Sea-Side Studies*, which was first serialized in *Blackwood's Edinburgh Magazine* (1856–7), then published in book format, the cheapest edition of which could be purchased for 6*s*. 6*d*. He makes particularly vigorous use of narrative flexibility to (like Gosse) execute a series of dizzying shifts of scale and perspective in his observation of the squid, linked along a thread of suspense as he seeks the answer to a puzzling physiological mystery. His discussion even begins with such a shift, when the narrator (ostensibly Lewes himself) finds a dead specimen along the shoreline and notices an unusual feature of its appearance. "Yet what is this?" he says. The specks puzzlingly vanish and reappear: "the colour-specs are coming out on the skin, like stars appearing at night, and now the whole surface, which was pearly-white, is of a variegated hue."[29] Lewes here turns from the setting of a real beach to an imaginary starscape, laying out both a scenario and a research question in the first of several swings between macroscopic and microscopic, real and imagined settings. He then places himself back in a real-life macroscopic scene (now working at the microscope), then swiftly dives into microscopic space. He says, "On placing a small strip of the skin under the Microscope, I was surprised to find two or three of these colour-specs expanding and contracting with great vigour." This description of the skin both connects its microscopical features to the previously observed macroscopic ones ("these colour-specs") and transforms them; in their vigorous pulsing they look little like the tiny specs that had gently emerged "like stars appearing at night."

Lewes's prose shifts not only between macroscopic and microscopic settings and features but also between different modes of inquiry and different "tales" (explanations), in an example of how narrative investigation (as Beatty suggests) allows the observer to try out multiple branching hypotheses. Acting as a frame story, narrative investigation shows the investigator pursuing one possible route of explanation after another. At one point Lewes pauses to reflect, recalling that Alcide d'Orbigny argues that color change in the *Cephalopoda* "*always depends* on the will of the animal." His interest is piqued by the apparent evidence before him that contradicts d'Orbigny's claim. He eventually shuts down d'Orbigny's claim as "preposterous," decrying the general lack of investigative rigor in zoology. Returning to his own work, his narrative follows up multiple possibilities with a blend of articulated practice, minute description, and reasoning. He explains, "At first I thought [the pulses of color] must be an optical illusion, but on close attention it became too decided for doubt; and not suspecting the truth, I concluded that some animalculae were imbedded in the tissue, and that their movements produced this apparent activity of the globules." He here offers and rejects two possible explanations: optical illusion and parasites embedded in the sample. Again he sets himself in front of the microscope and describes at length what he sees, calling on minute description as he does so: "To settle these doubts, two other strips underwent examination; in both of these, all, or almost all, the specs were in activity, shooting out prolongations, and retracting again—two specs sometimes seeming to run into one, but really overlapping each other, and sometimes a point not bigger than a millet seed expanding to the size of a sixpence, growing fainter in colour as it expanded." Satisfied, Lewes concludes, "This was decisive. If the globules in a strip of skin taken from the dead animal manifest precisely the same contractions and expansions which they manifest on the living animal, it is clear that their activity does not depend upon the 'will.'"[30] Narrative form here allows him to shift perspective and tactics repeatedly, modeling both investigative practice and theoretical argument. The observer prepares the specimen and apparatus, examines it, pauses to adjust the apparatus, continues on a new path, errs, pauses for reconnaissance and course correction, backs up to a lower magnification, moves forward again in surveying the object, and pauses to survey the scene with description and awe, or assessment and analysis.[31]

Narrative permits Lewes to report his diverse actions as investigator—his physical and logical movements and pauses—by drawing attention to them as a connected process, explaining for example that the observation of an oddity "leads" him onward "to further investigation." He constructs this passage as both argument, leading the reader through his alternating observations and conclusions using what Alan Gross, Joseph Harmon, and Michael Reidy have called "inferential explicitness," and narrative, telling the story of how the investigator (Lewes himself) pursued the mysterious source of these "stars."[32] Lewes is tempted into some wrong turns (optical illusion, embedded animalcules), but he regains momentum by appropriately skeptical double-checking. He places his movements and thoughts along a chronological chain of action, with temporal anchors such as "on placing a small strip," "At first I thought," "on close attention it became too decided for doubt," "not suspecting the truth, I concluded" (a wrong move!); then, "to settle these doubts, two other strips underwent examination" and finally, "This was decisive." At the beginning of the passage, Lewes himself is the protagonist as he pursues his investigation. At the critical moment, he shifts back offstage and lets the facts speak for themselves in the passive voice, with a pause for elaboration of what was now "clear."

Lewes's investigative narratives in *Sea-Side Studies* were not entirely successful; the book sold well but did not succeed in establishing him as a naturalist. T. H. Huxley's devastating critique in the *Westminster Review* of Lewes' 1853 book *Comte's Philosophy of the Sciences* is famously cutting, and *Sea-Side Studies* likewise attracted negative reviews for its brash claims and many errors.[33] Lewes struggled to claim an authoritative narrative voice in *Sea-Side Studies*, zigzagging between whimsy, travel narrative, claims from his own original research, and bold attacks on earlier workers. He was not yet widely accepted as a physiologist; he had attended medical classes as a young man but had abandoned medicine for a career in literature and criticism instead. These did not offer much preparation for his physiological investigations later in life. However, they allowed him to hone his fluency in multiple languages, facilitating his deep reading of Continental researchers, and they lent him skill in framing narrative as a form of argument. His *Physiology of Common Life* (1859–60) was better received, and Lewes eventually gained acceptance as an interlocutor on these topics—he was, for example, a founding member of the Physiological Society in 1876 alongside Sanderson, Huxley, Galton, Foster, and others. The hefty *Problems of Life and Mind* (1874–9) still exhibits traces of the narrator of his earlier work—nimble, persuasive, straightforward—but now paired with a judiciousness and gravitas acquired through years of microscopical work.

Affect: Dallinger

It is not surprising to find narrative investigation in Lewes's book, as he overtly adopts the travel narrative to structure his popularizing essays. However, it can frame research texts also. William Henry Dallinger uses narrative investigation in a research report presented to the Royal Microscopical Society in December 1876 and published as the lead article in the *Monthly Microscopical Journal* of January 1877. Dallinger was a Wesleyan Methodist minister and respected microscopist, who would later become President of the Society. His article on the *Navicula rhomboides* shows how even an elite microscopist addressing a research report to his peers may frame his investigative work in narrative form. Dallinger's use of narrative investigation allows him to infuse his descriptions with his own affective experience, an emotional charge that implies his own presence actually within the microscopic scene. This article examines the relation of three species of diatoms (single-celled algae with intricate glassy shells): *Navicula crassinervis*, *Frustulia Saxonica*, and *Navicula rhomboides*. The first two, originally misidentified as unique species, had recently been shown to be simply other forms of *N. rhomboides*. However, the range of size and scale now demonstrated by this more broadly defined species meant that "*N. rhomboides* is a most uncertain and unreliable *test*," and Dallinger wished to determine the precise value of each sub-species as a test object.[34]

Dallinger begins the report of his observations by alternating between articulated practice and minute description. He uses articulated practice to detail his use of Powell and Lealand's "new formula" 1/8-inch immersion lens and a "broad-wicked paraffin lamp with the flame turned edgeways to the condenser" to illuminate the specimen of *N. crassinervis*. He takes a leading role in such passages as he records his interventions with the apparatus and slide. Equipment in place, he shifts into minute description, recording the view for readers to observe. He continues this oscillation with each successive specimen, marking each new phase of his investigation with terms like "next," "now," and "finally."[35] This preliminary section of his narrative relies almost entirely on skeptical practice.

Dallinger signals the climax of this discovery narrative by coloring his minute description with heightened (more extreme and poetic) language. He projects his own affective response *at* the microscopical image *into* the image. If magnification brings the eye "closer" to the object, as microscopists insisted when explaining optical theory to their readers, then Dallinger's emotional investment in the image, coupled with his exquisitely minute description, implies that he has extended his persona as appreciative observer to an almost unimaginable proximity to the object—actually into microscopical space. Dallinger procures "some remarkable specimens of *Rhomboides*" and exclaims, "The result is given at E, Plate CLXVI., which may be taken at once as the interpreter of all the rest, and the witness of their oneness." These last samples of *N. rhomboides* demonstrate that all these specimens belong to the identical species, but the phrase "witness of their oneness" attributes an absolute truth and even, perhaps, spirituality to their resemblance. At this triumphant moment in his argument, Dallinger encourages admiration of the beautiful image, pausing to elaborate upon the view. He explains,

> The striation in this case is everywhere resolved into rows of hemispheres, placed in rectangular order; and I know of few more beautiful objects in "still" microscopy than this. The frustule [shell] is of exquisitely perfect form, slightly tinted with yellow-brown: the hemispheres are everywhere sharply and clearly separate, and the continuity of this is unbroken from end to end.[36]

Dallinger's close description places both observer and reader right in the midst of these peaks. This passage serves to establish the authority of his observation with its "sharply and clearly separate," "perfect" forms but also to infuse this scene with his own admiration for the beauty of the specimen. Dallinger moves on from this rhetorical pinnacle, continuing to alternate between articulated practice and minute description. He directs the reader through a tricky paragraph bristling with instructions on how to change the objective, stage, illumination, and eyepiece for the next observation. Then he turns again from practice to description and gazes upon another view of the diatom, which he describes with such an emphasis on clarity and color that it seems as if he and the reader overlook it directly. He says, "the frustule itself, especially if seen amongst brilliantly white diatoms, assumes a pale sapphire tint, and the hemispheres with skilful management are beautifully developed."[37] This scene again establishes his own "skilful management" in gaining such an image even as he inflects the scene with his own sense of the beauty of the frustule, seen close up. Here the physical manipulation of the object and apparatus respond to, express, and shape the investigator's cerebral and affective work. Overall, a narrative frame such as this enables an elegant interpellation of visual perception, mental deliberation, and the material work of practice. It accommodates diverse approaches to the work and allows the observer's affective experiences to suffuse it.

Movement and Stasis

Narrative investigation provides a useful model for reading microscopical texts because so often, when sharing or circulating the microscopical image, practitioners embedded narratives in that image. They could overlay the story of their work as investigators onto a tour of the microscopic world of the object; their record of events builds an alternating rhythm of practice and discovery using articulated practice and minute description. Articulated practice is already formatted as narrative and can easily drop into a larger narrative frame along with the minute description that marks close study of the object. When a text enters into either of these textual modes, time slows down and the focus of the text narrows in as the speaker lingers over details of action and material. This slower pace can feel like a pause but in fact both articulated practice and minute description present a record of movement, even if that movement is only in the shifting attention of the observer, which may move "forward" or dive "down" into stronger magnification, track "laterally" across different structures or portions of the object, or step "backward" to resume an earlier perspective. These moments of deliberate, elaborated process can appear as isolated set-pieces or can launch a new stage in the speaker's overarching investigative process. They help the observer model appropriately skeptical practice, but they also appear at times interlaced with wonder.

While microscopical narratives sometimes—as in lifecycle studies—follow a story arc centered on a natural object, narrative investigation centers on the physical and intellectual work of the investigator and models his or her entwined processes of observation, experimentation, reasoning, and discovery.[38] Skepticism and wonder call for different approaches to the observer's story. In skeptical passages, observers press ahead with action and assessment, eschewing speculative leaps; wonder-filled passages, on the other hand, are more likely to describe a pause for admiration or awe. For this reason, wonder appears more frequently threaded through moments of minute description, a textual strategy that already privileges elaborate description achieved by a static or slowly surveying mental focus.

In the examples above, in contrast to the many elements of narrative investigation that depict intellectual work as movement, the passages invoking wonder and the sublime often represent the moment of awe as a stasis—a pause in narrative action. Although the admiring gaze may survey different aspects of the image much as minute description does, the tone of the language suggests not movement so much as rest, enabling a deeper and more intent gaze upon the details of the lovely scene. Such moments invite the reader to accompany the observer in visually and virtually tracing through a microscopical image, share the rush of emotion generated by close proximity to such a beautiful object, and affirm the legitimated pleasures of microscopical work.

Similar moments occur macroscopically as well; Lewes, about to dip specimens out of a rock-pool, exclaims,

> Really this pool is enchanting! How gracefully the Polypes wave from its sides, like fairy fir-trees in the summer air. The longer we look, the more beauties and wonders we discover. I have just detected an *Ascidian* standing up like an amphora of crystal, containing strange wine of yellow and scarlet; and crawling about the root of that Oar-weed, I see various Annelids of great beauty.[39]

Some passages of bestilled wonder at the natural world represent the agency of the narrative voice as puny in the face of beauty and infinitude. Microscopical texts often embed these brief moments of wonder and inaction between passages that represent intellectual movement and agency.

Amy King identifies a strategy in naturalist writing that she calls *dilation*—the "accumulation of detail" that slows down or pauses "narrative momentum" in an act of "reverent attention." She identifies the "pleasure in looking" that moments of wonder express and solicit.[40] Though not particularly reverent, Lewes's description of the charming rock-pool, above, offers a secular version of King's dilation, a narrative pause to appreciate the view. Of course, Lewes is not a "reverent naturalist describ[ing] with especial attention what is humble and common" but a would-be elite microscopist engaging in a theoretical struggle with other investigators. Still, his appreciative pause allows him to isolate and exalt this moment of his practice. He celebrates the beauty of this little world (implicitly inviting lay readers to take up such work) even as he also casually references the scientific nomenclature of various organisms (thus marking his own expertise). For the evangelical Gosse, "*Description comes from the desire to be reverent*," but for observers like Lewes a pause for elaborated description may serve secular personal and professional goals.[41]

Revelation: Gosse

Victorian microscopical culture developed many of its new practices around the need to compensate for the inherent difficulties of microscopical vision—to mitigate skepticism and enable reliable investigation. But skeptical systems also enabled one of the most powerful and persistent tropes of wonder: the *revelation*. As Ernst Abraham Hart wrote in *Household Words* about the great botanist Matthias Jakob Schleiden, he "travelled through a tangled forest of prickly and entwined facts, till at last he saw the light, and could proclaim it."[42] Isobel Armstrong identifies a pattern of successive revelation in Victorian microscopy; she notes how Gosse "works through a series of corrective descriptions of each organism, or rather, each part of an organism," at each step "recalibrating the visual meaning of an object by discovering further detail and complexity." She identifies "a familiar expositional move" in which Gosse describes a sight, then explains that "what we see is not *in fact* the case; there is a further revelation still that reorganises the prior knowledge."[43] (This technique drives his exploration of the Broad Hornwrack, discussed in Chapter 19.) Gosse, a fervent member of the nonconformist Plymouth Brethren, was fully committed to Revelation theologically; the Brethren are biblical literalists, fixed on the Revealed Word of God. It is not surprising, then, to see Gosse using the trope of revelation frequently in his work as a naturalist, leveraging microscopical revelations as signs of God's promises to us.

Armstrong finds in Gosse an "authoritarian stance" of heavy-handed correction,[44] but in the movement of the "reveal" I see rather an elegant example of how narrative investigation figures knowledge work as a many-branched narrative filled—like a Victorian novel—with a wealth of protagonists and plots. Narrative investigation productively manages any potential friction between skepticism and wonder by accommodating many illusions, many revelations. For Gosse each revelation leads to another. His continued corrections rely upon and extend the cautious, self-correcting stance of the skeptical observer, always aware that any particular view through the microscope must be considered provisional, may shift with any change in objective, illumination, or observer. The observer must accept that "true vision" on this earth may be an ever-receding, perhaps unattainable ideal. Gosse's microscopical revelations are not themselves final; the observer must follow their lead onward to the one truth of God.

In texts, this approach powers that quintessentially wonder-filled move, the "reveal": with a quick twiddle of the apparatus, *this* is revealed as *that*, A is really B, a cuttlefish shell is actually an iceberg—no, actually, a railroad station—no, it's a cuttlefish shell after all. The novelist George Eliot comments, in this famous passage in *Middlemarch*,

> Even with a microscope directed on a water-drop we find ourselves making interpretations which turn out to be rather coarse; for whereas under a weak

> lens you may seem to see a creature exhibiting an active voracity into which other smaller creatures actively play as if they were so many animated tax-pennies, a stronger lens reveals to you certain tiniest hairlets which make vortices for these victims while the swallower waits passively at his receipt of custom.[45]

Eliot here, in a classic statement of realist skepticism, reminds readers of their own fallibility in interpreting what seems clear to the eye but is in fact constrained by the choice of one objective over another. The narrator endorses the pleasure and wonder of the "reveal" but goes on to suggest that the forms revealed by the "stronger lens" might themselves prove to be utterly different under a yet stronger objective.

Eliot's description plays on microscopical skepticism and offers a mordant critique of societal misreadings, reminding readers that (to quote her 1859 novel *Adam Bede*), "Falsehood is so easy, truth so difficult." Yet Victorian microscopists celebrated their fragile revelations, acknowledging the error and the marvel in a single turn of the lens. Narrative investigation enables microscopists to express and manage both skepticism and wonder as complementary approaches to the object; to place themselves both outside the image, managing it, and within the image, admiring it. This textual strategy persists through the seismic changes in physical microscopy that characterized the Victorian era. Microscopists writing from a range of contexts turn again and again to this rhythm of juxtaposed skepticism and wonder to narrate difficult passages of microscopical practice and express peak experiences of curiosity and awe.

Writing Embodiment in Victorian Microscopy: Beautiful Mechanism. Meegan Kennedy, Oxford University Press.
 DOI: 10.1093/9780198940623.003.0020

APPENDIX I

List of Abbreviations

BAAS	British Association for the Advancement of Science
BJHS	*British Journal for the History of Science*
EMWS	*English Mechanic and World of Science*
JMNS	*Journal of Microscopy and Natural Science: The Journal of the Postal Microscopical Society*
JPMS	*Journal of the Postal Microscopical Society*
JQMC	*Journal of the Quekett Microscopical Club*
JRMS	*Journal of the Royal Microscopical Society*
MMJ	*Monthly Microscopical Journal*
MSL	Microscopical Society of London
NMMN	*Northern Microscopist and Microscopical News*
PASM	*Proceedings of the American Society of Microscopists*
PBNS	*Proceedings of the Bristol Naturalists' Society*
PMS	Postal Microscopical Society
QJMS	*Quarterly Journal of Microscopical Science*
QJSLA	*Quarterly Journal of Science, Literature, and Art*
QMC	Quekett Microscopical Club
RCS	Royal College of Surgeons
RMS	Royal Microscopical Society
RTS	Religious Tract Society
SDUK	Society for the Diffusion of Useful Knowledge
SPCK	Society for Promoting Christian Knowledge
TMSL	*Transactions of the Microscopical Society of London*

APPENDIX II

Microscope History Timeline

1802 Thomas Wedgwood and Humphry Davy report Davy's experiments with fixing the image from a solar microscope.

1807 William Hyde Wollaston develops a version of the camera lucida with a Wollaston prism, which will prove important in the later development of polarizing microscopy.

1815 Giovanni Battista Amici develops his reflecting (catadioptric) microscope on the model of a reversed telescope.

1817 Joseph von Fraunhofer publishes an essay on dark lines in the solar spectrum. His work would be valuable in assessing the quality of optical lenses.

1818 Christian Gottfried Ehrenberg establishes that fungi derive from spores rather than spontaneous generation.

1821 September. John Herschel visits Fraunhofer in Bavaria to study his methods. He learns how to view Fraunhofer's spectral Iines but is not permitted to know his secret techniques for making optical glass.

1824 C. R. Goring selects a series of natural objects to serve as tests for the microscope.

1825 George Birkbeck gives first demonstration of the oxyhydrogen microscope.

1826 Joseph Jackson Lister commissions William Tulley to build a compound microscope that Lister had designed to accompany a new achromatic lens system. Tulley asked James Smith (later of Smith and Beck) to make the brass stand, completed in May 1826.

1827 William Nicol produces sections of fossilized wood.

1827 Joseph Jackson Lister and Tulley collaborate on a compound microscope. Lister and Thomas Hodgkin use it to examine the microscopic structure of tissues (histology).

1828 Robert Brown, a Scottish botanist, publishes a pamphlet on Brownian motion (the random movement of tiny particles suspended in a fluid).

1828 Nicol uses Iceland spar and Canada balsam to construct a polarizing prism, facilitating polarized light microscopy.

1829 Wollaston's doublet and Wollaston's condenser are described (after his death).

1829 Goring and Andrew Pritchard publish *The Natural History of Several New, Popular, and Diverting Living Objects for the Microscope*, the first of their many volumes on microscopy and the first major (illustrated) publication on microscopy in the nineteenth century.

1830 January. Joseph Jackson Lister publishes "On Some Properties in Achromatic Object-Glasses Applicable to the Improvement of the Microscope" in *Philosophical Transactions of the Royal Society* 120 (1830), 187–200. This article communicates a new design of compound microscope with greatly minimized spherical and chromatic aberration.

1831 Brown names and recognizes the importance of the plant cell nucleus.

1832 Slide-makers John New and James. R. Bond are the first to use Canada balsam in their work.

1835 Ehrenberg coins the term "bacillus."

1835–7 Adoption of thin glass cover slips.

1837 Andrew Ross suggests the addition of an adjustable collar to correct for thickness of cover glass.

1837 Joseph Bancroft Reade's method of dark field illumination published in Goring and Pritchard's *Micrographia.*

1837 Godefroy Engelmann (France) patents chromolithography.

1837 Matthias Jakob Schleiden argues for attention to the life-history of the individual plant cell.

1838 Ehrenberg argues that infusoria enjoy the use of organs (stomach, eggs, sexual organs) analogous to those of larger animals. Proven wrong by Felix Dujardin in 1841.

1838 Ehrenberg shows that chalk strata are composed of immense tranches of fossilized foraminifera.

1839 Theodor Schwann argues for the basic doctrine of modern cell theory, the homology between plant and animal cells as fundamental building blocks of the organism.

1839 January. Louis Daguerre exhibits daguerreotypes to the Académie des Sciences, including one of a spider taken through the solar microscope.

1839 January. Henry Fox-Talbot exhibits two photomicrographs (a section of wood and an insect wing) to the Royal Society, both taken using a solar microscope.

1839 September 3. Founding of the Microscopical Society of London (MSL): seventeen gentlemen including Edwin Quekett, Joseph Jackson Lister, Reade, Nathaniel Bagshaw Ward, James Scott Bowerbank, George Loddiges, and Richard Horsman Solly meet at Quekett's house (50 Wellclose Square, London). First full meeting, December 20, 1839. First President, Richard Owen.

1840 Friedrich Gustav Jakob Henle publishes a series of criteria by which to identify the causation of disease. These were revised by Robert Koch in 1890 to become the Henle–Koch Postulates of Infectious Diseases.

*c.*1840 Johann Evangelista Purkinje coins the term "protoplasm."

1840 January 29. The MSL establishes standard sizes for microscopical glass slides (1 × 3 inches and 1.5 × 3 inches).

1840 February. Alfred Donné takes photomicrographs, displays them to the Académie des Sciences.

1840s Amici experiments with oil and (mostly) water immersions.

1841 First issue of the *Microscopic Journal* by the MSL.

1843–7 Richard Marsh Hoe (New York City) develops the lithographic rotary press (printer), increasing the speed of print production. Patterned papers to be used for slide covers and labels become popular.

1844 *Microscopic Journal* becomes *Transactions of the Microscopical Society of London.*

1844 April. Donné and Léon Foucault demonstrate the photoelectric projection microscope to the Académie des Sciences.

1844–5 Donné publishes his *Cours de Microscopie* with accompanying *Atlas*—the first published medical photomicrographs.

1844–64 Bowerbank hosts a microscopical salon at his residence at Park Street, Islington (afterward Highbury Grove) every Monday evening.

1845 Repeal of Glass Excise Tax in Britain, reducing the price of glass. Microscopists increasingly use glass slides.

1845 F. A. Nobert produces his first test plate.

1846 Arthur Hill Hassall publishes *The Microscopic Anatomy of the Human Body, in Health and Disease.*

1851 G. E. Blenkins, teaching at Lane's School of Anatomy and Medicine (near St. George's Hospital, London), begins a course on practical histology, in which each medical student learned to use a microscope and prepare his own specimens. Each student had access to a single instrument during the course.

1851 Frederick Scott Archer invents the wet collodion process on glass plates—allowing prints with fine resolution, accelerating the publication of print photomicrographs, and spurring more widespread use of photomicrography.

1853 Edwin Lankester and George Busk launch the *Quarterly Journal of Microscopical Science.* The first volume includes the first known photomicrograph published in a scientific journal.

1853 Smith & Beck advertise the double nosepiece designed by Charles Brooke

1854 John Riddell, Tulane professor of chemistry, publishes on the binocular microscope.

1855 Rudolf Virchow, following work by Robert Remak, articulates the proposition *omnis cellula e cellula* ("every cell is derived from a [pre-existing] cell").

1856 William Henry Perkin, a chemistry student, discovers mauveine, the first aniline dye. Aniline dyes, with their intensity and staying power, would later greatly accelerate the practice of staining samples for better contrast under the microscope.

1857 The MSL designates a screw size to be recommended as a standard ("universal screw") for fittings to the microscope.

1858 Rudolf Virchow publishes the influential treatise *Cellular Pathology* (*Die Cellularpathologie in ihrer Begründung auf physiologische und pathologische Gewebelehre*).

1858 Joseph von Gerlach, a German anatomist and microphotographer, develops an effective staining technique on brain cells using carmine (a natural dye).

1858 Henry Clifton Sorby publishes "On the Microscopical Structure of Crystals" in the *Quarterly Journal of the Geological Society*, helping inaugurate the science of microscopic petrography (study of the structure of rocks).

1861 William Henry Olley publishes *The Wonders of the Microscope Photographically Revealed*, the first known English book of photomicrographs.

1862 Louis Pasteur develops the pasteurization process for milk, theorizing that the fermentation process is analogous to disease causation by microscopic organisms (germ theory).

1863 Sorby's work sparks another new field, microscopic metallurgy (study of the structure of metals).

1865 June 14. Mordecai Cooke, mycologist and editor of *Science-Gossip*, following a suggestion put forth by W. Gibson in a letter to that periodical, founds the Quekett Microscopical Club with Thomas Ketteringham and Witham Bywater. First full meeting, 7 July 1865. First President, Edwin Lankester.

1865 Lionel Beale experiments with a glycerin immersion.

1866 Ernst Abbe joins Zeiss.

1866 November 1. The MSL receives a Royal Charter to become the Royal Microscopical Society. Its journal is renamed the *Monthly Microscopical Journal* (*MMJ*).

1867 Joseph Lister publishes "On the Antiseptic Principle in the Practice of Surgery" in *The Lancet.*

1868 First issue of the *Quekett Journal of Microscopy.*

1869 Powell & Lealand issue their "No. 1" microscope in its canonical form.

1869 Col. J. J. Woodward, with Major F. Curtis, produces a photomicrographic image of his resolution of the Nobert "19th band," with lines spaced 0.28 microns apart.

1870 January. John Tyndall gives lecture on "Dust and Disease," supporting germ theory.

1870 Ferdinand Julius Cohn classifies bacteria into bacilli, cocci, spirochaetes, and terponemes. (He publishes *Untersuchungen uber Bacterien* in 1872.)

1870 Abbe develops the Abbe Condenser.

1870s The Battle of the Glasses (Aperture Wars) rage for much of this decade.

1871 Royal College of Surgeons requires its licentiates to be trained in general anatomy and physiology, practical anatomy, and histology.

1871 Adolf von Bayer develops fluorescein, the first fluorescent stain.

1872 Zeiss starts making lenses guided by Abbe's calculations.

1873 Abbe publishes his diffraction theory (on the limitations of optical microscopy) in "Beiträge zur Theorie des Mikroskops und der mikroskopischen Wahrnehmung" (Contributions to the Theory of the Microscope and to Microscopic Perception), *Max Schultze's Archiv für mikroskopische Anatomie*, IX (Dec. 1873), 413–68.

1873 Joseph Lister presents *A Contribution to the Germ Theory of Putrefaction* to the Royal Society of Edinburgh (published 1876).

1873 Anton Schneider recognizes mitosis in *Mesostomum ehrenbergii.*

1874 Robert Tolles, Zeiss, Abbe, and John Ware Stephenson dispute the origin of the homogeneous-immersion formula. John Mayall, Jr. gives the palm to Tolles (using balsam in 1873) due to priority of publication.

1874 Hermann von Helmholtz independently publishes his theory on the limitations of optical microscopy in "Die theoretische Grenze fiir die Leistungsfähigkeit des Mikroskops" (On the Theoretical Limits of the Optical Capacity of the Microscope), *Poggendorff's Annalen der Physick, Jubelband*, 136 (1874), 557–84.

1875 Eduard Strasberger identifies mitosis as a universal process of cell division and multiplication.

1875 Henry Edward Fripp publishes his translation of Abbe's article in *Proceedings of the Bristol Naturalists' Society* (n.s. 1, no. 2), reprinted in the October and November 1875 issues of the *MMJ.*

1876 Koch establishes that anthrax is caused by *Bacillus anthracis* (published 1877).

1876 Fripp publishes his translation of Helmholtz's article in *Proceedings of the Bristol Naturalists' Society* (n.s. 1, no. 3), reprinted in the July 1876 issue of the *MMJ.*

1876 Oscar Hertwig discovers meiosis in sea urchin eggs.

1877 The *MMJ* becomes the *Journal of the Royal Microscopical Society* (*JRMS*).

1877 Zeiss starts producing its first oil immersion lens, developed by Abbe on the suggestion of Stephenson (published 1879).

1878 Koch publishes on the microorganism causing wound infections—observations enabled by Abbe's new lenses and condenser.

1879 Pasteur establishes that puerperal fever is caused by *Streptococcus pyogenes* (named by Rosenbach in 1884).

1879 Walther Flemming shows the role of chromosomes in mitosis, specifically their replication by splitting longitudinally (published full account in 1882).

1879 Otto Schott develops a lithium-based optical glass. Zeiss would make the first microscopical lens with this glass in 1883 and take it to market in 1885.

1880 Alexander Ogston and Pasteur establish that *Staphylococcus aureus* (named by Ogston) causes disease.

1881 Pasteur successfully grows *Bacillus anthracis* in culture, producing anthrax in sheep when injected.

1881 *JRMS* publishes "Notes on Aperture, Microscopical Vision, and the Value of Wide-angled Immersion Objectives" (April 1881), 303–60, a definitive answer to the aperture question, by Frank Crisp (see "Proceedings of the Society," April 1881, 365).

1881 *MMJ* publishes Abbe, "On the Estimation of Aperture in the Microscope" (June 1881), 388–423.

1881 Koch demonstrates his techniques for culturing and inoculating specimens and for microscopical observation at the International Medical Congress in London.

1882 Koch uses differential staining to establish that tuberculosis is caused by *Mycobacterium tuberculosis.*

1882 Elie Metchnikoff proposes that macrophages (white blood cells) combat infection and inflammation by enveloping germs at the site.

1883 Edouard van Beneden describes the chromosomal effects of meiosis.

1884 Friedrich Augustus Loeffler establishes that diphtheria is caused by a bacterium (Edwin Klebs worked on this as well).

1884 Koch establishes that cholera is caused by *Vibrio cholerae* (the "comma bacillus").

1884 Friedrich Julius Rosenbach identifies *Streptococcus pyogenes* as the source of scarlatina and erysipelas.

1884 Arthur Nicolaier and Carl Flugge identify *Clostridium tetani* as the cause of lockjaw.

1884 Hans Christian Gram develops his method of differential staining.

1885 Wilson "Snowflake" Bentley, in Vermont, takes the first photomicrograph of a snowflake.

1886 Zeiss, consulting with Abbe, produce the first apochromatic lenses (not just red/blue but red/green/blue).

1886 Theobald Smith identifies the bacterial cause of typhoid.

1887 Anton Weichselbaum identifies that meningitis is caused by the bacterium now called *Neisseria meningitides.*

1888 Julius Richard Petri develops the Petri dish.

1889–90 Sergei Nikolai Winogradsky identifies the nitrogen-fixing bacteria in soil.

1890 Koch presents the refined form of the Henle–Koch Postulates.

1890 August. Weismann describes the importance of meiosis for reproduction and inheritance.

1894 Shibasaburo Kitasato and Alexandre Yersin independently show that bubonic plague is caused by *Pasteurella pestis* (*Yersinia pestis*).

1898 Kiyoshi Shiga identifies *Shigella dysenteriae* as the cause of dysentery.

1899–1904 Ernst Haeckel publishes *Kunstformen der Natur* (Artforms in Nature), with its spectacular chromolithographed images of microscopic forms.

APPENDIX III

List of Microscopical Tropes

Chapter 1

The microscope's decline in the eighteenth century
Microscope as a mere toy
Mechanical objectivity
Metaphor as lens
Metaphor as tool
Body as machine
Microscopy as scrutiny (visual/mental)
Microscopy as reading

Chapter 2

Telescope and microscope
Microscope = petty, vain, vermin
Seeing the invisible
Microscopist as Peeping Tom/voyeur

Chapter 4

Instructive and amusing

Chapter 5

Emotional contagion
Body as machine
Nervous currents and waves
Nervous system is like a telegraph
Microscope as a new sense
Body of the microscope

Chapter 6

Hidden worlds
Modern disenchantment
The explained supernatural
Revelations of the microscope

Mechanical objectivity
Beauty gives pleasure
Analogy and metaphor
Microscopical richness (jewels, crystal, gold, velvet…)
Hybrid forms
Visual desperation
Beautiful = small, intricate
Aesthetic of detail
Diatom as spiderweb, engine-turned watch, rose window
Elaborate and fanciful aesthetic descriptions
Scientific nomenclature

Chapter 7

The sublime/microscopical sublime
Dizzying vista of vastness/infinite minuteness
Vastness of the very small
Mystery
Magnificence
Counting the uncountable
Myriads
Natura in minima maxima (Nature is greatest in the smallest things)
Worlds within worlds
Multiple worlds
Beautiful AND sublime (diatoms)

Chapter 8

Narrative of natural theology
Theologies of nature
Watchmaker God
Argument from analogy
Argument from beauty
Argument from design
Microscope as metaphor for mental scrutiny
Created vs. constructed objects
Body as machine or mechanism
Beautiful mechanism (natural object or microscope)
Brass and glass
Diatomaniacs
Gimcrack
Bugs and slugs
Contrivance
Narrative of natural history
(Narrative of nature, narrative of science)
Reverent empiricism: commonness, detail, dilation, minuteness

Chapter 9

Wonder everywhere
The divine in the commonplace
The common wonder
The repulsive revealed to be wonderful
Microscope as scrutiny
Natura in minima maxima (Nature is greatest in the smallest things)
Reverent empiricism

Chapter 10

The romance of microscopy
Microscopical worlds
Virtual storyspace
Microscopical landscapes/vistas
Microscopical fairylands
Nature's laboratory
Peering into nature's secrets
Microscopist as Peeping Tom/voyeur
Life in a water-drop
Invisible life surrounds us
Sympathetic identification with microscopic objects
Oriental tale
Magic tube
Magician's wand
Phantasmagoria (spectacular, Gothic microscopy)
Anthropomorphism

Chapter 11

"Oh law" objects
Goodness gracious box
The pleasures of microscopy

Chapter 12

Many-sided comprehension/sight
Seeing is believing
Mitigated skepticism

Chapter 13

Tact (trained practice, hands and body)
Training for tact (manual dexterity)

Personified contrivance
Trained hand/educated hand
Dirt is matter out of place
The royal road to dexterity
Patient labor
Seeing is believing
Microscope as companion
Microscope as pet
A dancing stand (dancer)
Ascetic (disembodied) philosopher
The body of the microscope/of the observer
Embodiment relations
A new sense

Chapter 14

Pencil of light
Observer as an eye
Eye of the microscope
Personal equation
Eye as an optical instrument
Eye as a camera obscura
Eye as a screen
Eye (or retina) as a lens
Created vs. constructed objects
Eye as a wonderful contrivance
Counterbalancing
List of general rules of action
Microscopy injures the eye
Scrutinize vision
Mental scrutiny as a kind of vision
Natural/unnatural vision
Naked eye
Unarmed eye
Trained eye
Eyes or no eyes
Muscae volitantes (flying flies)
Pearly spectrum
Many-sided comprehension
Magnification as bringing the eye closer, as if walking across the street to read a bill
Play the jackal to the lion
Microscopical army/corps
Microscope as weapon against disease
Medicine as war with disease
Bewildered student
Picture of an object, not the object itself
Natural/unnatural object
Reagents (any intervention)
True structure

Chapter 15

Appreciation
Educated judgment
Duty to care
Seeing is believing
Innocent, unerring microscope
Contrivance
Brain (and eye) as optical contrivance
Body as machine/mechanism
Multiple views
Many-sided comprehension
Microscopical vision as reading
Ghostly parasites (diffraction lines)
Unnatural vision (again)
Counterfeit revelations

Chapter 16

Minute description
Taxonomic character
Inductive description
Defined description

Chapter 17

Articulating practice
Virtual witnessing
Narrative knowing
Experimental report
Narrative of science
Drop of water, slide, or live-box as a cage or prison

Chapter 18

Wonder and skepticism
Counterbalancing
Separation of wonder and skepticism
Juxtaposition of wonder and skepticism
Combination of wonder and skepticism

Chapter 19

The soirée as genre
Enthusiastic description
Description as invitation

Many-sided comprehension
Microscopical richness
Divine in the commonplace
Common objects
Magic lantern show as genre

Chapter 20

Narrative investigation
Argument as a journey
Investigation as a journey
The object tells its own tale
Dilation

APPENDIX IV

Graphs of Microscopical Titles in Britain, by Publication Date

Figs. AIV.1 and AIV.2 have been derived using the Nineteenth-Century Short Title Catalog accessed via Gale on June 24, 2023. I searched for publications published 1800–99 with microscope, microscopic, microscopical, microscopist, or microscopy in the title. I sorted them by date of publication and eliminated duplicates, titles published outside Britain or its imperial lands, and those using "microscope" metaphorically. My thanks to Catherine Hanson and Jeff Hanson for technical assistance with the figures.

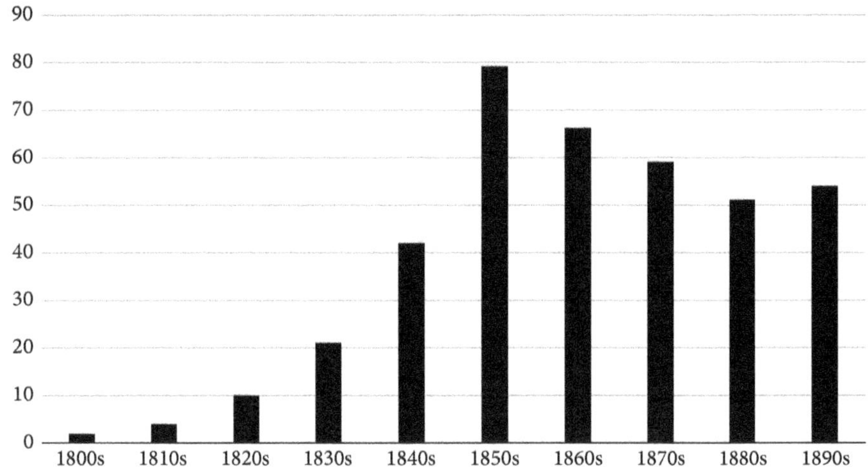

Fig. AIV.1 British microscopy titles published, by decade, 1800–1900.

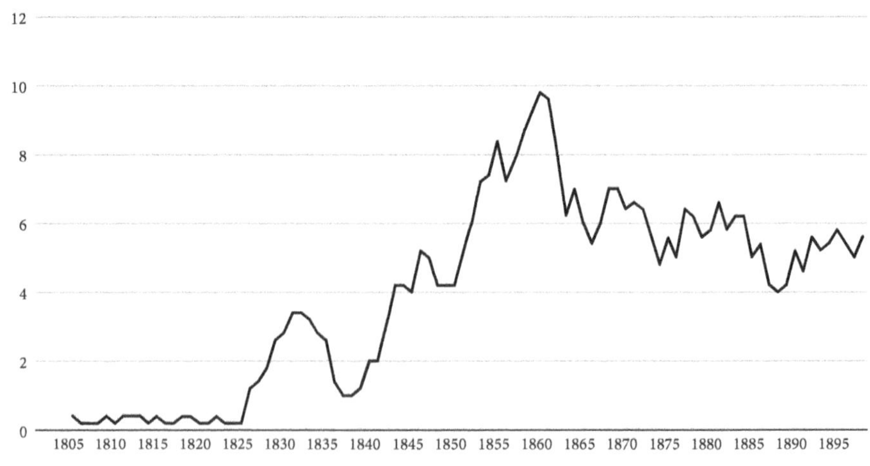

Fig. AIV.2 British microscopy titles, by year (five-year moving average), 1800–1900.

Glossary

Microscopic So small it can only be seen clearly through a microscope. Describes the object or specimen under observation.

Microscopical Having to do with microscopy.

Types of Microscope Construction

Compound microscope A microscope that uses separate multiple lenses in a system. The worker observes a magnified image of the object, not the object itself. Although Robert Hooke and others used compound microscopes, the early versions of this technology magnified not only the object and its image but also the aberration. Simple (single-lens) microscopes generally provided a clearer image until Lister and Tulley published on their achromatic compound microscope in 1830.

Engiscope C. R. Goring used this term to describe any microscope that uses an eyepiece to magnify an image produced by an earlier mirror (reflecting microscope) or lens (refracting microscope). The term did not catch on.

Light microscope (optical microscope) Uses properties of light to magnify an object. All microscopes before the twentieth century are light microscopes.

Reflecting microscope A microscope that uses a mirror to produce an enlarged image of the object.

Refracting microscope A microscope that uses a lens system to produce an enlarged image of the object.

Simple microscope (single microscope) A microscope that uses a single lens, which could be constructed of two or more lenses stacked together. The worker observes the object directly. Anton van Leeuwenhoek constructed hundreds of small simple microscopes, each one adjusted for its own particular object.

Parts of the Microscope and Its Apparatus

Condenser A lens that collects and focuses the illumination.

Cover-glass (cover-slip) A thin piece of glass of the same size as the slide, placed over the specimen. These became more common in the Victorian period, since they help hold and protect the specimen and provide a more uniform surface.

Diaphragm Beneath the stage, cuts off outside rays from entering the objective, effectively narrowing the angle of aperture.

Erector Some microscopes add this in the body-tube above the object-glass, so that the image will not be inverted.

Eye-glass The lens in the eyepiece.

Eyepiece The component of the microscope combining eye-glass and field-glass, which mounts onto the body-tube at the end nearest the eye. Microscopists can switch out "shallow" (low-magnification) and "deep" (high-magnification) eyepieces (A, B, C, D, E, F are mentioned). Great magnification is not useful in the eyepiece, as it will only magnify any errors in the image provided by the object-glass. A more perfect object-glass will "bear" deeper eyepiecing.

Field-glass A lens near the eye-glass, within the body-tube, refracting the (rays of the) enlarged virtual image (within the body-tube) so that it is the correct size for view by the eye-glass. It also helps to correct chromatic and spherical aberration (??) and thus to clarify the image. The term *field-glass* can also refer to a travel telescope used to survey a landscape.

Front The portion of the object-glass that is closest to the slide. Lenses corrected for immersion are mounted with a special front; some objectives offer both a wet and dry front.

Live-box (animalcule-cage) A slide providing a reservoir to hold live animalcules for observation.

Nosepiece This attachment to the body of the microscope can hold multiple objectives; a double nosepiece was first introduced by Charles Brooke in the 1850s. Carpenter discusses various nosepieces in his 1868 *Revelations* (1868, 102–3). The operator rotates the nosepiece to more easily select different magnifications.

Object The specimen under observation, usually mounted on a slide.

Object-glass (objective) The interchangeable component that mounts onto the body-tube of the microscope at the far end (farthest from the eye). Usually contains multiple lenses.

Slide The glass rectangle on which a specimen is mounted.

Stage The flat surface on which the slide is secured for viewing

Types of Lenses and Objectives

Achromatic Without chromatic aberration.

Aplanatic Without spherical aberration or coma.

Compound lens A lens formed of more than one piece of glass (doublet or triplet).

Doublet Lens formed of two pieces of different kinds of glass (for example, flint glass and crown glass) with different dispersion of refractive index, cemented together. This is usually to correct chromatic aberration by bringing two different wavelengths of light (red and blue) into focus together. First popularized by John Dolland in 1750s for telescopes, 1830 by Joseph Jackson Lister (with instrument-maker William Tulley) for microscopes.

Immersion lens An object-glass designed to be used in contact with a substance with refractive index closer to glass, eliminating refraction through airspace (usually water or oil); improves resolution at high magnifications. Both Hooke and Brewster had suggested the principle used in immersion lenses. Amici started experimenting with immersion lenses in 1840. Tolles developed water immersion lenses (duplex, dry/wet) in 1858; Hartnack followed in 1859, with Nachet and others in 1860. English instrument-makers were cautious about this technology; Powell and Lealand produced a water immersion in 1865. Tolles produced a homogeneous immersion lens for use with Canada balsam in 1873, and in 1877 Abbe (Zeiss) produced a homogeneous oil immersion lens.

Triplet Lens formed of three pieces of glass; improves color correction relative to an achromatic lens by adding, for example, fluorite and brings three different wavelengths of light (red, blue, and green) into focus together. Designed by Peter Dolland in 1763 for telescopes; developed for microscopes by Ernst Abbe in 1886. Bears deep eyepiecing better than other lenses.

Aspects of the Microscope and Its Lenses

Aberrations Error in the lens or lens system that distorts the image. There are a number of these potential flaws. Those most concerning to Victorian microscopists were chromatic and spherical aberrations.

Angle of aperture (angular aperture) The definition of this term was debated. The angle of aperture is the angle *a b c* from the edges of the object-glass to the object and back; the angle within which "pencils" (light/image) from the object can get into the objective. Wide angles are high angles of 100 degrees or higher. Microscopists differed on whether angular aperture over 180 degrees is really possible, sparking the "Battle of the Glasses" (the "Aperture Wars") in 1871–8. Abbe's work in 1873 introduced the notion of "numerical aperture," which also accounts for the refractive index of the immersion fluid (or air) and the wavelength of light. Abbe showed that, at any magnification, greater aperture improves resolution.

Chromatic aberration The image of the object is blurred by a colored fringe around its outline and markings of structure. This occurs when the different color wavelengths within a ray of light are refracted (bent) through the lens at different angles; the various colors splay out, somewhat misaligned, in the final image. This was the last major error to be corrected by nineteenth-century microscope-makers.

Coma A flaring-out of the light point made by an oblique ray on one side of the field. An element of Abbe's experimental demonstration of his diffraction theory.

Diffraction The bending of light around sharp edges or fine lines or minute objects, or splaying out through tiny holes. These diffracted light waves, interfering with each other, can create spurious lines or markings in the image. The object is shown both by direct light (transmitted or absorbed/in shadow) and by diffracted light, overlaying and sometimes obliterating true markings with spurious ones. It is especially common in lenses of wide aperture because a diaphragm or narrower aperture will cut off some of the diffraction rays. Abbe's 1873 theory of the light microscope sets the limit of resolution at half the wavelength of the light waves used, due to diffraction. Although the undulatory (wave) theory of light was not yet fully accepted, his work on the diffraction limit proved influential. Helmholtz independently discovered and derived this theoretical limit using a different approach, publishing his findings in 1874.

Flatness of field If the microscope is focused at a particular layer/depth of an object, a flat visual field will be equally distinct not only at the focal point but at the edges of the field as well. See Carpenter 1856, 195.

Focal length The distance at which the objective's point of focus occurs. A lens with long focal length brings a larger field of view (more of the object on the slide is visible) at a particular working distance than a lens with short focal length.

Polarization The restriction/flattening of a light wave through a filter. The resulting wave undulates only in one plane (orientation).

Refraction The bending of a light ray as it passes from one kind of substance (various kinds of glass, water, oil, Canada balsam) to another. It can be seen, for example, by the broken image of an oar dipping into a clear pond (passing from air to water). The amount of bend is the refractive angle. Longer-wavelength light (red) bends less than shorter-wavelength light (blue or violet). Different substances can be ranked by their refractive index. Objects with a similar refractive index will show less bend as the light passes from one to the other. Abbe argues that the laws of prismatic or lenticular refraction apply only to coarser objects: "they do not hold good unless these objects represent *large* multiples of the wavelength" of the rays (1881b, 421).

Spherical aberration The image of the object is distorted. This occurs when the light passes through a lens and is refracted unevenly. Rays at the edges of a curved lens reach a focus at different points than those in the center of the lens, creating a blurring or bulging effect.

Working distance Distance between the front lens and the slide, when the image is in focus. This decreases as magnification increases.

Powers

The "power" of a lens, if referenced without qualifiers (defining power, resolving power, etc.) generally refers to magnifying power, although clear vision requires many other qualities as well. Powers will differ based on the size, shape, and types of glass; on the cement used in an objective, on whether it is used dry or wet (oil or water immersion); and on the distance between the lenses and the order in which the lenses appear in relation to one another. Resolving power is also particularly affected by type of illumination (natural or artificial; central, oblique, dark field, polarized). Victorian texts demonstrate some confusion over terms (defining, penetrating, and resolving power) when different authorities defined them differently. See Phin 1877, 45; Schickore 2007, 110; Schickore 2009, 119–20. In defining the terms here, I try to identify a consensus for each.

Defining power (Nägeli: "outline power" [1887, 91]) Goring defines this as "*a destitution of both kinds of aberration*" (Goring in Pritchard and Goring 1832, 173); Carpenter defines it as clearness of the object and its outline (1856, 192). A well-defined image lacks spherical or chromatic aberration and shows elements sharply. See **Correction (of objective)** below. Abbé specifies that defining power describes "the purely dioptrical perfection of a system" and is independent of aperture or the microscope's ability to manage diffraction (1881b, 418).

Delineating power Defined by Frank Crisp as "the power of the Microscope to show things *as they are*" (1881, 355). Whereas a microscope with great *defining* power may show structure sharply etched, and one with great *resolving* power may separate fine lines well, some or all of the structure imaged may be spurious due to the effects of diffraction and the limitations of the objective. A microscope with great *delineating* power (for example, a wide-aperture objective that will not cut off many of the diverging spectra) will not just show the image sharply etched and with fine elements distinguished from one another, but also show that structure more accurately.

Magnifying power Early, measured in diameters at a distance of 10 inches, as if projected. See Carpenter 1856, 206. Increasing the magnifying power of an objective (for example, by using a stronger ["deeper" or "higher"] eyepiece) increases resolution to a certain point. After this, any increase of magnifying power simply magnifies the image (and any errors in the image) without improving the resolution, and perhaps actually impairing the resolution of what had been seen at a lower power. See "Summary of Current Researches: Highest Magnifying Powers," 130. The term "high-powered" was first used in microscopy to describe magnifying power, and only later applied to guns, engines, business success, etc.

Penetrating power (associated with *depth of field, focal depth*) (Nägeli: "piercing power" [1887, 91]) The range of depth through which the structure of the object remains in focus or nearly in focus, even if it is not in the focal point. A greater penetrating power is useful for gaining a sense of the relations of structure. Penetrating power usually decreases as angle of aperture increases (Carpenter 1856, 193). Goring arrives at *penetrating power* by analogy with the telescope; he defines it simply as a "*large angle of aperture*" (Goring in Pritchard and Goring 1832, 173). Griffith and Henfrey argue that the term "should be laid aside, as tending to cause confusion, the properties of object-glasses being reducible simply to their defining power and their angular aperture" (1856, 636).

Resolving power (resolution) Carpenter defines this as the ability to clearly distinguish distinct small elements in the object rather than running them together (for example, seeing dots instead of lines in a test object). See Carpenter 1856, 195, 196–7 on resolving power vs. defining power. Resolving power usually increases as angle of aperture increases. Nägeli points out that this power of differentiation should be what we compare to the double-star-resolving test in astronomy (1887, 92). Abbe showed that resolving power depends on how the objective manages diffraction (Abbe 1875, 251).

On these definitions, see also Quekett 1848, 423; Nägeli and Schwendener 1887, 91; Abbe 1881b, 418.

Manipulation of the Microscope

Bright field illumination The simplest and most elementary form of microscopical illumination. A full cone of white light passes through the specimen and the transmitted light is received by the objective, with contrasting areas of shadow where a dense or thick structure has absorbed the light. The specimen appears darker than the bright background. Used where the object has good contrast already (for example, it has inherent color or has been sectioned and stained) or where its refractive index is quite different from the surrounding mounting medium.

Correction (of objective) Adjusting (arrangement and type of lenses) to compensate for spherical aberration, centering, achromatism in the lens. A lens can be over- or under-compensated and this can be tested using an "artificial star," which is a tiny mercury globule observed out of focus, a micrometer, or Abbe's test slide with marked lines and differing thicknesses of cover-glass. See **Defining power** above. Some microscopes (Ross first) included a collar adjustment ring to allow microscopists to account for the cover-glass on top of the object (the glass refracts the rays—more of an issue with high-powered objectives).

Dark light illumination The cone of illumination produced by a circular condenser is hollowed out by a circular central stop, blocking direct rays from reaching the specimen or the objective. Indirect rays—forming the outside of the hollow cone of light, converging from an oblique angle to a point on the slide—brightly illuminate the specimen by light reflecting, refracting, or diffracting through different densities. The background appears dark, however, because the oblique rays diverge again at an angle excluded from the objective (the numerical aperture of the condenser must be greater/wider than that of the objective). This technique increases contrast of refractive objects, although it requires a strong beam. It was used, for example, to display diatoms with their intricate silica structure brilliantly lighted. Victorian microscopists had experimented with using oblique light to better resolve these complex objects; dark field illumination provides oblique light from all around the object rather than just from one side. (Oblique light improves the detail and accuracy of the image when viewing fine structure, but this is not—as Victorians assumed—because it highlights the prominences and other forms of the object, as when the setting sun edges every form with shadow, but rather—as Ernst Abbe showed in 1873—because it admits a greater number of diffractive rays.)

Mounting Placing a prepared object on a slide. Usually this entails affixing it to the slide, preserving it with a substance like Canada balsam, and cementing a protective cover-glass on top. Some objects, however, are best mounted dry or on the cover-glass.

Sectioning Slicing an object very thin so that its structure becomes more visible through the microscope.

Staining Treating an object with a dye that is taken up unevenly by different parts of the structure, thus rendering that structure more readily visible, especially when objects are transparent or have a refractive index very similar to the mounting medium. Although staining is described by Robert Hooke in *Micrographia* (1665), an effective technique was not developed until Joseph von Gerbach in 1858 described his work staining brain cells with a solution of carmine (crushed cochineal), thus inaugurating the field of histology (study of the structure of cells and tissues). The development of vibrant, colorfast, synthetic aniline dyes from coal tar by William Henry Perkin, beginning in 1856 with mauveine, proved a great aid to researchers.

For an online gallery illustrating a range of nineteenth-century microscopes, please consult Michael Davidson's online "Museum of Microscopy" at Florida State University (https://micro.magnet.fsu.edu/primer/museum/index.html).

Notes

Notes to Chapter 1

1. Hadjiafxendi and Plunkett 2017, 181.
2. See the Glossary of microscopical terms at the end of the book.
3. Gatty 1872, xxix, xix, viii. Gatty uses the term "algologist" to describe her role (xiii).
4. "L'Envoy," 1850, iv. The catalogue of the Museum Collection at the Royal College of Surgeons includes a Ross microscope from 1839–41 (RCSIC/O 12.22) and comments, "Microscopes of this design were used in the College following the appointment of Quekett in 1840." See "Surgicat," Royal College of Surgeons of England, last modified 2022, http://surgicat.rcseng.ac.uk/Details/collect/50580. The frontispiece to Quekett's 1855 (third) edition of *Practical Treatise on the Microscope* shows the Histological Theatre apparently equipped with these instruments.
5. An advertisement for the 1851 volume of the *Medical Times* puffs Quekett as "well known, not only in this country but throughout the scientific world, as the most able microscopist of the day." "*The Medical Times* [advertisement]," n.p. For an example of Quekett's consulting work, see, for example, Smith and Jordan 1850, 653.
6. Bennett 1841, 15.
7. Quekett 1848, viii, 420.
8. Bonner 1995, 194, 221.
9. Quekett 1851, 1.
10. [Owen] 1851, 433.
11. "Microscopical Memoranda: Owen on the Application of the Microscope," 1842, 64.
12. Owen 1840, xi.
13. Quekett 1852, vol. 1, 1. Marc Ratcliff (2009) refutes this dismissal of eighteenth-century microscopy, but most of the work he discusses occurred on the Continent. My focus here is on Victorian perceptions of British microscopical history and culture; the narrative of decline was widely accepted at the time.
14. Quekett 1852, vol. 1 (1–2).
15. Schaffer 1998, 194. The term "literary technology" used in this sense originates with Steven Shapin (Shapin 1984).
16. Buckland 2013, 2–4.
17. Barad 2007, 35. Barad argues that laboratory apparatuses themselves are "material-discursive practices" and that bodies (human and non-human) are "material-discursive phenomena" (146, 153).
18. For example, see Flint 2000, Armstrong 2008, Willis 2011, and Potter 2018, who all approach microscopy as one of a range of visual tools and toys.
19. T. Smith 2014, 10, 5.
20. Schmitt 2017, 48.
21. Hodder 2012, 96–7.
22. See, for example, the work of Bruno Latour, Don Ihde, and Karen Barad. On mechanical objectivity, see Daston and Galison 2007.
23. Morgan 2017, 4, 22, 6.
24. Daston and Galison 2007, 203.
25. See Kennedy 2019.
26. I speak here to Victorian microscopical wonder specifically; as Dennis Quinn remarks in reviewing nineteenth-century attitudes toward wonder, "The age was one of controversy on many issues, and wonder was one of them" (2002, 290).
27. Fisher 1998, 55, 40, 35.
28. Vasalou 2012, 33.
29. Daston and Park 1998, 304–5.

30. On the affective work of whimsy and other forms of comic microscopy, see Kennedy, "Water-tigers," forthcoming.
31. Notcutt 1859, 1.
32. Klein et al. 1873, 41, 86, 232.
33. Jacyna 2003, 68, quoting "Anti Spiral," 1844, 477.
34. Carpenter 1874, 497. For more on the mid-century distinction between uncontrolled Fancy and disciplined Imagination, see Potter 2018, 110.
35. Jacyna 2003, 71, quoting Bennett MSS, Edinburgh University Library, Gen. 2007/1–4 (1), p. 170.
36. Tyndall 1871, 157.
37. Tyndall 1871, 130–1.
38. See Lightman 2004, 200; Potter 2018, 110–11.
39. Carpenter 1874, 487.
40. Carpenter 1874, 506fn.
41. Beale warns, "Every false statement regarded as an observed fact, forms a terrible barrier to onward progress, since, before the slightest useful advance can be made, it is necessary to retrace our steps, it may be for a considerable distance, before we can hope to recommence our onward course.... Scientific enquiry ought continually to advance, and we should be able to extend our researches from the point where they have been left by our predecessors, adding successively to what they had discovered; but the observations which we owe to them should not require correction." He remarks specifically that, "led away by their imagination," some observers "think they see everything which has been delineated or which they have heard described." These careless and gullible observers then publicize their supposed confirmation of others' erroneous observations. He points out, "In this manner errors have been propagated and strengthened to an extent almost incredible, and years of laborious investigation have been spent in overthrowing statements which had never resulted from actual observation in the first place" (1857b, 97–8).
42. Fisher 1998, 55, 31.
43. Fisher 1998, 118.
44. Carpenter 1856, 8.
45. On Victorian research into perception and cognition, see, for example, Crary 1990, Garratt 2010, Potter 2018.
46. Carpenter 1856, 25.
47. Daston 2008, 99.
48. Sprat 1667, 331.
49. Daston 1995, 20.
50. See Myers 1991, 142.
51. Gates and Shteir 1997, 11–12; Gates 1998, 39–41.
52. Lynn Merrill emphasizes the "lively, visual, particularized language" of natural history, its "discursive" style, and its promise of "'opening up' new worlds" to spark an affective response in the observer (Merrill 1989, viii, 23, 260, 217). Lightman notes the close observation of curious, marvelous natural objects in a familiar, inviting narrative that finds meaning and value in the secret wonders of nature (Lightman 2000, 7–9, 13–15, 25).
53. King 2019, 11, 112–13.
54. Ketabgian 2011, 3, 44.
55. Ketabgian 2011, 13.
56. Lakoff and Johnson 2008 [1980], 22, 24.
57. See Webster 2020, 58–60 for a discussion of this suggestion.
58. "Preface" in Boyle 1690, [n.p.].
59. Reynolds 2018, 152.
60. Reynolds 2018, 147, 156, 13.
61. Reynolds 2018, 157, 160, 4.
62. Maurette 2018, 21.
63. [Tuthill] 1820, 3–4.
64. Swinburne [1872], 3. Appropriately, the dispute arose over the charge that the work of Swinburne and the Pre-Raphaelite poets was *too embodied*; for the critic prompting this publication was Robert Buchanan and the review was his notorious essay on the "fleshly school of poetry."
65. [Elliott] 1861, 14–15, 23–4, 111.
66. Bain 1859, 199–201.
67. Bain 1859, 202.

68. Bain 1859, 640.
69. Carpenter 1874, 257.
70. Lewes 1877, 59.
71. On Lewes and metaphor, see Rylance 2000, 306–7; Otis 2001, 59–78.
72. Lewes 1877, 224.
73. Lewes 1877, 224–5.
74. Lewes 1874, 315.
75. Lewes 1877, 276.
76. Reade 1870, 113–14.
77. Lankester and Busk 1853, 2.
78. "Microscope, Microscopic, Microscopical," 1886, 339.
79. Armstrong 2002, 43–4.
80. Pater 1873, 208–10. On the influence of Lewes and Spencer, see Inman 1981, 14–15. The passage from the *Fortnightly Review* that she quotes made its way into Lewes's *Physical Basis of Mind* (1877); the Spencer text she cites is *Principles of Biology* (1864–7). See also Morgan 2017 on the role of physiological psychology in Pater's work.
81. Baudelaire 1863, 13.
82. McGrath 2003, 151.

Notes to Chapter 2

1. The narrative of the general history of microscopy has stayed relatively stable across different moments in historiography; see Mayall 1886, Bradbury 1967a, and B. Jardine 2016. See Appendix II for a timeline of microscopical history.
2. See the Glossary of microscopical terms at the end of the book.
3. Turner 1967, 177–8.
4. Lister 1830. Lister reviews prior work in the field. See also Arthur Hughes on challenges to Lister's influence (Hughes 1956, 9).
5. See Bradbury 1967b, 151–2, 170.
6. Pritchard and Goring 1832, 14.
7. L. S. Jacyna's articles on this period offer useful case studies in these fields (Jacyna 1984, 2001, 2003).
8. On the MSL, see Turner 1989b.
9. Paget 1842, [4].
10. G. T. F. [Fisher] 1846, 30–1.
11. Fisher 1846. This was third in a series of Willats' Scientific Manuals. Fisher also wrote for Willats on medical electricity, for George Knight (a dealer of chemical and philosophical apparatus) on photography, and for a cigar dealer on the history of smoking. Louisa Lane Clarke's 1858 *The Microscope* was also published "solely to accompany the Object-box and explain its contents" (iii).
12. These figures have been derived using the Nineteenth-Century Short Title Catalog accessed via Gale on June 24, 2023. I searched for publications published 1800–99 with microscope, microscopic, microscopical, microscopist, or microscopy in the title. I sorted them by date of publication and eliminated duplicates, titles published outside Britain or its imperial lands, and those using "microscope" metaphorically only, to denote close study—as in *The World and Its Two Faces: A Microscopic Mirror of the World We Live In* (William Kidd, 1854) and *The Microscope of the New Testament* (William Sewell and W. J. Crichton, 1878). Periodicals or multi-year books were counted only upon their first appearance, a limitation of the data. It is instructive to compare these totals to a graph of overall book and periodical publication in Britain. There, periodical titles appear as a steady rise throughout the century; book titles show an only modest increase just before 1850 followed by a plateau with a temporary dip in the 1870s. In comparison, I have found that microscopical titles fare better than general titles, increasing sharply from 1840 to 1860, then settling into a plateau at a somewhat lower level. The microscopical list does include periodicals, but those are counted only in the first year of publication, which reduces their effect on the graph. For the graph of general titles, which also derives its data from the Nineteenth-Century Short Title Catalogue, see Dawson et al. 2004, 10. Both line graphs use a five-year moving average.
13. Schickore 2007, 220–39 focuses on this period of formalization, albeit mostly in German-language sources. See also Feffer 1996, 29–32. On cheap print and the 1850s, see J. Secord 2000, 2, 30; Fyfe 2012, especially 1, 4–5, 33–5.

14. Goring 1827a, 413, 415. This apparent deference to the reader's own preferences is somewhat undercut by his mockery of observers who are "besotted" with large fields of view or the "half-taught booby" who values sheer magnification over anything else (413, 419).
15. On the relation between medicine and microscopical study in Edinburgh, see Jacyna 2001.
16. On distribution, see, for example, J. Secord 2000, 138–42.
17. Plumer 1865, 8.
18. On the vitalist debates, see Strick 2000; for germ theories, see Worboys 2000.
19. Worboys 2000, 54, 59–60, 63, 289–90.
20. Warner 1982, 9.
21. Turner 1967, 186–7.
22. Carpenter 1856, 25.
23. Turner 1967, 187–91.
24. Reiser 1978, 73–81; Bonner 1995, 245–50, 259–64; Worboys 2000, 213–14.
25. Chakrabarti 2012, 5, 15.
26. Samuel J. M. M. Alberti notes a general and national decline in philosophical societies at this time; intriguingly, Diarmid Finnegan references a growth in natural history societies in Scotland between 1870 and 1900 (Alberti 2003, 351; Finnegan 2005, 54–5). It is not uncommon to find, in the meeting reports of smaller microscopical societies in the late century, mentions of a general falling-off in membership. Even the RMS suffered; Gerard L. E. Turner says, "The 1890s were a difficult time for the Society.... [R]ecruitment of new Fellows failed to keep pace with deaths and resignations" (1989a, 57). John Harley Warner chronicles a similar trend in American popular microscopy (1982, 32). Nathan Edward Charles Smith argues that "the collapse of [popular] microscopy can best be understood" as a result of "amateur differentiation"—a trend toward specializing on minority taxa like cryptogams—rather than the take-up of microscopy by scientific disciplines generally (N. Smith 2023, 270).
27. Smiley 1893, 33–4.
28. Armstrong 1996, 125.
29. G. T. F. [Fisher] 1846, 31. Fisher refers to Richard Owen and Gideon Mantell.
30. Hogg 1854, 335.
31. Hogg 1854, viii–ix.
32. Laycock 1860, 83.
33. It is likely that Dollond was simply building on the ideas of Chester Moor Hall, a lawyer who had employed opticians to make similar lenses in the 1730s (Dunn 2009, 78).
34. Campbell 1821, 129.
35. See Nicolson 1935, 1–2.
36. Chalmers 1817, 112.
37. Whewell 1853, 25.
38. Buckley 1890, 4.
39. Chalmers 1817, 114, 112.
40. See Timbs 1868, 114–15.
41. Quekett 1848, vii.
42. Goring and Pritchard 1829, 9.
43. Mantell 1846, 7.
44. Brocklesby 1851, 7.
45. Maudsley 1868, 64.
46. [Brewster] 1856, 466.
47. Crisp 1878, 126.
48. Pritchard and Goring 1838, 239–40.
49. See Thwaite 2002, 320.
50. Goring 1827b, 279.
51. Pritchard and Goring 1838, 239–40.
52. Hogg 1854, 3.
53. Willis 2011, 61.
54. Armstrong 2008, 10; Henchman 2014, 167.
55. McGrath 2003, 167.
56. Omar Nasim notes that nebulae cannot be sectioned or stained (2014, 6).
57. Wells 1897, 238, 244–5.
58. Turner 1967, 193.
59. See Nasim 2014, 38.
60. Bachelard 2014 [1958], 274.
61. Stewart 1993, 44.

Notes to Chapter 3

1. Hudson and Gosse 1886, vi.
2. Bennett 1850, 199.
3. See A. Secord 1994.
4. Beale 1880a, 4.
5. See Hamer 2023; Bracegirdle 2016 on the debates over this issue.
6. See Beiermann 2021 on the strategies used in the American Postal Microscopical Society (modeled on the British one) to train members in microscopical slide preparation despite their geographic dispersal.
7. Moore 1896, 2.
8. See, for example, Senior 2021 on Samuel Campbell's work with Gosse in Jamaica.
9. Schickore 2009, 118.
10. See Rudwick 1982, 190; Fleck 1979; Topham 2009, 14, 16.
11. Lightman uses *practitioners* to refer to this group (2007, 13).
12. Bradbury 1996, 293.
13. See Desmond 2001, 13–16. As James Bennett says, "The serious and influential amateur is a figure characteristic of Victorian science" (J. A. Bennett 1989, 275). Bennett specifically notes the elite, exclusionary nature of the gentlemanly amateur in Victorian microscopy, although makers such as Ross and Powell (and Beck) were active in the MSL. Similarly, Endersby notes that colonial collectors of botanical specimens wished to have a good compound microscope (not just a field or dissecting microscope) to use in their work and to brandish as a badge of skill, but Hooker was still reluctant to admit them to full recognition as naturalists (2008, 72–9). Samuel Alberti argues for the "heterogeneity of amateurs" and their role in constructing professional biology in Yorkshire (2001, 116).
14. Glaisher 1866, 45. It was, ironically, this address in which Glaisher proposed that the MSL should seek a Royal Charter (61).
15. Beale 1880b, 204.
16. Quekett 1848, 26; Lentz 2017, 305.
17. McCalla's remarks were given in his presidential address to the American Society of Microscopists and reprinted in McCalla 1883a, 767.
18. Kingsley 1884, 261–2.
19. Kitton 1876, 5.
20. Kingsley 1884, 261.
21. Bousfield 1887, 18, emphasis in original; Hogg 1898, 108.
22. Lankester and Busk 1853, p. 1.
23. For changes in how books were marketed to investigators and hobbyists over the course of the century, and how hobbyists might contribute to science, see J. Secord 2009, 471.
24. Rudwick 1982, 191–2.
25. Hogg 1854, 131.
26. Goring 1827a, 419.
27. Notcutt 1859, 1.
28. Dixon 1857, 132.
29. Carpenter 1856, vi.
30. "The Substage Condenser," 1890, 116–36; reprinted in "Summary of Current Researches Relating to Microscopy: Illuminating and Other Apparatus," 1891, 90, 91. This self-deprecating sequence may have been Nelson's attempt to lighten the mood before launching into a controversial correction of Abbe's work. The RMS supposedly banned such critiques for a while; see "Obituary: Edward Milles Nelson," 1938, 200. However, Nelson's important *JQMC* article on the substage condenser, which challenged Abbe on the question of a narrow or wide cone, was reprinted in the *JRMS* in 1891. Nelson went on to serve as President of the QMC in 1893–5 and the RMS in 1897–9.
31. See Part III on articulated practice.
32. Belknap 2016, 395, 421.
33. Adrian Desmond notes that Richard Proctor, founder of *Knowledge*, referred to "Huxley, Tyndall, Darwin, and Spencer" as "popularizers" due to their income from book sales (2001, 22).
34. P. Gosse 1859, 434, 164. Gosse gives the nomenclature of these as *Sagartia miniata* (now a variety of *Sargartia elegans*) and *Tenthredinoidea.* At the time, he was serializing publication of his massive *Actinologia Britannica* on sea-anemones and corals.
35. Lightman 2016, 355–6. See also Lightman 2007.
36. Carpenter 1856, v.
37. Wood 1861, 5. *Common Objects of the Microscope* was sold in two versions. The fine edition is listed at either 2*s.* 6*d.* or 3*s.* 6*d.*, the cheap edition at 1*s.*—the difference largely in the coloring of the plates.

38. Wood 1861, 7–8, 6.
39. On producers and receivers, see J. Secord 2004, 661. On scientific performers' sensitivity to audience responses, see, for example, Morus 2010; T. Smith 2014.
40. Topham 2004b, 437.
41. Csiszar 2018, 1, 15.

Notes to Chapter 4

1. Landecker 2007, 21–2.
2. J. Secord 2007, 25–30.
3. N. Jardine 2000, 401–2.
4. See Kennedy, "Circulation and Civility: Mid-Victorian Botany and Microscopical Method," 2024.
5. See, for example, Jackson 1873, 97.
6. In a letter dated March 1875, Hickie draws readers' attention to "Dr. J. Schumann's 'Diatomaceae of the High Tatra'" [*Die Diatomeen der hohen Tatra*, Wien, 1867] and "Otto Müller's 'Recent Microscopic Objectives Compared.'" In a later issue he translates a portion of Schumann's work, citing "a general and strongly expressed wish" for such a translation (Hickie 1875b, 265; Hickie 1875c, 6).
7. Lidwell-Durnin 2020, 81. He further notes that the 1856 and 1857 editions of *Revelations* "saw 5000 copies printed—one of the highest in Churchill's records (MS1393/M8/201–202, 307)" (81).
8. William St. Clair argues that, by the mid-nineteenth century, the great variability in print runs due to stereotyping meant that "the very concept of 'an edition' loses much of its meaning" (2004, 21–2). See also Altick 1969, 197–200; Altick 1986, 235–6. On sales patterns for popularizers' books, see Lightman 2007, 33–4.
9. Hassall 1857, 49. In citing "Beale" in these lists I refer to Beale's general handbook *How to Work with the Microscope* (1857), not his earlier book for medical readers, *The Microscope and Its Application to Clinical Medicine* (1854).
10. [Dixon] 1856, 380; Clarke 1858, iii; Ward 1864, 8, 65, 103; and (for Carpenter) 6–7 and passim; "Microscope," 1874, 443.
11. Suffolk 1870, 182.
12. For example, David Brewster's review essay in the *North British Review* examines Carpenter 1856, Ehrenberg 1854, Griffith and Henfrey 1856, Lardner 1856, Pritchard and Goring 1832, Pritchard 1834, Goring and Pritchard 1837, Pritchard 1852, and Quekett 1848—notably critiquing Carpenter and excluding Hogg.
13. Translations suggest not only domestic market success but also the publishers' influence and reach. Clarke and Hogg benefitted from being published by Routledge, which arranged joint publication in New York. Publications of the Religious Tract Society (RTS) were also likely to make an early journey to an American market. Thomas Dick's *The Telescope and Microscope* (RTS, 1848) was reprinted in Philadelphia in 1857, and Charles Williams' *Curiosities of Animal Life, with the recent discoveries of the microscope* (RTS, 1848) was reprinted in New York in 1855 (with a minor title change). Beale's *How to Work with the Microscope* did not appear in an American edition until 1880.
14. *Catalogue of New and Standard Works in Circulation at Mudie's Select Library* (1857).
15. *Catalogue of the Principal Books in Circulation at Mudie's Select Library* (1865). At this point Mudie's carried eleven books by Gosse, almost all aimed at broad audiences.
16. *Catalogue of the Books in the Library of the Quekett Microscopical Club* (1915).
17. Searson 1860, 13–28.
18. Only Lardner and *Chambers's Papers for the People* are missing from the later list, and some popularizers, like George Henry Lewes and Mary Ward, were added (see [Tierney] 1929). In the 1888 catalogue of Lewis's Medical and Scientific Library, a specialized circulating library, Beale, Carpenter, Hogg, Pritchard, and Quekett are well represented. However, Lewis's eschews many of the popularizing texts, largely but not exclusively privileging those who (like Gosse, Lankester, and Lewes) had earned reputations as workers (see [Tierney] 1929).
19. Plumer 1865, 6; Barnard 1941, 7. Plumer's little book was an expanded version of an 1864 *QJMS* article.
20. [Brewster] 1856, 472, 468–70. Ironically, in his 1842 *Principles of Human Physiology*, Carpenter explicitly "claim[s] a somewhat higher character than that of the mere Compiler" (xi).
21. Brock 1980 provides a useful history of nineteenth-century microscopical periodicals. See also Turner 1974.
22. "Quekett Microscopical Club," 1884, 48; "Illumination by Daylight," 1884, 622–4.

23. Sample publications consulted for this project include *Chambers's Edinburgh Journal*; *Chambers's Miscellany of Useful and Entertaining Tracts*; *Chambers's Papers for the People*; *Chambers's Journal of Popular Literature, Science, and Arts*; *Chambers's Useful Hand-Books*; *Chambers's Wonder Book of Light.*
24. See Ford 2001 on the microscope and the Royal Society. Charles Kingsley, indeed, says of one particularly inventive description in *Water Babies* that if it is "not all, every word, true, then there is no faith in microscopes, and all is over with the Linnaean Society" (Kingsley 1863, 198–9).
25. Csiszar 2018, 17, 200, 205.
26. Cooter and Pumfrey 1994, 248.
27. In the early decades of the nineteenth century, eminent men of science like Charles Lyell endorsed what is now identified as a diffusionist view, where popular discussions of science would be guaranteed by a "gentlemanly, disinterested, and apolitical" character; but an even earlier tradition had been more participatory. J. Secord 2014, 144.
28. Cooter and Pumfrey 1994, 256, 251, 249, 252.
29. Brock 1980; J. Secord 2000, 437–9.
30. Endersby 2008. See also Barton 1998, Barton 2003, Desmond 2001, and Belknap 2018.
31. Lightman 2016; Topham 2016.
32. O'Connor 2009, 341.
33. J. Secord 2014, 238–9.
34. A. Secord 1994, 277–8.
35. Topham 2004b, 440. See Topham 2009, 17 for an example of unintended reading of Buckland's *Bridgewater Treatise.*
36. Topham 2004b, 436. A paradigmatic case study of this sort is J. Secord 2000, but see also Topham 1998.
37. Topham 2009, 19; James Secord uses the term "knowledge in transit" to urge the study of "science as a form of communication" (2004, 655).
38. Clarke 1858, 1.
39. "The Third Annual Quekett Dinner," 1878.
40. Lewes 1859–60, vol. I, v–vi.
41. Fyfe 2004, 56.
42. Hogg 1854, v, his emphasis.
43. Hogg 1854, viii.
44. "Reviews," 1855, 36. Ironically, some of the most-cited, most-recommended authors—Quekett, Hogg, Carpenter—were called out for plagiarism, although they could also be lavishly praised in the same review. See, for example, Cassedy 1976 on Hogg.
45. "Reviews," 1856, 30–1.
46. In the Preface to his fourth edition (1859), Hogg discloses that he had sold "three editions, each of *five thousand copies*" (n.p.). Lightman considers a print run of up to ten thousand to be "moderately successful" and of up to twenty thousand to be "very successful" (2007, 34). On the sixth edition of Hogg's book and its history of successful sales, see "The Microscope," 1868, 84–5. The review was probably written by Edwin Lankester, an editor of *QJMS*, who was a prolific reviewer of books on science and medicine.
47. Carpenter 1856, vi, v.
48. "Carpenter on the Microscope," 1863, 352.
49. Quekett 1848, viii.
50. Hogg 1854, 130–1.
51. Hogg 1854, 182.
52. Carpenter 1856, vi–vii.
53. Pratt 1850, vi, 124.
54. Fyfe 2001, vi.
55. Rudwick 1982, 191. Rudwick's diagram is inspired by the exoteric/esoteric circles of Ludwik Fleck (1979).
56. The reader signs himself "W. T. Porter, 19 Otley Road, Headingley, Leeds" (Porter 1888, 48). *Science-Gossip* was a popular periodical founded by the mycologist Mordecai Cubitt Cooke.
57. Beale 1857b, 4.
58. Beale 1865, 264–5.
59. Lewes 1858, 54, 97.
60. Lewes 1860, 12.
61. [Brewster] 1856, 471.
62. White 1893, 1.

63. *Catalogue of the Books* (1915), 16, 21, 25, 29, 34–6, 39, 52, 55, 62–3. Somewhat surprisingly, Wood's *Common Objects of the Microscope* is not in the QMC list.
64. Darnton 1982, 68. Darnton later discusses the circulation of "the same anecdotes recounted in nearly the same words everywhere" and calls for study of "the system itself... intertextual combinations rather than the book as a self-sufficient unit" (2007, 507).
65. On anthologies, see L. Price 2000; on reviewing practices, see Dames 2010.
66. Topham 1998 provides an influential early example of this kind of analysis.
67. Dawson et al. 2004, 30.
68. O'Connor 2009, 336.
69. O'Connor 2009, 337, 339.
70. Shuttleworth and Charnley 2016, 298.
71. O'Connor 2009, 337.
72. Lightman 2007, 9–12.

Notes to Chapter 5

1. However, some idealist elements can be perceived in this work; Peter Garratt discusses how, despite their generally empiricist leanings, Spencer and Lewes align in some ways with a Kantian tradition (Garratt 2010, 46–7).
2. Lightman 2004, 200.
3. On Bain's readings but not experiments in physiology, see Rylance 2000, 173–4.
4. Rylance 2000, 173. The other authors named here are not particular boosters of microscopy. Laycock vied with Carpenter over the discovery of reflex action but hardly mentions microscopy in *Mind and Brain*; in *Lectures on the Principles and Methods of Medical Observation and Research* (1857) he warns medical students against too much reliance on, or interest in, microscopy (73–6). Holland and Maudsley cite microscopical research but were disappointed in its results at the time of writing. Holland was more interested in the effects of mental attention on bodily functions. He considered "minute anatomy" interesting and suggestive but, given the state of microscopy and chemistry at the time, doubted that analysis along those lines could provide useful knowledge about nervous function (1852, 250–3). Fifteen years later, Maudsley, who studied mental illness, directed readers to the work of Bain, Spencer, and Carpenter but, like Holland, he did not emphasize neural anatomy, considering it of only limited use (1867, vii). He remarks that "intimate special differences" in brain cells might underlie the variousness in neural functions, but these physiological distinctions "are not such, indeed, as the microscope is ever likely to reveal" (56–7). Even pathological changes in brain tissue are not, he says, particularly helpful: they "certainly do not admit of very definite description" (391).
5. Rylance 2000, 252–3.
6. R. Smith 2004, 81.
7. Rylance 2000 provides useful surveys of the reception of Bain, Spencer, and Lewes. Roger Smith argues that a "typical" response in mid-century periodicals was "a temperate articulation of middle ground" that he calls an "expressive reassurance" that these new ideas would not significantly change Victorian values (R. Smith 2004, 89–90).
8. *Catalogue of the Principal Books in Circulation at Mudie's Select Library*, 1865, 9, 73, 116. Lewes's Problems of Life and Mind was not yet published.
9. *Catalogue of Lewis's Medical and Scientific Library*, 1888, 9, 35–6, 119, 184.
10. [Tierney] 1929, 22. As late as 1929, the RMS library catalogue still lists Carpenter's *Animal Physiology* (1859) and *Principles of Human Physiology* (6th ed., 1864).
11. Carpenter 1856, 158, 160, 185–6.
12. Carpenter 1874, 195, 287.
13. Spencer 1855, 584–6, 613.
14. Bain 1859, 8.
15. Bain 1859, 340.
16. Bain 1859, 475–6.
17. Bain 1859, 215.
18. Rylance 2000, 170, 208.
19. Bain 1859, 475–7, his emphasis.
20. Spencer 1855, 221, 223, 236.
21. Carpenter 1842, 75.
22. Carpenter 1874, 1. See also Cohen 2009, xiii, xvi.
23. Carpenter 1874, x, 8–9, 676.
24. Bain 1859, 475–7.

25. See, for example, Carpenter 1856, 155.
26. Lewes 1879, 6, 112.
27. Lewes 1879, 74.
28. Lewes 1879, 74.
29. Bain 1855, 61–2.
30. Lewes 1859–60, vol. II, 4–6, his emphasis; E. Price 2014.
31. Lewes 1877, 166, 200, 191.
32. Schneider 1899, 105.
33. Bain 1855, 57.
34. Bain 1855, 57, 61.
35. Bain 1855, 57.
36. Bain 1855, 60, 62.
37. Bain 1859, 618.
38. Lewes 1877, 303–4, 366, 324.
39. Rylance 2000, 179–80.
40. See Garratt 2010, 151–60 on the model of a "continuous energy network" (157); Rylance 2000, 169, 174–6 on its acceptance.
41. Bain 1855, 359.
42. Rylance 2000, 10–1.
43. Rylance 2000, 222–3.
44. Rylance 2000, 275.
45. Bain 1859, 210, 213.
46. Bain 1855, 333–4, his emphasis.
47. Bain 1859, 339.
48. Barad 2007.
49. Bain 1859, 219.
50. Bain 1859, 216.
51. Bain 1859, 194, 216.
52. Lewes 1879, 92.
53. Bain 1859, 210, 219.
54. Lewes 1879, 98.
55. Lewes 1879, 160–1.
56. Rylance 2000, 279.
57. On "granule" and exactness of description, see Beale 1857b, 92–3.
58. Reddy 2001, 128–9.
59. Rosenwein 2010, 11.
60. Brennan 2004, 9.
61. Brennan 2004, 70, 9.
62. Massumi 2021 [2002], xxxix, xliii.
63. Ahmed 2014, 10.
64. Scheer 2012, 199, 209.
65. Ahmed 2014, 2, 8–11, 45.
66. Dames 2004, 210.
67. Vanessa Ryan emphasizes the extent to which these writers saw this training as a means to improve society; she argues that Victorian fiction provided a vehicle for such training (2012, 79, 86–7).
68. Carpenter 1856, 165.
69. Hermann von Helmholtz famously formulated this law in 1847, but others contributed to its establishment. It became one of the main tenets held by Victorian scientific naturalists like Huxley.
70. Rabinbach 1992, 53, 52, 55; Paterson 2021, 224.
71. Paterson 2021, 224.
72. Carpenter 1856, viii, his emphasis.
73. Rabinbach 1992, 6, 20. On fatigue in the factory setting, see also Sussman 2009, 49–53; Ketabgian 2011, 52–3.
74. Rabinbach 1992, 38, 22.
75. Carpenter 1846, 221–2.
76. Carpenter 1853, 844.
77. Bain 1855, 58–9.
78. Bain 1859, 193.
79. Bain 1859, 608.

80. Bell 1833, 206.
81. Rabinbach 1992, 48.
82. Ryan 2012, 4, 10.
83. Spencer 1855, 461–3.
84. Lewes 1858, 54.
85. "Proceedings of Societies: Brighton and Sussex Natural History Society," 1870, 118.
86. "E. B. P." [Edward Bagnell Poulton], "Elementary Instruction," 1888, 505–6, reviewing the new edition of Thomas Henry Huxley and Henry Newell Martin's *Course of Elementary Instruction in Practical Biology*. Quoted in "Summary of Current Researches: Microscopy: Learning to See with the Microscope," 1888, 495.
87. Winslow 1854, 121. The case was *Morgan* v. *Boys* (1838).
88. Maurette 2018, 6–7.
89. Downes et al. 2018, 11.
90. Barad 2003, 815.
91. Taylor 1889, 221–4.
92. For scratched slides, see Hannover 1853, 47.
93. Clarke 1863, 231.

Notes to Chapter 6

1. Leeuwenhoek to Henry Oldenburg, Sept. 7, 1674, p. 111.
2. Robert Hooke, "A Letter of the Ingenious and Inquisitive Mr. Leeuwenhoeck [sic] of Delft, sent to the Secretary of the Royal Society, 5 October 1677." Quoted in Gest 2004, 195.
3. Letter from Leeuwenhoek to Hooke, Nov. 12, 1680, p. 200.
4. On the formalization of scientific writing, see Gross et al. 2002. Nineteenth-century "scientific" writing was, of course, a literary style of its own, but its literary elements have not always been recognized. See, for example, Merrill 1989; Dear 1991.
5. Gross et al. also note the more extravagant language enabled in Introductions and Conclusions (2002, 136); and O'Connor in fact focuses upon the "pictorial" moments that, he says, "tended to cluster around beginnings and endings" (2008, 25).
6. For example, see Durham 1868–9, 95.
7. Scott 1830, 34, 40.
8. Scott 1830, 389–91.
9. Scott 1830, 35, 173.
10. Brewster 1832, 98–106.
11. Brewster 1832, 5.
12. Carlyle 1840, 10–11.
13. Carlyle 1827, 96; the translation is of *The Life of Quintus Fixlein*.
14. Ruskin [1887] (35.2.10), 430.
15. Weber 1958, 117, 120–1.
16. Levine 2008a, 24, 4.
17. Shapin 2019, 11.
18. Weber 1958, 116–17.
19. Jane Bennett 2016, 37.
20. "The Microscope and Its Marvels," 1850, 14.
21. "The Microscope and Its Marvels," 1850, 14.
22. Stafford and Terpak 2001, 53.
23. The final two editions were completed by the Rev. W. H. Dallinger and E. M. Nelson (unattributed).
24. Carpenter 1856, v.
25. "New Books with Short Notices," *MMJ*, 1873a, 67.
26. See Lightman 2007, 20.
27. "The Microscope and Its Marvels," 1850, 6–7, 9.
28. Hogg 1854, 335, my emphasis.
29. G. T. F. [Fisher] 1847, 35. Fisher was reviewing, among other books, his own *Microscopic Manipulation* (1846).
30. On the laboratory, see Gooday 1991 and Gooday 2008.
31. On depersonalization, see Gross et al. 2002, 122–4, 158–9; on explanation and theory, 141, 158–9.
32. Csiszar 2018, 14.
33. Daston and Galison 2007, 122.
34. Daston and Galison 2007, 321.

35. Daston and Galison 2007, 120.
36. Willis 2011, 8.
37. J. Smith 1994, 13. Smith's first chapter offers a thoughtful historical review of nineteenth-century revisions or rejections of strict Baconian induction, including the work of Whewell, Herschel, Tyndall, and Lewes, among others.
38. Elkins 2011, 89.
39. Bain 1859, 67–9, his emphasis.
40. Bain 1859, 69–70.
41. Bain 1859, 46–9, 70.
42. Bain 1859, 72, 623.
43. Bain 1859, 70–1.
44. Bain 1859, 189.
45. Bain 1859, 606.
46. Lightman notes that William Houghton and the Rev. J. G. Wood emphasize the beauty and wonder of the microscopic world (Lightman 2007, 84, 192–3).
47. Levine 2008a.
48. Beale 1854, 84, 133, 153.
49. Longinus provides an earlier source who was important to the Romantics, but his version of the sublime as an aesthetic of nobility and grandeur is less relevant to the nineteenth-century microscopical tradition.
50. Burke 1757, 95–104.
51. Kant [1790], §10–11, 14.
52. Burke 1757, 102–3; Kant [1790], §14.
53. Royston-Pigott 1869. The "beading" was eventually shown to be illusory; see Chapter 18.
54. "Proceedings of Societies: Croydon Microscopical Club," 1871, 54.
55. Lang 1870, 308. Lang was President of the Reading Microscopical Society at the time.
56. Merrill 1989, 122–4; Braga 2021.
57. Schor 1987.
58. See, for example, Chico 2018, 31–2.
59. Somerville 1869, vol. II, 203–4.
60. Clarke 1858, 11.
61. Butler 1886, 96–7. The current nomenclature for this moth is *Nymphula nitidulata.*
62. Clarke 1858, 11.
63. Wood 1861, 25, 46, 77. On *Common Objects of the Microscope*, see Lightman 2009, 201.
64. [Dixon] 1856b, 292.
65. [Dixon] 1856b, 293.
66. [Dixon] 1856b, 294.
67. See Stafford 1996. Rebecca Stott argues that natural theology texts tend to present analogies that retain an organic sense of unity, whereas some in the natural history tradition embrace a comic sense of hybridity that borders on the grotesque (Stott 2002, 169).
68. [Dixon] 1856b, 293.
69. Elkins 1992. Elkins focuses on the metaphorical description of "animalcules" as wild animals, which was common in describing lively infusoria for general audiences; but authors also turned to a strategy of figuration in describing many other species as well.
70. See, for example, Endersby 2008, 154–69 on the distinction between "lumpers" and "splitters" of categories, and Joseph Hooker's efforts to reduce the proliferating number of botanical species.
71. G. T. F. [Fisher] 1846, 43. He quotes from Arthur Hill Hassall, *A History of the British Freshwater Algae* (1845).
72. Carpenter 1856, 6.
73. See Mershon 2016, 473–4; Strick 2000.
74. See Worboys 2000, 2–3.
75. J. Secord 2009, 470.
76. Pratt 1850, 95.
77. Bennett 1850, 197.
78. Clarke 1858, 44–5.
79. Carpenter 1856, 318.
80. Carpenter 1868, plate X.
81. White 1887, 25–6.
82. Jane Bennett 2016, 169.
83. Morehouse 1874, 19.

84. Morehouse 1874, 21, 25.
85. William Saville Kent was the brother of Constance Kent, who in 1865 confessed to the 1860 murder of her younger half-brother Francis, in a sensational case that received wide coverage in Victorian media; some suspected William was an accomplice. His readers would have been aware of this shadow on his family history as well as his important work on ocean life.
86. Saville Kent 1880, vol. I, viii.
87. Saville Kent 1882, vol. II, 668.
88. Carpenter 1856, v–vi. The use of tropes of beauty is not materially different in the 1901 edition of *Revelations* revised by Dallinger.
89. Carpenter 1856, 206, 318.
90. Quekett 1848, 364.
91. Carpenter 1856, 395.
92. P. Gosse 1859, 272.
93. J. Smith 2006, 84.
94. P. Gosse 1859, 275–7.
95. P. Gosse 1887, 862–4, 868–9.
96. Stokes 1887, 37, 35.
97. Beale 1857b, 76.

Notes to Chapter 7

1. Burke 1757, 96–7.
2. For a discussion of the history of the microscopical sublime in the eighteenth century, see Nicolson 1935, especially 45–6, 67–8; Wilson 1995, 82.
3. See, for example, Fisher 1998, 1–2.
4. Burke 1757, 41–2.
5. Kant [1790], §23, 25, 26. Considering the effect in telescopes and microscopes of disclosing a nested series of worlds growing infinitely large in one direction and infinitely small in the other, Kant argues that "there is nothing in nature," no matter how small, that wouldn't look large to us in relation to a still smaller scale of existence (§25). Thus no single object can be considered in itself sublime. This argument does not preclude microscopic infinity from evoking in us a judgment of the sublime.
6. Kant [1790], §29.
7. Burke 1757, 43, 51–2, 60.
8. The nineteenth-century microscopical sublime corresponds to Kant's "mathematical" sublime. His "dynamical" sublime, on the other hand, occurs when we observe greatness not in extent but in force (§24). James Elkins examines another kind of microscopical image that could be considered sublime, although he resists using the term. He studies twentieth-century images of microscopic objects as examples of the "limits of representation" because they show how "how some objects resist depiction" (2008, xv). He describes this as the "last moments of ordinary representation" (16).
9. Jason Lindquist finds a "post-Kantian pattern" among mid-century Victorian scientists such as Tyndall and Lewes, who, Lindquist argues, reverse Kant's narrative of the mind-boggling sublime. He specifies, "where Kant had given the name 'imagination' to the collapsing faculty, Victorian writers such as Tyndall assign that name to the faculty that resists or recuperates meaning from surfeit" (2008, 192). Lindquist defines the Victorian visual imagination in roughly Kantian terms—as that which takes in multitudinous diverse particulars via sensory experience and negotiates between these and an overarching unity, pattern, or system. But he also acknowledges conflicting uses of the term and "a continuing ambiguity in British definitions of the word at mid-century" (174, 183).
10. Pascal 1859, 159–60.
11. Henchman 2014, 2.
12. Jacques Derrida remarks that Kant, describing the sublime, associates it with the absolutely large but not the absolutely small. Derrida comments that Kant here reintroduces the notion of comparison and thus measurable size into the supposedly immeasurable sublime (Derrida 1987, 136–7). This argument aside, Victorian observers did find the sublime in the vastness of the very small, as the examples for this section demonstrate.
13. Carpenter 1856, 4.
14. Carpenter 1856, 37.
15. "The Horror of Astronomy," 1891, 285.
16. Carpenter 1856, 4.
17. Carpenter 1856, 36–7.

18. Dick 1851, 152–3.
19. Harvey 1849, 173–4.
20. O'Connor 2008, 22.
21. Hogg 1854, 230, 131.
22. Roget 1834, 12.
23. [Hart] 1857, 511.
24. Roach 1991, 131.
25. Jones 1845, 99. This passage circulated widely in nineteenth-century texts and periodicals.
26. Kerr 1895, 133.
27. Kerr 1895, 135–6. John Tyndall had offered a somewhat different view of the respective roles of the microscope and the imagination, in his 1870 Address to the British Association for the Advancement of Science on "The Scientific Use of the Imagination." He warned of areas where the microscope fails: it "can have no voice in the real question of germ structure" nor in questions of molecular form, which present "a world of matter and of motion to which the microscope has no passport." Regardless, he argues, "beyond the present outposts of microscopic enquiry lies an immense field for the exercise of the speculative power" (Tyndall 1870, 40–1).
28. [Joyce] 1806, 17, 12. Joyce was a Unitarian minister and radical activist who had been imprisoned for treason in 1794, although charges were eventually dropped. He drew heavily on the work of Robert Hooke for his popular book on the "wonders of the microscope," published anonymously in 1806; he published "dialogues on the microscope" (for children) under his own name in 1811.
29. Lewes 1860, 27.
30. De Morgan 1872, 377 ("The Siphonaptera"), based upon a verse in Jonathan Swift's long satirical poem "On Poetry: A Rhapsody" (1733).
31. Lewes 1860, 28, 13.
32. Hume 1777, ¶12.18.
33. Hume 1777, ¶12.20.
34. Among the nineteenth-century authors promoting the plurality of worlds were the enthusiastic preacher Thomas Chalmers, in his *Astronomical Discourses* (1817), and the "Christian Philosopher" Thomas Dick in most of his books, especially *Celestial Scenery; or, the Wonders of the Planetary System Displayed* (1838). William Whewell, in *Of the Plurality of Worlds. An Essay* (1853), rejected the possibility of extraterrestrial intelligence while admitting the possibility of animal life on other planets.

    The doctrine of successive acts of special creation of species became more hotly debated in the Victorian period due to arguments for deep time, derived from the fossil record, in James Hutton's *Theory of the Earth* (1789) and Charles Lyell's *Principles of Geology* (1830–3); and the challenges to special creation in Robert Chambers' *Vestiges of the Natural History of Creation* (1844) and Charles Darwin's *Origin of Species* (1859). Whewell juxtaposes the theories of special creation and plurality of worlds as analogous models, rejecting the existence of many worlds inhabited like Earth by reference to the "doctrine of the successive creation of species" over "immense intervals of time" like the interstellar "intervals of space" (77–8).
35. Churchill 1989, 191.
36. Kennedy 2019, especially 104–11.
37. Freedgood 2019, xv, xvii.
38. Freedgood 2019, xv.
39. Pritchard 1841, 8. Translated by "Dr. Willshire" (v).
40. Pritchard 1841, iv–vi. This somewhat haphazard inclusion of important material in appendices might have reflected not just late-breaking news but also a lack of organization; a later edition of this text was chastised by a reviewer as being "disfigured and injured by its slovenly arrangement." "Miscellanies," 1857, p. 389.
41. Marx 2000, 195.
42. Burke 1757, 97, 115, 161; Kant [1790], §23, 28.
43. Burke 1757, vi, 60–1.
44. Kant [1790], §23, 24, 27.
45. Kant [1790], §23.
46. Kant [1790], §29.
47. Ward 1864, viii, 153.
48. Ward 1864, 153–4.
49. Ward 1869, vol. 1, 206.
50. Kerr 1895, 125–6.
51. Burke 1757, 84–90.

Notes to Chapter 8

1. Brooke 2014, 285–9.
2. Brooke 1974, 8–9. Aileen Fyfe discusses evangelicals' role in defining theologies of nature (2004, 6–8, 94, 111).
3. Brooke 1974, 9.
4. Topham 2004a, 38.
5. Brooke and Cantor 1998, 179.
6. Brooke and Cantor 1998, 184–95. *Theologies of nature*, the *discourse of design*, and the genre of natural theology as discussed by Brooke and Cantor are distinct from the *narrative of natural theology* identified by Barbara Gates and Ann Shteir. This is usually a charming pedagogical story in a domestic setting, in dialogue, with a Paleyite moral. See Gates and Shteir 1997, 11–12; Gates 1998, 39–41.
7. Natural history refers to naturalists' study of the natural world (including botany, geology, biology, zoology, and entomology) via observation, often examining objects through magnification. For more on the *narrative of natural history*, a conversational genre emphasizing curiosity, wonder, and admiration, see Chapter 1.
8. King 2019, 12.
9. Goring, "Exordium," in Goring and Pritchard 1829, 10.
10. Goring and Pritchard 1829, 11.
11. The father instructing his son in Joyce's 1812 *Dialogues on the Microscope* calls the mosquito's proboscis a "fine sharp trunk." When the son protests that it is "not easily seen," the father replies, "It is with the microscope" (vol. 1, 80–1).
12. Olley 1857, ii.
13. Olley 1857, 8–9, 13.
14. Ferguson 1858, iii.
15. "Art. III.—The Microscope: And Its Relations [sic]," 1856, 349.
16. "Art. III.—The Microscope: And Its Relations [sic]," 1856, 349.
17. Clarke 1858, iv.
18. Clarke 1858, 39.
19. Carpenter 1856, 37.
20. Hogg 1854, 429. The original passage appears in Baker 1742, 300–1.
21. Hogg 1854, 430. The original passage appears in Baker 1742, 301.
22. Hogg 1854, 430. Hogg slightly misquotes the passage.
23. Hogg 1854, 433.
24. This is a portion of a passage Hogg encloses in quotation marks in the 1854 edition but not in the 1856 edition. The passage originates in Baker 1742, 310–11.
25. Baker 1742, 302.
26. Hogg 1856, vi.
27. Hogg 1856, 450–1.
28. Marjorie Nicolson surveys a range of seventeenth- and eighteenth-century versions of this trope (1935, 63–6). See also Schickore 2007, 25.
29. Joyce 1806, 115–16. Joyce makes minor changes to Baker's words. See Baker 1742, 297–8.
30. Joyce 1806, 117; Baker 1742, 296–7.
31. Joyce 1806, 117; Baker 1742, 299.
32. Joyce 1806, 27; Baker 1742, xii.
33. Hogg 1854, 138.
34. Gatty, "Microscopic Objects," reprinted in Gatty 1871, 98–107, p. 106.
35. On the eye as a beautiful mechanism, see Schickore 2007, 29–34.
36. T. Brown 1974.
37. P. Gosse 1859, 60.
38. P. Gosse 1859, iii.
39. Hogg 1854, 300.
40. Dick 1851, 166.
41. Hogg 1854, 272, 279.
42. Reade 1849, 21, 23.
43. "Obituary Notices: Joseph Bancroft Reade," 1871, 94.
44. Goring 1827b, 273.
45. Ross 1839, 182.
46. "The Microscope and Its Marvels," 1850, 6. The article attributes this statement to Carpenter; it is likely to derive from his series of microscopical lectures at the Royal Manchester Institution in 1845. See "Recent Revelations of the Microscope," 1846, 114.

47. Chico 2018, 50.
48. Pritchard and Goring 1832, 192. Glassmakers use the term *metal* to describe the mixture of materials used to make glass, especially in molten form. Here and elsewhere Goring uses the term, metonymically, to refer to what is made from this mixture, the individual glass lenses that are then combined to make doublets or triplet object-glasses. See, for example, Pritchard and Goring 1832, 205. The phrase "exact figures" here denotes the carefully worked physical shape of the finished lenses.
49. Welborn, "An Inexpensive but Efficient Microscope," 1882, 495.
50. Quekett 1848, 203, 31, 16, 24.

Notes to Chapter 9

1. King 2019, 11, 10.
2. G. T. F. [Fisher] 1846, 54.
3. Jackson 1835, 158–9.
4. See Kennedy 2010.
5. See Lightman 2007, 372.
6. Browning 1864, 219.
7. [White] 1887, 27.
8. Daston and Park 1998, 17; Fisher 1998, 1.
9. Kerr 1895, 134.
10. Gatty 1871, 104–5.
11. Goring 1827a, 433.
12. "The Microscope and Its Marvels," 1850, 22–3.
13. The reference was in Shadwell [1676], I.ii.12–13. Cole 2016 and Chico 2018 discuss this tradition.
14. Bacon [1620], 106.
15. Nicolson 1935, 50–6, 65; Armintor 2011, 67–8, 72.
16. "The Microscope and Its Marvels," 1850, 1–2.
17. Quigley 1819, 8.
18. Swinburne 1899 [1872], 83.
19. Goring 1827b, 271.
20. Goring and Pritchard 1829, 8. The passage derives from the "Exordium" penned by Goring.
21. Goring and Pritchard 1829, 9–10.
22. Quekett 1848, 379–80.
23. Kingsley 1855, 88.
24. See, for example, P. Gosse 1859, 60; "The Microscope and Its Marvels," 1850, 3; "J. P. P." (likely J. P. Philips) 1859, 5; title page epigraph to Somerville 1869, vol. 1. The trope is a simplification of a passage from Pliny the Younger (Hist. Nat. XI. 1); Augustine renders it in a sacred mode, as "*Deus magnus in magnis, maximus in minimis*" (God is great in great things, most great in the smallest things).
25. Carlyle 1837, 3.
26. Jane Bennett 2016, 4.
27. Dickens 1850, 1.
28. Ngai 2005, 22.
29. Ngai 2005, 335.
30. Ngai 2005, 336.
31. [White] 1887, 47–8.
32. Hogg 1854, 310, 313–15.
33. Pirie 1864, 112, 119–20.
34. Pirie 1864, 115.
35. Notcutt 1859, 100.
36. Dick 1851, 126–7.
37. "The Microscope and Its Marvels," 1850, 22.
38. Marsh 1878, vi, 72, 78, 79, 82.

Notes to Chapter 10

1. The oxyhydrogen microscope was a projection microscope that used brilliant oxyhydrogen light (limelight) to create a kind of hybrid of the solar microscope and magic lantern. It was also known as the lantern microscope or gas microscope. Later in the century, electric arc lighting was used for the same purpose.
2. Tresch 2012, 1–2.

3. Merrill 1989, 13.
4. Merrill 1989, 12.
5. Ursula Seibold-Buitmann relates the aesthetic of lantern microscopy to Victorian fairy paintings; Melanie Keene and Laura Forsberg examine the pedagogical role of fairy-talk in science writing for children or general audiences (Seibold-Buitmann 2013; Keene 2015, especially 82–109; Forsberg 2021, 109).
6. These fairy references ask readers not to believe in fairies outright but rather, as Forsberg says, to "imagine[e] believing in them" (Forsberg 2015, 109).
7. The story may have been plagiarized, either from William North or Vladimir Odoevsky; see Whitley and Weidman 2009, 41–2.
8. O'Brien 1858, 364, 354.
9. O'Brien 1858, 363–7.
10. Carole Silver identifies 1880–1910 as a period when Victorians were especially interested in fairy culture (2000, 6).
11. Badcock 1883, 5–7.
12. Badcock 1883, 6.
13. Dwight Atkinson identifies a similar concentration of "author-centered style" at the beginning of eighteenth- and nineteenth-century scientific articles from the *Philosophical Transactions*; and Gross, Harmon, and Reidy note that introductions and conclusions could more often "deviat[e] into the personal and the polemical" than the body of scientific articles. Atkinson 1998, 79; Gross et al. 2002, 136.
14. See, for example, Durham, "President's Address," p. 95. Gerard L'E. Turner notes that the Presidential Addresses of the RMS became more like a scientific lecture beginning in 1871, when Henry J. Slack steps in to deliver the address in the place of the recently deceased Joseph Bancroft Reade (Turner 1989a, 47).
15. Piers Hale argues that William Carpenter's 1872 Presidential Address to the BAAS shows how the members of the Metaphysical Society could "define the terms of key debates within that community" (2019, 219).
16. Hudson 1890, 132. The address was reprinted in *Nature*.
17. "Obituary. Charles Thomas Hudson," 48; Wootton 2011.
18. G. T. F. [Fisher] 1846, 35.
19. P. Gosse 1853, 240.
20. P. Gosse 1859, iii.
21. Blackham 1882, 25.
22. [Hart] 1857, 511.
23. Harvey 1849, 169. Quotations are from Spenser's *Faerie Queene*, Book IV, Canto xii, and Newman's "The Invisible World," Sermon XIII, *Parochial Sermons* (1849). Newman discusses the world of visible "brute animals" but Harvey turns this passage toward the wonder of microscopic ones instead.
24. Gould 1835, 26–7. The "ditiscus" refers to the larvae of *Dytiscus*, a diving-beetle; "Hydropholus" refers to the larvae of *Hydrophilus*, a water-beetle.
25. Quekett 1848, 340.
26. See Keene 2015, 82–104; Keene focuses on fairylands but examines a range of examples of this trope.
27. Catlow 1851, x–xii.
28. Hudson 1889, 179.
29. See Keene 2015, 91–9 on "water-panics" and Kennedy 2019, 95 on the contribution of the oxyhydrogen microscope to these.
30. "Microscopic View of a Drop of Water," 1833, 177; "Mr. New's Grand Exhibition!"
31. Leech 1850, 188; Morley 1850.
32. Mathews 1839, 176–7.
33. Catlow 1851, x–xii, 4.
34. Brewster 1864, 169.
35. Brewster 1864, 176.
36. Hogg 1854, 122.
37. Kennedy 2019, 97.
38. Clarke 1863, 231.
39. Warner 1982, 32; Benjamin 1996, 122.
40. A. Secord 2002, 51.

41. Hassall 1850; originally published in *The Lancet*. See Hamlin 1990, 322 for other notable analyses of water and air.
42. Hassall 1850, 5.
43. Hassall 1850, 1–2.
44. Zieger 2018, 104–5.
45. Castle 1988, 27, 29, 33–40, 45, 47, 51.
46. Crary 1990, 132–3; see also Zieger 2018, 104–5.
47. Hunt 2008.
48. Armstrong 2008, 319, 321, 329–31.
49. McGrath 2003, 165.
50. Willis 2011, 12, 25.
51. Kennedy, 2019, 91–2, 94–101.
52. Willis 2011, 27. McGrath comments that "Crucially, and this remains the case, the manipulating hand always remained out of the frame" of the microscopic image (McGrath 2003, 165). The impossibility of portraying the operator's hand in a microscopic image aside, hands did indeed appear in illustrations of the microscope and its tools; see Part III of this book on difficulty.
53. Carpenter 1856, 158.
54. See Part III on difficulty.
55. Wood 1861, 1–2.
56. Hogg 1854, xi.
57. See Part III on difficulty.
58. Clarke 1858, 3.
59. Clarke 1858, iii.
60. Clarke 1858, 8.
61. Goring and Pritchard 1830, 6–7.
62. See Radick 2007, 50–83; Harley 2014.
63. Lewes 1858, 255.
64. Lewes 1858, 341, 365.
65. Lewes 1858, 369–70.
66. Lewes 1858, 364–5.
67. Lewes 1858, 213.
68. Lewes 1858, 51.
69. Lewes 1858, 57.
70. Lewes 1858, 136.
71. Lewes 1858, 136.
72. Wood 1861, 161.
73. Wood 1861, 166–7.

Notes to Chapter 11

1. Sprat 1667, 111–13, 334–5.
2. Hume [1777], ¶12.25.
3. Daston and Park 1998, 19.
4. Whewell 1840, xxxviii, 281.
5. Gross et al. 2002, 69, 244.
6. Kennedy 2010.
7. Wells 1894, 301.
8. Lightman 2007, 364, 366–7.
9. A. Secord 2002, 31–2.
10. Goring and Pritchard 1829, 12.
11. Goring and Pritchard 1829, 32.
12. Goring and Pritchard 1829, 20–2, Goring's emphasis.
13. Clarke 1858, iv (NB while this page is marked "iv," the previous pages are unpaginated, and the following page is marked "vii"). She mentions this again in the discussion of *Echinus* (201).
14. Clarke 1858, iv.
15. Clarke 1858, 7.
16. Lang 1870, 308.
17. Michael 1895, 3, 10.
18. Michael 1895, 10.
19. Carpenter 1856, 428.

20. Carpenter 1856, 25.
21. McCalla 1883a, 769–70.
22. Cole 1895, 163, 173.
23. Bernard Lightman explains that the scientific naturalists were "naturalistic in the sense that they ruled out recourse to causes not present in empirically observed nature, and...scientific in that they interpreted nature in accordance with three major midcentury scientific theories, the atomic theory of matter, the conservation of energy, and evolution" (Lightman 2007, 6).
24. Opposition to scientific naturalism simmered for various reasons among parson-naturalists, gentleman naturalists, north British energy physicists (who wanted to blend scientific naturalism with the Christian faith), scientific popularizers (many of whom retained the language of natural theology), a new group of scientific "Anglicans with an interest in spiritualism," and socialist writers on science (see Lightman 2007, 6–9; Lightman 2009; Lightman and Dawson 2014, 14–15).
25. Harrison 2006, 28.
26. Joseph Hooker, Charles Darwin, and John Tyndall were not regular members of the MSL/RMS or the QMC. John Tyndall appears in the annals of the RMS via republished material in the ongoing dispute in the 1870s over spontaneous generation, but not as the author of any paper submitted to the Society. Charles Darwin worked for eight years at the microscopic examination of barnacles, and Fellows of the RMS—such as William Benjamin Carpenter, William Henry Dallinger, Henry James Slack, and Henry Clifton Sorby—were among his extensive correspondents. However, he declined to publish within its (or indeed any) journals due to poor health (Darwin to H. J. Slack, Dec. 2 [1867]).
27. Desmond 2001, 40–1; Lightman 2001, 347.
28. Huxley 1853, 2. The paper was delivered on Dec. 31, 1851.
29. Huxley 1880, vi.
30. Levine 2008a.
31. Reidy 2014.
32. Tyndall 1871, 102, 104.
33. Tyndall 1871, 111.
34. Barton 1998, 418; Lightman 2019.
35. Lightman 2004, 200.
36. Desmond 2001, especially 35–40. Of the twenty-six Presidents of the RMS between 1840 and 1901, fifteen (nearly two-thirds) were either surgeons or physicians. Huxley was, of course, himself trained as a surgeon.
37. Nicolaas Rupke notes that "Owen published on the ideal vertebra as early as 1840" (1993, 233).
38. Beale 1880b, 202.
39. Beale 1869; Beale 1880b.
40. Hudson, "Preface," in Hudson and Gosse 1886, vol. 2, vi.
41. Desmond 1989, 210–22.
42. Barton 1998, 411–12. On the X-Club, see also Desmond 2001; Barton 2017.
43. Hale 2019, especially 225–6. On Carpenter's navigation of the materialism debate, see Winter 1997, 36–42.
44. Scholars have interrogated Carpenter's complex alignments. Shannon Delorme notes that Carpenter framed his neurophysiological researches as supporting "an expanded and revised natural theology," acknowledging a "divine mind" as a first cause behind all natural force and causation. Piers Hale argues that Carpenter "was instrumental in forging a new post-Darwinian scientific orthodoxy around a theistic account of the natural world that was not based in traditional natural theology." Delorme 2014, 60–1, 64; Hale 2019, 227.
45. E. Lankester 1863.
46. See Strick 2000, 119–23.
47. Toal 2013.
48. "Summary of Current Researches...Zoology," 1884, pp. 721–2. The article quotes Dallinger's address as reported in *The Times* of Sept. 2 and 4, 1884.
49. Toal 2013.
50. Marsh 2019, 4.
51. Gerard L'E. Turner remarks, "The high proportion of medical men among the early members of the Society bears witness to the importance of the improved, post-Lister microscope in all branches of medicine" (Turner 1989a, 9). The value of the microscope in medicine was not widely accepted until later in the Victorian period, but microscopists who were medical men were often advocates for its use in their profession.

52. See Daston and Galison 2007, 120–3. Whether these are bourgeois virtues or Christian ones, they provide moral guidance; however, in the context of microscopy these virtues are not "non-instrumental" (Levine 2008b, 72).
53. On difficulty and error in microscopy, see Part III.
54. Hooke 1665, n.p., his emphasis.
55. Brooke and Cantor 1998, 181–3.
56. King 2019, 16.
57. Darwin to J. M. Herbert, Nov. 21, 1872.
58. Darwin to Joseph Hooker, May 1, 1847; Darwin to Hooker, Nov. 14, 1846.

Notes to Chapter 12

1. On scientific error and its mitigation, see Gooday 1991, Schickore 2007, and Rheinberger 2010.
2. Flint 2000, 8, 30.
3. Armstrong 2008, 325, 327.
4. Willis 2011, 5.
5. Huxley 1880, 2.
6. "Obituary. Christian Gottfried Ehrenberg," 1877, 167.
7. Rylance 2000, 259.
8. Hume [1777], ¶12.4, ¶12.6. For the influence of Hume on nineteenth-century skepticism, see Garratt 2010, 43–70.
9. Hume [1777], ¶5.8.
10. Hume [1777], ¶12.9.
11. Hume [1777], ¶5.5.
12. Fieser 2005, 428, 437, 439, 429.
13. See Garratt 2010, 43–70. However, these thinkers also rejected Hume's anti-materialism (Potter 2018, 143).
14. Fieser 2005, 437; Potter 2018, 54, 85–6, 114, 120.
15. Herschel 1830, 83. He presumably refers here to the optical telegraph (semaphore signaling).
16. Lewes 1859–60, vol. II, 335–8, 341–3, his emphasis.
17. Garratt 2010, 35.
18. The surgeon Joseph Lister, a pioneer of antisepsis, was a young child at the time of his father's experiments in microscopy.
19. Quekett 1848, 146.
20. Using his terms and definitions, penetrating power (the ability of the instrument to show elements lying outside the plane of focus) is always inversely related to angle of aperture and to resolving power (definition of image). Carpenter 1856, 193–5.
21. Schleiden 1849, 584.
22. Schleiden 1849, 585.
23. Chico 2018, 33.
24. Lankester, "Translator's Preface," in Schleiden 1849, iii.
25. Schleiden 1849, 589–91, my emphasis. The original passages appear at Schleiden 1842, 151, 154; I use Lankester's translations of these terms.
26. For the uneven British uptake of mid-century German botanical technique, see Kennedy, "Circulation and Civility: Mid-Victorian Botany and Microscopical Method," 2024.
27. Whewell 1840, xxxix. He clarifies, "The act of the mind, by which it converts facts into theories, is of the same kind as that by which it converts impressions into facts. In both cases there is a new principle of unity introduced by the mind" where "that which was many becomes one; that which was loose and lawless becomes connected and fixed by rule. And this is done by induction…by superinducing upon the facts, as given by observation, the conception of our own minds" (1840, 44).
28. Whewell 1840, 42.
29. Herschel 1830, 118–19.
30. Herschel notes that Whewell sees the line connecting the dots as a pre-existing idea, "already self-traced on [the mind's] own tablets" before experience comes into play; for Herschel that connective thread arises out of experience instead. Herschel 1857 [1841], 172–3.
31. Nasim 2014, 161. Nasim shows that Herschel's model was influenced by James Douglas, a Scottish philosopher of the Common Sense school (2014, 164). It is curious that the astronomer Herschel chooses a metaphor evocative of the constellations, which are essentially connect-the-dot puzzles that do *not* reflect any necessary relation between the dots.

32. Carpenter 1856, 185–91.
33. Hacking 1983, 207–8.
34. Hume [1777], ¶12.24.
35. Hume [1777], ¶12.4.
36. Hume [1777], ¶12.24, ¶12.25.
37. Michels 1875, 177.
38. Schickore 2007, 242, 247, 250.
39. Schickore 2007, 147.
40. Hickie 1876, 124.
41. Schickore 2007, 84.
42. Schickore 2007, 92–3.
43. Schickore 2007, 146, 100–1.
44. Lewes 1858, 37–8.
45. Jacyna 2001, 231.
46. For other examples of this use of the term, see Carpenter 1856, 88, 97, 109, 116, 184.
47. Beale 1854, 7.
48. Willis cites the work of Jonathan Crary and Terry Castle on these earlier moments of mystified vision. He points to microscopy as an instance of Crary's phantasmagoric, mystified vision, arguing, "The visual experience afforded by the microscope was very precisely phantasmagoric in both its mystification of the means by which the extraordinary images were produced, and in the ways in which it gave rise to questions of accurate or imaginative seeing" (2011, 26).

Notes to Chapter 13

1. Fripp 1877, 306.
2. On the role of the hand in scientific observation, see also Nasim 2014.
3. See the Glossary for definitions of these terms. In the 1920s the upper portion with eyepiece begins to be named as the "head."
4. Buckley 1890, 37.
5. Quekett 1848, vii–viii.
6. Hogg 1854, v.
7. Hogg 1854, 61.
8. Hogg 1854, xiv.
9. Wilson and Geikie 1861, 121–2.
10. Hogg 1854, 86.
11. [Broderip] 1858, 455.
12. On this tradition, see Briefel 2015, 2–16.
13. Schleiden 1849, 590. Indeed, the image is reversed in the compound microscope also, albeit to less effect. Some instruments were provided with an "erector" to turn it round again. See E. Lankester 1859, 9–10.
14. Schleiden 1849, 589–90.
15. Wood 1861, 3, 23.
16. Worboys 2000, 62, 138.
17. The sociologist Mary Douglas is often cited as the author of the saying "dirt is matter out of place," but she attributes it to Lord Chesterfield; Richard Fardon traces it to Lord Palmerston, discussing sewage and fertilizer during the sanitation debates of the 1850s (2013).
18. Quekett 1848, 448.
19. Carpenter 1856, 257.
20. Koch 1880 [1878], 30.
21. Ward 1864, viii. Groombridge published the book in cloth for *7s. 6d.*
22. Ward 1864, 35.
23. Phin 1877, 176; Carpenter, 1856, 166.
24. Wood 1861, 64.
25. Pillischer 1873, 18–19.
26. Pillischer 1873, 36, 37. These prices are not far off what Gideon Mantell reports in 1846 (111–12).
27. Rheinberger 2010, 219.
28. Quekett 1848, 146.
29. "Wonders of the Microscope," 1847, 4.
30. Griffith and Henfrey 1856, x, iv.
31. Otter 2008, 182.
32. Woodward to John Mayall, Sept. 8, 1875, 113, his emphasis.

33. Pritchard and Goring 1832, 141–2.
34. Carpenter 1874, 417–18.
35. Carpenter 1856, 164–5.
36. Illustrated hands most commonly accompany discussions of the dissecting microscope, a simple (single-lens) microscope especially useful in botanical or medical investigation.
37. Otis 2001, 17.
38. Durham 1868–9, 107–8.
39. Bennett 1850, 190–1.
40. Cole 1895, 7.
41. Smith 1880, 392.
42. "An Amateur," 1890, 270.
43. Wenham 1854, 137. *Pleurosigma angulatum*, a diatom (form of unicellular algae), often served as a test object due to its intricate structure.
44. Wenham 1854, 138.
45. Wenham 1854, 139.
46. B. Brown 2001, 4.
47. Beale 1857b, vii.
48. Beale 1857b, 104–11.
49. For example, the report notes that during the second class meeting, "Bread crumbs were . . . examined in various ways, as recommended for a preliminary lesson by Dr. Beale." "Proceedings at Mr. Suffolk's Class," 1869, 211.
50. Holmes 1877, 606.
51. Beale 1857b, vii.
52. Beale 1865a, 249; Glaisher 1866, 55, my emphasis.
53. Worboys 2000, 36.
54. Beale 1857b, 4.
55. Beale 1857b, 4, 2.
56. Beale 1857b, 2.
57. Beale 1857b, 98–9.
58. Beale 1868, 1–2.
59. Beale was actively engaged on these debates around the time of his 1868 and 1880 editions of *How to Work with the Microscope* (see Beale 1880a, 3).
60. Beale 1880a, 2.
61. Beale 1868, 1–2.
62. Beale 1880a, 2.
63. Durham 1868–9, 109.
64. Durham 1868–9, 109, 19–20.
65. Beale 1865b, 268.
66. Lakoff and Johnson 2008 [1980], 134–5.
67. Hickie 1875a, 262.
68. Siesser 1981, 173–4.
69. Hermann von Helmholtz leveraged this sensitivity to vibration to develop a "vibration microscope" allowing him to study sound waves as visual patterns.
70. Carpenter 1856, 80; White 1893, 109.
71. Carpenter 1856, 67, 162, 164.
72. Wood 1861, 22.
73. Carpenter 1856, 80.
74. Pocklington 1872, 621; "An Amateur," 1890, 374.
75. See, for example, Hogg 1854, 44.
76. Bennett 1841, 6.
77. Bennett 1850, 188.
78. Plumer 1865, 7, 14.
79. Plumer 1865, 16.
80. Carpenter 1883, 40, his emphasis.
81. "Obituary: Edward Milles Nelson," 1938, 202.
82. For cautions about the breath, see "Wonders of the Microscope," 1847, 5.
83. Agassiz 1863, 297–8.
84. Similarly, as Omar Nasim explains, specialized observing chairs were a feature of mid-Victorian astronomy. They supposedly "increased the quality and quantity of observations because they regulated discomforts and distractions that inevitably fatigued the observer's body and mind" (2021, 194).

85. Hogg 1854, 50.
86. Hack 1821, 90–1.
87. Wood 1861, 20–1.
88. Carpenter 1878, 260, 275.
89. Brewster 1837, 98, 168.
90. Brewster 1837, 165.
91. Bennett 1850, 188–9.
92. Frey 1872, 97.
93. Bennett 1850, 189.
94. Hannover 1853, 60.
95. Van Heurck 1893, 134.
96. Schneider 1899, 105.
97. See Shapin 1998.
98. Schneider 1899, 108, 110.
99. Schneider 1899, 181–3.
100. Schneider 1899, 111.
101. Gatty 1872, viii, x, vii.
102. The term *ergonomics* was coined by the Polish botanist Wojciech Jastrzebowski in 1857, but his essay was not available in English until the twentieth century. Frederick Winslow Taylor first published on his principles in 1895.
103. Bennett 1850, 188–9. Impecunious students often bought relatively basic and affordable microscopes by Parisian opticians such as Oberhauser.
104. Hartley 1749, 205.
105. Bain 1859, 478–9.
106. Spencer 1855, 569.
107. Carpenter 1874, 213.
108. Carpenter 1874, 217.
109. Ryan 2012, 79–81.
110. Lewes 1877, 197fn.
111. Wells 1896, 244–5.
112. Ihde 1991, 74–5, 88.
113. Wilson and Geikie 1861, 120.

Notes to Chapter 14

1. See Castle 1988, Crary 1990, Armstrong 2008, Willis 2011, and Potter 2018.
2. Hogg 1854, 75.
3. Fripp 1874–6, 462. Chris Otter has shown how this notion functioned in debates over illumination technology generally; see Otter 2008, 205–6.
4. On Donders' measurement of refraction, see Almond-Brown 2023, 84.
5. Schickore argues that Joseph Jackson Lister influentially explored this model of vision in the 1820s (2007, 123).
6. Brewster 1837, 165.
7. Carpenter 1868, 28.
8. Ward 1864, 85. The eminent ophthalmologist Haynes Walton, discussing eye examinations, notes that the retina also reflects rays back through the lens of the eye (1875, 267–8).
9. Fripp 1874–6, 462, 464–5, 475.
10. Helmholtz 1874–6, 409. The article "On the Limits of Optical Capacity of the Microscope" was originally published as "Die theoretische Grenze für die Leistungsfähigkeit der Mikroskope," *Poggendorff's Annalen* 136 (1874); it was reprinted in the *MMJ* in 1876. The quoted phrase appears in Fripp's prefatory remarks.
11. In some ways this model resembles the trope, discussed by Mark Paterson, of vision as *touching with the eyes*; but the trope of the *eye as part of the microscope* goes further in declaring the human eye to be an integral part of an optical system. Paterson 2007, 37–57.
12. Brewster 1831, 47.
13. For example, the Welsh popularizer J.P.P. [J. P. Philips] 1859, 40.
14. Schickore 2007, 122–4.
15. Fripp 1874–6, 464–5.
16. Royston-Pigott 1876, 181.
17. "Proceedings of Societies: Quekett Microscopical Club," 1875, 95.
18. See Almond-Brown 2023, 80–94.

19. See Daston 1995, 21–3.
20. Brewster 1832, 8–36; quoted terms are on 11, 15, 21. On nineteenth-century skepticism on the eye, see also Armstrong 2008, 324–7.
21. Crary 1990, 90, 92. Müller's book was published in English as *Elements of Physiology* (1837–43).
22. Crary 1990, 91.
23. Schickore 2007, 96.
24. Otter 2008, 28, 32, 46.
25. Helmholtz 1873, 202, 219.
26. Tyndall 1873, 16–17.
27. Carpenter 1842, 264.
28. Carpenter 1874, 203–4, 206, his emphasis.
29. Carpenter 1874, 546.
30. Youmans 1894, 512.
31. Lardner 1855, vol. 5, 50, 65–80.
32. Lardner 1854, vol. 1, 82–4.
33. [Scoffern] 1865, 186–7.
34. Helmholtz 1873, 202.
35. Tyndall 1873, 16–17.
36. Quekett 1848, 187–8; Hogg 1854, 51. Hogg, following Quekett, quotes only the first six rules, omitting Brewster's recommendations for an intense light of small diameter when using high powers, and a homogeneous light at all times.
37. Brewster 1837, 138–9, quoting from "Microscope," *The Edinburgh Encyclopaedia* vol. 14.1 (1830), 215–32, p. 228. He misdates his original article as having been published in 1820 (Brewster 1837, 135).
38. Brewster 1837, 165–6.
39. Brewster 1837, 165.
40. Brewster 1837, 166. He quotes the earlier rules on pp. 138–9.
41. Helmholtz 1874–6, 426.
42. Phin 1877, 59.
43. Frey 1872, 97. Although Frey's book was translated by an American, the physician George Cutter, it was known in England. A review in the *MMJ* claimed that "to the medical man and student no better work has been offered to the English reader." "New Books with Short Notices," 1873b, 87.
44. Letter from Nicholl to William Henry Fox Talbot, June 23, 1816.
45. Armstrong 2008, 305.
46. Bennett 1850, 192.
47. Carpenter 1856, 158; see also 154.
48. Durham 1868–9, 98.
49. Brewster 1837, 167–8.
50. Carpenter 1856, 157–8.
51. Carpenter 1856, 158, his emphasis.
52. Lewes 1858, 36–7.
53. "Summary of Current Researches: Microscopy: Advantage of the Binocular," 1881, 835. Quoted from *Zeitschrift für Mikroskopie*, ii (1880), 20.
54. Van Heurck 1893, 133–4.
55. Fripp 1877, 316–17.
56. Frey 1872, 98.
57. Carpenter 1856, 639.
58. See, for example, Schleiden 1842, 139.
59. Bennett 1850, 6–7.
60. [P.] Lankester 1861, 121, 124. Lankester cites "Eyes and No Eyes" from John Aikin and Anna Laetitia Barbauld's children's book *Evenings at Home* (1792–6) as the origin of this trope.
61. See Schaffer 1998, 190–1.
62. Carpenter 1856, 35.
63. See, for example, nos. 4, 7, and 9 in Klaus Hentschel's list of iconic insights of visual studies (2014, 70–1). However intuitive it feels to us, three-dimensional vision with the unassisted eye also requires experience and to some degree training. "Molyneux's question" about this (1688), made famous by Locke in the second edition of his *Essay Concerning Human Understanding* (1694), was answered empirically (in the affirmative) by cataract operations.
64. Goring in Pritchard and Goring 1832, 185.
65. Clarke 1858, iv–vii. The Preface skips directly from page number iv to vii.
66. Carpenter 1854, 625.

67. Schleiden 1849, 585–9; Carpenter 1856, 188–9. Brownian motion was first observed in 1827 by the botanist Robert Brown, then explained in 1905 by Einstein as evidence of molecular motion upon the particles.
68. Balfour 1860, 21–2. Balfour derives the second phrase from William Mackenzie, *A Practical Treatise on the Diseases of the Eye* (London: Longman, Green, Brown, and Longmans, 1854), 951.
69. Carpenter 1856, 186.
70. Beale 1868, 70.
71. Beale 1857b, 99–101.
72. Gage 1886, 9–14.
73. Schleiden 1849, 589, 591.
74. [Ross] 1839, 177.
75. Goring in Pritchard and Goring 1832, 185.
76. Carpenter 1856, 119–23.
77. For example, Carpenter 1856, 160. He also recommends the use of an object-finder (119–20). These changes in focus due to optical depth of field and shifts in magnification arguably affected fictional depictions of mental perspective as well; see Kennedy 2010, chapter 5.
78. [Dixon] 1856a, 378–9, my emphasis.
79. [Dixon] 1856a, 380. An erecting eyepiece could be added that would reverse the reversed image.
80. Cartwright 1995, 84.
81. Cartwright argues that an "aesthetic of abstraction" associated with microscopy and with science in general, which features "flatness, segmentation, and planar division of space," sparked Modernist artistic movements such as cubism (1995, 91).
82. Armstrong 2008, 317–18.
83. Griffith and Henfrey 1856, xxxix.
84. Other scholars have noted this defamiliarizing effect. For instance, Marjorie Hope Nicolson discusses the alienation the tiny Gulliver experiences when the world around him suddenly presents itself to his eye as if greatly magnified. Nicolson 1935, 51–2.
85. Armstrong 1996, 143.
86. Jacyna 2001, 238. The quotation derives from William Waring Johnston, who had studied with Bennett as a postdoctoral student in the late 1860s. Johnston, "John Hughes Bennett—His Services to Medicine," Address to the Medical and Chirurgical Society of Maryland, 1899, British Library, BL 10600.ff.9 (3), p. 5. See also Jacyna 1997 on Bennett's efforts to discipline medical students in microscopy. On Bennett's interest in balancing disciplined clinical observation with medical art, see Kennedy 2010, 70, 123–4.
87. Rutherford 1872, 3.
88. For the war with disease, see Servitje 2021.
89. Jacyna 2001, 239; the phrase is quoted from an undated and untitled fragment, EUL, MS Gen. 2007, folder 1.
90. Beale 1857b, 98.
91. Beale 1868, 189–90.
92. Wood 1861, 27.
93. Wood 1861, 55.
94. Beale 1878, 38. He credits the histologist William Rutherford for the idea.
95. Pritchard 1827, 15, 18. Goring had seemed unconcerned about the distinction Pritchard makes in this passage; compound lenses "show objects more satisfactorily than single lenses of the most exquisite workmanship, though the former shows only a simulacre, or representation of the object, while the latter shows the original in nature itself, from which the picture was formed. Such is the effect of destroying aberration!" (Goring 1827b, 280). But Pritchard reiterated his view in the *Microscopic Cabinet* (Pritchard and Goring 1832, 106–7).
96. Abbe 1881a, 205.
97. Fripp 1877, 308.
98. Tuckwell 1908, 46–7.
99. Gooday 1991, 327.
100. Reiser 1978, 37.
101. Schacht 1853, 2.
102. Schacht 1853, 18.
103. Julien 1893, 32.
104. Lewes 1877, 226–7.
105. Gooday 1991, 333, his emphasis.
106. Julien 1893, 33.
107. Rheinberger 2010, 220.

108. Rheinberger 2010, 219.
109. Rheinberger 2010, 222.
110. Beale 1857b, 49–50; see also 1868, 116, his emphasis.
111. Schleiden 1842, 154.
112. Garratt 2010, 115.
113. Griffith and Henfrey 1856, xxvii–xxviii.
114. Schacht 1853, 2.
115. Pritchard and Goring 1832, 189.

Notes to Chapter 15

1. Gatty 1872, xxix. For the "List of Rules," see xxi–xxvii.
2. On Gatty, see Lightman 1997, 192–5; Bryant et al. 2016.
3. Koch 1880 [1878], 22, 25.
4. Carpenter 1856, 187.
5. Carpenter 1856, 8. Italics in Carpenter are his original emphasis.
6. Carpenter 1856, 35.
7. Carpenter 1856, 185, his emphasis.
8. Carpenter 1856, 36.
9. Carpenter 1856, 307–8.
10. Carpenter 1856, 35.
11. Carpenter 1856, 185.
12. Beale 1857b, 97.
13. Beale 1868, 187.
14. Beale 1857b, 98.
15. Beale 1857b, 97–8. In the 1868 fourth edition of *How to Work with the Microscope*, exercised by debates over spontaneous generation, Beale further points to "how difficult it is to prove that what [the observer] sees is really the thing as it is in nature" and argues that erroneous observations introduce as many or more errors as speculative theories do (Beale 1868, 189; see also 194–5).
16. Willis 2011, 5.
17. Garratt 2010, 115.
18. Spiers 1890b, 223–4.
19. Fripp 1877, 305.
20. Beale 1857b, 102.
21. Schleiden 1849, 583.
22. Carpenter 1856, 10.
23. Fripp 1877, 327.
24. The OED sites the first instance of this usage in 1787: "appreciate, v." *OED Online*.
25. Bain 1859, 623.
26. Crisp 1878, 126–7.
27. Carpenter 1856, 189.
28. Carpenter 1856, 162.
29. Beale 1857b, 51.
30. Carpenter 1856, 184.
31. Carpenter 1874, 137; Lewes 1877, 446.
32. Carpenter 1855, 727.
33. Carpenter 1874, 204.
34. Bain 1859, 505.
35. Bain 1859, 517.
36. Carpenter 1874, 213, 350. He also described mental pathology and deficiency as governed by automaticity and reflex action; see, for example, Carpenter 1853, 672, 843.
37. Carpenter 1853, 819–20. See also Carpenter 1874, 137, 195, 287, 538. Carpenter's theory of unconscious cerebration, related to his work on ideomotor action, was influenced by the work of Thomas Laycock, who claimed priority in his work on reflex action. See Laycock 1860, vol. 2, 465–80.
38. Carpenter 1853, 819.
39. Carpenter 1874, 538–9.
40. Helmholtz 1874–6, 16. The quoted phrase appears in Fripp's prefatory remarks.
41. Carpenter 1874, 7.
42. Carpenter 1874, ix–x.
43. Carpenter 1874, xiii.
44. Carpenter 1874, 12–13.

45. Carpenter 1874, 206.
46. Carpenter 1874, 259.
47. Carpenter 1874, 14.
48. McCalla 1883a, 767. The *JRMS* excerpted largely the latter portion of McCalla's address (the practical suggestions), omitting his introductory section on the microscope as "the vigilant sentinel that stands guard over our health and safety" or "an unerring detective," and the microscopist as a "votary of the tube" (McCalla 1883b, 2, 3, 5).
49. Griffith and Henfrey 1856, xxxi. They continue with a reference to other difficulties in ascertaining true structure: "Are they granules of pigment, or rows of them? Do the lines represent a true lined structure, or are they optical illusions?"
50. F. H. Wenham's stereoscopic binocular microscope held the field for many years after his publication of the technology in 1860, favored for its ease of use and adaptability. See Part I on Wenham's development of this design. Beck's "Popular" model instrument, built on Wenham's model, offered portability as well (Beck 1865, 48–9).
51. Plumer 1865, 13.
52. Phin 1877, 39.
53. *How to Choose a Microscope. By a Demonstrator*, 1876, 16.
54. Schacht 1853, 2, his emphasis.
55. Elkins 2008, 118, his emphasis.
56. Lewis to James Lorimer Graham, Mar. 18, 1859, p. 4; Apr. 3, [1859], p. 2.
57. Bain 1855, 411.
58. Crary 1999, 92.
59. Schleiden 1849, 590–1.
60. Beale 1857b, 57–8 and table II, 105.
61. Beale 1880a, plate XXIII, between pp. 80–1.
62. Woodward to Richard Leach Maddox, Aug. 4, 1871, p. 27.
63. Griffith and Henfrey 1856, xxxii.
64. Griffith and Henfrey 1856, xxxiv.
65. Stephenson 1873, 1, 4. Stephenson's technique undercut the then-common adherence to "beads" or "bosses" in diatom structure.
66. Carpenter 1856, 652.
67. Clarke 1863, vi, 231.
68. Carpenter 1856, 652.
69. Dancer 1869, 243.
70. Schacht 1853, 23.
71. Royston-Pigott 1875, 153–4. Royston-Pigott wrote frequently, insistently, and voluminously to the *MMJ* about the supposed claims of his "Aplanatic Searcher" (an additional lens system meant to improve penetration, magnifying power, and definition) and the "beads" on the *Podura*-scale (which are not beads).
72. Carpenter 1856, 165, 196, his emphasis.
73. *How to Choose a Microscope. By a Demonstrator*, 1876, 16.
74. "Essex," Letter 12790, June 22, 1877, p. 363.
75. Schleiden 1849, 585.
76. Riegl 1999, 281–2.
77. Schacht comments, "Many longitudinal sections in different determinate directions, must be carefully examined and compared with one another before the observer can be satisfied that he has made out the construction of the body under observation" (1853, 23). Later in the century, advances in the microtome allowed investigators to produce precise serial sections from one side to the other of a three-dimensional specimen (even a relatively large one such as a human organ). Identifying and following structures from slide to slide, with the aid of differential staining, enables the observer to mentally build up a three-dimensional model of the specimen's anatomical structures, their relation, and the differences in their cellular makeup. As *The American Naturalist* explained in 1883, "the value of the series…will depend" in part "on our ability to grasp all the topographical relations of each section." This feat requires that "the sections must be made of *uniform thickness*, arranged in *serial order*, and all *similarly disposed*." Thus the careful investigator can "make a single series of sections tell the whole story" of the object. See "Orientation in Microscopic Sections," 1883, 109–11, emphasis in original. On the development of serial sections, see Bracegirdle 1978, 77, 81. In the twentieth century, the optical sectioning and image stacking of microscopical images was adapted into the "serial section film" (Cartwright 1995, 96–7).
78. See Crary 1999, 95–7.
79. Brewster 1837, 165.

80. Schneider 1899, 105.
81. Carpenter 1868, 41, his emphasis.
82. Willis 2011, 13.
83. See Koch 1880 [1878], 26–30.
84. Carpenter 1856, 186.
85. Carpenter 1856, 210; 1881, 185.
86. Smith 1880, 388–9.
87. Abbe's article, "A Contribution to the Theory of the Microscope and of Microscopic Vision" (Abbe 1874–5), was originally published as "Beitrage zur Theorie des Mikroskops und der mikroskopische Wahrnehmung," *Max Schultze's Archiv for mikroskopische Anatomie* 9 (1873), 413–68; it was reprinted in *MMJ* 14.4 (Oct. 1875), 191–201; 14.5 (Nov. 1875), 245–54. Stuart Feffer argues that "at the time Abbe's work was made public, [the practical] consequences were not yet apparent" (1996, 51). Fripp's translation was widely circulated.
88. Abbe 1881b.
89. Abbe 1881b, 418.
90. Helmholtz 1874–6, 16, 20–1.
91. Hacking 1983, 187.
92. Bradbury 1996.
93. Stephenson 1877.
94. Dallinger explicitly claims authorship of the Abbe section: "I summarized [Abbe's works], not without anxiety" (Dallinger 1891, vii–viii). However, the *JRMS* insists this was Nelson's work ("it is a fact") and I accordingly credit him here ("Obituary: Edward Milles Nelson," 1938, 201).
95. Nelson in Carpenter and Dallinger 1891, 62, his emphasis; Hacking 1983, 187. Abbe phrases it somewhat differently: "The very first step of every understanding of the Microscope is to abandon the gratuitous assumption of our ancestors, that microscopical vision is an *imitation* of macroscopical, and to become familiar with the idea that it is a thing *sui generis*, in regard to which nothing can be legitimately inferred from the optical phenomena connected with bodies of large size" (Abbe 1881b, 423).
96. Nelson in Carpenter and Dallinger 1891, 72.
97. Nelson in Carpenter and Dallinger 1891, 70. These passages draw largely from Crisp's elucidation of Abbe's theory (Crisp 1881, 354).
98. Nelson in Carpenter and Dallinger 1891, 72. As early as 1874, the *MMJ* warns, "According to the researches of Professor Abbe…the limit of capability of our best instruments is almost reached," with the editorial comment "[!!]." Readers are also informed of the possibility "that different structures give the same microscopical image, and, on the other, that like structures give different images" ("Progress of Microscopical Science," 1874, 29).
99. Nelson in Carpenter and Dallinger 1891, 72. This passage recirculates phrases from Fripp's translation of Abbe (Abbe 1881b, 250) and Crisp's version of Fripp's text (Crisp 1881, 354).
100. McCalla 1883a, 767, 770.
101. Mantell 1846, 8.
102. Fripp 1877, 308, 304.
103. McCalla 1883, 767.
104. Fripp 1877, 333.
105. Daston and Galison 2007, 311, 314, their emphasis.
106. Daston and Galison 2007, 313, 314, 311, 322.
107. Rylance 2000, 76.
108. For mechanical objectivity, see Daston and Galison 2007, 115–90.
109. Fripp 1877, 331.
110. See, for example, Nasim 2014, 91, 113–14.

Notes to Chapter 16

1. Schickore 2007, 154.
2. Nasim 2014, 1.
3. J. Secord 2000, 382.
4. On the participatory ideal, see Sheets-Pyenson 1985, Barton 1998, and Lightman 2016.
5. See Topham 2020.
6. Johnston 1838, 255.
7. Endersby 2008, 193.
8. Schacht 1853, 4.
9. "Reviews," *Natural History Review*, 1855, 39.

10. Winter 1997, 25.
11. Schacht 1853, 44.
12. Schacht 1853, 45.
13. Schacht 1853, 49.
14. Schacht 1853, 46.
15. Schacht 1853, 47.
16. Schacht 1853, 45–6.
17. Schacht 1853, 47–8.
18. Knorr-Cetina 2005, 195.
19. Schacht 1853, 48.
20. See Endersby 2008, 60–72, 82, 91 on the depredations suffered by collected botanical specimens; 125–30 on the difficulty of drawing such a specimen and the usefulness of field sketches for information.
21. Schacht 1853, 60, 67.
22. Schacht 1853, 65.
23. On Schacht's endorsement of the trained eye, see Kennedy, "Circulation and Civility: Mid-Victorian Botany and Microscopical Method," 2024.
24. Daston and Galison 2007.
25. Beale 1857b, vii.
26. Beale was not the first to do so. He cites John Hughes Bennett's 1853 *Lectures on Clinical Medicine*, which contains a sizable chapter on microscopical observation for the medical man (Beale 1854, 159). Bennett lists eight "physical characters which distinguish microscopic objects," essentially a checklist guiding the medical microscopist's observation and description. They are shape, color, edge or border, size, transparency, surface, contents, and effects of reagents. Bennett provides an extended description of what to examine for each of these elements (1853, 70–3).
27. Strick, writing about the vitalism debates, characterizes Beale as "quite offensive" in both public and private and "an extremist who criticized just about everybody," including influential men of science such as Huxley and Tyndall (Strick 2000, 137, 184). However, as a microscopist, Beale was well respected for his tact (in physical manipulation) and discernment.
28. Beale 1857b, 3.
29. Beale 1857b, 93.
30. Beale 1857b, 92.
31. Beale 1857b, 94.
32. Beale 1857b, 92.
33. Some two decades later, the physician H. E. Fripp criticizes this approach, saying, "the advice to abstain from noting or describing what is *not clearly seen* is at best but of temporary and negative value" because "few persons are so distrustful of their sight or judgment as to think the advice necessary in their case." Furthermore, this advice misdirects the observer's skepticism: "Optical illusions are as clearly seen as material objects, and the microscope image though perfectly seen may be an illusion in so far as any inference of material structure is founded upon it" (Fripp 1877, 327).
34. Beale 1857b, 93.
35. Beale 1857b, 93–4.
36. Beale 1857b, 94.
37. Beale 1857b, 92.
38. Beale 1865a, 240–1.
39. This somewhat differs from the advice Beale gives a few years earlier in the volume *The Microscope and Its Application to Clinical Medicine*: "Every observation should be carefully recorded in a notebook at the time it is made, and drawings added, if necessary; and it is very important that the student should take every opportunity of describing, in the simplest manner possible, the appearances which he has observed under the microscope" (1854, 124). Here we see the familiar exhortation to keep description simple, but it is concurrent with the observation, and not subordinated to the drawings, which are only included in the notebook "if necessary." This passage largely persists in the second (1858) edition but disappears from the third (1867) edition of the book, which was published after Beale's 1865 advice on retrospective note-taking.
40. Beale 1857a, 3.
41. *Medico-Chirurgical Review* no. 41, 132; quoted in an advertisement for *Archives of Medicine* on the back cover of *The Use of the Microscope in Clinical Medicine* (1857).
42. Beale 1854, 1.
43. Beale 1854, 166.
44. Beale 1854, 173.
45. Beale 1854, 130.

46. Beale 1854, 125.
47. Beale 1854, v–vi, 36 (fig. 38).
48. Beale 1863, 95–6.
49. Beale 1863, 100.
50. Beale 1854, 94fn; 1857, 4fn.
51. Daston 2004, 445.
52. Beale 1863, 98.

Notes to Chapter 17

1. Morgan 2017, 21.
2. Holmes 1987, 229.
3. See, for example, Fyfe et al. 2015, 227.
4. The remit for this series directed that, for scientific topics, the author should adopt a "general and philosophical manner" rather than a "technical and professional" one. See Fyfe 2012, 152.
5. "A Microscopic Entertainment," 1854.
6. Daston 2004, 444 notes this longstanding association.
7. Wood 1861, 36.
8. P. Gosse 1859, 210–12.
9. J. Secord 2000, 238, 350.
10. "The Story of Microscopical Discovery," 1859, 519.
11. E. Lankester 1880, 789.
12. Beale 1865a, ix.
13. Carpenter and Dallinger 1891, 341–2.
14. Topham 2004b, 433. For an instructive comparison to this, see Nasim 2014, 123–70, where two case studies in astronomy show a similar staging of the process of observation but with a focus on sketching rather than note-taking.
15. Bazerman 1988, 70–1, 73–5.
16. Dear 1985, 152–4.
17. Shapin and Schaffer 1985, 25, 77. The concept of the literary technology originates in Shapin 1984.
18. Shapin and Schaffer 1985, 60–1, 63.
19. Morgan and Wise 2017, 4.
20. Dear 1985, 153.
21. Shapin and Schaffer 1985, 60.
22. Gross et al. note that replication was not a major concern of investigators writing in the seventeenth or early eighteenth centuries, but it becomes important later in the eighteenth century as a way to broaden the base on which an experimental report might be deemed trustworthy (2002, 55, 98).
23. Dear 1991, 137.
24. Holmes 1991, 170.
25. Rheinberger 2010, 219.
26. Jacyna 2003, 55.
27. Jonathan Topham argues that historians should be more careful about distinguishing "science *for* the people" from the "science *of* the people," suggesting that the outreach encouraging a broad uptake of science was not always in earnest (2009, 2). Some instances of this outreach, as with the oxyhydrogen microscope shows designed for mass audiences, may indeed be offered as a perfunctory gesture of respectable instruction cloaking sensational amusement; but many of the textual examples seem entirely sincere, albeit anxious to discipline the novices in community norms.
28. Gross et al. 2002; see also Atkinson 1998.
29. On the rise of depersonalized scientific reporting and the methods section, see Atkinson 1998, xxiii–xxiv, 89; Gross et al. 2002, 118, 122. On citation, see Gross et al. 2002, 131–2.
30. Bazerman 1988, 8.
31. Articulated practice is distinct from both the *narrative of nature* and the *narrative of science* (see Part II). Victorian articulated practice presents a strongly chronological narrative (like the narrative of nature) centered on the mutual interaction of the investigator, the apparatus, and the object, and framed by its theoretical implications (like the narrative of science).
32. Gross et al. 2002, 69, 122–4.
33. Atkinson 1998, 144.
34. Beale 1861, "Preface to the Illustrated Edition," n.p.
35. Beale 1868, 114.
36. Beale 1857, 81–2.
37. Beale 1857, vii.

38. Beale 1857b, 110. By 1868, the items in these tables were correlated with plates and specific figures as well.
39. Beale 1861, "Preface to the Illustrated Edition," n.p.
40. "How to Work with the Microscope" (review), 1857, 459.
41. "Pathological Laboratory," a four-page brochure included in Beale 1857b between plates 14 and 15. Quotations appear on pp. 1–2.
42. Suffolk 1870, 175.
43. Edwards 1874, 228.
44. Edwards 1874, 228–36, at p. 234.
45. "Sensory Nerve-endings in Skin of Insects," 1881, 235, reprinted from *Comptes Rendues.*
46. Sommer 1881, 49, reprinted from the *Zeitschrift für wissenschaftliche Zoologie.*
47. "Chemical Composition of the Ink of Cephalopods," 1881, 877, reprinted from *Comptes Rendus.*
48. There are 618 references to "method" or "process" over 1,275 pages.
49. Wood 1861, 109–11.
50. Wood 1861, 86–7.
51. "Professor Morton's Lantern Lectures," 1875, 101.
52. "Professor Morton's Lantern Lectures," 1875, 100.
53. "Professor Morton's Lantern Lectures," 1875, 100.
54. Kennedy 2019, 94.
55. Schickore 2007, 154.
56. Hughes 1956, 11.
57. Csiszar 2018, 150–1.
58. Many of these roles were not fully developed until later in the period. For instance, for much of the century, periodical publishing was not universally accepted as a means of establishing priority, as the disputes over photographic technologies demonstrate. To point to one example among several, J. B. Reade's photomicrographs had been displayed, circulated, and even sold in 1839, before Talbot heard about Reade's fixing process from Andrew Ross, but a court ruled that Reade's work had not been sufficiently published to claim priority.
59. Carpenter 1856, 762–3; Carpenter 1862, 767.
60. Carpenter 1862, 261.
61. [P.] Lankester 1861, 124.
62. [P.] Lankester 1861, 125.
63. Goring and Pritchard 1830, 7.
64. Pritchard and Goring 1832, 235–6.
65. Pritchard and Goring 1832, 25.
66. Bates 2014, 31–3.
67. Anderson 2012, 50.
68. [P.] Lankester 1861, 127.
69. Quekett 1848, 338–9.
70. Anderson 2012, 55–6.
71. See Lewes 1858, 330; Bates 2014, 33.
72. [P.] Lankester 1861, 124.
73. See, for example, Dick 1851, 145–7.
74. "Carey's Oxy-Hydrogen Microscope," 1835, 3.

Notes to Chapter 18

1. Eliot 1868, 119.
2. Michael 1895, 10.
3. Michael 1884, 2.
4. Ross 1839, 182.
5. Hogg 1854, v.
6. J. Secord 2014, 10–23; Lightman 2007, 5–9.
7. Hogg 1854, 212. This species is now known as *Plumatella fungosa.*
8. Dixon 1856a, 377.
9. Dixon 1856a, 378–80.
10. Dixon 1856a, 379.
11. Clarke 1858, 42.
12. Clarke 1858, 23–4.
13. "Proceedings of Societies: Microscopical Society of London," 1853, 147.
14. Martineau 1877, 15.

15. Lyons 1851, 8–10.
16. Lyons 1851, 45.
17. Buckley 1890, 53–4.
18. Hogg 1854, 204.
19. This diatom is now known not as *Navicula rhomboides* but as *Frustulia rhomboides*.
20. Reade 1869, 8–9.
21. Wood, 1861, 22.
22. Abbe 1874–5. Abbe identifies the limits as "the wave-length of violet rays when direct illumination is used" or half that with oblique (245).
23. Abbe 1874–5, 257.
24. Abbe 1874–5, 250, my emphasis.
25. Abbe says, "zwar kann aus dem mikroskopischen Befund mit Sicherheit auf Nichts weiter geschlossen werden, als auf das Vorhandensein solcher Structurbedingungen, als zur Erzeugung des die Abbildung vermittelnden Beugungsphänomens nothwendig und hinreichend sind" (Abbe 1873, 452–3). This was just one of multiple changes by Fripp; Barry Masters, who retranslated Abbe's essay, characterizes Fripp's translation as "faulty," involving both "omissions and changes" (Masters 2020, 93; see also 67).
26. "Report," 1879–82, 76.
27. Goring and Pritchard 1832, 187.
28. Goring and Pritchard 1832, 190.
29. Goring and Pritchard 1832, 11.

Notes to Chapter 19

1. On soirée culture, see J. Secord 2000, 410–16; Alberti 2003; J. Secord 2007.
2. "Notices of Meetings: Liverpool Microscopical Society," 1882, 68.
3. E. Gosse 1907, 136.
4. Merrill 1989, 197.
5. Samuel Alberti has compared "the experience of a conversazione visitor" to "the reader of a general periodical" (2003, 215).
6. Fyfe 2000, 276, 288.
7. Gates 1993, 292–4.
8. Lightman 2007, 20. On children's science books, see also Fyfe 2000 and Topham 2007.
9. Gates notes that by the mid-Victorian period this catechismal style was being parodied (1993, 294).
10. Thwaite 2002, 288.
11. Belknap 2018, 398.
12. For "Science Gossips," see Belknap 2018, 413. On Bowerbank, see Michael 1895, 3, 9–10.
13. P. Gosse 1859, iv–v, 132.
14. P. Gosse 1859, iv.
15. Spiers 1890a, 256. The *Journal of Microscopy and Natural Science* was the official organ of the Postal Microscopical Society; Spiers was listed as a Fellow of the Geological Society (FGS) and Fellow of the Royal Microscopical Society (FRMS).
16. P. Gosse 1859, 398.
17. E. Gosse 1890, 337; 1907, 136.
18. J. Smith 2006, 82.
19. P. Gosse 1859, 272, 170.
20. Topham 2004a, 59.
21. P. Gosse 1859, 417.
22. P. Gosse 1859, v.
23. P. Gosse 1859, 25.
24. P. Gosse 1859, v.
25. Lightman 2007, 362.
26. "Evenings at the Microscope," 1859, 570–1.
27. "The Story of Microscopical Discovery," 1859, 525.
28. Later chapters present fewer or shorter fanciful passages, perhaps because the reader would be assumed to have by that point developed a self-sustaining interest in the topic.
29. P. Gosse 1859, 2–3.
30. White 1887, 9–10.
31. White 1887, 17.
32. White 1887, 24.
33. E. Lankester 1859, 49.

34. E. Lankester 1859, 58.
35. Catlow 1851, x–xi.
36. Catlow 1851, xiv.
37. Catlow 1851, xi.
38. Spiers 1890b, 179–80.
39. Spiers 1890b, 183–6.
40. P. Gosse 1859, 327–9.
41. P. Gosse 1859, 336, 339–40.
42. Schleiden 1849, 589; 1842, 151.
43. P. Gosse 1859, 47. Lynn Merrill classifies Gosse's prose into *passages of approach* such as this one (macroscopic framing), *passages of aesthetics* (delving into and admiring the microscopic world), and *passages of animation* (narrating the animalcules' lively actions). The techniques I discuss here—personal address, chatty tone, admiration of beauty, and comparison to "common objects"—apply in all these modes.
44. P. Gosse 1859, 71–2.
45. P. Gosse 1859, 86.
46. P. Gosse 1859, 467.
47. P. Gosse 1859, 63.
48. P. Gosse 1859, 62.
49. See Part II.
50. See J. Secord 2000, 98–100.
51. Hogg 1854, 126.
52. P. Gosse 1859, 3–4.
53. Armstrong 2002, 45–6.
54. P. Gosse 1859, 72–4.
55. Gosse must have been unaware that the botanist Edward Forbes had identified this form in 1846 as *Willsia stellata*; it is now accepted as *Proboscidactyla stellata* (Forbes 1846). See *World Register of Marine Species* (WORMS), http://www.marinespecies.org/aphia.php?p=taxdetails&id=158169. Accessed May 29, 2021.
56. P. Gosse 1859, 417–23.
57. P. Gosse 1859, 74–5.
58. P. Gosse 1859, 56.
59. Gosse offers the same remarks in his *Naturalist's Rambles on the Devonshire Coast*, where he adds a quotation from a French authority making a similar comparison (P. Gosse 1853, 392–3).
60. See Merrill 1989 on the "romance of natural history," especially 13. See also Keene 2015.
61. P. Gosse 1859, 395–6.
62. Merrill 1989, 208.
63. King 2019, 127.
64. On pleasure, Merrill 1989, 196–7; King 2019, 131.
65. Merrill 1989, 202, 206–7, 211.
66. King 2019, 112–13.
67. King 2019, 112, 130, 113, 136. On dilation specifically, see 130–3.
68. P. Gosse 1859, iii.
69. Armstrong 2002, 36.
70. Lukacs 1983 [1937], 230–50 on the decadence of the arbitrary and inessential detail; Schor 1987, 42–64.
71. Schor 1987, 43, 61, 59, 4.
72. King 2019, 129. Even apart from his bizarre argument in *Omphalos*, Gosse presented a distinct version of natural theology. He criticized the notion that faith could be founded solely on the Book of Nature without knowing or accepting the revelations of the Bible as the Word of God. See J. Smith 2019.
73. P. Gosse 1859, 77.
74. Ward 1864, vii.
75. Ward 1864, viii.
76. See Kennedy 2019, 91, 101–4.

Notes to Chapter 20

1. Vasalou (2012, 51).
2. (Fisher 1998, 40–1).
3. McCalla 1883b, 182.
4. McCalla 1883a, 772.

5. This is where the *JRMS* reprint ends. McCalla's original concludes with another spatialization of microscopical work—giving the instrument itself agency in "pushing its researches into many fields... finding out truth in realms remote from ordinary human gaze"—and an even clearer call to skepticism as a route to spiritual revelation: "Let us resolve to use this wonderful agent of investigation more earnestly, with greater carefulness against error and reverence for the truth, and so seek more and more to know things as they are, and as we shall know them some day, I trust, when we shall no longer see through a glass, darkly; but shall know even as we are known" (McCalla 1883b, 187).
6. Morgan 2017, 87.
7. In narratology, following the Russian formalists Vladimir Propp and Victor Shklovsky, the order of events as they happened is referred to as *fabula* and the order of events as they are told is *syuzhet*.
8. Influential early work on narrative in science includes Shapin 1984, Dear 1985, and Holmes 1991. See also O'Connor 2008 and Buckland 2013.
8. Beatty 2017, 32.
10. Morgan and Norton Wise 2017, 2.
11. Hurwitz 2017, 65.
12. Mussell 2007, 77. He cites Wells 1894.
13. Bain 1859, 196, 60, 185.
14. Bain 1859, 196.
15. Bain 1859, 193.
16. Lewes 1858, 34.
17. L. Price 2000, 4–5.
18. Dames 2007, 82.
19. Lakoff and Johnson 2008 [1980], 89–91.
20. Bain 1859, 147–8.
21. Bain 1859, 184.
22. Bain 1859, 156.
23. Bain 1859, 185.
24. Bain 1859, 196.
25. White 1887, 63.
26. West 1883, 236–7.
27. Herrick 1883, 166–8. Herrick's given name was Sophia, misrepresented on the title page as Sophie.
28. Smith 1880, 300–1. On the history of the *Podura* test, see Walker 2011.
29. Lewes 1858, 94.
30. Lewes 1858, 95–6.
31. This movement resembles the "tuning" that Andrew Pickering identifies as a process of "resistance and accommodation" but on a much smaller timescale—hours instead of months or years—and with the important addition of the moment of "description and awe." Pickering describes a more or less cleanly delineated stop-and-start pattern to tuning, which he terms a "dance of agency" or "dialectic of resistance and accommodation," but he also portrays the mangle as proceeding through a complex and overlapping process (1995, 21–2).
32. The term "inferential explicitness" derives from Gross et al. 2002, 160, where it describes a deliberate leading of the reader from step to step in an argument from observed facts to mechanical explanation and/or theory.
33. On Huxley's critique of Lewes, see White 2002. Harsh reviews of *Sea-side Studies* appeared in, for example, the *Natural History Review* ("Sea-side Studies at Ilfracombe, 1858) and the *Westminster Review* ("Contemporary Literature. Science," 590–603, esp. 599–600).
34. Dallinger 1877, 7.
35. Dallinger 1877, 5–6.
36. Dallinger 1877, 6–7.
37. Dallinger 1877, 7.
38. Terrall discusses these two different approaches to narrative, both common in natural history (2017, 51–2).
39. Lewes 1858, 23.
40. King 2019, 130–1.
41. King 2019, 132.
42. [Hart] 1857, 511.
43. Armstrong 2002, 37.
44. Armstrong 2002, 37.
45. Eliot 1997 [1872], 55.

Works Cited

Primary Sources

Abbe, Ernst. "Beiträge zur Theorie des Mikroskops und der mikroskopischen Wahrnehmung." *Archiv für Mikroskopische Anatomie* 9.1 (1873): 413–68.

Abbe, Ernst. "A Contribution to the Theory of the Microscope and of Microscopic Vision." Trans. H. E. Fripp. *PBNS* n.s. 1.2 (1874–5): 200–61.

Abbe, Ernst. "Extracts from Mr. H. E. Fripp's Translation of Professor Abbe's Paper on the Microscope." *MMJ* 14.4 (Oct 1875): 191–201; *MMJ* 14.5 (Nov. 1875): 245–63.

Abbe, Ernst. "On the Conditions of Orthoscopic and Pseudoscopic Effects in the Binocular Microscope." *JRMS* 1.2 (Apr. 1881a): 203–11.

Abbe, Ernst. "On the Estimation of Aperture in the Microscope." *JRMS* 1.3 (June 1881b): 388–423.

Agassiz, Louis. *Methods of Study in Natural History* (Boston, MA: Ticknor and Fields, 1863).

"An Amateur." "Notes on the Microscope Stand and on Some of Its Accessories." *The Microscope* 10 (1890): 369–74.

"Anti Spiral." "Transcendental Anatomy, Strictures of Professor Valentin on the Views of Dr. Martin Barry." *Lancet* 42.1062 (1844): 477.

"Art. III.—The Microscope: And Its Relations [sic]" (review). *Buffalo Medical Journal and Monthly Review of Medical and Surgical Science* 12.6 (Nov. 1856): 348–50.

Bacon, Francis. *Novum Organum* [1620]. In *The Works of Francis Bacon*, ed. Robert Leslie Ellis et al., vol. 4 of 14 (London: Longman, 1858).

Badcock, John. *Vignettes from Invisible Life* (London: Cassell, 1883).

Bain, Alexander. *The Senses and the Intellect* (London: John W. Parker and Son, 1855).

Bain, Alexander. *The Emotions and the Will* (London: John W. Parker and Son, 1859).

Baker, Henry. *The Microscope Made Easy* (London: R. Dodsley, 1742).

Balfour, [John Hutton]. *The Botanist's Companion* (Edinburgh: Adam and Charles Black, 1860).

Baudelaire, Charles. "The Painter of Modern Life." In *The Painter of Modern Life and Other Essays*, trans. Jonathan Mayne (London: Phaidon, 1965): 1–40.

Beale, Lionel S. *The Microscope and Its Application to Clinical Medicine*, 3d ed. (London: Samuel Highley, 1854).

[Beale, Lionel S.] "Pathological Laboratory." In Lionel Beale, *The Use of the Microscope in Clinical Medicine* (London: John Churchill, 1857): advertising matter included between plates 14 and 15.

Beale, Lionel S. "*Archives of Medicine*...Advertisement." In Lionel S. Beale, *The Use of the Microscope in Clinical Medicine* (London: John Churchill, 1857a), after plate marked "Urine—IV."

Beale, Lionel S. *How to Work with the Microscope* (London: John Churchill, 1857b).

Beale, Lionel S. *The Use of the Microscope in Clinical Medicine* (London: John Churchill, 1857c).

Beale, Lionel S. *How to Work with the Microscope*, 2d ed. (London: John Churchill, 1861).

Beale, Lionel S. "Of the Formation of the So-Called Intercellular Substance of Cartilage, and of Its Relation to the So-Called Cells; with Observations upon the Process of Ossification." *Transactions of the Microscopical Society of London (Quarterly Journal of Microscopical Science)* 11.1 (1863): 95–104.

Beale, Lionel S. *How to Work with the Microscope*, 3d ed. (London: Harrison, 1865a).

Beale, Lionel S. "On the Ultimate Nerve-Fibres Distributed to Muscle and Some Other Tissues." *Proceedings of the Royal Society of London* 14 (1865b): 229–68.

Beale, Lionel S. *How to Work with the Microscope*, 4th ed. (London: Harrison, 1868).

Beale, Lionel S. "Protoplasm and Living Matter." *MMJ* 1.5 (May 1869): 277–88.
Beale, Lionel S. *The Microscope in Medicine*, 4th ed. (London: J. and A. Churchill, 1878).
Beale, Lionel S. *How to Work with the Microscope*, 5th ed. (London: Harrison, 1880a).
Beale, Lionel S. "The President's Address." *JRMS* 3.2 (Apr. 1880b): 202–24.
Beck, Richard. *Treatise on the Construction, Proper Use, and Capabilities of Smith, Beck, and Beck's Achromatic Microscopes* (London: John Van Voorst, 1865).
Bell, Charles. *The Hand; Its Mechanism and Vital Endowments, as Evincing Design* (London: Pickering, 1833).
Bennett, John Hughes. *On the Employment of the Microscope in Medical Studies* (Edinburgh: Maclachlan, Stewart, 1841).
Bennett, John Hughes. *An Introduction to Clinical Medicine.* (Edinburgh: Sutherland and Knox, 1850).
Bennett, John Hughes. *An Introduction to Clinical Medicine. Six Lectures on the Method of Examining Patients; Percussion; Auscultation; the Use of the Microscope; and the Diagnosis of Skin Diseases*, 2d ed. (Edinburgh: Sutherland and Knox, 1853).
Blackham, George E. "President's Address: The Evolution of the Modern Microscope." *PASM* 4 (1882): 25–47.
Bousfield, Edward Collins. *A Guide to the Science of Photo-micrography* (London: W. Kent and H. H. Sharland, 1887).
Boyle, Robert. *The Christian Virtuoso Shewing that by Being Addicted to Experimental Philosophy, a Man is Rather Assisted, Than Indisposed, to be a Good Christian*, 2 vols. (London, 1690).
Brewster, David. *A Treatise on Optics* (London: Longman, Rees, Orme, Brown, and Green, 1831).
Brewster, David. *Letters on Natural Magic: Addressed to Sir Walter Scott, Bart.* (London: John Murray, 1832).
Brewster, David. *A Treatise on the Microscope, forming the Article under that Head in the Seventh Edition of the Encyclopedia Britannica* (Edinburgh: Adam and Charles Black, 1837).
[Brewster, David]. "The Microscope and Its Revelations" (review essay). *North British Review* 25 (Aug. 1856): 437–76.
Brewster, David. "Life in a Drop of Water." *Good Words* 5 (1864): 169–76.
Brocklesby, John. *Views of the Microscopic World* (New York: Pratt, Woodford, 1851).
[Broderip, William John]. "The Rambles of a Naturalist [Rev. of A. de Quatrefages, *The Rambles of a Naturalist on the Coasts of France, Spain, and Sicily*, trans. E. C. Otté]." *Fraser's Magazine* 57.340 (Apr. 1858): 441–58.
Brooke, Charles. "The President's Address for the Year 1865." *TMSL* n.s. 13.1 (Jan. 1865): 19–23.
Browning, Robert. "Mr. Sludge, "The Medium." In *Dramatis Personae* (London: Chapman and Hall, 1864): 171–236.
Buckley, Arabella Burton. *Through Magic Glasses and Other Lectures* (London: Edward Stanford, 1890).
Burke, Edmund. *A Philosophical Enquiry into the Origin of our Ideas of the Sublime and Beautiful* (London: R. and J. Dodsley, 1757).
Butler, Edward A. *Pond Life: Insects* (London: Swan Sonnenschein, 1886).
Campbell, Thomas. "Lecture I. Part 2. Lectures on Poetry." *New Monthly Magazine and Literary Journal* 1 (1821): 129–41.
"Carey's Oxy-Hydrogen Microscope." *The Morning Post* (Nov. 6, 1835), 3. Gale British Library Newspapers, Part II: 1800–1900.
Carlyle, Thomas. *Tales by Musaeus, Tieck, Richter* (London: Chapman and Hall, 1827).
Carlyle, Thomas. "The Diamond Necklace." *Fraser's Magazine for Town and Country* 15.85–6 (Jan.–Feb. 1837): 1–19, 172–89.
Carlyle, Thomas. *On Heroes, Hero-worship and the Heroic in History* (London: Chapman and Hall, 1840).
"Carpenter on the Microscope." *Intellectual Observer* 2 (1863): 352.
Carpenter, William B. *Principles of Human Physiology* (London: John Churchill, 1842).
Carpenter, William B. *Principles of Human Physiology*, 3d ed. (London: John Churchill, 1846).

Carpenter, William B. *Principles of Human Physiology*, 4th ed. (London: John Churchill, 1853).
Carpenter, William B. "On the Study of Natural History as a Branch of Education." *Journal of the Society of Arts* 2.88 (July 28, 1854).
Carpenter, William B. *Principles of Human Physiology. With Their Chief Applications to Psychology, Pathology, Therapeutics, Hygiène, and Forensic Medicine*, 5th ed. (London: John Churchill, 1855).
Carpenter, William B. *The Microscope: And Its Revelations* (London: John Churchill, 1856).
Carpenter, William B. *The Microscope and its Revelations.* (London: John Churchill, 1862).
Carpenter, William B. *The Microscope and Its Revelations* (London: John Churchill and Sons, 1868).
Carpenter, William B. *Principles of Mental Physiology* (London: Henry S. King, 1874).
Carpenter, William B. "Microscope." In *Encyclopedia Britannica*, 9th ed., vol. 16 (New York: Samuel L. Hall, 1878): 258–78.
Carpenter, William B. *The Microscope and Its Revelations.* 6th British ed. (London: J. & A. Churchill, 1881).
Carpenter, William B. *The Microscope and Its Revelations.* 6th American ed. (New York: William Wood, 1883).
Carpenter, William B. and W. H. Dallinger. *The Microscope and Its Revelations*, 7th ed. (London: J. and A. Churchill, 1891).
Catalogue of the Books in the Library of the Quekett Microscopical Club (London, 1915).
Catalogue of Lewis's Medical & Scientific Library: Including a Classified Index of Subjects (London: Lewis's Library, 1888).
Catalogue of New and Standard Works in Circulation at Mudie's Select Library (London, 1857).
Catalogue of the Principal Books in Circulation at Mudie's Select Library (London, 1865).
Catlow, Agnes. *Drops of Water: Their Marvelous and Beautiful Inhabitants Displayed by the Microscope* (London: Reeve and Benham, 1851).
Chalmers, Thomas. *A Series of Discourses on the Christian Revelation, Viewed in Connection with the Modern Astronomy* (Glasgow: J. Smith, 1817).
"Chemical Composition of the Ink of Cephalopods." *JRMS* 1.6 (Dec. 1881): 877.
Clarke, L[ouisa] Lane. *The Microscope* (London: G. Routledge, 1858).
Clarke, L[ouisa] Lane. *Objects for the Microscope*, 3d ed. (London: Groombridge and Sons, 1863).
Cole, Arthur C. *The Methods of Microscopical Research: A Practical Guide to Microscopical Manipulation* (London: Baillière, Tindall and Cox, 1895).
"Contemporary Literature. Science." *Westminster Review* n.s. 13.2 (April 1858): 590–603.
Crisp, Frank. "On the Present Condition of Microscopy in England." *JRMS* 1.3 (July 1878): 121–32.
Crisp, Frank. "Notes on Aperture, Microscopical Vision, and the Value of Wide-Angled Immersion Objectives." *JRMS* 1.2 (Apr. 1881): 303–60.
Dallinger, W. H. "On 'Navicula crassinervis,' 'Frustulia Saxonica,' and 'Navicula rhomboides,' as Test-objects." *MMJ* 17 (Jan. 1877): 1–7.
Dallinger, W. H. "Preface." In William B. Carpenter and W. H. Dallinger, *The Microscope and Its Revelations*, 7th ed. (London: J. Churchill, 1891): v–x.
Dancer, J. B. "On the Markings on the Pleurosigma Angulatum and on the Lepisma saccharina." *MMJ* 1.4 (Apr. 1869): 243–4.
Darwin, Charles to Henry Slack. Dec. 2 [1867]. "Letter no. 5703F." Darwin Correspondence Project, 2020. https://www.darwinproject.ac.uk/letter/DCP-LETT-5703F.xml
Darwin, Charles to J. M. Herbert. Nov. 21, 1872. "Letter no. 8639." Darwin Correspondence Project, 2020. https://www.darwinproject.ac.uk/letter/?docId=letters/DCP-LETT-8639.xml
Darwin, Charles to Joseph Hooker. May 1, 1847. "Letter no. 1085." Darwin Correspondence Project, 2020. https://www.darwinproject.ac.uk/letter/?docId=letters/DCP-LETT-1085.xml
Darwin, Charles to Joseph Hooker. Nov. 14, 1846. "Letter no. 1024." Darwin Correspondence Project, 2020. https://www.darwinproject.ac.uk/letter/?docId=letters/DCP-LETT-1024.xml
De Morgan, Augustus. *A Budget of Paradoxes* (London: Longmans, Green, 1872).
Dick, Thomas. *The Telescope and Microscope* (London: RTS, 1851).
Dickens, Charles. "A Preliminary Word." *Household Words* 1.1 (Mar. 30, 1850): 1–2.
[Dixon, Edmund Saul]. "Microscopics." *Household Words* 14.345 (1856a): 377–81.

[Dixon, Edmund Saul]. "The World Unseen." *Household Words* 14.342 (Oct. 11, 1856b): 291–6.
[Dixon, Edmund Saul]. "Microscopic Preparations." *Household Words* 16.385 (1857): 132–8.
Durham E. "President's Address." *JQMC* 1 (1868–9): 95–110.
"E. B. P." [Edward Bagnell Poulton]. "Elementary Instruction in Practical Biology" (review). *Nature* 37 (1888): 505–6.
Edwards, A. Mead. "How to Prepare Specimens of Diatomaceae for Examination and Study by Means of the Microscope." *MMJ* 12.5 (Nov. 1874): 225–37.
Ehrenberg, Christian Gottfried. *Mikrogeologie* (Leipzig: L. Voss, 1854).
Eliot, George. *The Works of George Eliot: The Spanish Gypsy* (Edinburgh: William Blackwood and Sons, 1868).
Eliot, George. *Middlemarch.* [1872]. Ed. David Carroll. (Oxford: Oxford UP, 1997).
[Elliott, Emily Steele]. *Under the Microscope; or, "Thou shalt call me My Father"* (London: T. Nelson and Sons, 1861).
"Essex." Letter 12790. *English Mechanic* 25.639 (June 22, 1877): 363.
"Evenings at the Microscope." *Saturday Review* 184.7 (May 7, 1859): 570–1.
Ferguson, John. *The Microscope, Its Revelations and Application in Science and Art* (Edinburgh: Thomas Constable, 1858).
Fisher, George Thomas, Jr. *Microscopic Manipulation; containing the Theory and Plain Instructions for the use of the Microscope, including the best methods for the mounting of objects, and a review of the important discoveries effected by this instrument* (London: Thomas and Richard Willats, Opticians, 1846).
Frey, Heinrich. *The Microscope and Microscopical Technology.* Trans. George R. Cutter (New York: William Wood, 1872).
Fripp, H. E. "On the Physiological Limits of Microscopic Vision." *PBNS* n.s. 1 (1874–6): 457–75.
Fripp, H. E. "Is There a Science of Microscopy?" *PBNS* n.s. 2.1 (1877): 301–37.
G. T. F. [George Thomas Fisher, Jr.]. "Thoughts on Animalcules [The Microscope and Its Revelations]" (review essay). *The Westminster and Foreign Quarterly Review* 46.1 (Oct. 1846): 29–60.
Gage, Simon Henry. *Notes on Microscopical Methods* (Ithaca, NY: Andrus and Church, 1886).
Gatty, Mrs. Alfred (Margaret). *Waifs and Strays of Natural History* (London: Bell and Daldy, 1871).
Gatty, Mrs. Alfred (Margaret). *British Sea-weeds*, vol. 1. (London: Bell and Daldy, 1872).
Glaisher, James. "The President's Address for the Year 1866." *TMSL* n.s. 14.1 (1866): 45–64.
Goring, C. R. "On Achromatic Microscopes with a Description of Certain Objects for Trying Their Definition." *QJSLA* n.s. 23, no. 2 (1827a): 410–34.
Goring, C. R. "On Mr. Tulley's Thick Aplanatic Object-Glasses, for Diverging Rays." *QJSLA* 22.44 (1827b): 265–84.
Goring, C. R. and Andrew Pritchard. *The Natural History of Several New, Popular, and Diverting Living Objects for the Microscope* (London: Whittaker, Treacher, 1829).
Goring, C.R. and Andrew Pritchard. *Microscopic Illustrations of a Few New, Popular, and Diverting Living Objects, with their Natural History, &c.; Conjoined with Accurate Descriptions of the Latest Improvements in the New Microscopes, the Best Methods of Constructing their Mountings, Apparatus, &c and Complete Instructions for Using Them* (London: Whittaker, Treacher, 1830).
Goring, C. R. and Andrew Pritchard. *Micrographia: Containing Practical Essays on Reflecting, Solar, Oxy-hydrogen Gas Microscopes; Micrometers; Eye-pieces, &c. &c.* ([London]: Whitaker and Company, 1837).
Gosse, Edmund. *The Life of Philip Henry Gosse, FRS* (London: Kegan Paul, Trench, Trübner, 1890).
Gosse, Edmund. *Father and Son: A Study of Two Temperaments* (London: William Heinemann, 1907).
Gosse, Philip Henry. *A Naturalist's Rambles on the Devonshire Coast* (London: John van Voorst, 1853).
Gosse, Philip Henry. *Evenings at the Microscope; or, Researches among the Minuter Organs and Forms of Animal Life* (London: SPCK, 1859).
Gosse, Philip. "Twenty-Four More New Species of Rotifer." *JRMS* 7.6 (Dec. 1887): 861–71.

Gould, C. *The Companion to the Compound, Ox-Hydrogen [sic], and Solar Microscopes*, 10th ed. (London: W. Cary, 1835).
Griffith, John William and Arthur Henfrey. *The Micrographic Dictionary; A Guide to the Examination and Investigation of the Structure and Nature of Microscopic Objects*, 3d ed. (London: John Van Voorst, 1856).
Hack, Maria. *Harry Beaufoy, Or, The Pupil of Nature* (London: Harvey and Darton, 1821).
Hannover, Adolphe. *On the Construction and Use of the Microscope*, ed. John Goodsir (Edinburgh: Sutherland and Knox, 1853).
[Hart, Ernst Abraham]. "Nature's Greatness in Small Things." *Household Words* 16.401 (1857): 511–13.
Hartley, David. *Observations on Man: His Frame, His Duty, and His Expectations* (London: S. Richardson, 1749).
Harvey, W. H. *The Sea-Side Book* (London: John van Voorst, 1849).
Hassall, Arthur Hill. *A Microscopic Examination of the Water Supplied to the Inhabitants of London and the Suburban Districts.* (London: Samuel Highley, 1850).
Hassall, Arthur Hill. *Adulterations Detected, or, Plain Instructions for the Discovery of Frauds in Food and Medicine* (London: Longman, Brown, Green, Longmans, and Roberts, 1857).
Helmholtz, Hermann von. "The Recent Progress of the Theory of Vision" [1868], trans. Pye-Smith. In *Popular Lectures on Scientific Subjects* (New York: D. Appleton, 1873): 197–316.
Helmholtz, Hermann von. "On the Limits of the Optical Capacity of the Microscope." Trans. H. E. Fripp. *PBNS* 1.3 (1874–6): 407–40.
Herrick, Sophie [Sophia] Bledsoe. *The Wonders of Plant Life under the Microscope* (New York: G. P. Putnam's Sons, 1883).
Herschel, John Frederick William. *A Preliminary Discourse on the Study of Natural Philosophy*, ed. Dionysius Lardner (London: Longman, Rees, Orme, Brown, and Green, 1830).
Herschel, John Frederick William. "Whewell on the Inductive Sciences." In *Essays from the Edinburgh and Quarterly Reviews: With Addresses and Other Pieces.* (London: Longman, Brown, Green, Longmans, and Roberts, 1857): 142–256.
Hickie, W. J. "Correspondence: Professor Hasert's New Objective." *MMJ* 14.5 (Nov. 1875a): 260–3.
Hickie, W. J. "Correspondence: Recent Works on the Diatomaceae, and on Objectives." *MMJ* 13.6 (June 1875b): 265–6.
Hickie, W. J. "On Dr. Schumann's Formulae for Diatom-lines." *MMJ* 14.1 (July 1875c): 6–15.
Hickie, W. J. "Further Notes on Frustulia Saxonica." *JRMS* 15.3 (Mar. 1876): 122–7.
Hogg, Jabez. *The Microscope: Its History, Construction, and Applications.* (London: W. S. Orr, 1854).
Hogg, Jabez. *The Microscope: Its History, Construction, and Applications. Being a Familiar Introduction to the Use of the Instrument and the Study of Microscopical Science* (London: Herbert Ingram & Co., 1856).
Hogg, Jabez. *The Microscope: Its History, Construction, and Application*, 4th ed. (London: Routledge, Warnes, and Routledge, 1859).
Hogg, Jabez. *The Microscope: Its History, Construction, and Application*, 15th ed. (London: George Routledge and Sons, 1898).
Holland, Henry. *Chapters on Mental Physiology* (London: Longman, Brown, Green, and Longmans, 1852).
Holmes, Oliver Wendell. "An Address Delivered at the Annual Meeting of the Boston Microscopical Society." *Boston Medical and Surgical Journal* 96.21 (May 24, 1877): 601–12.
Hooke, Robert. *Micrographia: or, Some Physiological Descriptions of Minute Bodies Made by Magnifying Glasses* (London: J. Martyn and J. Allestry, 1665).
"The Horror of Astronomy." *The Spectator* 67.3296 (Aug. 29, 1891): 285.
How to Choose a Microscope. By a Demonstrator (London: Hardwicke and Bogue, 1876).
"How to Work with the Microscope" (review). *The Edinburgh Medical and Surgical Journal* 3 (Nov. 1857): 459.
Hudson, C. T. "Transactions of the Society: The President's Address." *JRMS* 9.2 (Apr. 1889): 169–79.
Hudson, C. T. "Transactions of the Society: The President's Address on Some Needless Difficulties in the Study of Natural History." *JRMS* 10.2 (Apr. 1890): 129–41.

Hudson, Charles Thomas and Philip Henry Gosse. *The Rotifera or Wheel-animalcules* (London: Longmans, Green, 1886).
Hughes, Arthur. "Studies in the History of Microscopy. 1. The Influence of Achromatism." *JRMS* 76.1–2 (1956): 47–60.
Hume, David. "Enquiry Concerning Human Understanding." In *Essays and Treatises on Several Subjects* [1777]. Eds. Amyas Merivale and Peter Millican. Hume Texts Online, accessed Aug. 28, 2020. https://davidhume.org/texts/e
Huxley, T. H. "*Lacinularia socialis*: A Contribution to the Anatomy and Physiology of the Rotifera." *TMSL* n.s. 1.1 (Jan. 1853): 1–19.
Huxley, T. H. *The Crayfish: An Introduction to the Study of Modern Zoology* (New York: Appleton, 1880).
"Illumination by Daylight and Artificial Light—Paraboloids and Lieberkühns." *JRMS* n.s. 4.2 (1884): 622–4.
"J. P. P." [J. P. Philips]. *The Microscope and Its Revelations* (Solva, Wales: John Williams, 1859).
Jackson, B. Daydar. "Inexperienced Artists *v.* Experienced Ones." *MMJ* 10.2 (Aug. 1873): 97–8.
Jackson, James. *A Memoir of James Jackson, Jr., M. D.* (Boston: I. R. Butts, 1835).
Johnston, George. *History of the British Zoophytes* (Edinburgh: W. H. Lizars, 1838).
Jones, Thomas Rymer. *Natural History of Animals* (London: John van Voorst, 1845).
[Joyce, Jeremiah]. *The Wonders of the Microscope* (London, 1806).
Julien, Alexis. "Suggestions in Microscopical Technique." *Journal of the New-York Microscopical Society* 9.2 (Apr. 1893): 23–43.
Kant, Immanuel. *Critique of Judgement* [1790]. Trans. Werner Pluhar. (Indianapolis: Hackett, 1987).
Kerr, Richard. *Hidden Beauties of Nature* (London: RTS, 1895).
Kidd, William. *The World and Its Two Faces: A Microscopic Mirror of the Land We Live In* (London: Groombridge and Sons, 1854).
Kingsley, Charles. *Glaucus; or, The Wonders of the Shore* (Cambridge: Macmillan, 1855).
Kingsley, Charles. *The Water-Babies: A Fairy Tale for a Land-Baby* (London: Macmillan, 1863).
Kingsley, J. S. "Workers and Their Instruments." *Science Record* 2 (1884): 261–2.
Kitton, F[rederic]. "The Microscope and Microscopic Work." *Science-Gossip* 12.133 (1876): 3–5.
Klein, E[manuel Edward], John Burdon-Sanderson, Michael Foster, and T. Lauder Brunton. *Handbook for the Physiological Laboratory*, ed. John Burdon-Sanderson (London: J. and A. Churchill, 1873).
Koch, Robert. *Investigations into the Etiology of Traumatic Infective Diseases* [1878], trans. W. Watson Cheyne (London: New Sydenham Society, 1880).
Lang, Fred H. "On Selecting and Mounting Diatoms." *MMJ* 4.6 (Dec. 1870): 308–12.
Lankester, Edwin. *Half-hours with the Microscope* (London: Robert Hardwicke, 1859).
Lankester, Edwin, "Notes on Raphides." *TMSL* 2–3.12 (1863): 243–51.
Lankester, E. Ray. "The Crayfish" (review). *Popular Science Monthly* 16.46 (Apr. 1880): 789–95.
Lankester, Edwin and George Busk, eds. "Preface." *QJMS* 1 (1853): 1–2.
Lankester, Mrs. [Phoebe]. "Eyes or No Eyes." *St. James's Magazine* 2.1 (1861): 121–8.
Lardner, Dionysius. *The Museum of Science and Art*, vols. 1-2, 4, 5 (London: Walton and Maberly, 1854–5).
Lardner, Dionysius. *The Microscope: From "The Museum of Science and Art"* (London: Walton and Maberly, 1856).
Laycock, Thomas. *Lectures on the Principles and Methods of Medical Observation and Research, for the Use of Advanced students and Junior Practitioners* (Philadelphia: Blanchard and Lea, 1857).
Laycock, Thomas. *Mind and Brain; or, The Correlations of Consciousness and Organisation; with their Applications to Philosophy, Zoology, Physiology, Mental Pathology, and the Practice of Medicine.* (Edinburgh: Sutherland and Knox, 1860).
Leech, John. "The Wonders of a London Water-Drop." *Punch* (May 11, 1850): 188.
Leeuwenhoek, Antony van to Henry Oldenburg. Letter of Sept. 7, 1674. In *Antony Van Leeuwenhoek and His Little Animals*, trans. and ed. Clifford Dobell (New York: Harcourt, Brace, 1932): 109–11.

Leeuwenhoek, Antony van to Robert Hooke. Letter of Nov. 12, 1680. In *Antony Van Leeuwenhoek and His Little Animals*, trans. and ed. Clifford Dobell (New York: Harcourt, Brace, 1932): 200–4.
"L'Envoy." *Medical Times* n.s. 1.1 (July 6, 1850): iii–iv.
Lewes, George Henry. *Sea-Side Studies at Ilfracombe, Tenby, the Scilly Isles, and Jersey* (Edinburgh: William Blackwood and Sons, 1858).
Lewes, George Henry. *The Physiology of Common Life*, 2 vols. (Edinburgh: William Blackwood and Sons, 1859–60).
Lewes, George Henry. *Studies in Animal Life* (New York: Harper and Brothers, 1860).
Lewes, George Henry. *Problems of Life and Mind* (*Foundations of a Creed*, 1874; *The Physical Basis of Mind*, 1877; *The Study of Psychology*, 1879) (London: Trübner, 1874–9).
Lewis, Francis West to James Lorimer Graham. Letter of Mar. 18, [1859], HM 81256. Huntington Library, San Marino, California.
Lewis, Francis West to James Lorimer Graham. Letter of Apr. 3, [1859], HM81267. Huntington Library, San Marino, California.
Lister, Joseph Jackson. "On Some Properties in Achromatic Object-Glasses Applicable to the Improvement of the Microscope." *Philosophical Transactions of the Royal Society of London* 120 (1830): 187–200.
Lyons, Robert D. *Apology for the Microscope* (Dublin: Fannin, 1851).
Mantell, Gideon Algernon. *Thoughts on Animalcules: or, a Glimpse of the Invisible World Revealed* (London: John Murray, 1846).
Marsh, Sylvester. *Section-Cutting: A Practical Guide to the Preparation and Mounting of Sections for the Microscope* (London: J. and A. Churchill, 1878).
Martineau, Harriet. *Autobiography*, ed. Maria Weston Chapman, vol. 1 (London: Smith, Elder, 1877).
Mathews, Charles. *Memoirs of Charles Mathews, Comedian*, ed. Mrs. Mathews (London: Richard Bentley, 1839).
Maudsley, Henry. *The Physiology and Pathology of the Mind* (London: Macmillan, 1867).
Maudsley, Henry. *The Physiology and Pathology of the Mind*, 2d ed. (London: Macmillan, 1868).
Mayall, John. *Cantor Lectures on the Microscope* (London: 1886).
McCalla, Albert. "Summary of Current Researches: Microscopy: Verification of Microscopical Observation." *JRMS* n.s. 3.5 (Oct. 1883a): 766–72.
McCalla, Albert. "The Verification of Microscopical Observation": *PASM* 5 (Oct. 1883b): 1–19.
"*The Medical Times*" (advertisement). *Medical Times* n.s. 1.25 (Dec. 21, 1850): n.p.
Michael, Albert D. *British Oribatidae*. Vol. 1 of 2. London: Ray Society, 1884.
Michael, A. D. "The President's Address: The History of the Royal Microscopical Society." *JRMS* 15.1 (Feb. 1895): 1–15.
Michels, John. "The Microscope and Its Misinterpretations." *Popular Science Monthly* 7 (June 1875): 177–83.
"Microscope." In *Chambers's Encyclopedia* (Edinburgh: W. and R. Chambers, 1874): 441–3.
"The Microscope and Its Marvels." *Chambers's Papers for the People* 6.41 (1850): 1–32.
"The Microscope: Its History, Construction, and Application" (review). *QJMS* 8 (1868): 84–5.
"Microscope, Microscopic, Microscopical." *Journal of the New York Microscopical Society* 1 (1885): 209.
"A Microscopic Entertainment." *The Leisure Hour: An Illustrated Journal of Instruction and Recreation* 157 (1854): 826–7.
"Microscopic View of a Drop of Water." *The Tourist; or, Sketch Book of the Times* 1.22 (Jan. 28, 1833): 177–8.
"Microscopical Memoranda: Owen on the Application of the Microscope." *Microscopical Journal and Structural Record* 2 (1842): 63–4.
"Microscopy." *American Naturalist* 10.8 (Aug. 1876): 503.
"Miscellanies: Notes on Microscopic Literature." *Canadian Naturalist and Geologist* 2.5 (Nov. 1857): 387–92.
Moore, R. H. "Presidential Address." *JMNS* 15 (1896): 1–7.
Morehouse, G. W., "On the Structure of Diatoms." *MMJ* 12.1 (July 1874): 19–25.

Morley, Henry. "The Water-Drops." *Household Words* 1.21 (Aug. 17, 1850): 482–9.

"Mr. New's Grand Exhibition!" Undated advertising material. Bodleian Library, University of Oxford: John Johnson Collection: Scientific Instruments 2.

Nägeli, Karl Wilhelm von, and Simon Schwendener. *The Microscope in Theory and Practice* trans. Frank Crisp and John Mayall, Jr. (London: Swan Sonnenschein, Lowery, & Co., 1887).

"New Books with Short Notices." *MMJ* 9.2 (Feb. 1873a): 67–9.

"New Books with Short Notices." *MMJ* 10.2 (Aug. 1873b): 85–9.

Nicholl, Jane Harriot to William Henry Fox Talbot. Letter of June 23, 1816, no. 702. In *The Correspondence of William Henry Fox Talbot*, ed. Larry J. Schaaf (2012), http://foxtalbot.dmu.ac.uk/letters/transcriptName.php?bcode=Nich-JH&pageNumber=20&pageTotal=68&referringPage=1

Notcutt, W. L. [William Lowndes]. *A Handbook of the Microscope and Microscopic Objects* (London: Edward Lumley, 1859).

"Notices of Meetings: Liverpool Microscopical Society." *NMMN* 2.15 (Mar. 1882): 67–70.

"Obituary. Charles Thomas Hudson." *JRMS* 24.1 (Feb. 1904): 48–9.

"Obituary. Christian Gottfried Ehrenberg." *MMJ* 17.3 (Mar. 1877): 166–7.

"Obituary: Edward Milles Nelson." *JRMS* 58.3 (Sept. 1938): 200–3.

"Obituary Notices: Joseph Bancroft Reade." *MMJ* 5.27 (Mar. 1871): 92–6.

O'Brien, Fitz-James. "The Diamond Lens." *Atlantic Monthly* 1.3 (Jan. 1858): 354–67.

Olley, William Henry. *The Wonders of the Microscope, Photographically Revealed by Olley's Patent Micro-photographic Reflecting Process.* (London: W. Kent, 1857).

"Orientation in Microscopic Sections," *American Naturalist* 17.1 (Jan 1883), 109–12.

Owen, Richard. *Odontography; Or, A Treatise on the Comparative Anatomy of Teeth: Their Physiological Relations, Mode of Development, and Microscopic Structure in the Vertebrate Animals* (London: Hippolyte Bailliere, 1840).

[Owen, Richard], "Lyell—On Life and its Subsequent Development." *Quarterly Review* 89.178 (Sept 1851): 412–51.

Paget, James. *Report on the Chief Results Obtained by the Use of the Microscope, in the Study of Human Anatomy and Physiology* (London: John Churchill, 1842).

Pascal, Blaise. *The Thoughts, Letters, and Opuscules of Blaise Pascal*, trans. O. W. Wight (New York: Derby and Jackson, 1859).

Pater, Walter. *Studies in the History of the Renaissance* (London: Macmillan and Co., 1873).

Phin, John. *Practical Hints on the Selection and Use of the Microscope. Intended for Beginners.* 2d ed. (New York: Industrial, 1877).

Pillischer, M. *Illustrated Descriptive Catalogue of Achromatic Microscopes, Telescopes, Opera, Race, and Field Glasses* (London: 1873).

Pirie, Mary. *Familiar Teachings of Natural History: A Book for the Use of Schools and Families* (Aberdeen: Lewis Smith, 1864).

Plumer, J. J. *A Few Words on the Choice of a Microscope* (London: John Churchill and Sons, 1865).

Pocklington, Henry. "The Microscope in Pharmacy." *Pharmaceutical Journal and Transactions* 3d series, vol. 2 (Feb. 3, 1872): 621–2.

Porter, W. T. "Exchanges." *Hardwicke's Science-Gossip* 24.278 (Feb. 1888): 48.

Pratt, Anne. *Chapters on the Common Things of the Sea-Side* (London: SPCK, 1850).

Pritchard, Andrew. "On the Art of Forming Diamonds into Single Lenses for Microscopes." *QJSLA* 23 (1827): 15–21.

Pritchard, Andrew. *The Natural History of Animalcules, with Instructions for Procuring and Viewing Them* (London: Whittaker, 1834).

Pritchard, Andrew. *A History of Infusoria, Living and Fossil* (London: Whittaker, 1841).

Pritchard, Andrew. *A History of Infusorial Animalcules, Living and Fossil* (London: Whittaker, 1852).

Pritchard, Andrew and C. R. Goring. *The Microscopic Cabinet of Select Animated Objects* (London: Whittaker, Treacher, and Arnot, 1832).

Pritchard, Andrew and C. R. Goring. *Microscopic Illustrations of Living Objects* (London: Whittaker, 1838).

"Proceedings at Mr. Suffolk's Class." *JQMC* 1.7 (1869): 211–12.
"Proceedings of Societies: Brighton and Sussex Natural History Society: Microscopical Section." *MMJ* 4.2 (Aug. 1870): 118–20.
"Proceedings of Societies: Croydon Microscopical Club." *MMJ* 6.1 (July 1871): 52–6.
"Proceedings of Societies: Microscopical Society of London. 1852." *QJMS* 1 (1853): 147.
"Proceedings of Societies: Quekett Microscopical Club." *MMJ* 13.2 (Feb 1875): 95–6.
"Professor Morton's Lantern Lectures." *The Magic Lantern* 1.12 (Aug 1875): 99–101.
"Progress of Microscopical Science," *MMJ* 12.1 (July 1874): 26–38.
Quekett, John. *Practical Treatise on the Microscope*, 2d ed. (London: H. Bailliere, 1848).
Quekett, John. "Lectures on Histology. Delivered at the Royal College of Surgeons, London." *Medical Times* n.s. 2.27 (Jan. 4, 1851): 1–2.
Quekett, John. *Lectures on Histology* (London: Hippolyte Bailliere, 1852).
Quekett, John. *Practical Treatise on the Microscope, Including the Different Methods of Preparing and Examining Animal, Vegetable, and Mineral Structures*, 3d ed. (London: H. Bailliere, 1855).
"Quekett Microscopical Club." *English Mechanic* 39.991 (Mar. 21, 1884): 48.
Quigley, Catharine. *The Microscope; Or Village Flies: In Three Cantos* (Monaghan: Nathaniel Greacen, 1819).
Reade, J. B. "On Animals of the Chalk Still Found in a Living State in the Stomachs of Oysters." *TMSL* 2 (1849): 20–4.
Reade, J. B. "On the Diatom Prism, and the True Form of Diatom Markings." *MMJ* 2 (July 1869): 5–11.
Reade, J. B. "The President's Address." *MMJ* 3.3 (Mar. 1870): 113–33.
"Recent Revelations of the Microscope." *Chambers's Edinburgh Journal* 5.112 (Feb. 21, 1846): 114–16.
"Report." *PBNS* n.s. 3.1 (1879–82): 76–8.
"Reviews [Review Essay on the Microscope]." *Natural History Review* 2 (1855): 28–45.
"Reviews [Review Essay on the Microscope]." *New Quarterly Review* 5.17 (1856): 28–32.
Roget, Peter Mark. *Animal and Vegetable Physiology Considered with Reference to Natural Theology* (London: William Pickering, 1834).
[Ross, Andrew]. "Microscope." In *The Penny Cyclopaedia of the Society for the Diffusion of Useful Knowledge*, vol. 15 (London: Charles Knight, 1839): 177–88.
Royston-Pigott, G. W. "On High-Power Definition: with Illustrative Examples." *MMJ* 2.6 (Dec. 1869): 295–305.
Royston-Pigott, G. W. "On the Principle of Testing Object-Glasses by Miniatures..." *MMJ* 13.4 (Apr. 1875): 147–55.
Royston-Pigott, G. W. "The Present Limits of Vision." *MMJ* 16.4 (Oct. 1876): 175–90.
Rupke, Nicolaas. "Richard Owen's Vertebrate Archetype." *Isis* 84.2 (1993): 231–51.
Ruskin, John. *Praeterita* [1887], eds. Edward Tyas Cook and Alexander Wedderburn (Cambridge: Cambridge University Press, 2015).
Rutherford, William. "Notes of a Course of Practical Histology for Medical Students." *QJMS* n.s. 12.45 (1872): 1–21.
Saville Kent, William. *A Manual of the Infusoria* (London: David Bogue, 1880).
Schacht, Hermann. *The Microscope in Its Special Application to Vegetable Anatomy*, trans. and ed. Frederick Currey (London: Samuel Highley, 1853).
Schleiden, Matthias Jacob. *Grundzüge der wissenschaftlichen Botanik* (Leipzig: Wilhelm Engelmann, 1842).
Schleiden, Matthias Jakob. *Principles of Scientific Botany*, trans. and ed. Edwin Lankester (London: Longman, Brown, Green, and Longmans, 1849).
Schneider, Albert. *Microscopy and Micro-Technique* (Chicago, IL: Chicago Medical Books, 1899).
[Scoffern, John]. "The Senses, Their Limits and Their Fallacies." *St. James's Magazine* 14 (Sept. 1865): 177–87.
Scott, Sir Walter. *Letters on Demonology and Witchcraft, addressed to J. G. Lockhart, Esq.* (London: John Murray, 1830).
Searson, W. G. "Catalogue of the Library of the Microscopical Society of London, 1859." *TMSL* 8.1 (Jan. 1860): 13–28.

"Sea-side Studies at Ilfracombe." *Natural History Review* 5 (1858): 98–102.
"Sensory Nerve-Endings in Skin of Insects." *JRMS* 1.2 (Apr. 1881): 235.
Sewell, William. *The Microscope of the New Testament*, ed. William Crichton (London: Rivingtons, 1878).
Shadwell, Thomas. *The Virtuoso* [1676], eds. Marjorie Hope Nicolson and David Stuart Rodes (Lincoln, NE: University of Nebraska Press, 1966).
Smiley, Charles W. "American Societies Interested in Microscopy, Part II." *The Microscope* n.s. 1.3 (Mar. 1893): 33–9.
Smith, Henry and Robert C. R. Jordan. "King's College Hospital: Cases of Cancer of the Stomach." *Medical Times* 1.25 (Dec. 21, 1850): 651–3.
Smith, J. Edwards. *How to See with the Microscope* (Chicago, IL: Duncan, 1880).
Somerville, Mary. *On Molecular and Microscopic Science*, 2 vols. (London: John Murray, 1869).
Sommer, F. "Anatomy of the Liver-Fluke." *JRMS* 1.1 (Feb. 1881): 49.
Spencer, Herbert. *The Principles of Psychology* (London: Longman, Brown, Green, and Longmans, 1855).
Spiers, William. "Dips into My Aquarium." *JMNS* n.s. 3 (1890a): 256–9.
Spiers, William. *Rambles and Reveries of a Naturalist* (London: Charles H. Kelly, 1890b).
Sprat, Thomas. *The History of the Royal Society of London* (London, 1667).
Stephenson, J. W. "Observations on the Optical Appearances Presented by the Inner and Outer Layers of Coscinodiscus When Examined in Bisulphide of Carbon and in Air." *MMJ* 10.1 (July 1873): 1–5.
Stephenson, J. W. "Observations on Professor Abbe's Experiments Illustrating His Theory of Microscopic Vision." *MMJ* 17.2 [Feb. 1877], 82–8.
Stokes, A. C. "Notices of New American Fresh-Water Infusoria." *JRMS* 7.1 (July 1887): 35–40.
"The Story of Microscopical Discovery." *Universal Review* 1.4 (June 1859): 512–26.
"The Substage Condenser: Its History, Construction, and Management; and Its Effect Theoretically Considered." *JQMC* 4 (1890): 116–36.
Suffolk, W. T. *On Microscopical Manipulation* (London: Henry Gillman, 1870).
"Summary of Current Researches: Microscopy: Advantage of the Binocular." *JRMS* n.s. 1.5 (Oct. 1881): 835.
"Summary of Current Researches: Microscopy: Highest Magnifying Powers." *JRMS* n.s. 1.1 (Feb. 1881), 130.
"Summary of Current Researches: Microscopy: Learning to See with the Microscope." *JRMS* 8.3 (June 1888): 495–6.
"Summary of Current Researches Relating to Microscopy: Illuminating and Other Apparatus." *JRMS* 11.1 (Feb. 1891): 90–105.
"Summary of Current Researches…Zoology. Lowest and Smallest Forms of Life as revealed by the modern Microscope." *JRMS* 4.5 (Oct. 1884): 721–4.
Swinburne, Algernon Charles. *Under the Microscope* (Portland, ME: Thomas B. Mosher, 1899 [1872]).
Taylor, John Ellor. *The Playtime Naturalist* (London: Chatto and Windus, 1889).
"The Third Annual Quekett Dinner, on Saturday, the 23rd November, 1878, (With an account of the origin and progress of the Quekett Microscopical Club) with Songs Written for the Excursionists' Annual Dinners, Q.M.C. by Dr. M. C. Cooke." (London: Keating, 1878).
[Tierney, Clarence]. *Catalogue of the Printed Books and Pamphlets in the Library of the Royal Microscopical Society* (London: Royal Microscopical Society, 1929).
Timbs, John. *Wonderful Inventions: From the Mariner's Compass to the Electric Telegraph Cable* (London: George Routledge and Sons, 1868).
Tuckwell, William. *Reminiscences of Oxford* (New York: E. P. Dutton, 1908).
Tuthill, Cornelius. "The Microscope, edited by a Fraternity of Gentlemen." *The Microscope, edited by a Fraternity of Gentlemen* 1.1 (1820): 3–8.
Tyndall, John, "Scientific Use of the Imagination" (1870 Address to the British Association for the Advancement of Science) in John Tyndall, *Essays on the Use and Limit of the Imagination in Science* (London: Longmans, Green, 1870), 13–51.

Tyndall, John. *Fragments of Science for Unscientific People: A Series of Detached Essays, Addresses, and Reviews* (London: Longmans, Green, 1871).
Tyndall, John. *Six Lectures on Light* (New York: D. Appleton, 1873).
Van Heurck, Henri. *The Microscope: Its Construction and Management*, trans. Wynne E. Baxter (London: Crosby, Lockwood and Son, 1893).
Walton, [Henry] Haynes. *A Practical Treatise on the Diseases of the Eye* (London: J. and A. Churchill, 1875).
Ward, The Hon. Mrs. [Mary]. *Microscope Teachings* (London: Groombridge and Sons, 1864).
Ward, The Hon. Mrs. [Mary]. *The Microscope, or Descriptions of Various Objects of Especial Interest and Beauty, Adapted for Microscopic Observation* (London: Groombridge and Sons, 1869).
Welborn, G. "An Inexpensive but Efficient Microscope [Letter 19659]." *English Mechanic* 34.879 (Jan. 27, 1882): 495.
Wells, H. G. "Popularising Science." *Nature* 50 (1894): 300–1.
Wells, H. G. "A Slip under the Microscope." *The Yellow Book* 8 (1896): 229–51.
Wells, H. G. "Through the Microscope: Some Moral Reflections." In *Certain Personal Matters: A Collection of Material, Mainly Autobiographical* (London: William Heinemann, 1897): 238–45.
Wenham, F. H. "Memoranda: On Measuring the Aperture of Object-Glasses; and Remarks on Their Adjustment." *QJMS* 2 (1854): 134–9.
West, Tuffen. "Half-an-hour at the Microscope." *JPMS* 2 (Oct 1883): 231–7.
Whewell, William. *The Philosophy of the Inductive Sciences: Founded upon Their History*, 2 vols. (London: John W. Parker, 1840).
Whewell, William. *On the Plurality of Worlds* (London: John W. Parker, 1853).
White, T. L. Charters. *A Manual of Elementary Microscopical Manipulation, for the Use of Amateurs* (London: Roper and Drowley, 1887).
White, T. Charters. *The Microscope and How to Use It: A Handbook for Beginners* (London: Sutton, 1893).
Wilson, George and Archibald Geikie. *Memoir of Edward Forbes* (Cambridge: Macmillan, 1861).
Winslow, Forbes. *Lettsomian Lectures on Insanity* (London: John Churchill, 1854).
"Wonders of the Microscope." In *Chambers's Miscellany of Useful and Entertaining Tracts*, ed. (Edinburgh: William and Robert Chambers, 1847): 1–32.
Wood, J. G. *Common Objects of the Microscope*. (London: Routledge, Warne and Routledge, 1861).
Woodward, J. J., to Dr. Maddox. Letter of Aug. 4, 1871 (OHA28-005-00001-034 to -038). OHA 28 Woodward Letterbooks. Otis Historical Archives, National Museum of Health and Medicine.
Woodward, J. J., to John Mayall. Letter of Sept. 8, 1875 (OHA28-006-00002-113 to -114). OHA 28 Woodward Letterbooks. Otis Historical Archives, National Museum of Health and Medicine.
Youmans, E. A. "Tyndall and His American Visit." *Popular Science Monthly* 44 (Feb. 1894): 502–14.

Secondary Sources

Ahmed, Sara. *Cultural Politics of Emotion* (Edinburgh: Edinburgh University Press, 2014).
Alberti, Samuel J. M. M. "Amateurs and Professionals in One County: Biology and Natural History in Late Victorian Yorkshire." *Journal of the History of Biology* 34.1 (2001): 115–47.
Alberti, Samuel J. M. M. "Conversaziones and the Experience of Science in Victorian England." *Journal of Victorian Culture* 8.2 (2003): 208–30.
Almond-Brown, Gemma. *Spectacles and the Victorians: Measuring, Defining and Shaping Visual Capacity* (Manchester: Manchester University Press, 2023).
Altick, Richard D. "Nineteenth-Century English Best-Sellers: A Further List." *Studies in Bibliography* 22 (1969): 197–206.
Altick, Richard D. "Nineteenth-Century English Best-Sellers: A Third List." *Studies in Bibliography* 39 (1986): 235–41.
Anderson, Nancy. "Facing Animals in the Laboratory: Lessons of Nineteenth-Century Medical School Microscopy Manuals." In *The Educated Eye: Visual Culture and Pedagogy in the Life Sciences*, eds. Nancy Anderson and Michael R. Dietrich (Hanover, NH: Dartmouth College Press, 2012): 44–67.

"appreciate, v." *OED Online*; Oxford University Press; June 2022, https://www-oed-com.proxy.lib.fsu.edu/view/Entry/9787?redirectedFrom=appreciate.
Armintor, Deborah Needleman. *The Little Everyman* (Seattle, WA: University of Washington Press, 2011).
Armstrong, Isobel. "Transparency: Towards a Poetics of Glass in the Nineteenth Century." In *Cultural Babbage: Technology, Time and Invention*, eds. Francis Spufford and Jenny Uglow (London: Faber and Faber, 1996): 123–48.
Armstrong, Isobel. "The Microscope: Mediations of the Sub-visible World." In *Transactions and Encounters: Science and Culture in the Nineteenth Century*, eds. Roger Luckhurst and Josephine McDonagh (Manchester: Manchester University Press, 2002): 30–54.
Armstrong, Isobel. *Victorian Glassworlds: Glass Culture and the Imagination 1830–1880* (Oxford: Oxford University Press, 2008).
Atkinson, Dwight. *Scientific Discourse in Sociohistorical Context: The Philosophical Transactions of the Royal Society of London, 1675–1975* (London: Routledge, 1998).
Bachelard, Gaston. *The Poetics of Space* (New York: Penguin, 2014).
Barad, Karen. "Posthumanist Performativity: Toward an Understanding of How Matter Comes to Matter." *Signs: Journal of Women in Culture and Society* 28.3 (2003): 801–31.
Barad, Karen. *Meeting the Universe Halfway: Quantum Physics and the Entanglement of Matter and Meaning* (Durham, NC: Duke University Press, 2007).
Barnard, J. E. "President's Address: The Centenary of the Royal Microscopical Society 1839–1939: Some Personal Recollections and Impressions." 61.1–2 (1941): 1–15.
Barton, Ruth. "'Huxley, Lubbock, and Half a Dozen Others': Professionals and Gentlemen in the Formation of the X Club, 1851–1864." *Isis* 89.3 (1998): 410–44.
Barton, Ruth. "'Men of Science': Language, Identity and Professionalization in the Mid-Victorian Scientific Community." *History of Science* 41.1 (2003): 73–119.
Barton, Ruth. "Scientific Authority and Scientific Controversy in Nature: North Britain against the X Club." In *Culture and Science in the Nineteenth-Century Media*, eds. Louise Henson, Geoffrey Cantor, et al. (London: Routledge, 2017): 223–35.
Bates, A. W. H. "Vivisection, Virtue Ethics, and the Law in 19th-Century Britain." *Journal of Animal Ethics* 4.2 (2014): 30–44.
Bazerman, Charles. *Shaping Written Knowledge: The Genre and Activity of the Experimental Article in Science* (Madison, WI: University of Wisconsin Press, 1988).
Beatty, John. "Narrative Possibility and Narrative Explanation." *Studies in History and Philosophy of Science*. Part A, 62 (April 2017), 31–41.
Beiermann, Lea. "'A Method for Safe Transmission': The Microscope Slides of the American Postal Microscopical Club." *British Journal for the History of Science* 54.4 (2021): 403–22.
Belknap, Geoffrey. *From a Photograph: Authenticity, Science and the Periodical Press, 1870–1890* (London: Bloomsbury, 2016).
Belknap, Geoffrey. "Illustrating Natural History: Images, Periodicals, and the Making of Nineteenth-Century Scientific Communities." *British Journal for the History of Science* 51.3 (2018): 395–422.
Benjamin, Marina. "Sliding Scales: Microphotography and the Victorian Obsession with the Miniscule." In *Cultural Babbage: Technology, Time and Invention*, eds. Francis Spufford and Jenny Uglow (London: Faber, 1996): 99–122.
Bennett, J. A. "The Social History of the Microscope." *Journal of Microscopy* 155.3 (1989): 267–80.
Bennett, Jane. *The Enchantment of Modern Life* (Princeton: Princeton University Press, 2016).
Bonner, Thomas Neville. *Becoming a Physician: Medical Education in Britain, France, Germany, and the United States, 1750–1945* (London: Oxford University Press, 1995).
Bracegirdle, Brian. *A History of Microtechnique* (Ithaca, NY: Cornell University Press, 1978).
Bracegirdle, Brian. *The Quekett Microscopical Club 1865–2015* (London: Quekett Microscopical Club, 2016).
Bradbury, Savile. *The Evolution of the Microscope* (Oxford: Pergamon, 1967a).
Bradbury, Savile. "The Quality of the Image Produced by the Compound Microscope: 1700–1840." In *Historical Aspects of Microscopy*, eds. Savile Bradbury and Gerard L'E Turner (Cambridge: W. Heffer and Sons, for the Royal Microscopical Society, 1967b): 151–73.

Bradbury, Savile. "The Reception of Abbe's Theory in England." *Proceedings of the Royal Microscopical Society* 31.4 (1996): 293–301.

Braga, Ariane Varela. "Microorganisms, Microscopes, and Victorian Design Theories." *British Art Studies* 21 (2021), https://doi.org/10.17658/issn.2058-5462/issue-21.

Brennan, Teresa. *The Transmission of Affect* (Ithaca, NY: Cornell University Press, 2004).

Briefel, Aviva. *The Racial Hand in the Victorian Imagination* (Cambridge: Cambridge University Press, 2015).

Brock, William H. "The Development of Commercial Science Journals in Victorian Britain." In *The Development of Science Publishing in Europe*, ed. A. J. Meadows (Amsterdam: Elsevier, 1980): 95–122.

Brooke, John Hedley. "Natural Theology in Britain from Boyle to Paley." In *New Interactions between Theology and Natural Science*, eds. John Hedley Brooke et al. (Milton Keynes: Open University Press, 1974).

Brooke, John Hedley. *Science and Religion: Some Historical Perspectives* (Cambridge: Cambridge University Press, 2014).

Brooke, John Hedley and Geoffrey Cantor. *Reconstructing Nature: The Engagement of Science and Religion* (Edinburgh: T. and T. Clark, 1998).

Brown, Bill. "Thing Theory." *Critical Inquiry* 28.1 (2001): 1–22.

Brown, Theodore M. "From Mechanism to Vitalism in Eighteenth-Century English Physiology." *Journal of the History of Biology* 7 (1974): 179–216.

Bryant, J. A., H. Plaisier, L. M Irvine, et al., "Life and Work of Margaret Gatty (1809–1873), with Particular Reference to British Sea-weeds (1863)." *Archives of Natural History* 43.1 (2016): 131–47.

Buckland, Adelene. *Novel Science: Fiction and the Invention of Nineteenth-Century Geology* (Chicago, IL: University of Chicago Press, 2013).

Cartwright, Lisa. *Screening the Body: Tracing Medicine's Visual Culture* (Minneapolis, MN: University of Minnesota Press, 1995).

Cassedy, James H. "The Microscope in American Medical Science, 1840–1860." *Isis* 67.1 (1976): 76–97.

Castle, Terry. "Phantasmagoria: Spectral Technology and the Metaphorics of Modern Reverie." *Critical Inquiry* 15.1 (1988): 26–61.

Chakrabarti, Pratik. *Bacteriology in British India: Laboratory Medicine and the Tropics* (Rochester, NY: University of Rochester Press, 2012).

Chico, Tita. *The Experimental Imagination: Literary Knowledge and Science in the British Enlightenment* (Stanford, CA: Stanford University Press, 2018).

Churchill, Frederick B. "The Guts of the Matter. Infusoria from Ehrenberg to Bütschli: 1838–1876." *Journal of the History of Biology* 22.2 (1989): 189–213.

Cohen, William A. *Embodied: Victorian Literature and the Senses* (Minneapolis, MN: University of Minnesota Press, 2009).

Cole, Lucinda. *Imperfect Creatures: Vermin, Literature, and the Sciences of Life, 1600–1740* (Ann Arbor, MI: University of Michigan Press, 2016).

Cooter, Roger and Stephen Pumfrey. "Separate Spheres and Public Places: Reflections on the History of Science Popularization and Science in Popular Culture." *History of Science* 32.3 (1994): 237–67.

Crary, Jonathan. *Techniques of the Observer: On Vision and Modernity in the Nineteenth Century* (Cambridge: MIT Press, 1990).

Crary, Jonathan. *Suspensions of Perception: Attention, Spectacle, and Modern Culture* (Cambridge: MIT Press, 1999).

Csiszar, Alex. *The Scientific Journal: Authorship and the Politics of Knowledge in the Nineteenth Century* (Chicago, IL: University of Chicago Press, 2018).

Dames, Nicholas. "Wave-Theories and Affective Physiologies: The Cognitive Strain in Victorian Novel Theories." *Victorian Studies* 46.2 (2004): 206–16.

Dames, Nicholas. *The Physiology of the Novel: Reading, Neural Science, and the Form of Victorian Fiction* (Oxford: Oxford University Press, 2007).

Dames, Nicholas. "On Not Close Reading: The Prolonged Excerpt as Victorian Critical Protocol." In *The Feeling of Reading: Affective Experience and Victorian Literature*, ed. Rachel Ablow (Ann Arbor, MI: University of Michigan Press, 2010): 11–26.

Darnton, Robert. "What Is the History of Books?" *Daedalus* 111.3 (1982): 65–83.

Darnton, Robert. "'What Is the History of Books?' Revisited." *Modern Intellectual History* 4.3 (2007): 495–508.

Daston, Lorraine. "The Moral Economy of Science." *Osiris* 10 (1995): 3–24.

Daston, Lorraine. "Type Specimens and Scientific Memory." *Critical Inquiry* 31 (2004): 153–82.

Daston, Lorraine. "On Scientific Observation." *Isis* 99.1 (2008): 97–110.

Daston, Lorraine and Peter Galison. *Objectivity* (New York: Zone Books, 2007).

Daston, Lorraine and Katherine Park. *Wonders and the Order of Nature, 1150–1750* (New York: Zone Books, 1998).

Dawson, Gowan, Richard Noakes, and Jonathan R. Topham. "Introduction." In *Science in the Nineteenth-Century Periodical: Reading the Magazine of Nature*, eds. Geoffrey Cantor et al. (Cambridge: Cambridge University Press, 2004): 1–34.

Dear, Peter, "*Totius in verba*: Rhetoric and Authority in the Early Royal Society." *Isis* 76 (1985): 145–61.

Dear, Peter, ed. *The Literary Structure of Scientific Argument: Historical Studies* (Philadelphia, PA: University of Pennsylvania Press, 1991).

Delorme, Shannon. "Physiology or Psychic Powers? William Carpenter and the Debate over Spiritualism in Victorian Britain." *Studies in History and Philosophy of Biological and Biomedical Sciences* 48 Pt A (2014): 57–66.

Derrida, Jacques. *The Truth in Painting*, trans. Geoffrey Bennington and Ian MacLeod (Chicago, IL: University of Chicago Press, 1987).

Desmond, Adrian. *The Politics of Evolution: Morphology, Medicine, and Reform in Radical London* (Chicago, IL: University of Chicago Press, 1989).

Desmond, Adrian. "Redefining the X Axis: 'Professionals,' 'Amateurs' and the Making of Mid-Victorian Biology—A Progress Report." *Journal of the History of Biology* 34.1 (2001): 3–50.

Downes, Stephanie, Sally Holloway, and Sarah Randles, eds.. "Introduction." In *Feeling Things: Objects and Emotions through History* (Oxford: Oxford University Press, 2018): 8–24.

Dunn, Richard. *The Telescope: A Short History* (London: Conway, 2009).

Elkins, James. "On Visual Desperation and the Bodies of Protozoa." *Representations* 40 (1992): 33–56.

Elkins, James. *Six Stories from the End of Representation* (Stanford, CA: Stanford University Press, 2008).

Elkins, James. "Against the Sublime." In *Beyond the Finite: The Sublime in Art and Science*, eds. Roald Hoffmann and Iain Boyd Whyte (New York: Oxford University Press, 2011): 75–90.

Endersby, Jim. *Imperial Nature: Joseph Hooker and the Practices of Victorian Science* (Chicago, IL: University of Chicago Press, 2008).

Fardon, Richard. "Citations out of Place." *Anthropology Today* 29.1 (2013): 25–7.

Feffer, Stuart M. "Ernst Abbe, Carl Zeiss, and the Transformation of Microscopical Optics." In *Scientific Credibility and Technical Standards in 19th and early 20th-Century Germany and Britain* (Dordrecht: Kluwer, 1996): 23–66.

Fieser, James. *Early Responses to Hume's Life and Reputation: Series Bibliography*, vol. 10 (Bristol: Thoemmes Press, 2005).

Finnegan, Diarmid A. "Natural History Societies in Late Victorian Scotland and the Pursuit of Local Civic Science." *British Journal for the History of Science* 38.1 (2005): 53–72.

Fisher, Philip. *Wonder, the Rainbow, and the Aesthetics of Rare Experiences*. Cambridge, MA: Harvard University Press, 1998.

Fleck, Ludwik. *Genesis and Development of a Scientific Fact*, trans. Fred Bradley and Thaddeus J. Trenn (Chicago, IL: University of Chicago Press, 1979).

Flint, Kate. *The Victorians and the Visual Imagination* (Cambridge: Cambridge University Press, 2000).

Ford, Brian J. "The Royal Society and the Microscope." *Notes and Records of the Royal Society of London* 55.1 (2001): 29–49.

Forsberg, Laura. "Nature's Invisibilia: The Victorian Microscope and the Miniature Fairy." *Victorian Studies* 57.4 (2015): 638–66.

Forsberg, Laura. *Worlds Beyond: Miniatures and Victorian Fiction* (New Haven: Yale University Press, 2021).

Freedgood, Elaine. *Worlds Enough: The Invention of Realism in the Victorian Novel* (Princeton, NJ: Princeton University Press, 2019).

Fyfe, Aileen. "Young Readers and the Sciences." In *Books and the Sciences in History*, eds. Marina Frasca-Spada and Nicholas Jardine (Cambridge: Cambridge University Press, 2000): 276–90.

Fyfe, Aileen. "Introduction." In *Science for Children*, ed. Aileen Fyfe (Bristol: Thoemmes Press, 2001): xi–xxii.

Fyfe, Aileen. *Science and Salvation: Evangelical Popular Science Publishing in Victorian Britain* (Chicago, IL: University of Chicago Press, 2004).

Fyfe, Aileen. *Steam-Powered Knowledge: William Chambers and the Business of Publishing, 1820–1860* (Chicago, IL: University of Chicago Press, 2012).

Fyfe, Aileen, Julie McDougall-Waters, and Noah Moxham, "350 years of Scientific Periodicals." *Notes and Records: The Royal Society Journal of the History of Science* 69.3 (2015): 227–39.

Garratt, Peter. *Victorian Empiricism: Self, Knowledge, and Reality in Ruskin, Bain, Lewes, Spencer, and George Eliot* (Madison, NJ: Fairleigh Dickinson University Press, 2010).

Gates, Barbara T. "Retelling the Story of Science." *Victorian Literature and Culture* 21 (1993): 289–306.

Gates, Barbara T. *Kindred Nature: Victorian and Edwardian Women Embrace the Living World* (Chicago, IL: University of Chicago Press, 1998).

Gates, Barbara T. and Ann B. Shteir Ann. *Natural Eloquence: Women Reinscribe Science.* (Madison, WI: University of Wisconsin Press, 1997).

Gest, Howard. "The Discovery of Microorganisms by Robert Hooke and Antoni Van Leeuwenhoek, Fellows of the Royal Society." *Notes and Records of the Royal Society of London* 58.2 (2004): 187–201.

Gooday, Graeme, "Placing or Replacing the Laboratory in the History of Science." *Isis* 99.4 (2008): 783–95.

Gooday, Graeme. "'Nature' in the Laboratory: Domestication and Discipline with the Microscope in Victorian Life Science." *British Journal for the History of Science* 24.3 (1991): 307–41.

Gross, Alan G., Joseph E. Harmon, and Michael Reidy. *Communicating Science: The Scientific Article from the 17th Century to the Present* (Oxford: Oxford University Press, 2002).

Hacking, Ian. *Representing and Intervening: Introductory Topics in the Philosophy of Natural Science* (Cambridge: Cambridge University Press, 1983).

Hadjiafxendi, Kyriaki, and John Plunkett. "Science at the Seaside: Pleasure Hunts in Victorian Devon." In *Coastal Works: Cultures of the Atlantic Edge*, eds. Nicholas Allen, Nick Groom, and Jos Smith (Oxford: Oxford University Press, 2017): 181–204.

Hale, Piers. "Between Intuition and Empiricism: William Benjamin Carpenter on Man, Mind, and Moral Responsibility." In *The Metaphysical Society (1869–1880): Intellectual Life in Mid-Victorian England*, eds. Catherine Marshall et al. (Oxford: Oxford University Press, 2019): 204–27.

Hamer, Pam. "The Election of Women as Fellows of the Royal Microscopical Club." *Journal of Microscopy* 291.1 (2023): 8–15.

Hamlin, Christopher. *A Science of Impurity: Water Analysis in Nineteenth Century Britain* (Oakland, CA: University of California Press, 1990).

Haraway, Donna J. "A Cyborg Manifesto: Science, Technology, and Socialist-Feminism in the Late Twentieth Century." In *Simians, Cyborgs, and Women: The Reinvention of Nature* (New York: Routledge, 1991): 149–82.

Harley, Alexis. "Darwin's Ants: Evolutionary Theory and the Anthropomorphic Fallacy." In *Representing the Modern Animal in Culture*, eds. Jeanne Dubino, Ziba Rashidian, and Andrew Smyth (New York: Palgrave Macmillan, 2014): 103–18.
Harrison, Peter. "'Science' and 'Religion': Constructing the Boundaries." *Journal of Religion* 86.1 (2006): 81–106.
Henchman, Anna. *The Starry Sky Within: Astronomy and the Reach of the Mind in Victorian Literature* (Oxford: Oxford University Press, 2014).
Hentschel, Klaus. *Visual Cultures in Science and Technology: A Comparative History* (Oxford: Oxford University Press, 2014).
Hodder, Ian. *Entangled: An Archaeology of the Relationships between Humans and Things* (Chichester: John Wiley and Sons, 2012).
Holmes, Frederic L. "Scientific Writing and Scientific Discovery." *Isis* 78.2 (1987): 220–35.
Holmes, Frederic L. "Argument and Narrative in Scientific Writing." In *The Literary Structure of Scientific Argument: Historical Studies*, ed. Peter Dear (Philadelphia, PA: University of Pennsylvania Press, 1991): 164–81.
Hughes, Arthur. "Studies in the History of Microscopy. 1. The Influence of Achromatism." *JRMS* 76.1–2 (1956): 47–60.
Hunt, Verity. "Raising a Modern Ghost: The Magic Lantern and the Persistence of Wonder in the Victorian Education of the Senses." *Romanticism and Victorianism on the Net* 52 (2008).
Hurwitz, Brian. "Narrative Constructs in Modern Clinical Case Reporting." *Studies in History and Philosophy of Science, Part A* 62 (2017): 65–73.
Ihde, Don. *Instrumental Realism: The Interface between Philosophy of Science and Philosophy of Technology* (Bloomington, IN: Indiana University Press, 1991).
Inman, Billie Andrew. "The Intellectual Context of Walter Pater's 'Conclusion.'" *Prose Studies* 4.1 (1981): 12–30.
Jacyna, L. S. "The Romantic Programme and the Reception of Cell Theory in Britain." *Journal of the History of Biology* 17.1 (1984): 13–48.
Jacyna, L. S. "John Hughes Bennett and the Origins of Medical Microscopy in Edinburgh." *Proceedings of the Royal College of Physicians of Edinburgh* 27 (1997): 12–21.
Jacyna, L. S. "'A Host of Experienced Microscopists': The Establishment of Histology in Nineteenth-Century Edinburgh." *Bulletin of the History of Medicine* 75.2 (2001): 225–53.
Jacyna, L. S. "Moral Fibre: The Negotiation of Microscopic Facts in Victorian Britain." *Journal of the History of Biology* 36.1 (2003): 39–85.
Jardine, Boris. "Microscopes." In *A Companion to the History of Science*, ed. Bernard Lightman (Chichester, UK: John Wiley and Sons, 2016): 515–29.
Jardine, Nick. "Books, Texts, and the Making of Knowledge." In *Books and the Sciences in History*, eds. Silvia De Renzi and Nicholas Jardine (Cambridge: Cambridge University Press, 2000): 393–407.
Keene, Melanie. *Science in Wonderland: The Scientific Fairy Tales of Victorian Britain* (Oxford: Oxford University Press, 2015).
Kennedy, Meegan. *Revising the Clinic: Vision and Representation in Victorian Medical Narrative and the Novel* (Columbus, OH: Ohio State University Press, 2010).
Kennedy, Meegan. "'Throes and struggles…witnessed with painful distinctness': The Oxy-hydrogen Microscope, Performing Science, and the Projection of the Moving Image." *Victorian Studies* 62.1 (2019): 85–118.
Kennedy, Meegan. "Circulation and Civility: Mid-Victorian Botany and Microscopical Method." In *The Victorian Revolution: A Botanical Perspective*, ed. Luís Mendonça de Carvalho (New York: Springer, 2024): 131–65.
Kennedy, Meegan. "Water-tigers, Jam-pots, and 'Ye Mikroskopiker's Arms': Boundary-Work and Boundary Objects in Comic Microscopy." In *Humor across Victoriana*, eds. Mou-Lan Wong and James Whitehead (New York: Routledge, 2025).
Ketabgian, Tamara. *The Lives of Machines: The Industrial Imaginary in Victorian Literature and Culture* (Ann Arbor, MI: University of Michigan Press, 2011).

King, Amy M. *The Divine in the Commonplace: Reverent Natural History and the Novel in Britain* (Cambridge: Cambridge University Press, 2019).

Knorr-Cetina, Karin. "Objectual Practice." In *The Practice Turn in Contemporary Theory*, eds. Karin Knorr Cetina et al. (London: Routledge, 2005): 184–97.

Lakoff, George and Mark Johnson. *Metaphors We Live By* (Chicago, IL: University of Chicago Press, 2008).

Landecker, Hannah. *Culturing Life: How Cells Became Technologies* (Cambridge, MA: Harvard University Press, 2007).

Lentz, Thomas L. *Lentz Microscopy and Technology Collection, Yale*. Peabody Museum of Natural History, Yale University, 2022, https://peabody.yale.edu/sites/default/files/documents/history-science-technology/Lentz%20Collection%20Guide%203-3-2022.pdf

Levine, George. *Darwin Loves You* (Princeton, NJ: Princeton University Press, 2008a).

Levine, George. *Realism, Ethics and Secularism* (Cambridge: Cambridge University Press, 2008b).

Lidwell-Durnin, John. "William Benjamin Carpenter and the Emerging Science of Heredity." *Journal of the History of Biology* 53.1 (2020): 81–103.

Lightman, Bernard. "'The Voices of Nature': Popularizing Victorian Science." In *Victorian Science in Context*, ed. Bernard Lightman (Chicago, IL: University of Chicago Press, 1997): 187–211.

Lightman, Bernard. "The Story of Nature: Victorian Popularizers and Scientific Narrative." *Victorian Review* 25.2 (2000): 1–29.

Lightman, Bernard. "Victorian Sciences and Religions: Discordant Harmonies." *Osiris* 16 (2001): 343–66.

Lightman, Bernard. "Scientists as Materialists in the Periodical Press: Tyndall's Belfast Address." In *Science Serialized: Representations of the Sciences in Nineteenth-Century Periodicals*, ed. Geoffrey Cantor and Sally Shuttleworth (Cambridge, MA: MIT Press, 2004): 199–237.

Lightman, Bernard. *Victorian Popularizers of Science: Designing Nature for New Audiences* (Chicago, IL: University of Chicago Press, 2007).

Lightman, Bernard. "Science, Scientists and the Public: The Contested Meanings of Science in Victorian Britain." In *Evolutionary Naturalism in Victorian Britain: The "Darwinians" and Their Critics* (Farnham, UK: Ashgate, 2009): 1–40.

Lightman, Bernard. "Popularizers, Participation and the Transformations of Nineteenth-Century Publishing: From the 1860s to the 1880s." *Notes and Records of the Royal Society of London* 70.4 (2016): 343–59.

Lightman, Bernard. "The Theology of Victorian Scientific Naturalists." In *Science without God? Rethinking the History of Scientific Naturalism*, eds. Peter Harrison and John H. Roberts (Oxford: Oxford University Press, 2019): 235–53.

Lightman, Bernard and Gowan Dawson. *Victorian Scientific Naturalism: Community, Identity, Continuity* (Chicago: University of Chicago Press, 2014).

Lindquist, Jason H. "'The Mightiest Instrument of the Physical Discoverer': The Visual 'Imagination' and the Victorian Observer." *Journal of Victorian Culture* 13.2 (2008): 171–99.

Lukács, Georg. *The Historical Novel*, trans. Hannah and Stanley Mitchell (Lincoln, NE: University of Nebraska Press, 1983).

Marsh, David. "Sir Frank Crisp." *GT [Gardens Trust] News* 10 (2019): 4–5.

Marx, Leo. *The Machine in the Garden: Technology and the Pastoral Ideal in America* (New York: Oxford University Press, 2000).

Massumi, Brian. *Parables for the Virtual: Movement, Affect, Sensation* [2002]. (Durham, NC: Duke University Press, 2021).

Masters, Barry. *Superresolution Optical Microscopy: The Quest for Enhanced Resolution and Contrast* (Cham, Switzerland: Springer Nature, 2020).

Maurette, Pablo. *The Forgotten Sense: Meditations on Touch* (Chicago, IL: University of Chicago Press, 2018).

McGrath, Roberta. *Seeing Her Sex: Medical Archives and the Female Body* (Manchester: Manchester University Press, 2003).

Merrill, Lynn L. *The Romance of Victorian Natural History* (Oxford: Oxford University Press, 1989).
Mershon, Ella. "Ruskin's Dust." *Victorian Studies* 58.3 (2016): 464–92.
Morgan, Mary S. "Narrative Ordering and Explanation." *Studies in History and Philosophy of Science, Part A* 62 (2017): 86–97.
Morgan, Mary S. and M. Norton Wise. "Narrative Science and Narrative Knowing." *Studies in History and Philosophy of Science, Part A* 62 (2017): 1–5.
Morus, Iwan Rhys. "Worlds of Wonder: Sensation and the Victorian Scientific Performance." *Isis* 101.4 (2010): 806–16.
Mussell, James. *Science, Time and Space in the Late Nineteenth-Century Periodical Press* (Aldershot, UK: Ashgate, 2007).
Myers, Greg. *Writing Biology: Texts in the Social Construction of Scientific Knowledge* (Madison, WI: University of Wisconsin Press, 1991).
Nasim, Omar W. *Observing by Hand: Sketching the Nebulae in the Nineteenth Century* (Chicago, IL: University of Chicago Press, 2014).
Nasim, Omar W. *The Astronomer's Chair: A Visual and Cultural History* (Cambridge, MA: MIT Press, 2021).
Ngai, Sianne. *Ugly Feelings* (Cambridge, MA: Harvard University Press, 2005).
Nicolson, Marjorie Hope. *The Microscope and English Imagination* (Northampton, MA: Smith College, 1935).
O'Connor, Ralph. *The Earth on Show: Fossils and the Poetics of Popular Science, 1802–1856* (Chicago, IL: University of Chicago Press, 2008).
O'Connor, Ralph. "Reflections on Popular Science in Britain: Genres, Categories, and Historians." *Isis* 100.2 (2009): 333–45.
Otis, Laura. *Networking: Communicating with Bodies and Machines in the Nineteenth Century* (Ann Arbor, MI: University of Michigan Press, 2001).
Otter, Chris. *The Victorian Eye: A Political History of Light and Vision in Britain, 1800–1910* (Chicago: University of Chicago Press, 2008).
Paterson, Mark. *The Senses of Touch: Haptics, Affects and Technologies* (Oxford: Berg, 2007.
Paterson, Mark. *How We Became Sensorimotor: Movement, Measurement, Sensation* (Minneapolis, MN: University of Minnesota Press, 2021).
Pickering, Andrew. *The Mangle of Practice: Time, Agency, and Science* (Chicago, IL: University of Chicago Press, 1995).
Potter, Jonathan. *Discourses of Vision in Nineteenth-Century Britain: Seeing, Thinking, Writing* (Cham, Switzerland: Springer, 2018).
Price, Elfed Huw. "George Henry Lewes (1817–1878): Embodied Cognition, Vitalism, and the Evolution of Symbolic Perception." In *Brain, Mind and Consciousness in the History of Neuroscience*, eds. C. U. M. Smith and Harry Whitaker (Dordrecht, Netherlands: Springer, 2014): 105–23.
Price, Leah. *The Anthology and the Rise of the Novel: From Richardson to George Eliot* (Cambridge: Cambridge University Press, 2000).
Quinn, Dennis. *Iris Exiled: A Synoptic History of Wonder.* Lanham, MD: University Press of America, 2002.
Rabinbach, Anson. *The Human Motor: Energy, Fatigue, and the Origins of Modernity* (Oakland, CA: University of California Press, 1992).
Radick, Gregory. *The Simian Tongue: The Long Debate about Animal Language* (Chicago, IL: University of Chicago Press, 2007).
Ratcliff, Marc. *The Quest for the Invisible: Microscopy in the Enlightenment* (Aldershot, UK: Ashgate, 2009).
Reddy, William M. *The Navigation of Feeling: A Framework for the History of Emotions* (Cambridge: Cambridge University Press, 2001).
Reidy, Michael S. "Evolutionary Naturalism on High: The Victorians Sequester the Alps." In *Victorian Scientific Naturalism: Community, Identity, Continuity*, eds. Gowan Dawson and Bernard Lightman (Chicago, IL: University of Chicago Press, 2014): 55–79.

Reiser, Stanley Joel. *Medicine and the Reign of Technology* (Cambridge: Cambridge University Press, 1978).

Reynolds, Andrew S. *The Third Lens: Metaphor and the Creation of Modern Cell Biology* (Chicago, IL: University of Chicago Press, 2018).

Rheinberger, Hans-Jörg. *An Epistemology of the Concrete: Twentieth-Century Histories of Life* trans. and ed. Timothy Lenoir (Durham, NC: Duke University Press, 2010).

Riegl, Alois. *The Group Portraiture of Holland*, trans. Evelyn M. Kain (Los Angeles, CA: Getty, 1999).

Roach, Joseph, "The Artificial Eye: Augustan Theater and the Empire of the Visible." In *The Performance of Power*, ed. Sue-Ellen Case and Janelle G.Reinelt (Iowa City: University of Iowa Press, 1991): 131–45.

Rosenwein, Barbara H. "Problems and Methods in the History of Emotions." *Passions in Context* 1.1 (2010): 1–32.

Rudwick, Martin J. S. "Charles Darwin in London: The Integration of Public and Private Science." *Isis* 73.2 (1982): 186–206.

Rupke, Nicolaas. "Richard Owen's Vertebrate Archetype." *Isis* 84.2 (1993): 231–51.

Ryan, Vanessa L. *Thinking without Thinking in the Victorian Novel* (Baltimore, MD: JHU Press, 2012).

Rylance, Rick. *Victorian Psychology and British Culture, 1850–1880* (Oxford: Oxford University Press, 2000).

Schaffer, Simon. "The Leviathan of Parsonstown: Literary Technology and Scientific Representation." In *Inscribing Science: Scientific Texts and the Materiality of Communication*, ed. Timothy Lenoir (Stanford, CA: Stanford University Press, 1998): 182–222.

Scheer, Monique. "Are Emotions a Kind of Practice (and is that what makes them have a history)? A Bourdieuian Approach to Understanding Emotion." *History and Theory* 51.2 (2012): 193–220.

Schickore, Jutta. *The Microscope and the Eye: A History of Reflections, 1740–1870* (Chicago, IL: University of Chicago Press, 2007).

Schickore, Jutta. "Test Objects for Microscopes." *History of Science* 47 (2009): 117–46.

Schmitt, Cannon. "Victorian Beetlemania." In *Victorian Animal Dreams*, eds. Martin A. Danahay and Deborah Denenholz Morse (London and New York: Routledge, 2017): 35–51.

Schor, Naomi. *Reading in Detail: Aesthetics and the Feminine* (New York: Methuen, 1987).

Secord, Anne. "Science in the Pub: Artisan Botanists in Early Nineteenth-Century Lancashire." *History of Science* 32 (1994): 269–315.

Secord, Anne. "Botany on a Plate: Pleasure and the Power of Pictures in Promoting Early Nineteenth-Century Scientific Knowledge." *Isis* 93.1 (2002): 28–57.

Secord, James. *Victorian Sensation: The Extraordinary Publication, Reception, and Secret Authorship of Vestiges of the Natural History of Creation* (Chicago, IL: University of Chicago Press, 2000).

Secord, James. "Knowledge in Transit." *Isis* 95.4 (2004): 654–72.

Secord, James. "How Scientific Conversation Became Shop Talk." In *Science in the Marketplace: Nineteenth-Century Sites and Experiences*, eds. Aileen Fyfe and Bernard Lightman (Chicago, IL: University of Chicago Press, 2007): 23–59.

Secord, James. "Science, Technology, and Mathematics." In *History of the Book in Britain 6 (1830–1914)*, ed. David McKitterick (Cambridge: Cambridge University Press, 2009): 443–74.

Secord, James. *Visions of Science* (Chicago, IL: University of Chicago Press, 2014).

Seibold-Buitmann, Ursula. "Monster Soup: The Microscope and Victorian Fantasy." *Interdisciplinary Science Reviews* 25.3 (2013): 211–19.

Senior, Emily. "'Glimpses of the Wonderful': The Jamaican Origins of the Aquarium." *Atlantic Studies* (2021): 1–25.

Servitje, Lorenzo. *Medicine Is War: The Martial Metaphor in Victorian Literature and Culture* (Albany: State University of New York, 2021).

Shapin, Steven. "Pump and Circumstance: Robert Boyle's Literary Technology." *Social Studies of Science* 14 (1984): 481–520.

Shapin, Steven. "The Philosopher and the Chicken: On the Dietetics of Disembodied Knowledge." In *Science Incarnate: Historical Embodiments of Natural Knowledge*, eds. Christopher Lawrence and Steven Shapin (Chicago, IL: University of Chicago Press, 1998): 21–50.

Shapin, Steven. "Weber's Science as a Vocation: A Moment in the History of 'Is' and 'Ought'." *Journal of Classical Sociology* 19.3 (2019): 1–18.

Shapin, Steven and Simon Schaffer. *Leviathan and the Air-Pump: Hobbes, Boyle, and the Experimental Life* (Princeton, NJ: Princeton University Press, 1985).

Sheets-Pyenson, Susan. "Popular Science Periodicals in Paris and London: The Emergence of a Low Scientific Culture, 1820–1875." *Annals of Science* 42.6 (1985): 549–72.

Shuttleworth, Sally and Berris Charnley. "Science Periodicals in the Nineteenth and Twenty-First Centuries." *Notes and Records of the Royal Society of London* 70.4 (2016): 297–304.

Siesser, William G. "Christian Gottfried Ehrenberg: Founder of Micropaleontology." *Centaurus* 25.2 (1981): 166–88.

Silver, Carole G. *Strange and Secret Peoples* (Oxford: Oxford University Press, 2000).

Smith, Jonathan. *Fact and Feeling: Baconian Science and the Nineteenth-Century Literary Imagination* (Madison, WI: University of Wisconsin Press, 1994).

Smith, Jonathan. *Charles Darwin and Victorian Visual Culture* (Cambridge: Cambridge University Press, 2006).

Smith, Jonathan, "Philip Gosse and the Varieties of Natural Theology." In *Reinventing Christianity*, ed. Linda Woodhead. London: Routledge, 2019: 251–62.

Smith, Nathan Edward Charles. "Fertile Substrate: The Rise, Fall, and Succession of Popular Microscopy in Great Britain." *Annals of Science* 80.3 (2023): 268–92.

Smith, Roger. "The Physiology of the Will: Mind, Body, and Psychology in the Periodical Literature, 1855–1875." In *Science Serialized: Representations of the Sciences in Nineteenth-Century Periodicals*, eds. Geoffrey Cantor and Sally Shuttleworth (Cambridge, MA: MIT Press, 2004): 81–110.

Smith, Tiffany Watt. *On Flinching: Theatricality and Scientific Looking from Darwin to Shell Shock* (Oxford: Oxford University Press, 2014).

St. Clair, William. *The Reading Nation in the Romantic Period* (Cambridge: Cambridge University Press, 2004).

Stafford, Barbara Maria, "Images of Ambiguity: Eighteenth-Century Microscopy and the Neither/Nor." In *Visions of Empire: Voyages, Botany, and Representations of Nature*, eds. David Phillip Miller and Peter Hanns Reill (Cambridge: Cambridge University Press, 1996): 230–57.

Stafford, Barbara Maria and Frances Terpak. *Devices of Wonder: From the World in a Box to Images on a Screen* (Los Angeles, CA: Getty, 2001).

Stewart, Susan. *On Longing: Narratives of the Miniature, the Gigantic, the Souvenir, the Collection* (Durham, NC: Duke University Press, 1993).

Stott, Rebecca. "Darwin's Barnacles: Victorian Natural History and the Marine Grotesque." In *Transactions and Encounters: Science and Culture in the Nineteenth Century*, eds. Roger Luckhurst and Josephine McDonagh (Manchester: Manchester University Press, 2002): 151–81.

Strick, James E. *Sparks of Life: Darwinism and the Victorian Debates over Spontaneous Generation* (Cambridge, MA: Harvard University Press, 2000).

Sussman, Herbert L. *Victorian Technology: Invention, Innovation, and the Rise of the Machine* (Santa Barbara, CA: Praeger, 2009).

Terrall, Mary. "Narrative and Natural History in the Eighteenth Century." *Studies in History and Philosophy of Science, Part A* 62 (2017): 51–64.

Thwaite, Ann. *Glimpses of the Wonderful: The Life of Philip Henry Gosse, 1810–1888* (London: Faber and Faber, 2002).

[Tierney, Clarence]. *Catalogue of the Printed Books and Pamphlets in the Library of the Royal Microscopical Society* (London: Royal Microscopical Society, 1929).

Toal, Ciaran. "Science, Religion and the Geography of Speech at the British Association: William Henry Dallinger (1839–1909) under the Microscope." *Endeavour* 37.2 (2013): 64–70.
Topham, Jonathan R. "Beyond the 'Common Context': The Production and Reading of the Bridgewater Treatises." *Isis* 89.2 (1998): 233–62.
Topham, Jonathan R., "Science, Natural Theology, and the Practice of Christian Piety in Early-Nineteenth-Century Religious Magazines." In *Science Serialized: Representations of the Sciences in Nineteenth-century Periodicals,* ed. Geoffrey Cantor and Sally Shuttleworth (Cambridge: MIT Press, 2004a): 37–66.
Topham, Jonathan R. "A View from the Industrial Age." *Isis* 95.3 (2004b): 431–42.
Topham, Jonathan R, "Publishing 'popular science' in early nineteenth-century Britain." In *Science in the Marketplace: Nineteenth-Century Sites and Experiences,* ed. Aileen Fyfe and Bernard Lightman (Chicago: University of Chicago Press, 2007): 135–68.
Topham, Jonathan R. "Rethinking the History of Science Popularization/Popular Science." In *Popularizing Science and Technology in the European Periphery*, ed. Faidra Papanelopoulou et al. (London: Routledge, 2009): 1–20.
Topham, Jonathan R. "The Scientific, the Literary and the Popular: Commerce and the Reimagining of the Scientific Journal in Britain, 1813–1825." *Notes and Records of the Royal Society of London* 70.4 (2016): 305–24.
Topham, Jonathan R. "Redrawing the Image of Science: Technologies of Illustration and the Audiences for Scientific Periodicals in Britain, 1790–1840." In *Science Periodicals in Nineteenth-Century Britain*, eds. Gowan Dawson et al. (Chicago, IL: University of Chicago Press, 2020): 65–102.
Tresch, John. *The Romantic Machine* (Chicago, IL: University of Chicago Press, 2012).
Turner, Gerard L'E. "The Microscope as a Technical Frontier in Science." In *Historical Aspects of Microscopy*, eds. Savile Bradbury and G. L'E. Turner (Cambridge: W. Heffer and Sons, 1967): 175–97.
Turner, Gerard L'E. "Microscopical Communication." *Journal of Microscopy* 100.1 (1974): 3–20.
Turner, Gerard L'E. *God Bless the Microscope! A History of the Royal Microscopical Society over 150 Years* (London: RMS, 1989a).
Turner, Gerard L'E. *The Great Age of the Microscope* (Bristol: Hilger, 1989b).
Vasalou, Sophia. "Wonder: Toward a Grammar." In *Practices of Wonder: Cross-Disciplinary Perspectives*, ed. Sophia Vasalou. Eugene, Oregon: Pickwick, 2012. 16-63
Walker, David. "Exploring Classic Insect Test Objects for the Microscope: II—Scales of 'Podura' (Springtail)." *Micscape* (Feb. 2011). http://www.microscopy-uk.org.uk/mag/artfeb11/dw-insect-test2.html
Warner, John Harley. "'Exploring the Inner Labyrinths of Creation': Popular Microscopy in Nineteenth-Century America." *Journal of the History of Medicine and Allied Sciences* 37.1 (1982): 7–33.
Weber, Max. "Science as a Vocation." *Daedalus* 87.1 (1958): 111–34.
Webster, Erin. *The Curious Eye: Optics and Imaginative Literature in Seventeenth-Century England* (Oxford: Oxford University Press, 2020).
White, Paul, "Cross-Cultural Encounters: The Co-production of Science and Literature in Mid-Victorian Periodicals." In *Transactions and Encounters: Science and Culture in the Nineteenth Century,* ed. Roger Luckhurst and Josephine McDonagh (Manchester: Manchester University Press, 2002): 75–95.
Whitley, Edward and Robert Weidman. "The (After) Life of William North among the New York Bohemians." *Victorian Newsletter* 115 (Spring 2009): 29–45.
Willis, Martin. *Vision, Science and Literature, 1870–1920: Ocular Horizons* (London: Pickering and Chatto, 2011).
Wilson, Catherine. *The Invisible World: Early Modern Philosophy and the Invention of the Microscope* (Princeton, NJ: Princeton University Press, 1995).
Winter, Alison. "The Construction of Orthodoxies and Heterodoxies in the Early Victorian Life Sciences." In *Victorian Science in Context*, ed. Bernard Lightman (Chicago, IL: University of Chicago Press, 1997): 24–50.

Wootton, Robin. "The Hudson Transparencies: A Set of Remarkable Visual Aids by a Distinguished Victorian Microscopist." *Report and Transactions of the Devonshire Association* 143 (2011): 61–90.

Worboys, Michael. *Spreading Germs: Disease Theories and Medical Practice in Britain, 1865–1900* (Cambridge: Cambridge University Press, 2000).

Zieger, Susan. *The Mediated Mind: Affect, Ephemera, and Consumerism in the Nineteenth Century* (New York: Fordham University Press, 2018).

Index

Note: locators in *italics* denote illustrations.

Since the index has been created to work across multiple formats, indexed terms for which a page range is given (e.g., 52–53, 66–70, etc.) may occasionally appear only on some, but not all of the pages within the range.